Cosmetics
SCIENCE AND TECHNOLOGY

Cosmetics

SCIENCE AND TECHNOLOGY

Second Edition

Volume 2

Edited by

M. S. BALSAM

and

EDWARD SAGARIN

WILEY-INTERSCIENCE

a division of John Wiley & Sons, Inc.

New York • London • Sydney • Toronto

To the Advisory Editors who originally made possible the launching of this publication:

H. D. GOULDEN
EMIL KLARMANN
DONALD H. POWERS

Library of Congress Catalog Card Number: 75-177888

ISBN 0-471-04647-7

Printed in the United States of America.

10 9 8 7 6 5 4 3 2 1

Preface

The second edition of Cosmetics: Science and Technology is now a reality, and it appears in three volumes, of which this is Volume Two.

Volume One contains descriptive material on the manufacture and formulation of fifteen different products and classes of products. These are, in the order of appearance in that book, the following: cleansing creams and lotions, emollient creams and lotions, hormone creams, baby toiletries, hand creams and lotions, skin lighteners and bleach creams, suntan preparations, beauty masks, foundation makeup, face powders, rouge, lipsticks, eye makeup, dentifrices, and mouthwashes. An additional eighteen products or classes of products are described in Volume Two.

The third and last volume in the series will follow shortly. It will contain chapters on the physiology of the skin and hair, the physiology of sweat, the use of colors in cosmetics, problems of quality control, preservation, and psychological testing, historical information, legal information: in short, all that the editors deem important on the subject described in our title, Cosmetics: Science and Technology, other than the manufacture and formulation of the individual products, as covered in the first two volumes.

To all those who have collaborated in this venture, we express our gratitude.

THE EDITORS

M. S. Balsam	E. Sagarin
S. D. Gershon	S. J. Strianse
M. M. Rieger	

March, 1972

Notice

The information in this publication is to the best of our knowledge reliable. However, nothing herein is to be construed as a warranty or representation as to the safety, effectiveness, or utility of the raw materials or formulas presented. Users should make their own tests to determine the applicability of the information or the suitability of any products for their own particular purpose. Nor is anything to be construed as an expression of opinion concerning scope of protection afforded by any patent referred to herein.

Preface to the First Edition

The science of cosmetics is as old as recorded history, and perhaps older, and it is as new as any branch of chemistry in this age of rapid technological developments. But whether one dips back into antiquity, or examines the most recent product to come from the laboratory of the cosmetic chemist, one becomes aware that the recorded literature of cosmetic science lags far behind its actual progress in research and technology.

In the fifteenth to seventeenth centuries, the cosmetic literature was limited to the "books of secrets," devoted not only to bodily embellishment but also to medicine, in addition to the care of the home and sundry other subjects. Later, there began to appear formulation and instruction books, and no doubt many a usable formula was set forth in such works. But very few efforts had been made (until a short time ago) to gather the most advanced scientific information on all phases of cosmetic formulation and manufacture, and the books that did appear could not be taken too seriously as compendia of scientific and technological data. Among the exceptions one might cite certain German works, as well as some of the books dealing with perfumery and perfume materials; but the former were little used in English-speaking countries, and the latter seldom dealt with matters outside of the narrow realm of fragrance.

Thus, when two cosmetic compendia were published in English in the early 1940s, they filled a gap that had been felt seriously by all those engaged in this field. The two books, by M. G. de Navarre and R. G. Harry, did more than furnish up-to-date and pertinent information to cosmetic chemists; they demonstrated that most advances in cosmetics—including formulation—could be described without endangering any proprietary interests which individuals or companies held in their own products. With the books by de Navarre and by Harry available to them, the cosmetic chemists and their companies found themselves better informed rather then embarrassed. The discussion of the different cosmetic formulas and procedures actually served to raise the standards of an entire industry, without affecting the privileged position of those companies which, either through patents or trade secrets, controlled formulas or processes superior to those generally known.

vii

The years that passed since the publication of these works have confirmed this position and have justified the enthusiasm with which the books were greeted upon their appearance by cosmetic chemists and technologists. They have also been years of fruitful cosmetic research, characterized by many important developments, e.g., in synthetic detergents and emulsifiers, in antiperspirants, in pressure-packaged cosmetics, and in countless other new products or improvements of older ones. Fortunately for the science and industry which they serve, the authors of these two works are striving to keep their books up to date by means of new editions, either published or in preparation.

By contrast, the editors of the present volume have felt the need for a work of a collective nature. It is well known that the formulation and production of cosmetics actually involves familiarity with numerous materials and processes which vary considerably from one another in many respects, even though the diverse end products are intended ultimately for use on the human body for cosmetic purposes. Consider, for example, face powder and hair lacquer, toilet water and antiperspirant, sunscreen and shampoo, tooth paste and depilatory, lipstick and hand lotion, nail lacquer and shaving cream; here are groups of products which are as dissimilar one from the other as if they had originated in entirely unrelated industries. This is why some chemists who are well versed in certain types of cosmetics may have comparatively little knowledge of others.

Out of this realization grew the determination to bring together a number of authorities each of whom has acquired a thorough familiarity with some special phase of cosmetics or of some related science or technique. It is the editors' sincere hope that this has been accomplished successfully; and so we offer this—the first collective volume in the history of cosmetic literature—with a deep sense of gratification.

With sixty-one men and women engaged in the writing of the fifty-three separate chapters, the problem of duplication and redundancy was a serious one. Certainly, almost every chapter in this book contains material related to that found in some other chapters. The editors have striven to reduce overlapping and repetition to a minimum, while at the same time retaining the essential unity of each individual contribution.

It is not only to the authors who have contributed the several chapters that we are deeply grateful, but also to their respective companies who encouraged them to participate in this project. For a collective volume which presents scientific and technical data—formerly guarded or scattered, and often ill-publicized—symbolizes the coming of age of both a science and an industry. The book demonstrates not only that there is no reluctance to disclose significant details of technological advances in cosmetics, but also that numerous firms were willing to cooperate in order to gain the benefits of this undertaking.

Such an industry cannot but go forward. Furthermore, the contributions to this work tend to illustrate a fact which is but inadequately realized not only by the public, but also by many specialists in other fields, *viz.*, what a remarkable body of research supports each cosmetic product in its several aspects, and notably in regard to its formulation, manufacture, application, and safety in use.

A book of this type owes so much to so many that it is virtually impossible to thank all of them. In addition to the contributors, we are indebted to the following who have reviewed the several chapters or have in other ways assisted the editors:

Mr. Marvin S. Balsam of Standard Aromatics, Inc.; Dr. A. Bass of Plough, Inc.; Hans Behrendt, M.D.; Mr. Martin Brookins of Rayette, Inc.; Dr. Abraham Cantor of West Laboratories; Dr. E. J. Casselman of American Safety Razor Co.; Dr. Dan Dahle (retired) of Bristol-Myers Co.; Mr. M. G. de Navarre of Cosmetic Laboratories, Inc.

Also Mr. Milton G. Eckstrom of Rayette, Inc.; Mr. Walter Edman of Evans Research and Development Corp.; Mr. J. C. Ervin of Procter & Gamble Company; Dr. George Fiero of Esso Standard Oil Co.; Dr. A. L. Fishback of Oxzyn Co.; Dr. O. S. Gibbs of Gibbs Medical Research Laboratory; Mr. Robert Goldemberg of Coty, Inc.; Dr. J. Heyman of The Mennen Co.; Dr. Paul Jewel of Max Factor & Co.; Dr. G. W. Johnston of Warner-Lambert Pharmaceutical Corp.; Mr. George S. Kachajian of Dow Corning Corp.; Mr. Kazmaier of The Mennen Co.

Also Mr. Steve Mayham of The Toilet Goods Association, Inc.; Dr. E. G. McDonough of Evans Research and Development Corp.; Mr. Paul C. Olsen of Topics Publishing Co.; Dr. Sholom Pearlman of the American Dental Association; Dr. A. W. Radike of Proctor & Gamble Co.; Dr. George Reddish of Warner-Lambert Pharmaceutical Corp.; Dr. Kenneth Russell of Colgate Palmolive Co.

Also Dr. Maurice Sage of Sage Laboratories, Inc.; Dr. Robert Sanders of Warner-Lambert Pharmaceutical Corp.; Mr. Milton H. Schwarz of Revlon, Inc.; Dr. Morris Shelanski of Industrial Biological Testing Laboratories; Mr. Frazier Sinclair of Drug and Cosmetic Industry; Mr. Richard Sommers of Caryl Richards Co.

Also F. H. Theodore, M.D.; Mr. Paul E. Traum of A. C. Nielsen Co.; Dr. Emery I. Valko of Onyx Oil & Chemical Co.; Mr. William Wellman of Colgate Palmolive Co.; Mr. Ross C. Whitman of Rayette, Inc.; Mr. Henry J. Wing of Chesebrough-Ponds Co.

Also members of the staffs of E. I. du Pont de Nemours and Co., Inc., Freon Products Division; of General Chemical Co., Genetron Division; and of Toni Company.

Once again, the editors tender their sincere thanks to the contributors as well as to all others who have aided in making this book possible.

THE EDITORS

June 1957

Contents

Note to Readers

*on patents, trade names, abbreviations, and other
information of interest to users of this book*

Trade names: Mention of proprietary products by their trade names
has been permitted, since many of these products are not generally known
by other nomenclature. An index of the trade names used in the text,
giving their description and source, is furnished as an appendix to
this book.

Temperatures: An effort was made to give all temperature data in
degrees Centigrade, in conformity with standard practice of scientific
literature. In some instances, this may be contradictory to the practice in
certain areas of the industry in the United States; in a few cases, therefore,
Fahrenheit temperatures are given, either alone or together with Centigrade.

Abbreviations: Abbreviations of journals listed in the bibliographies
are in almost all instances those used by *Chemical Abstracts.* The full
titles of these journals, their addresses, subscription prices, and the
libraries in which they are found are given in *Chemical Abstracts* for
December, 1956. A few journals cited quite frequently by the contributors
were assigned shorter abbreviations. These periodicals and abbreviations are:

American Perfumer and Aromatics (formerly *American Perfumer and Essential Oil Review*)	*Am. Perf.*
Drug and Cosmetic Industry	*DCI*
Journal of the Society of Cosmetic Chemists	*JSCC*
Proceedings of the Scientific Section of the Toilet Goods Association	*Proc. Sci. Sec. TGA*
Perfumery and Essential Oil Record	*PEOR*
Soap and Chemical Specialties (formerly *Soap and Sanitary Chemicals*)	*Soap*
Soap Perfumery and Cosmetics	*SPC*

Chapter 16

SHAVING PREPARATIONS: SOAPS, CREAMS, OILS, AND LOTIONS

Warren R. Schubert

The origin of shaving, that is, some form of cutting the hair on and about the face, undoubtedly predates recorded history. One of the earliest records attesting to the existence of barbers is mention of them in the Bible, and some have been traced back as early as 400 B.C. in Greece (1). Shaving soaps or aids as such were first described over 125 years ago (2) and, as would be supposed, were prepared by the saponification of cooking fats with potash and soda lye. Since these early attempts, man has been trying to make this process of shaving a more pleasant, convenient, and comfortable necessary ritual.

The total shaving cream market has been constantly growing because of an ever-expanding population and also, especially in the last decade, a vastly growing awareness of personal appearance, as evidenced by the fact that more American men are shaving more frequently now than they did in the last century (3).

Even though many men look upon shaving as a chore, and at best a necessary evil, they are rather particular about the shaving preparation they will use. Although some men have definite preferences and will use only one brand or type of shaving soap or cream, most are willing to try another product or type of preparation if it can offer any advantages, such as greater speed or more comfort. This can be seen by the evolution of various types of products offered by manufacturers, and the fact that they are constantly working to improve their existing products and create new ones which will appeal to or satisfy the shaving male.

Preferences in shaving preparations have undergone definite changes. For the longest time soap dominated the market, first in the form of the cake or bar, then in the mug, stick, or powdered form. These in turn gave way to

1

the cream-type products, both lather and brushless. Again these products encountered competition from the invention of the dry electric shavers with their claimed advantage over the old-fashioned wet methods. Then the advent of the aerosol age and the aerosol shaving creams caused another change in men's shaving preferences, one that to this day accounts for the majority of shaving preparation sales.

In the year 1965, aerosol shaving cream accounted for over one-half (57%) of all shaving creams sold in the United States (4). By 1966 this figure was increased to almost two-thirds (65%) (5). Interestingly, the profile of users shows that convenience was the biggest factor in the growth of aerosols and that the greatest users of the brushless and lather type of products were mostly men in the age bracket of 45 and over.

Regardless of the type of product chosen, most must possess certain attributes which will make them useful, necessary, and salable. Some of the more important properties of shaving preparations are the easy and rapid production of a copious lather; resistance to rapid drying or collapse while on the face during the normal length of time needed for shaving; freedom from the possibility of causing skin irritation, especially to the mucosa of the eyes, nose, etc.; rapid softening of the beard and sufficient viscosity to hold the hairs erect to facilitate cutting the individual hairs; and lubrication to make the razor glide over the face more easily and painlessly. Also, as with any consumer product, the stability of the preparation must be adequate at all temperatures corresponding to all possible geographical areas where one might be expected to use it. These preparations should not cause rusting or dulling of razor blades, although with the advent of the stainless steel blades this is no longer as important a consideration.

The evaluation and testing for many of these criteria are largely subjective and ultimately rest with the user and shaver. However, some researchers have attempted to devise methods of evaluation. Hair-softening studies were conducted by Valko and Barnett (6) and by Hollander and Casselman (7). These were primarily measurements of the swelling of hair by water and its effect on the softening or cutting strength. Hollander and Casselman determined that with the use of 120°F water for presoftening the beard, a minimum of $2\frac{1}{2}$ to 3 min would be needed to attain satisfactory softening prior to shaving. This preparation time increased with a decrease in water temperature.

Ross and Miles (8) devised an apparatus and published a method for measuring the foaming capabilities of soaps and detergents, and Scott and Thompson (9) were able to measure the consistency or viscosity of foams.

Many shavers have found that the passage of the razor edge across the face was invariably accompanied by the production of superficial cutaneous abrasion of the skin. These abrasions more often than not were invisible but resulted in subsequent irritation (10). A study of the skin trauma caused by shaving (11) was made by careful examination of the scrapings and residues obtained from the faces of men who had shaved. The skin scrapings were found to be composed of hair with large amounts of varied epithelial components, both nucleated and nonnucleated. This localized trauma could be increased with the factors that tend to promote a closer shave. These factors would be the use of a sharp new blade, the use of thin lathers or plain water, excessive stretching of the skin or the raising of the hair follicles as after the use of a pilomotor agent, or shortened preparation times. It is to prevent, or at least reduce, this trauma that the use of thicker lathers is preferable, and that emollients, lubricants, and other additives are added to most shaving preparations.

Shaving Soaps, Sticks, and Powders

Shaving Soaps

Shaving soaps originally were prepared in bar or cake form, then in the shaped mug or bowl form. They all had to be applied with a shaving brush to produce a sufficient quantity of lather. Although the bar shaving soap may appear physically to resemble some ordinary bar toilet soaps, there are changes that must be made in the formulation in order to meet the requirements of a shaving soap.

A shaving soap must lather quickly and copiously, which is a property of the coconut oil soaps. However, the lather must be thick or dense in texture and reasonably long lasting, which is more characteristic of the fatty acid soaps of palm oil or tallow derivation. Soaps of this second group lather more slowly than the coconut soaps but have better lathering properties, and so some combination will usually give a better product than will the soap of any single fatty acid or oil alone. The ease of lathering also depends on the solubility of the soaps, and potassium soaps in general are more soluble than the corresponding sodium soaps. Nevertheless, potassium soaps alone are usually softer, and a certain proportion of sodium soap must be present to give body to the mixture. In addition, various soaps have different potentials for causing irritation. Emery and Edwards (12) conducted a study based on the production of pure soaps and the use of patch testing and classified them in their order of irritancy. See Table I, which lists them in order of *decreasing* irritancy. Emery and Edwards concluded that in general the potassium soaps are more capable of producing irritation than are the sodium soaps, but that the differences may not be due to alkalis alone but also may be associated with solubility, double bonds, and pH.

A typical formula for the manufacture of a bar or bowl shaving soap is represented in Formula 1 (13).

TABLE I.   Irritancy of Soaps on the Skin

Sodium soaps		Potassium soaps	
Lauric	$(2)^a$	Lauric	(1)
Myristic	(4)	Myristic	(3)
Linoleic	(5)	Linoleic	(6)
n-Capric	(7)	n-Capric	(10)
Oleic	(8)	Oleic	(11)
Ricinoleic	(9)	n-Caprylic	(17)
Palmitic	(15)	Palmitic	(13)
n-Caprylic	(12)	Stearic	(14)
Stearic	(18)	Ricinoleic	(16)

^{a} (   ) Decreasing order for all types.

Formula 1.   Bar or Bowl Shaving Soap

Tallow	15 to 30 parts
Coconut oil	10 to 15
Stearic acid	10 to 50
Sodium hydroxide (40°Bé)	5 to 10
Potassium hydroxide (38°Bé)	10 to 15

Procedure: This soap may be manufactured by the semiboiled method of adding the sodium hydroxide to the mixture of tallow and coconut oil in a crutcher heated to 110 to 120°F, after which the potassium hydroxide is added, followed by the melted stearic acid. The crutcher is then heated to about 150°F and held until saponification is completed. Adjustments for free alkali or fatty acid are then made. The soap can then be poured into molds while still fluid or it can be chipped, dried, milled, plodded, and pressed into suitably shaped cakes.

Many additives and modifications have been suggested to improve upon the basic soap type of system. Doran (14) has suggested the use of triple-pressed stearic acid in place of tallow, provided it is used in conjunction with a vegetable oil to enhance lathering. Myers (15), on the other hand, claims that a 70:30 mixture of palmitic/stearic soaps, instead of coconut oil soap, is less irritating and lathers as well. German patent 421,490 (16) claims the use of superfatting agents for soap of the type of monodecyl, dodecyl, tetra-decyl, or hexadecyl ethers of diglycerol amines such as dodecyl (β-dihydroxy-propyl) amine. Albumen decomposition products (17) of high molecular weight, such as lysalbinic acid and protalbinic acid acylated at the nitrogen with higher fatty acids of carbon length 12 to 18, are claimed to aid in lime soap dispersion and also to help retain moisture in the lather, thereby re-tarding dry-out. Smith (18) claims that alginic acid or its salts of potassium or sodium added to shaving soap at 0.1 to 20% will soften the fatty material or oils in the hair better than soap alone and, because of their high viscosity, will create, in conjunction with soap, a firm, lasting foam.

Styptic or hemostatic materials, such as the lipid-soluble fraction of soybeans (19), and adrenalin and its salts (20), have been incorporated into shaving soaps so that they give the shaving preparation a mild astringent action which is sufficient to prevent or stop bleeding caused by accidental cutting by the razor blade. Also, the addition of 10 to 15% talc (21) is said to increase the persistence of the lather and reduce the danger of scratching and cutting by aiding the razor to glide over the face more readily than when ordinary shaving soap is used. In 1937, before the advent of the stainless steel razor blade, a United States patent (22) recommended as an additive to shaving preparations oxidizing compounds such as chromic acid, chromates, dichromates, chlorates, perchlorates, and perborates. The purpose was to produce a passive condition of the iron or steel of the blade, so that the blade would efficiently resist corrosion, thereby prolonging its usefulness through more shaves.

In order to obtain closer, cleaner-appearing shaves, the use of pilomotor agents has been suggested. These materials, when applied to the skin in concentrations of 0.1 to 10%, are supposed to cause contraction of the hair follicle muscle (*arrectus pilorum*), thereby projecting the hair fiber further out of the follicle by about 0.2 to 0.3 mm. If the beard hair is shaved off while in the extended position, the hair stub will withdraw lower into the skin after relaxation of the muscle and give a cleaner appearance. Several patents for alkali-stable materials have been granted. These are for the types 2-(phenyl-amino)-1,3-diazocyclopentene-(2) such as 2-(2′,6′-dichlorophenyl)-amino-diazocyclopentene-(2) (23), 2-(3′-hydroxyphenyl)-morpholine and its acid addition salts (24), and 2-aminoimidazolines and its salts such as 2-benzyl-aminoimidazoline hydrochloride (25).

Shaving Sticks

Shaving sticks are usually produced in a very dry, firm form which is rubbed onto the moistened skin, then worked into a lather with a brush. This product is not significantly different from the bar shaving soap and is much like the mug or bowl soap.

An example of a shaving stick is Formula 2 (26).

These products usually possess only small amounts of glycerol, are composed mainly of stearic acid, and contain larger percentages of potassium stearate than the bar soaps.

A nonsoap-containing, nonlathering shave stick was patented in 1939 (27). This product was to be applied directly to the moistened skin, at which time oxygen was released from an alkali peroxide such as magnesium peroxide. This peroxide was claimed to be capable of conditioning the beard

Formula 2. Shaving Stick

Stearic acid	10.5 parts
Coconut oil	3.0
Potassium hydroxide (50°Bé)	5.5
Sodium hydroxide (45°Bé)	0.4
Glycerol	1.0

Procedure: The method for the manufacture of this type of product is essentially the same as for the bar soap and mug soap. However, the moisture content must be less than 15% and, after being chipped, dried, and milled, must be treated with a soap plodder of great packing power to give it a rigid shape and maintain it in that form.

for shaving by its depilating ability in the decomposition or dissolution of the hair substance.

Shaving Powders

This type of product is relatively unknown now, but at one time was popular with barbers. This was because the powder could easily be dispensed into a shaving mug prior to the addition of water and the production of a lather using a shaving brush. The product was also more sanitary in that a fresh quantity of powder was used for each application. A typical product of this type is Formula 3 (28).

Formula 3. Shaving Powder

Stearic acid	5.5 parts
Coconut oil	1.1
Sodium hydroxide (35°Bé)	1.4
Potassium hydroxide (50°Bé)	1.7

Procedure: This is manufactured in the same fashion as the bar soap or mug soap, except that the moisture content must be no higher than about 3% in order to avoid lumping after the soap has been pulverized. In addition, other dry pulverized soaps may be added. The inclusion of about 5% talc or other inert mineral will help to prevent caking and keep the powder free flowing.

Lather Shaving Cream

Essentially, a lather-type shaving cream contains ingredients similar to those of the bar shaving soap. However, because of the inclusion of a greater amount of water and the creamy or pasty consistency desired, this type of product is subject to all of the usual stability problems associated with bar soaps and, in addition, the problems of viscosity stability and product separation.

As with the bar soap, the type of fatty acids and their ratio, as both the potassium and sodium soap, greatly affect the foam which is subsequently produced. In the lather creams these ratios affect not only the foam qualities, but the consistency and stability as well. Small changes in these ratios can greatly change the viscosity or consistency of a cream. Sodium soaps, for example, tend to produce a stringy and firmer cream, and potassium gives a cream that is softer and better-lathering but possibly not as stable. Here it is for the formulator to decide which qualities are most important to the product he wishes to make. A typical starting formula can be found in Formula 4.

Formula 4.   Lather Shaving Cream

Stearic acid	20 to 40%
Coconut oil or fatty acids	6 to 10
Glycerol	5 to 15
Potassium hydroxide	2 to 6
Sodium hydroxide	1 to 3
Vegetable or mineral oil	1 to 5
Water	q.s.

Procedure: The usual method of preparation calls for saponification of the melted stearic acid and coconut oil at about 180°F. This is followed by the addition of melted, excess superfatting fatty acids or oils. Then heated water and glycerol are added. One very important note is that the entrapment of air is very possible during the manufacture of this type of product. Great care must be taken, and the batch should be cooled with a minimum of agitation or, preferably, while under a vacuum. The latter method is most widely employed at this time.

The addition of borax in minor amounts of 0.1 to 1.0% has been found to have a marked effect on the viscosity of the creams and is sometimes employed for this purpose. Other ingredients can be added to the lather cream to help improve the lather, make the face feel better, retain moisture, or make the lather more lubricating. Most creams contain superfatting agents which serve a twofold purpose: first, to neutralize any free alkali that might be present and, second, to help stabilize both the cream and the lather. The most common superfatting agents are free stearic acid, free coconut oils or other vegetable oils, mineral oil, or lanolin. However, lanolin or its derivatives are normally used in lesser amounts for their emollient properties. Significant quantities of glycerol (5 to 10%) are usually used to help keep the cream soft and pliable and improve the lather by retaining moisture. Propylene glycol or sorbitol may be substituted for glycerol, if it is preferred. Menthol is sometimes added for the apparent cooling effect it imparts or because the odor is preferred by some men. Hexachlorophene has been added by some manufacturers for its contribution as a germicide. Triethanol-amine is usually avoided in lather creams because it will often discolor with age and, upon hydrolysis, can produce ammonia.

The use of fractionated acids from coconut or palm kernel oil has been claimed (29) to make nonirritating soaps with good lathering properties. The soaps of almost completely hydrogenated vegetable oils or animal oils (such as cottonseed, peanut, soybean, sesame, corn, sunflower, olive, and linseed) and tallow, lard, oleo, etc., have been recommended (30) for their improved high-temperature stability, in preference to soaps made with commercial stearic acid or unhydrogenated oils. As mentioned previously, alkali-stable pilomotor agents, such as derivatives of 2-aminoimidazoline (31), have also been utilized in lather creams. Since many lather creams today are packaged in collapsible aluminum tubes, 0.1 to 0.4% sodium silicate can be added to prevent corrosion (32).

Brushless Shaving Cream

The brushless shaving cream resembles lather shaving cream in appearance. There the similarity ends. Whereas the lather cream is a soap which is intended to produce foam, the brushless cream's purpose is not to produce a foam. Brushless shaving creams are, in essence, oil-in-water emulsions of the vanishing cream type. They possess certain possible advantages over shaving soaps, such as greater convenience and speed since they do not require a brush for application, and they give a more comfortable shave because of their greater lubricating ability and subsequent reduction in razor pull or drag. They also leave the face with a thin coating of oil or grease after shaving, which makes the skin feel less irritated and softer.

Since the brushless creams do not contain any major amounts of soap, they do not of themselves soften the beard, and this function must be accomplished by first washing the beard with soap and water before application of the cream. The brushless cream does keep the hairs of the beard erect, retains moisture well, and above all provides a maximum of lubrication between the razor blade and the skin.

A simple starting formula for the manufacture of a brushless shaving cream is found in Formula 5.

Many additives may be incorporated to improve upon the product's consistency, lubricity, wetting ability, or feel. Borax in concentrations of 0.1 to 0.5% may be used to vary consistency or viscosity. Humectants such as glycerol, propylene glycol, or sorbitol may be used for their moisturizing ability and their effect on softening the cream. Emollients, such as lanolin and its derivatives, or fatty alcohols, such as cetyl alcohol, may be added for the unique feel that they impart to the skin. The use of various wetting agents,

Formula 5.   Brushless Shaving Cream

Stearic acid	10 to 35%
Mineral oil and/or petrolatum	5 to 15
Soap or surfactant	1 to 3
Glycerol	1 to 10
Water	q.s.

Procedure: The manufacture of this type of product is in accordance with the making of any oil-in-water emulsion; namely, the hot oils at 180°F are added to heated water at 175°F or it is possible in some instances to add the water to the oils. Again, adequate mixing is necessary and the entrapment of air must be avoided.

such as the sulfonated aromatic hydrocarbons (33), sulphuric acid esters of lauryl alcohol (34), fatty acid amides (35), and phosphatides such as lecithin (36), has been suggested to increase beard softening. Gums, such as methyl cellulose, tragacanth (37), alginates, and carragheenates (38), have been suggested. These materials add stiffness or body to the cream, help retain moisture, and possibly provide more slip to the face.

Brushless creams have also been prepared with the use of major amounts of glyceryl or glycol stearates rather than stearic acid. These creams are somewhat more translucent in appearance than standard creams, but also are somewhat better in their emollient action.

An example of this type is seen in Formula 6 (39).

Formula 6.   Brushless Shaving Cream

Glyceryl monostearate	10.0%
Mineral oil	3.0
Lanolin	5.0
Glycerol	3.0
Stearic acid	2.0
Potassium hydroxide	0.1
Water	q.s.

Shaving Oils and Lotions

Another class of shaving preparations has existed which is neither soap nor brushless cream in the classical sense. These products are various compositions of oils and lotions which are applied before another shaving preparation, as a pretreatment, or may be used alone for the purpose of shaving with a razor blade. Some are merely intended to improve the shave by performing the function of beard softening. One of the earliest of this type was patented in 1927 (40) and suggested the use of a mixture of sugar, borax, turpentine, and water for treating the beard prior to the application of a shaving soap. A more up-to-date version of this is one based upon a 0.5 % solution of a polyoxyalkylene derivative of sorbitan monolaurate in witch hazel (41).

Other products have been claimed which depend on gums or other mucilaginous substances to provide additional lubrication on the skin other than that obtained with the regular soap-type shaving preparations. These materials are intended to act as primary lubricants, usually in combination with additives such as surfactants, mineral oil, and lanolin derivatives. The most often employed have been starch and gelatin (42), sodium carboxy-methyl cellulose (43), acrylamides and acrylates (44), and alkyl cellulose ethers as in Formula 7 (45).

Formula 7. Shaving Lotion

Cellulose alkyl ether	73.70%
Glycerol	5.00
Sodium lauryl sulfate	1.00
Mineral oil	15.00
Perfume	0.30
Water	q.s.

A number of people have also made claims for the use of various silicone fluids of the dimethylpolysiloxane type as a lubricant to be used alone or in addition to a regular shaving preparation. To these silicones have been added alcohols and surfactants as in Formula 8 (46), octadecanol and mineral oil (47), isopropyl myristate (48), and Carbopol (49).

Formula 8. Shaving Oil

Methyl phenyl polysiloxane	5 to 25%
Dimethyl polysiloxane	5 to 25
Nonionic surfactant	1 to 25
Ethyl or isopropyl alcohol	10 to 80
Quaternary ammonium salt	0.0001 to 0.0010
Water	10 to 80

REFERENCES

1. Barber: in *New standard encyclopedia*, Vol. 2, Funk & Wagnalls Company, New York, 1931, p. 457.

2. Guest, H. H.: Shaving soaps and creams, in Sagarin, E., *Cosmetics, science and technology*, Interscience Publishers, New York-London, 1955, p. 422.

3. Lesser, M. A.: Shaving preparations, *Soap*, **27**: 44 (Oct. 1951).

4. Anon: Carbide tells results of shave cream study, *Deterg. Age*, **3**: 34 (July 1966).

5. Anon: Consumer attitudes toward shaving creams, *Deterg. Age*, **4:** 40 (June 1967).

6. Valko, E. I., and Barnett, G.: A study of the swelling of hair in mixed aqueous solvents (1), *JSCC*, **3:** 108 (1952).

7. Hollander, L., and Casselman, E. J.: Factors involved in satisfactory shaving, *J. Am. Med. Assoc.*, **109:** 95 (1937).

8. Ross, J., and Miles, G. D.: An apparatus for comparison of foaming properties of soaps and detergents, *Oil and Soap*, **18:** 99 (May 1941).

9. Scott, G. V., and Thompson, W. E.: Measurement of foam consistency, *J. Am. Oil. Chem. Soc.*, **29:** 386 (1952).

10. Gettemuller, G. D.: U.S. Pat. 1,991,501 (1935).

11. Bhaktaviziam, C., Mescon, H., and Matoltsy, A.: Shaving, *Arch. Dermatol.*, **8:** 874 (1963).

12. Emery, B. E., and Edwards, I.. D.: The pharmacology of soaps, *J. Am. Pharm. Assoc., Sci. Ed.*, **29:** 251 (1940).

13. Day, D. I.: Making modern shaving creams, *Am. Perf.*, **4:** 47 (1942).

14. Doran, G. F.: U.S. Pat. 1,771,707 (1930).

15. Myers, L. D.: U.S. Pat. 2,298,019 (1942).

16. Henkel & Cie, G.m.b.H.: Ger. Pat. 421,490 (1934).

17. Sommer, F., and Nassau, M.: U.S. Pat. 2,100,090 (1937).

18. Smith, R. F.: U.S. Pat. 1,940,026 (1933).

19. Kröper, H., and Thomae, E.: U.S. Pat. 2,185,255 (1940).

20. Sulzberger, N.: U.S. Pat. 1,744,061 (1930).

21. Smith, R. F.: U.S. Pat. 1,667,993 (1928).

22. Fash, R. H.: U.S. Pat. 2,074,833 (1937).

23. Zeile, K., Hauptman, K. H., and Stahle, H.: U.S. Pat. 3,190,802 (1965).

24. Thomä, O.: U.S. Pat. 3,296,076 (1967).

25. Berg, A.: U.S. Pat. 3,296,077 (1967).

26. Lesser, M. A.: *Op. cit.* (ref. 3), p. 77.

27. Dzialoschinsky, V., and Deutschland, G.: U.S. Pat. 2,143,060 (1939).

28. Lesser, M. A.: *Op. cit.* (ref. 3), p. 106.

29. Dreger, E. E., and Ross, J.: U.S. Pat. 2,462,831 (1948).

30. Dehn, F. B.: Brit. Pat. 454,660 (1936).

31. Anon: Pre-shaving preparations containing pilomotor agents, *Schimmel Briefs*, No. 347 (February 1964).

32. Blough, E., and Churchill, H. V.: U.S. Pat. 1,912,175 (1933).

33. Kritchevsky, W.: U.S. Pat. 2,167,180 (1939).

34. Kritchevsky, W.: U.S. Pat. 2,144,884 (1939).

35. Kritchevsky, W.: U.S. Pat. 2,134,666 (1938).

36. Kritchevsky, W.: U.S. Pat. 2,164,717 (1939).

37. White, S.: U.S. Pat. 2,148,285 (1939).

38. Hilfer, H.: Shaving creams, *DCI*, **67:** 562 (Oct. 1950).

39. Kalish, J.: Cosmetic manual, *DCI*, **45:** 173 (1939).

40. Poole, W. E.: Brit. Pat. 274,659 (1927).

41. Sharawara, W. H.: U.S. Pat. 3,063,907 (1962).

42. Regard, G. L.: Brit. Pat. 260,268 (1927).

43. Mueller, A. J.: U.S. Pat. 3,072,535 (1963).
44. Pye, D. J.: U.S. Pat. 3,072,536 (1963).
45. Bird, J. C.: U.S. Pat. 2,085,733 (1937).
46. Harrison, B.: U.S. Pat. 3,136,696 (1964).
47. Erickson, R.: U.S. Pat. 3,178,352 (1965).
48. Kass, G. S.: U.S. Pat. 3,185,627 (1965).
49. Fainer, P.: U.S. Pat. 3,314,857 (1967).

Chapter 17

PRESHAVE AND AFTERSHAVE PREPARATIONS

SAUL A. BELL

The process of shaving, even under optimum conditions, involves significant skin trauma. In less favorable circumstances, this can amount to discomfort, irritation, and actual physical damage to the skin. The disagreeable and undesirable aspects of wet razor shaving can be mitigated to some extent by the intelligent choice of shaving instrument and the use of a soap or soap-based product that has been properly formulated (1). The shaving preparation alone, however, cannot be depended on always to prepare the face adequately for the shave, to keep it comfortable through the shave, and to leave it refreshed afterward. The shortcomings in this regard of shaving soaps, whatever their form, have encouraged the development of a significant number of useful preparations as accessories to the shave. The purpose of preshave and aftershave preparations is to supplement and complement the shaving preparation in these accessory but important functions. In the special case of the electric "dry" shaver, the preshave functions as the shaving preparation itself.

These accessories to the shave exist in a variety of forms. Lotions and creams are the more conventional; talc powders also have been used for many years; and solid (stick and gel) forms continue to be popular. The lotions usually are alcoholic or, less frequently, emulsified. Alcoholic aftershave lotions, by use and tradition, have been the classical form for a men's fragrance product. They continue to lead by far in sales despite the strong challenge of colognes in the men's toiletries explosion. The solid preparations generally are talc sticks and sometimes alcoholic gels, and include also styptic pencils and alum blocks. A relatively new form is the quick-breaking aerosol foam.

13

Preshave Preparations

The primary purpose of a preshave preparation is to prepare the beard and the skin of the face more completely and effectively than can be done with the shaving preparation alone. The product can accomplish this by such effects as better softening of the hair, increasing the lubricity of the shaving soap lather, and reducing the sensitivity of the skin to the mechanical and chemical effects of shaving. Products also can be developed that will harden or erect the beard hairs and tauten the skin, if these effects are useful.

Skin Conditioners

The conventional shaving soaps and lather shave creams seldom are lacking in beard-softening action. On the other hand, it is frequently desirable to increase the body and lubricity of their lather and minimize abnormal sensitivity of the skin to their mechanical and chemical effects. Preparations formulated specifically for this purpose were used in barber shops rather than in the home. These were of both the vanishing cream and cold cream types, the former being generally more useful before the shave, the latter after the shave.

The shaver who feels the need for a preshave preparation at home often finds a multipurpose "medicated" skin cream useful. Formula 1 is a modified brushless shave cream reformulated to increase its beard-softening, moisturizing, skin-lubricating, and skin-protecting properties. Menthol and camphor are added for their cooling effect on the skin. A suitable soap-compatible antiseptic can be incorporated into the product, if desired.

Beard Softeners

Nonlathering or brushless shaving creams are high in lubricating action but frequently do not soften the beard rapidly or adequately. This shortcoming is most evident when the brushless cream is applied to the unwashed face or when the preparation of the face before the shave is otherwise incomplete or curtailed as to time. Emulsification of the natural oil on the beard and suspension of the facial soil by a solution of soap or synthetic detergent, applied before the brushless cream, are highly effective in speeding up and making more complete the wetting and softening of the beard by the water contained in the shaving cream (2).

To expedite or improve beard softening, preshave products have been developed. A product of this type is shown in Formula 2. It is based on synthetic surfactants and would be more effective in hard-water areas than a soap-based preparation. A mild shampoo detergent, with more soaplike

Formula 1

Part A

Stearic acid, triple-pressed	20.6%
Diglycol stearate, self-emulsifying	2.5
Mineral oil 55/65	4.0
Lanolin, anhydrous	1.0
Sulfonated castor oil, 70%	1.0

Part B

Triethanolamine	1.3
Borax U.S.P.	0.9
Water	14.0
Propylene glycol	4.0
Sodium alginate, 2% mucilage (preserved)*	50.0

Part C

Menthol	0.1
Camphor	0.1
Perfume oil	0.5

Procedure: Heat A until melted and homogenous at 70°C. Prepare a 2% mucilage of sodium alginate, then add other ingredients of B, and heat to 70°C. Add A to B with good agitation and continue stirring down to 45°C. Stir in the menthol and camphor dissolved in the perfume oil. The cream can be poured at 45°C. If a softer consistency is desired, continue slow stirring (avoiding incorporation of air) until cream is cooled below 35°C. Remix briefly next day and package at room temperature.

Formula 2

Duponol WAT	20.0%
Aerosol OT-100%	0.1
Carbitol	3.0
Ethyl alcohol, specially denatured	8.0
Water	68.9

Procedure: Dissolve the Aerosol OT-100% in a mixture of the Carbitol, alcohol, and water. Add the Duponol WAT and mix until uniform. Other brands of triethanolamine lauryl sulfate and dioctyl sodium sulfosuccinate can be used in place of Duponol WAT and Aerosol OT-100%, respectively, if of equivalent content of active ingredient.

* When mucilages of sodium alginate and similar natural gums are prepared in advance of use, it is advisable to preserve with 0.1 to 0.2% of methyl *p*-hydroxybenzoate, or with a combination of 0.15% methyl *p*-hydroxybenzoate and 0.05% propyl *p*-hydroxybenzoate. Synthetic gums and thickeners, such as carboxymethyl cellulose and the Carbopols, can be used here to advantage.

feel, such as a fatty alcohol ethoxy sulfate, may be substituted for lauryl sulfate, and an octyl half-ester for the dioctyl surfactant.

Formula 3 is adapted from a soap shampoo containing a slight excess of amine. An alkyl aryl polyethylene glycol ether (Tergitol NPX) is added to disperse insoluble lime soaps and increase wetting action in hard water.

Formula 3

Coconut oil fatty acids, double-distilled	4.20%
Oleic acid, low linoleic content	5.60
Propylene glycol	5.00
Triethanolamine	2.85
Monoethanolamine	1.26
Tergitol NPX	2.00
Water, demineralized	79.09

Procedure: Mix the fatty acids and stir in the propylene glycol. Add the amines and stir until a clear solution is obtained. No heating is required. Add the Tergitol NPX and dilute with water.

The proportions of coconut oil fatty acid and oleic acid in Formula 3 are based on equivalent weights of 210 and 282, respectively. The proportions of amines are based on 142 for triethanolamine and 61 for monoethanolamine. These proportions allow for some uncombined or free amine to remain in the finished product and improve the beard-softening effect of the preparation without affecting significantly the bland action of the amine soaps.

Both these products can be colored, if desired, and can be perfumed with 0.25 to 0.5% of a suitable perfume oil. If a small amount of terpineol (not over 0.1%) is used to cover the soapy odor, the lower percentage of perfume will be adequate. The addition of about 0.1% menthol will impart a pleasant cooling effect to the shave. It is advisable to preserve with 0.15% of methylparaben or other preservative. The preparation, of course, can be made antiseptic by incorporating hexachlorophene, trichlorocarbanilide, or other soap-compatible antibacterial agent in sufficient amount. Such finished products should be tested, not only for antibacterial action but also for irritation and sensitization characteristics.

It should be obvious from this brief discussion that other effective beard softeners, i.e., liquid creams and jellies, can be formulated by modification of the many shampoo types available. Soaps, synthetic surfactants, and other materials suggested for shampoos should be screened carefully for eye irritation before use in beard softeners. Certain relatively innocuous materials have been found to be markedly irritating to the eyes when used in combination; hence the users of such materials should test the final preparation very carefully for such tendencies by the method of Draize *et al.* (3).

Preelectric Shave Preparations

A constantly increasing number of men now employ the "dry" electric shaver. A preshave preparation, in this instance, has a highly specialized purpose and function. Primarily, it must remove the film of moisture on the skin, or otherwise prevent it from interfering with the smooth passage of the cutting head of the shaver over the beard.

Two different solutions to this problem are popular. The first employs talc to absorb moisture and to leave its characteristic "slip" on the skin. The second depends on an alcoholic lotion to "dry" and tauten the skin by its astringency and, preferably, to leave a lubricating film on the skin. A third approach, the use of pilomotor active compounds to erect the beard hairs, has been suggested in several German patents. Preparations containing such compounds would probably be classified in the United States as "new drugs."

The logical way to provide talc powder in a form convenient for use to prepare the skin before an electric razor shave is to fashion it into a stick. This avoids the inconvenience of getting powder on the hands in order to apply it to the face and, if one is using an electric shaver when fully clothed, having it drop on the clothing.

A simple and obvious method for preparing a talc stick for this purpose is to mix a suitable binder, such as dried calcium sulfate (plaster of paris), with the talc and other ingredients, moisten with water, and then mold. An excellent binder for powdered sticks is colloidal magnesium aluminum silicate (Veegum). Only a relatively small amount is required, and the strength of the stick and the degree of rub-off can be controlled by varying the amount used. The addition of a metallic soap, such as zinc stearate, increases "slip" and improves adhesion. Emollients can be added, as well, but this must be done in such a manner as not to reduce the "slip" or interfere with the ability of the stick to dry the skin.

Obviously, all solid ingredients employed in this product must be soft and free from grit to minimize abrasive action on the cutting head of the electric shaver. These instruments differ significantly in design, and the efficiency of some may be reduced seriously by continued use of a talc-based product. Hence thorough use-testing of the finished product with a number of types of electric shavers is advisable. Formula 4 is an illustration of a talc stick that should be suitable for a variety of electric shaver designs.

A recent patent (15) teaches that the addition of a major proportion of finely divided mica (400 mesh) to wet-molded stick compositions will improve their usefulness as a dry shaving agent.

A method is available for manufacturing a talc stick without incorporating a binder in the talc (4). Instead, the powder is formed into a round stick by

Formula 4

Part A

Zinc stearate	5.0 parts
Iron oxide pigments	q.s.
Magnesium carbonate, light	2.0
Perfume	q.s.
Talc, to make	98.5

Part B

| Veegum | 1.5 |
| Water | 30.0 |

Procedure: Adsorb the perfume completely on the magnesium carbonate, add the zinc stearate and pigments, and disperse all thoroughly in the talc. Prepare an aqueous dispersion of Veegum by adding slowly to the water with good agitation until smooth. Add B to A and mull to a smooth paste. Add enough water, if required, to produce a flowable paste. Pour into molds, allow to dry until hard, and then completely oven-dry the sticks. (Less water should be employed to produce the proper consistency of the paste if the sticks are to be made by extrusion.)

the application of pressures varying from 450 to 600 psi, and the stick is then coated, except on the end, with a suitable material to protect it against cracking, crumbling, and chipping. The vinyl resins, such as polyvinyl chloride and polyvinyl acetate dissolved in ethyl acetate, are preferred, although other film-forming materials, either in solution or melted, can be applied by spraying, brushing, or dipping. The thickness of the coating must be sufficient for adequate protection, but it must not be so thick or so hard that it does not wear away in use at substantially the same rate as the powder itself.

The alcoholic preelectric shave lotion may be either astringent or oily (5). The former, theoretically, should make the hair stiff and dry, assisting in making it stand more upright. This is achieved by employing a high concentration of alcohol for its dehydrating effect, and by adding a mildly astringent material, such as lactic acid or zinc phenolsulfonate. Menthol and camphor serve to diminish the discomforts of minor trauma.

The oily type lubricates the beard and skin by depositing a thin film of oil on the face. This prevents drag and pull of the cutting head against the skin, most troublesome in warm, humid weather, and generally improves performance by reducing friction.

Formula 5 is an example of the astringent type, and Formula 6 of a lubricating lotion. In Formula 5, a more distinctive product can be achieved by replacing the witch-hazel extract with water and perfuming with an attractive bouquet. The quantities and relative proportions of menthol,

Formula 5

Zinc phenolsulfonate	1.8%
Ethyl alcohol, specially denatured	40.0
Menthol	0.1
Camphor	0.1
Witch-hazel extract, distilled	58.0

Procedure: Dissolve the zinc phenolsulfonate, menthol, and camphor in the alcohol and dilute with the distilled witch-hazel extract. Add color, if desired, and filter clear.

Formula 6

Isopropyl myristate	20.0%
Ethyl alcohol, specially denatured	80.0
Perfume oil	q.s.

Procedure: Dissolve the isopropyl myristate and the perfume oil in the alcohol, add color if desired, and filter clear.

camphor, and perfume oil should be adjusted to give the best fragrance effect. The alcohol content can be increased if necessary to dissolve the perfume oil completely.

The lower alcohol esters of the higher fatty acids, such as isopropyl myristate, are effective lubricants for this application, and the diesters of adipic and sebacic acid are particularly useful. Other alcohol-soluble, oily materials, as silicone fluids, can also be used if bland and of suitable solubility and viscosity. If the lubricant is chosen judiciously and employed in correct proportion, this type of preparation can be highly useful, enabling the user to get a close and comfortable shave with an electric razor even under conditions of high temperature and high humidity. Also, it can be said to have an advantage over talc-based preparations or astringent lotions in that it will lubricate the electric shaver rather than clog its moving parts or dull its cutting edges. A combination of esters with the right balance of viscosity, film strength, and surface tension, will lubricate well during the shave and leave little or no oiliness afterwards. The finished product should be checked for any effect on the plastic materials used in electric shavers because certain esters are solvents and plasticizers.

Aftershave Preparations

The general function of an aftershave preparation is to relieve the feeling of tautness and discomfort caused by shaving. Its purpose is to refresh and cool the skin, soothe minor irritations, and impart a feeling of well-being. In the formulation of such products, one or more special characteristics can be given emphasis. In some cases this will dictate the physical form of the product and its relative efficacy when used after different types of shaves.

Clear Lotions

The alcoholic lotion is by far the most widely used aftershave preparation. When properly formulated, it is especially effective after the usual wet-razor shave. In earliest form, aftershave lotions were simple fragrance products based on the use of floral waters, witch-hazel extract, bay rum, and similar materials. For many years such materials were listed in official and quasi-official pharmacopoeias and formularies and were widely used both as such and in simple compositions to finish off the shave. Their significant content of alcohol and aromatic oils made them particularly pleasurable for the user.

Monographs on the preparation of floral waters can be found in the earlier revisions of the *Pharmacopoeia of the United States*. The *National Formulary*, tenth edition, contained monographs on the preparation of orange flower water, distilled witch-hazel extract, and compound myrcia spirit (bay rum). The aromatic waters mentioned are saturated solutions of odoriferous principles prepared by distilling the respective plant materials with water, separating the excess volatile oil, if any, from the clear water portion of the distillate, and adding alcohol, if necessary, for preservation. Bay rum is a hydroalcoholic solution of volatile oils customarily chosen from among bay oil (myrcia), pimenta oil (allspice), clove oil, and cinnamon leaf oil.

Poucher (6) gives a basic formulation for bay rum, as shown in Formula 7.

Formula 7

Bay oil	0.20%
Pimenta oil	0.05
Ethyl alcohol	50.00
Jamaica rum	10.00
Water	39.75
Caramel	q.s.

Because rum rarely is used, he suggests adding a very small amount of ethyl acetate and a trace of heptaldehyde to simulate its bouquet when only ethyl alcohol and water are used as the vehicle.

Compound myrcia spirit, an official preparation of the National Formulary, tenth edition (7), is quite similar, although it does not bear the synonym of bay rum. It differs in having a higher concentration of bay oil and contains orange oil as well as pimenta oil. It is shown as Formula 8.

Examples of lotions of an early type, based on bay rum or on floral waters, are given in Formulas 9 and 10, both taken from Poucher (8).

Formula 8

Myrcia oil	8.0 ml
Orange oil	0.5
Pimenta oil	0.5
Ethyl alcohol	610.0
Water, to make	1000.0

Procedure: Mix the oils with the alcohol and gradually add water until the product measures 1000 ml. Set the mixture aside in a well-closed container for 8 days, then filter, using 10 g of talc, if necessary, to render the product clear.

Formula 9

Peppermint oil	1%
Glycerol	5
Bay rum	94

Formula 10

Potash alum	2%
Glycerol	3
Menthol, 1% in ethyl alcohol	5
Orange flower water	20
Rose water	20
Witch-hazel extract	50

Fragrance and feel continue to be primary considerations in the formulation of aftershave lotions. In the recent unprecedented growth of the men's toiletries market, colognes are threatening to supplant aftershaves as the dominant fragrance product. This can be said to reflect the current trend to stronger, longer-lasting fragrances. The proliferation of highly scented products has been not merely in the direction of sophisticated modern odors. Citrus blends, classical floral and herbal compositions, and traditional bay rums have reappeared in great numbers.

Before considering the question of fragrance, the modern formulator should endeavor to incorporate the following specific characteristics in his product, to at least some degree:

1. Relief of irritation and tension of the freshly shaven skin.

2. Cooling and refreshing action.

3. Mild astringency.

4. Neutralization of soap left on the skin, to help restore normal "acid mantle."

5. Antibacterial action.

The selection of a suitable perfume undoubtedly is a critical aspect of the formulator's task. The usual considerations of freedom from irritation and sensitization of course apply with special weight in a product used daily on the freshly shaven face, but the question of immediate and continued consumer acceptance of the odor is of primary importance. Among those popular for men's toiletries are the fougere, lavender, spice, sandalwood, and eau de cologne types. In addition, oriental, tobacco, and leather notes have proved their appeal. Because the perfume used is such an important factor in the commercial success or failure of a shaving accessory, its formulation should be placed in the hands of a competent perfumer of experience and originality. The task of the cosmetic chemist is to evaluate the perfumer's suggestions for compatibility and stability in the product. The decision as to whether a popular fragrance or a novel note is suitable for the product should be based on the degree of innovation in the product itself, the kind of merchandising planned for it, and the market in which it must compete. The usual range of concentrations employed is from 0.75 to 1.5 avoirdupois oz/gal, depending on the nature of the compound and the effect desired. The final selection of the odor can be made by controlled panel tests of suitable size.

The matter of "feel" also is subjective, but here the formulator is on somewhat firmer ground. Ethyl alcohol, when employed at the right concentration in an aftershave lotion, produces a mild astringency and refreshing coolness that is unique. Small amounts of menthol enhance this markedly, and the perfume oil likewise contributes greatly to the total effect on the skin. Special effects, such as increased astringency, greater emolliency, and antiseptic, hemostyptic, and anesthetic action, also can be achieved by the use of suitable ingredients.*

The popular aftershave lotions contain ethyl alcohol between 40 and 60% by volume. In the United States a Specially Denatured Alcohol Formula, authorized for such purpose, is employed. Generally one of the several S.D.A. No. 40 variants will be satisfactory, but other authorized formulas can be used if suitable to the lotion being formulated (9). Isopropyl alcohol sometimes is used in products of foreign origin. However, it is not acceptable in a quality product because its characteristic odor, or that of the strong odorants required to mask it, distorts the fragrance.

---

* Cosmetic and toilet preparations may fall under the drug provisions of the Federal Food, Drug, and Cosmetic Act of 1962, depending on the claims made for the product. Claims that a product is antiseptic, hemostatic, or anesthetic will result in its being classified as a drug. This will require that, among other things, the label show the active ingredients and give adequate directions for effective use. Such a product, if it contains active ingredients not generally recognized as safe and effective for this use, may be a "new drug" for which a New Drug Application must be filed with the Food and Drug Administration to obtain approval before marketing.

Concentrations of alcohol higher than 60% by volume cause excessive sting and smarting. Concentrations of alcohol below 40% by volume are difficult to perfume because many perfume oils contain materials with limited solubility. To overcome this, solubilizers can be employed to produce clear solutions (actually transparent emulsions) of perfume compounds in vehicles containing little or no alcohol. These materials belong to the class of nonionic surfactants. The most useful are the ethylene oxide condensation products either of monoesters of sorbitol anhydrides (Tweens) or of alkyl phenols (Igepals and Tritons). Generally, three to five times as much solubilizer (by weight) as perfume is necessary. With certain perfume oils, as much as eight or nine parts of solubilizer may be required for one part of oil.* This increase in the amount of oillike material in the product may cause it to leave an excessively oily feel on the skin. Such products also are lacking in the desirable mild astringency and cooling afterfeel on the face characteristic of lotions containing between 40 and 60% alcohol by volume.

Current products frequently contain some menthol or a menthol-like compound for its cooling effects. Its concentration can be as low as 0.005% or as high as 0.2%. About 0.1% menthol gives ample cooling to allay the pain of a "close" shave. When higher concentrations are used, menthol has an undesirable rubefacient effect and its odor becomes overpowering, causing difficulty in perfuming the product. Lower concentrations of menthol can be potentiated in action by the judicious selection of fragrance compound. Some volatile oils and aromatic chemicals have a mild topical anesthetic effect. In combination with menthol, they are highly effective but must be used with care because they can produce an uncomfortable feeling of numbness about the lips. A potent topical anesthetic is not needed, and use is not advisable because of the possibility of skin sensitization.

After a shave with soap, the skin of the face is quite alkaline and only slowly recovers its normal acidity (10,11). Small amounts of weak acids, such as boric, lactic, or benzoic, in an aftershave lotion will neutralize this alkalinity and help restore the normal, slightly acid, condition of the skin.

The addition of an aluminum or zinc salt will increase astringency and styptic action. The phenolsulfonates of aluminum and zinc are somewhat milder than the chloride and sulfates. Up to 2% of the former can be used, but the lattter are best kept below 1%. Aluminum chlorohydrate probably is the best choice because its higher pH reduces the tendency to cause irritation. The new alcohol-soluble aluminum chloride complexes could be suitable for lotions having a high alcoholic content. It should be noted that styptic

---

* In the selection of a solubilizer, due consideration must be given to its odor, color, and stability, and to its effect on the fragance of the product. Because the optimum effective concentration depends not only on the perfume oil but also on the alcohol content of the vehicle, it must be determined by experiment for each case.

action sufficient immediately to stanch bleeding from minor razor cuts is difficult to achieve in lotions of high alcoholic content because of the solvent action of alcohol on the blood clot.

Emolliency is imparted most readily by the use of humectants. Low concentrations (up to 3 %) of such polyols as glycerol, propylene glycol, and sorbitol generally are employed for this purpose. By controlling the moisture exchange between the skin and air, they assist in softening the skin through rehydration. They do not soften the skin in the manner of the lipophilic emollients, i.e., by complete or partial occlusion of the skin surface.

Glycerol has a long and favorable history of use in this type of product. Propylene glycol has also been used in cosmetics for many years without significant incidence of irritation or sensitization of normal skin. It is less humectant than glycerol, but has greater solvent action for perfume oils, a useful property if low alcohol content is used. Sorbitol, as a 70 % solution of noncrystallizing isomers, likewise is acceptable. Its humectancy falls between those of glycerol and of propylene glycol, but it has less solvent action than either. All these are considered innocuous. Propylene glycol has the lowest viscosity and the highest volatility. Hence it is most useful when one desires that the residual film be at a minimum. Sorbitol has the highest viscosity and its films are least sensitive to extremes of humidity in the atmosphere. Other bland materials that possess humectant properties can be used, such as the polyethylene glycols, provided the films they leave are not tacky or greasy, and they are mild in odor.

Lipophilic emollients such as lanolin and its derivatives, hydrocarbons, phospholipids, and the fatty acid alcohols and esters are difficult to incorporate in conventional alcoholic aftershave lotions because of their limited solubility. They help to counter the potential drying effect of alcohol in these preparations, but unless used judiciously, they tend to reduce the characteristic cooling and refreshing "afterfeel" of these preparations. They are, however, employed to good advantage in the emulsified creams and lotions. If one desires to incorporate an emollient of this type in a hydroalcoholic preparation, this can, of course, be done. The concentration of alcohol need not be increased materially if the emollient is chosen from among the more hydrophilic polyoxyalkylene derivatives of fatty acids, alcohols, and esters. The water-soluble derivatives of lanolin and lanolin alcohols are especially useful. Between 2 and 3 % is adequate. More than 5 % may prove to be unpleasantly oily.

The use of a bland antiseptic in shaving preparations has been recommended as a prophylactic measure (12). The minor cuts and other breaks in the continuity of the skin frequently caused by shaving can be infected by the pathogens present in "normal" skin flora. Infections are particularly common in hard-water areas because the insoluble soap residues left on the razor, towel, and face harbor viable organisms for a considerable time and enable these organisms to enter follicles and sebaceous glands and produce infection (13). An aftershave lotion containing alcohol and a bland, nonirritating, nonsensitizing antiseptic should be of value in minimizing such infections.

The quaternary ammonium salts have been suggested for this purpose. These cationic antiseptics are effective in very low concentrations but are inactivated by soap and other anionic surfactants. Hence the aftershave preparation must be formulated to be compatible with these materials, and they must be used in sufficient amount to compensate for loss of activity in use. Some of the commonly used quaternaries are benzalkonium chloride (Zephiran), cetyl trimethyl ammonium bromide (Cetab), cetyl pyridinium chloride (Ceepryn), diisobutyl cresoxy ethoxy ethyl dimethyl benzyl ammonium chloride (Hyamine 10-X), and N(acyl colamino formyl methyl) pyridinium chloride (Emcol E-607).

These antiseptics are usually employed at concentrations below 0.1 % of active ingredient, calculated on the basis of the pure quaternary ammonium salt contained in the material as available commercially.* Such concentrations generally are sufficient to ensure effectiveness in the presence of soap residues and other contamination on the surface of the skin, and are considered to be free from any irritating characteristics. The addition of about 0.1 % of a neutral or slightly acid salt of ethylenediamine tetraacetic acid, or equivalent concentration of a similar chelating agent, will improve performance and stability of the antiseptic. Finalized formulations containing any antiseptic should be tested carefully for safety and effectiveness in use.

The halogenated phenolic antiseptics are stated to be advantageous in products to be used before, during, and after shaving because they are soap-compatible and have been demonstrated to possess marked substantivity to the skin (14). The recommended concentrations are higher than for the cationic antiseptics. They have rather limited solubilities in the alcohol concentration employed in aftershave lotions and have the further disadvantage of discoloring to varying degrees in contact with iron or when exposed to direct sunlight. Hexachlorophene (G-11) is most commonly used.

Although other antiseptics and fungicides of this type have been proposed, their use in aftershave preparations is not favored, either because of their toxicity or because only insufficient published knowledge is available about their use. The possibility of photosensitization from the halogenated phenolic antiseptics must be considered. Finished formulations containing these compounds should be skin-tested thoroughly in this regard.

Formula 11 is a simple, modern aftershave lotion illustrating the use of

* The quaternary ammonium salts, in some cases, are commercially available only in the form of solutions or mixtures with inert ingredients.

Formula 11

Ethyl alcohol, specially denatured	50.0%
Sorbitol, 70% solution	2.5
Perfume oil	0.5
Menthol	0.1
Boric acid	2.0
Water, demineralized	44.9

Procedure: Dissolve all the ingredients completely in the alcohol and dilute with water, using good agitation to prevent occurrence of local concentrations of water high enough to throw the less water-soluble materials (i.e., perfume and menthol) out of solution. Allow to stand, preferably with adequate chilling, until the poorly soluble constituents of the perfume oil have agglomerated, and then filter clear. If the lotion is to be colored, add color solution to the clarified preparation after it has warmed back to room temperature, and refilter without further chilling to obtain a brilliant "polish."

mild acid to correct the alkaline reaction of the skin after shaving with soap. The sorbitol can be replaced with glycerol if a more active humectant is desired, or with propylene glycol if a less residual film is preferred. Combinations of two or more humectants are useful to obtain effects not possible with a single humectant. Menthol can be reduced or eliminated for less cooling action, or can be increased or potentiated with aromatic compounds having a topical anesthetic effect. Other mild acids can be used, in place of boric acid, at suitable low concentrations. The percentage of perfume will depend on the type of perfume oil used, its compatibility with menthol, and the fragrance effect sought.

The percentages presented in Formulas 11 and 12 are percentages by weight. It is common practice in large-scale operations also to manufacture on a weight-to-volume basis. Such a procedure is convenient because larger volumes of alcohol and demineralized water are most easily metered directly into the mixing tanks from their storage tanks, and the finished product is metered from the bulk-manufacturing to the bottle-filling area. By using recording meters, permanent records of alcohol consumed and product manufactured are obtained automatically. This simplifies record-keeping for alcohol tax purposes, because it is required that such records be kept and reports made on a volume basis. Meters used for this purpose, of course, must be compensated for temperature variations. In manufacturing on a volume basis, it is essential that all volumetric measurements be made at the standard temperature chosen or that suitable corrections be made for deviations from that temperature.

Formula 12 is an antiseptic aftershave lotion containing a quaternary ammonium salt, a surface anesthetic, and menthol. *In vitro* tests indicate that

Formula 12

Hyamine 10-X	0.25%
Ethyl alcohol, especially denatured	40.00
Menthol	0.05
Benzyl alcohol	0.25
Water	59.45
Perfume oil	q.s.

Procedure: Dissolve the menthol, benzyl alcohol, and perfume oil in the alcohol, dilute with the water, and clarify by chilling and filtering. Add the Hyamine solution and color, if desired, and polish the product. The cationic antiseptic is added after clarification because of the possibility of loss through adsorption on the insoluble perfume constituents that are filtered out. The addition of a low concentration of glycerol or other humectant to this formulation will increase its emolliency.

organic contamination on the surface of the skin will lower the germicidal efficiency of this formulation, but it will need to retain only 20% of its efficiency to be effective within 1 min. If a chlorinated phenolic antiseptic is used in place of the cationic antiseptic, efficiency is not affected significantly by soap and other contaminants on the skin. However, a higher concentration of antiseptic is necessary, and the alcohol content must be increased to effect its solution.

Stick Lotions and Gels

Whether, from a technical viewpoint, a product in stick form can properly be called a lotion is questionable. However, the term "stick lotion" is applied to solid aftershave products in stick form, and for a time they achieved some

Formula 13

Ethyl alcohol, specially denatured	80.5%
Perfume oil	1.4
Sodium stearate, purified	6.0
Glycerol	4.0
Propylene glycol	3.0
Menthol	0.1
Water, demineralized	5.0

Procedure: Place all ingredients, except the perfume oil, in a closed, stainless steel or glass-lined steam-jacketed kettle fitted with an agitator and a water-cooled condenser. Heat with stirring and, when the temperature reaches about 55°C, add the perfume oil, preferably through an addition funnel. Continue heating to reflux temperature and stir until completely dissolved. Adjust temperature to 71 to 74°C and pour into molds. Color, if desired, is added by dissolving in the water of the formula.

popularity, but not great success. This preparation is very convenient to use, particularly when traveling, and owed its limited acceptance to novel and functional form rather than to a marked superiority in effectiveness. It tends to leave a heavier residue on the skin than conventional aftershave lotions, and is somewhat alkaline, rather than acid, in reaction. Formula 13 is a typical example.

A more sophisticated product, which can be packaged in tubes, is the clear hydroalcoholic gel, illustrated in Formula 14. The carboxy vinyl

Formula 14

Ethyl alcohol, specially denatured	50.00%
Menthol	0.05
Water	45.40
Carbopol 940 resin	0.75
Diisopropanolamine	0.80
Solulan 98	3.00
Perfume	q.s.

Procedure: Dissolve the menthol in the alcohol and add the water. Disperse the Carbopol 940 with vigorous agitation until completely free from lumps. Reduce agitator speed and add the diisopropanolamine slowly to form a clear gel. Mix the perfume with Solulan 98 and add to the gel.

polymer, Carbopol 940 resin, neutralized with diisopropanolamine, will provide increased viscosity up to a moderately firm gel, depending on the amount used. The acetylated ethoxylated lanolin derivative, Solulan 98, improves application properties and provides residual emollience.

Creams and Emulsified Lotions

Aftershave products can also take the form of creams and emulsified lotions. At the present time, few such products are marketed specifically for aftershave use. Men who find an alcoholic aftershave lotion uncomfortable or irritating, particularly after overexposure to sun, wind, or inclement weather, frequently will use a simple emollient vanishing cream or hand lotion to finish off the shave.

The "witch-hazel foam" or "snow" of Formula 15 can be the basis of a suitable masculine preparation. The slightly more complex emollient vanishing cream of Formula 16 will be acceptable to men if formulated with an eau de cologne, lavender, lilac, or other fresh bouquet. Formula 17 is a hand lotion that can be utilized, as is or modified, to make an emulsified aftershave lotion.

Creams developed specifically for aftershave use indicate a new trend in toiletries for men. A cosmetically elegant, soft cream designed to relieve

Formula 15

Stearic acid, triple-pressed	18.0%
Potassium hydroxide U.S.P.	1.2
Glycerol	5.0
Water	25.8
Witch-hazel extract, distilled	50.0
Preservative	q.s.

Procedure: Dissolve the potassium hydroxide in the water, add the glycerol and preservative, and heat to 80°C. Melt the stearic acid in a separate vessel and heat to 75°C. Add the alkali solution slowly to the melted stearic acid with good agitation. When the mixture has cooled to about 50°C, add the witch-hazel extract slowly with good mixing and continue slow mixing until cool. Cover and let stand overnight. Remix briefly next day before packaging.

Formula 16

Stearic acid, triple-pressed	15.00%
Potassium hydroxide U.S.P.	0.50
Sodium hydroxide U.S.P.	0.18
Cetyl alcohol	0.50
Isopropyl myristate	3.00
Glycerol	5.00
Water	75.82
Preservative	q.s.
Perfume oil	q.s.

Procedure: Dissolve the potassium hydroxide and sodium hydroxide in the water, add the glycerol and preservative, and heat to 80°C. In a separate vessel, melt together the stearic acid, cetyl alcohol, and isopropyl myristate and heat to 75°C. Add the alkali solution slowly to the melted oily phase with good agitation. When the mixture has cooled to about 45°C, add the perfume and continue slow mixing until cool. Cover and let stand overnight. Remix briefly next day before packaging.

tenderness, dryness, and roughness of the skin after shaving has appeared in "prestige" lines. A leading "popular" line has designed a lotion cream for men who find the characteristic cooling astringency of alcohol irritating rather than refreshing on the freshly shaven skin. Both creams are modern, functional types. They are acid in reaction to correct the alkalinity of soap, and they are markedly emollient without oiliness.

Formulas 18 and 19 are basic examples from which such preparations can be developed. The first is a soft cream, the second a heavy lotion. Desired consistency is obtained by adjusting the concentrations and relative proportions of glyceryl monostearate, stearyl alcohol, salt, and polyol. The cationic emulsifiers impart a desirable acid reaction and a rather unique "feel."

Formula 17

Stearic acid, triple-pressed	3.0%
Cetyl alcohol	0.5
Glycerol	2.0
Methylparaben	0.2
Quince seed mucilage*	40.0
Triethanolamine	0.8
Water	48.5
Ethyl alcohol, specially denatured	5.0
Perfume oil	q.s.

Procedure: Heat together the glycerol, methylparaben, quince seed mucilage, triethanolamine, and water to 75°C. In a separate vessel, melt together the stearic acid and cetyl alcohol and heat to 75°C. Add the latter mixture to the aqueous phase with good stirring. When cooled to 40°C, mix the alcohol and perfume and add slowly to the emulsion, with stirring. Discontinue stirring when temperature reaches 30°C, and bottle.

	Formula 18	*Formula 19*
Glyceryl monostearate, pure	10.00%	3.00%
Stearyl alcohol	3.00	1.50
Sorbo	5.00	2.50
Emcol E-607-S	1.00	1.00
Emcol E-607	0.25	0.25
Sodium benzoate	0.10	0.10
Perfume oil	0.30	0.30
Water	80.35	91.35

Procedure: Heat glyceryl monostearate and stearyl alcohol together to 70°C. In another vessel, dissolve the Sorbo, Emcols, and sodium benzoate in the water and heat to 70°C. Add the oily phase to the aqueous phase with good agitation and continue mixing while cooling. Add the perfume at 40°C. Mix slowly until cooled to 25°C, and package.

* The following procedure from de Navarre (16) will give a suitable mucilage. Add the seeds to 20 times their weight of warm water (not over 65°C) containing a preservative (either 1:750 methylparaben, 1:5000 butyl *p*-hydroxybenzoate, or other suitable preservative). Set aside overnight to soak, and strain through a cloth or fine screen. Return seeds to the soaking vessel, add another 20 times their weight of preserved water heated to about 65°C, and again allow to stand overnight. Strain the following day and mix the two solutions. The mixture represents about 2% quince seed mucilage. If cold water is used in place of hot water, the mucilage will be somewhat light in color.

Emolliency and lubricity can be increased by adding lanolin, lanolin derivatives, mineral oil, isopropyl myristate and palmitate, silicone fluids,* and similar materials, either alone or in suitable combinations. To produce special soothing and cooling effects on the skin, these preparations also can contain menthol, camphor, and surface anesthetics. The addition of allantoin is claimed to aid the skin in repairing the superficial damage caused by shaving. If, because of the cationic emulsifiers used, antiseptic properties are to be claimed, the final product must be tested for antibacterial efficiency.

The current medical literature will suggest other useful topical materials for inclusion. Claims for more than cosmetic effects usually will make the product a drug within the meaning of the Federal Food, Drug, and Cosmetic Act of 1962.

A heavily pigmented cream or lotion, formulated in a manner comparable to foundation makeup, occasionally has been offered as a product for men. Its purpose, when used several hours after shaving, is to make unnecessary a second shave in one day, by hiding the beard stubble with flesh-tinted pigments. This product has not had any significant success to date.

By incorporating a significant proportion of alcohol in a relatively light pigmented lotion, an interesting and potentially more acceptable product can be developed, an aftershave "talc lotion." This is feasible because the

Formula 20

Carbopol 941 resin	0.25%
Water	43.25
Ethyl alcohol, specially denatured	50.00
Triethanolamine, 10% aqueous	2.50
Solulan 98	3.00
Talc	1.00
Perfume	q.s.

Procedure: Slowly add the Carbopol to the water with rapid agitation, mixing until free from undispersed particles. Add the alcohol, stirring to disperse. Add the triethanolamine solution slowly with good agitation. Wet the talc with the Solulan 98, mixing until uniform and add the Solulan 98/talc mixture. Add the perfume and mix well.

* The silicone fluids suitable for use in cosmetics are the dimethyl polysiloxanes available from Dow Corning Corporation as Dow Corning 200 Fluid and from General Electric Company as SF-96 Silicone Fluid. These are offered in a very wide range of viscosities, 100 centistokes and 350 centistokes being the viscosities generally employed. The methyl phenyl polysiloxane fluids known as "Dow Corning 556 Fluid" or "General Electric SF-1075" have better solubility and miscibility with alcohols, polyols, esters, and other materials commonly employed in the formulation of cosmetics and toiletries.

carboxy vinyl polymer resins are efficient suspending agents compatible with alcohol. Formula 20 is an example of this type of pigmented hydroalcoholic aftershave lotion. The carboxy vinyl polymer (Carbopol 941 resin), neutralized with triethanolamine, is capable of producing a stable suspension of iron oxide pigments as well as talc to achieve a flesh-colored product. The acetylated ethoxylated lanolin derivative (Solulan 98) imparts residual emollience.

Powders

Aftershave powders are traditional and still popular accessories to the shave. Their obvious function is to impart a smooth matte finish to the face, masking the shine left by a too-oily brushless shave or toning down the "too-scrubbed" look after a lather shave. In this masking function, they help to cover minor skin defects and to compensate for an inadequate shave. In certain dark-haired individuals with heavy beards, the use of an aftershave powder is almost a necessity in obtaining a clean-shaven appearance.

The less obvious but important functions are identical with those mentioned for aftershave lotions. The application of a correctly formulated powder relieves mechanically the irritation and tension of the freshly shaven skin by the smooth, cooling feel it imparts. Although of course not so effective as an alcoholic lotion, a finely divided powder will dissipate a significant amount of heat from the face, leaving it cooled and refreshed after the shave. If desired, this cooling effect can be augmented by the addition of menthol. Mild astringency, acidity, and antibacterial action are specific characteristics that can be incorporated in aftershave powders as well as in aftershave lotions. The choice of odor is important in this preparation also, particularly because of the changes that can occur as the perfume "matures" on the extensive surface area of the powder.

The basic properties desired are the same as those sought in a face powder (q.v.) although the relative importances of these properties are not the same: slip, adherence, absorbency, covering power, and bloom.* The materials used also are those employed in face powders: talc, cosmetic metallic soaps, and lipophilic materials for adherence; calcium and magnesium carbonates, kaolin, and starch for absorbency; oxides of titanium, magnesium, and zinc, and kaolin for covering power; chalk and starch for bloom. Other materials are used in aftershave powders that are not employed in cosmetic face powders: boric acid for mild acidity and antibacterial action; menthol and camphor for cooling effect; aluminum and zinc salts for astringency; antiseptics and healing agents for "medication." These ingredients, however, are sometimes incorporated in body dusting powders for their special effects.

---

* "Bloom" is described by Harry (17) as the ability "to give the smooth velvetlike appearance of a peach to the skin."

The formulation of an aftershave powder differs from that of a face powder in several significant respects. First, slip and adherence are relatively more important than covering power and bloom. High slip enhances the very desirable smooth, cooling feel on the skin. The powder must have good adherence; otherwise, it will dust off too easily on the clothing to the embarrassment of the user. Covering power is of secondary importance in the sense that it is necessary only that the preparation should mask shine. It must not be so opaque that its presence on the face is obvious. Absorbency in an aftershave powder does not have quite the same significance as in a cosmetic face powder. The aftershave powder should be able to absorb moisture from the skin without caking or streaking because the face usually is difficult to dry completely after the shave. It need not, however, have the same degree of oil absorbency required in a cosmetic face powder.

An aftershave powder contains a relatively large proportion of talc. If this talc is of the highly platy variety, it will give good, almost greasy, slip but will have low covering power because of the thinness and translucency of the plates. The more granular talcs do not have this pronounced greasy slip but will give better hiding power to the powder. Opaque pigments like titanium dioxide and zinc oxide will compensate for low covering power, and a small percentage of metallic stearates will make the slippery plates adhere better to the skin. The less platy talcs generally require less hiding pigments than the platy types. However, both require the metallic stearates: platy varieties, to improve adherence; granular varieties, to improve slip. Starch is seldom if ever employed in aftershave powders, but small amounts of kaolin and precipitated chalk are included for absorbency. Magnesium carbonate and precipitated chalk, of course, are ideal for the incorporation of perfume. Boric acid has had a long history of usage as a mildly acid antiseptic in aftershave powders. Despite some rather ill-advised publicity concerning its toxicity when ingested or absorbed in large amounts, boric acid still remains, in the writer's opinion, a very safe and useful material for this application. It is customary to use 5 % or less of boric acid (impalpable powder) to "borate" aftershave powders. There is no reason why the newer antiseptics of proved efficacy and blandness cannot be used, where compatible, in place of or to supplement boric acid in these preparations.

Coloring of aftershave powders is quite important. It is generally agreed that the so-called "invisible" powders are preferred. These are lightly tinted a neutral or, better still, a slightly sun-tanned flesh tone to make them less evident on the skin. Deep tones are unsuitable because they tend to look artificial on the skin and to rub off on shirt collars and other light-colored clothing. The cosmetic earth colors or mineral pigments, particularly the synthetic ochres and iron oxides, are best. The synthetic ochres should be diluted or extended with talc before use because of their greater strengths as compared to the natural pigments. Lakes usually are too bright for use in aftershave powders. The water-soluble and alcohol-soluble dyes are definitely unsuitable.

Formula 21 is a typical example of an aftershave powder. Many variations

Formula 21

Talc	80%
Kaolin, colloidal	10
Zinc stearate	5
Precipitated chalk	3
Boric acid	2
Yellow ochre	q.s.
Perfume oil	q.s.

could be listed, but it is questionable whether the average user can detect any but the grossest differences in slip, adherence, opacity, or absorbency among them. Hence it is up to the formulator himself to reach a suitable balance of these characteristics that will achieve and maintain popularity when correctly colored and attractively perfumed.

The commercial manufacture of aftershave powders does not differ materially from that of face and body powders. The ingredients must be of high quality and free from grit, alkalinity, and soluble impurities. Uniformity of the ingredients is especially important, particularly in regard to color, bulk density, particle size, and particle shape.

An aftershave powder can be prepared in stick form in the same manner as described for preshave talc sticks. The formula given for the preshave preparation (Formula 4) need not be modified unless greater adherence or covering power is desired.

Styptics

Styptic pencils and alum blocks are shaving accessories having rather special functions. The use of styptic pencils is limited solely to the stanching of bleeding from minor cuts produced in the course of shaving. Alum blocks offer a convenient method of applying a high concentration of aluminum salt to a limited area of the face.

The physical properties desired in a styptic pencil are adequate mechanical strength, stability, and good appearance. In their manufacture, alum (aluminum potassium sulfate dodecahydrate), or a combination of astringent and styptic salts with alum, is fused and cast into stick form. If a very translucent stick is desired, a small amount (5 % or less) of glycerol is added to the alum and care is taken to avoid loss of water of crystallization during fusing. For opaque sticks, approximately 5 % of talc is triturated with about 5 %

of glycerol and added to the liquefied alum. More talc will, of course, give a whiter stick. Other salts suggested for use with alum are the hydrated sulfates of copper, zinc, and iron. A simple alum stick is shown in Formula 22.

Formula 22

Aluminum potassium sulfate dodecahydrate	90%
Talc	5
Glycerol	5

Procedure: Heat the potash alum crystals until liquefied, avoiding overheating and loss of water of crystallization. Remove scum from the surface of the melt, and incorporate the talc, previously triturated to a smooth paste with the glycerol. Pour into molds lubricated with mineral oil. Remove the pencils when solidified and polish, if desired, with a moistened cloth. The talc may be omitted if a translucent stick is desired or may be increased to give a whiter stick.

Alum blocks of good quality are characterized by a high degree of translucency and freedom from checks, cracks, and flaws in the block. The surface should be smooth and polished. They are made in the same manner as clear styptic pencils. Crystal potash alum is liquefied in its own water of crystallization with heat, up to 5% of glycerol or a mixture of glycerol and water is added, and the melt is cast into well-oiled block molds. The solidified blocks are polished with a moistened cloth.

Aerosols

Aerosol shaving foams were developed almost two decades ago and now dominate the men's toiletries field, but there has been no significant use of the aerosol foam for preshave and aftershave products. Until the relatively recent development of quick-breaking foams, the aerosol form was not suited to the performance requirements of preshave and aftershave preparations. The quick-breaking foams, unlike the stable shaving foams, have inherent characteristics that make them especially adaptable for use as shaving accessories. They can be made with high alcohol content, in the range most acceptable for preshave and aftershave lotions. They require only minor amounts of emulsifier and surfactant. They tolerate useful levels of additives such as humectant, emollient, lubricant, and antiseptic. They are excellent vehicles for the fragrance types most popular in shaving products.

The quick-breaking property of these foams is produced by use of a foam-stabilizing emulsifier with good solubility in the alcohol–water–propellant system of the aerosol, but with limited solubility in the concentrate alone. When the foam is discharged, loss first of propellant and then of alcohol quickly reduces the solubility of the emulsifier and its ability to stabilize the foam. By selection of emulsifier and adjustment of the proportions of alcohol, water, and propellant, foams can be obtained that will break within a few seconds to several minutes. The longer foam life would be desirable for products that are to be applied with some friction.

Formulas 23 and 24 are examples of two quick-breaking aerosol foam aftershave preparations that use polyoxyethylene stearyl ether foam-stabilizing emulsifiers of limited water solubility. Formula 23 contains humectant and is acidified with boric acid. Formula 24 contains a lubricating emollient, diisopropyl adipate, and the antiseptic hexachlorophene.

	Formula 23	Formula 24
Ethyl alcohol, 95%, specially denatured	54.00%	54.0%
Brij 72, polyoxyethylene (2) stearyl ether	2.00	—
Polawax A-31, polyoxyethylene stearyl ether	—	2.5
Crodamol, diisopropyl adipate	—	1.5
Sorbitol, 70% solution	5.00	—
Boric acid	1.00	—
Hexachlorophene	—	0.1
Menthol	0.05	0.1
Perfume	0.50	0.3
Water	27.45	31.5
Propellant 12/114, 40:60	10.00	10.0

Procedure: Dissolve the polyoxyethylene stearyl ether in the alcohol with slight heat (35 to 38°C). Add other ingredients, except propellants, and mix warm until smooth and uniform. Replace any alcohol lost during mixing. Fill at room temperature or above into aerosol containers and pressurize with propellant.

A quick-breaking foam preelectric shave preparation can be made by increasing diisopropyl adipate in Formula 24 to a level that will provide lubrication of the beard hairs. This will require adjustments of the alcohol–water ratio and the concentration of emulsifier to maintain quick-breaking foam characteristics. Examples can be found in a recent patent (18).

REFERENCES

1. Hollander, L., and Casselman, E. J.: Factors involved in satisfactory shaving, *J. Am. Med. Assoc.*, **109:** 95 (1937).

2. Hilfer, H.: Shaving preparations, *DCI*, **67:** 482 (1950).

3. Draize, J. H., Woodard, G., and Calvery, H. O.: Methods for study of irritation and toxicity of substances applied topically to skin and mucous membranes, *J. Pharmacol. Exp. Ther.*, **82:** 377 (1944).

4. Teichner, R. W.: U.S. Pat. 2,390,473 (1941).

5. Harry, R. G.: *The principles and practice of modern cosmetics*, Vol. 1 (*Modern cosmeticology*), 5th ed., Chemical Publishing Co., New York, 1962, p. 461.

6. Poucher, W. A.: *Perfumes, cosmetics and soaps*. Vol. 3, 7th ed., Chapman and Hall, London, 1959, p. 49.

7. *National formulary*, 10th ed., American Pharmaceutical Association, Washinging, D.C., 1955, pp. 385–386.

8. Poucher, W. A.: *Op. cit.*, pp. 145–146.

9. Department of the Treasury, Internal Revenue Service: *Formulas for denatured alcohol and rum, Publication 368 (Rev. 3–71)*, U.S. Government Printing Office, Washington, D.C., 1971.

10. Burckhardt, W.: New investigations on skin sensitivity to alkalis, *Dermatologica*, **94:** 73 (1947).

11. Cornbleet, T., and Joseph, N. R.: Recovery of normal and eczematous skin from alcohol, *J. Invest. Dermatol.*, **23:** 455 (1954).

12. Bryan, A. H.: Hygienic shaving cuts blemishes, *Am. Perf.*, **57:** 33 (1951).

13. Jones, K. K., and Lorenz, M.: Relation of calcium soaps to staphylococcal infections in skin, *J. Invest. Dermatol.*, **4:** 69 (1941).

14. Sindar Corporation: G-11 (Hexachlorophene U.S.P.), Technical bulletins and bibliographies.

15. Rieger, M. M.: U.S. Pat. 3,429,964 (1969).

16. de Navarre, M. G.: *The chemistry and manufacture of cosmetics*, Van Nostrand, New York, 1941, p. 290.

17. Harry, R. G.: *Op. cit.*, p. 146.

18. Lanzet, M.: Brit. Pat. 1,096,753 (1967).

Chapter 18

DEPILATORIES

RICHARD H. BARRY

The desire for removal of hair for social, religious, or vengeful purposes, or for improving personal appearance, has led men and women since ancient times to try the most varied measures. It is recorded that the primitives used singeing, as well as mineral, vegetable, or animal matter in ointment and paste form, for the removal of unwanted hair or to prevent its regrowth (1). The Papyrus Ebers, which dates from about 1500 B.C. and is the richest source of ancient medical recipes, contains references to formulas for this purpose containing "burnt chaetopod boiled with balanites oil, burnt leaf of lotus in oil; shell of the tortoise with the fat of the hippopotamus; the blood of oxen, asses, pigs, hounds, and goats together with stibium and malachite." These concoctions were intended also for vengeful purposes, e.g., for application to the head of a hated rival, or to prevent regrowth of hair already pulled out (2).

Among the civilizations which flourished in the Mediterranean area and in the Orient between 4000 and 3000 B.C., the use of cosmetics for body adornment included various epilants and depilatories (3–5). Religious and historical treatises mention the customs and laws which decreed the removal of body hair.

The classical depilatory *Rhusma Turcorum* contained orpiment (natural arsenic trisulfide), quicklime, and starch and was made into a paste with water. This was used extensively by women in Oriental harems (6,7). It is still common practice among certain African tribes to use hair-removing agents (3,8), and the South American natives use the exudation of the tree *coco de mono* for this purpose (9).

Certain ethnographers make no distinction between "epilation" and "depilation" (3). In this review, the author prefers to call the process epilation when the relatively intact hair is removed by uprooting, e.g., by plucking, electrolysis, X-ray, or by topically applied thallium compositions, reserving the term "depilation" for removal as the result of chemical degradation of the human hair fiber, e.g., by inorganic sulfides and organic thiols.

The most frequent subject of inquiry received in 1967 by the American Medical Association Committee on Cutaneous Health and Cosmetics concerned superfluous hair (10). At some time during her lifetime, every woman has desired the removal of unwanted hair on the face, underarms, and legs. Good grooming requires that these areas be free from unsightly hair the year around if bathing and casual attire, sheer hosiery, and off-the-shoulder dresses are to be worn without embarrassment. Not only is unwanted hair a cosmetic problem, but to most women its presence is a typical masculine trait unacceptable psychologically. Excessive hair growth (hirsutism) is, however, a relative term; the amount of facial or axillary hair normally acceptable to women living in Mediterranean countries might be quite disturbing to someone living in an area where native women are commonly less hirsute (e.g., in Scandinavia). Although such genetic factors as familial or racial background influence the degree of hirsutism, its cause may have physiologic origin since hair growth is related to endocrine influences (11). High androgen level or decrease in estrogen production is prone to cause hirsutism in women. Occasionally hirsutism is observed during puberty or in pregnancy, but it is more common during the menopause. Hirsutism often results from the administration of testosterone or adrenocortico steroids. Since the taking of oral contraceptives may promote hirsutism, present package inserts accompanying these products list hirsutism as a possible side effect. Polycystic ovarian syndrome will produce hirsutism in half the women afflicted; similarly, adrenal tumors or hyperplasia is associated with excessive hair growth. Treatment of these conditions (usually by surgery) diminishes or eliminates hirsutism.

Shaving by women is usually a semiweekly procedure, and the process is made more convenient and efficient by specially designed razors and blades with improved cutting edges and by the use of elegantly formulated shaving creams. To some, however, the procedure is considered tedious and hazardous because of cuts or scratches. The inability to see all the areas being shaved and the alleged regrowth of stubble have also been cited as disadvantageous. It is often painful to pull out each hair with tweezers; extensive skin damage may result, with the risk of subsequent infection.

Bleaching of hair on upper lips and cheeks may help conceal unwanted hair. Rubbing the hairy area with an abrasive such as pumice (12) will break the hair shaft. Although somewhat inconvenient to use and often too abrasive to the skin, such compositions are still sold.

Hair can be permanently removed by electrolysis, which involves introduction of a needle into the hair follicle with resultant destruction of the papilla. Earlier electrolysis machines used a weak direct current, but newer methodology involves electrocoagulation with a high-frequency alternating current, which is faster. Both methods are not only costly but time-consuming, and there is the risk of possible infection and scarring. The overall result is largely operator-dependent. Even with the best technique, a recurrence of about 15 to 25 % is to be expected.

Ionizing radiation (e.g., X-ray) is capable of permanent destruction of the hair follicle and loss of hair, and is a useful modality in the treatment of certain skin disorders and for temporary epilation in cases of ringworm of the scalp. It should never be employed for the purpose of cosmetic removal of excessive hair.

Although the use of the razor is widespread, other methods of hair removal involve the use of epilating waxes or other adhesive compositions and the application of depilatory creams. Sales of these products were estimated at $7,670,000 in 1965, and 95 % of all drug outlets stocked them (13).

EPILATORY COMPOSITIONS

Epilants of the "hair-pull" type are essentially semisolid adhesive mixtures of two general types:

1. Wax-rosin compositions designed to be applied in the molten state to the hirsute area and allowed to solidify, so that the hair becomes enmeshed in the plastic mass. Subsequent removal of the waxy film from the site uproots and removes the hair.

2. Adhesive semisolid compositions which are permanently sticky at room temperature and applied usually on a flexible backing material, such as fabric. Removal of the adhesive composition by stripping removes the hair without injury to the skin.

A typical "hair-pull" composition granted a U.S. patent in 1910 (14) is shown in Formula 1.

Formula 1

Rosin	69%
Beeswax	20
Burgundy pitch	4
Gum camphor	3
Oil of bergamot	2
Oil of eucalyptus	1
Oil of skunk	1

The molten mass was poured into molds to form sticks, which were heated before being applied to the hairy area. On cooling, the mass was removed quickly from the skin, thereby removing the embedded hair. In Europe, Unna's epilatory stick (fused rosin and beeswax) was in use about the time the above patent was issued (15,16). An innovation patented in 1923 combined the "hair-pull" idea with a composition for preventing further hair growth. After epilation by a rosin-wax mixture, another mixture of lime-water, hydrogen peroxide, and oil of turpentine, with color and perfume, was applied to the area in an attempt to restrict further hair growth (17). Rosin or rosin-wax compositions were later modified to include synthetic amber (18), viscous plasticizers such as honey (19), and nondrying oils, such as mineral oil or olive oil, to make the composition usable without preliminary heating (20–22). Replacement of the wax-rosin mixture by glucose, molasses, or honey and with water-insoluble fillers in a ratio of 2:1, respectively, was proposed (23). Raw rubber in a volatile solvent was used to provide a cohesive coating of deposited rubber for impounding unwanted hair. This film was subsequently removed by affixing a sheet of raw rubber to the coating *in situ* and withdrawing the sheet and coating with the embedded hair (24). The wax-rosin compositions were also supplied with a backing material which formed a flexible mounting for the adhesive mixture (25). Adhesive compositions containing zinc oxide and various ketones similarly mounted were patented abroad (26,27). Encompassing all the previous art, there was patented in 1944 a "ready to use and instantly removable depilatory pad for application to the body, comprising in combination, a substantially impermeable base section and a coating permanently mounted thereon; said coating being made of plasticized rosin of which an amount in the order of 75 to 88 % is rosin and the balance is rosin plasticizer; said proportions producing a substance having the physical properties at room temperature of being substantially permanently sticky while mounted on said base section, of permanently securing itself to the base section and to any hair with which it comes in contact and of easily releasing itself from the skin when detached therefrom without injury thereto" (28).

Proponents of the "hair-pull" epilatories point out that the advantages of this method include their use by untrained persons with no apparatus, metering equipment, or pretreatment; all hair is removed instantly, completely, and painlessly, without shaving or chemically degenerating or discoloring hair tissue by uprooting each hair as a unit, without infection or mechanical injury to hair follicle or skin—all without either inhibiting regrowth of the hair or promoting such regrowth as a thick coarse stubble (20–22).

That the "hair-pull" method is painless will be disputed by many who have used it. Moreover, allergic reactions to the adhesives are not uncommon.

Epilatories of the "hair-pull" type generally give satisfactory results when the epilatory is applied and removed by a beautician. When a woman applies the material to her skin, she winces from the pain attendant to the epilation and frequently removes the composition slowly and carefully instead of sharply and firmly. When this is done, epilation is often incomplete. Consequently epilatories of this type are apt to be ineffective when applied at home.

Several brands of these mechanical hair-removing aids in small cakes, canisters, kits, or tubes are available, but their popularity has been largely superseded by that of the newer chemical depilatories. Simple rosin-based epilatory compositions (29) are given in Formulas 2 to 5.

Formulas for Rosin-Based Epilatories

	2	3	4	5
Rosin	42%	50%	64%	68%
Beeswax	37	24	8	22
Paraffin	—	20	—	—
Ceresin	—	—	—	10
Carnauba wax	6	—	24	—
Linseed oil	—	—	4	—
Mineral oil	15	—	—	—
Petrolatum	—	4	—	—
Benzocaine	—	2	—	—

More recently, a British patent (30) concerned a cold-application epilatory which resists oxidative hardening and maintains a viscosity suitable for packaging between plastic sheets. An emulsified wax composition on silicone-treated kraft paper containing butyl p-aminobenzoate for topical anesthetic effect with resinous and oily substances as perfuming agents was patented (31). Although the addition of local anesthetics to epilants has been recommended (32), it is doubtful that their inclusion minimizes the pain of the "hair-pull" process.

CHEMICAL DEPILATORIES

Great advances in the art of removal of superfluous hair have been made through the use of chemical agents in paste or cream formulas. Within the last two decades, cosmetic depilatory creams have become popular.

An ideal chemical depilatory formulation would possess the following attributes:

1. It should convert human hair completely in 2 to 5 min to a soft, plastic mass easily removed from the skin by wiping or rinsing.

2. It should be nontoxic systemically and nonirritating to the skin even on long contact.

3. It should be easily applied, economical to use, and stable in the tube or jar.

4. It should be cosmetically elegant, odorless or pleasantly perfumed, white or neutral in color, stainless to the skin, and noninjurious to clothing.

Public acceptance of such a depilatory would be assured. To date, no one has perfected a product with all of these desirable properties. One obstacle in the path of fulfilling the requirements listed above is due to the similarity in structure of the skin and hair. Both are proteinaceous in nature and both are subject to attack by the same chemicals to nearly the same degree. Logically, formulation studies involving chemical depilatories should be undertaken only after reviewing the structure of hair and acquiring an understanding of the nature of the depilation process.

Keratin, the protein to which hair largely owes its characteristic physical properties, is particularly rich in the sulfur-containing amino acid, cystine (about 17%). Cystine, in turn, is linked chemically with other nonsulfur amino acids, including aspartic and glutamic acids. The disposition of the sulfur in the hair fiber becomes such as to form a sulfur-to-sulfur bridge between polypeptide chains (R):

$$R—CH_2—S—S—CH_2—R$$

The kind and arrangement of these amino acid residues in juxtaposition to cystine seem to govern the degree of protection or the vulnerability of the S–S linkage to attack by chemical agents, since all S–S linkages in human hair (and in wool) are not equally reactive. Most of the chemical treatment of hair involves the rupture and reformation of S–S linkages which, under normal conditions, confer stability and flexibility to the hair fiber. Whereas it is most important to prevent complete breakdown of all cross-linking disulfide bonds in the permanent waving process, it is the object of chemical depilation to cleave sufficient S–S bonds indiscriminately so that the hair will readily disintegrate. Under the influence of such different chemical agents as alkali-metal sulfides and sulfites, cyanides, amines, mercaptans, etc., the S–S bond in keratin is affected; increasing osmotic pressure develops within the hair fiber, which swells, loses its tensile strength, and generally deteriorates. A mass of almost jelly-like consistency which can be easily removed by wiping or scraping is the final stage of such degradation. Both the outer layer (cuticle) and the inner color-bearing layer (cortex) are disintegrated (33). The resistance of the latter to chemical attack, especially when the hair is coarse and highly pigmented, may explain the reason for the easier dissolution of fine and lightly pigmented hair by chemical depilatories.

Effect of Alkali on Disulfides

The nature of the reaction between alkali and model disulfide compounds was studied as early as 1906, when it was suggested that alkaline hydrolysis formed an unstable sulfenic acid and a mercaptan (34). Schöberl and his co-workers (35–38), working first with a simple model disulfide such as dithiodiglycolic acid and later with many disulfides, supported the foregoing hypothesis, noting that many of these disulfides on treatment with alkali formed mercaptan, hydrogen sulfide, and a carbonyl-containing compound such as aldehyde or ketone:

$$R—CH_2—S—S—CH_2—R' \xrightarrow{OH^-} R—CH_2—SH^+[R'—CH_2SOH]$$

$$[R'—CH_2SOH] \xrightarrow{H_2O} R'—CHO + H_2S$$

Speakman (39) applied the hydrolytic cleavage ideas of Schöberl to explain the action of steam and alkali on wool. Despite some anomalous results that were noted with certain secondary and tertiary disulfides, the original postulates of Schöberl were generally accepted; they served to explain the reactions which wool and human hair underwent in processing. A review of the complexity of alkaline hydrolysis of hair in general is given by Stoves (34). Schöberl's views on the alkaline degradation of disulfide linkages have been challenged by Rosenthal and Oster (40) in their review and critical evaluation of S–S reactivity. Ionic (modified β elimination) and radical displacement mechanisms are offered to explain scission of the S–S bond without requiring postulation of sulfenic acid formation.

The use of lime in the dehairing of hides was probably the first instance of simple alkali depilation, and it is of interest to note that a British patent issued in 1912 disclosed a depilatory for cosmetic use containing lime and/or caustic alkali in a petrolatum base (41).

In a study of the effect of base strength on the dehairing process, Moore (42) found that the addition of organic bases such as methyl- and dimethyl-amines, monoethanolamine, ethylenediamine, hydroxylamine, hydrazine, guanidine, aminoguanidine, and piperidine accelerated the dehairing effect of calcium hydroxide suspensions. Depilation was retarded or prevented when a phenyl group was substituted for hydrogen, or when ($=$O) or ($=$S) replaced ($=$NH). Patents have been granted for dehairing compositions containing calcium hydroxide and guanidine sulfide (43) or a biguanide or guanylthiourea (44) as accelerators.

In general, alkaline degradation of hair involves first the rupture of the cystine linkages between main polypeptide chains. The process is dependent on concentration of hydroxyl ions, temperature, and time of reaction (45).

Certain agents like glycerol, phenol, tannin, glucose, and hydrolyzed protein are said to moderate the alkaline degradation of hair (46).

Action of Reducing Agents

The disulfide linkages in keratin can be broken by chemical reduction, which in simplest terms is analogous to the reduction of cystine to cysteine (47,48):

$$R—S—S—R \underset{}{\overset{H_2}{\rightleftharpoons}} 2R—SH$$

cystine   cysteine

This process is independent of R and in the above equation can be accomplished by tin or zinc in acid solution or by sodium in liquid ammonia. Reduction of the S–S groupings in keratin with aliphatic thiols, sulfides, bisulfites, or cyanides occurs in neutral or, better, in alkaline media through a nucleophilic attack (Schöberl) or by a process involving a radical displacement or an ionic mechanism (40):

$$(a) \quad R—S—S—R + R'S^- \xrightarrow{H_2O} RS^- + R—S—S—R'$$

$$(b) \quad R—S—S—R + HS^- \xrightarrow{H_2O} RSH + R—SS^-$$

$$(c) \quad R—S—S—R + HSO_3^- \xrightarrow{H_2O} RSH + R—SSO_3^-$$

$$(d) \quad R—S—S—R + CN^- \xrightarrow{H_2O} RS^- + RSCN$$

Azide, fluoride, and cyanate do not produce a mercaptan (40). Simple primary, secondary, and tertiary aliphatic thiols accelerate depilation in alkaline solution, the extent of hair softening depending on the structure and solubility of the thiol and the ease of oxidation of thiol to disulfide under operating conditions. Alicyclic thiols, sulfites, and thiosulfates are less active than the corresponding aliphatic compounds, and the aromatic thiols are inactive (49,50).

The sulfides of lithium, sodium, potassium, cesium, magnesium, calcium, strontium, barium, aluminum, arsenic[III], and tin[IV] accelerate dehairing in the presence of calcium hydroxide suspensions (51–53). Of the depilatory formulas containing sulfides, the most frequently employed were lime pastes based on orpiment, barium sulfide, strontium sulfide, and calcium sulfide or sulfhydrate. The alkali-earth sulfides probably function because of hydrolysis to sulfhydrate and hydroxide:

$$2CaS + 2HOH \rightarrow Ca(SH)_2 + Ca(OH)_2$$

$$2SrS + 2HOH \rightarrow Sr(SH)_2 + Sr(OH)_2$$

$$2BaS + 2HOH \rightarrow Ba(SH)_2 + Ba(OH)_2$$

Moreover, further hydrolysis of sulfhydrate produces hydroxide and hydrogen sulfide, and the final reaction products are intractable mixtures of alkali-earth sulfhydrate, sulfhydryl hydrate (R–OHSH), and hydroxide. Both alkali and sulfhydrate are capable of reacting with keratin. Calcium sulfide is sparingly soluble in water and is probably the least efficient depilatory of the four mentioned above; calcium sulfhydrate, made by passing hydrogen sulfide into milk of lime, is soluble and was used in early paste compositions. Depilatories containing 25 to 50% strontium sulfide soften hair rapidly, and certain preparations containing it are still popular abroad. Regarding the famous *Rhusma* depilatory, which contained slaked lime and mineral arsenic trisulfide (orpiment), it is possible that orpiment was tried primarily as a yellow coloring agent for body adornment and that its depilating properties were observed thereafter. Orpiment reacts with lime, producing $Ca(SH)_2$, $Ca_3(AsO_3)_2$ and $Ca_3(AsS_3)_2$ (54); probably calcium sulfhydrate is the active depilating agent. As expected, the use of orpiment paste is not without danger, especially if applied to the broken skin.

Selenium and sulfur show similar chemical properties, and, as might be predicted, aqueous solutions of sodium selenide (Na_2Se) remove hair as rapidly as those of sodium sulfide (55). Such solutions are unstable, however, producing poisonous hydrogen selenide and oxidizing to free selenium. The nauseating odor of alkyl and aryl selenides completely prevents their use in depilatory formulas.

Advantages and Disadvantages of Metal Sulfide Depilatories

The use of the alkali and alkaline-earth sulfides as dehairing agents represented the first major advance in the art in nearly 6000 years and was the inevitable result of progress in chemical knowledge in Europe in the nineteenth century. Compositions containing these sulfides are capable of depilating rapidly. Their low equivalent weights and low cost make them economical to use. It has been said that a properly compounded sulfide depilatory will remove hair in one-third to one-half the time required for a comparable calcium thioglycolate preparation (56).

Since barium sulfide is poisonous (57,58), its use as a depilatory (59) has largely been abandoned. Pastes containing calcium sulfhydrate or strontium sulfide have a strong odor of hydrogen sulfide because of hydrolysis. This odor may be controlled by adding alkali to the formulation, so that the pH is 11 or above. Moreover, sulfide preparations have the propensity to turn to shades of green, blue, or gray, especially if the pH is lower than 11, owing to oxidation and to reaction with trace metals, such as iron. A problem with strontium sulfide preparations is that strontium hydroxide often crystallizes out soon after the product has been manufactured. Milling may control this problem initially, but temperature fluctuation is likely to cause the reappearance of hard, gritty crystals. The best formulas containing the metallic sulfides appear to be oil-in-water emulsions compounded with emulsifying agents which are not affected by the high alkalinity and electrolyte concentration.

Many of the sulfide depilatories have little or no odor when freshly made but develop a nauseating smell on standing. Moreover a strong odor of hydrogen sulfide is always produced when the preparation is removed from the skin with water. This can be largely avoided, however, if the depilatory is scraped off with a spatula or wiped off with a cloth or cleansing tissue before rinsing with water.

Formulations with Metallic Sulfides

The early American patent literature on depilatories pertained to tablet, powder, soap, or paste compositions containing barium, strontium, and sodium sulfides or polysulfides. A depilatory tablet, patented in 1885 (60), contained barium polysulfide. It was intended to be used to make a paste, and was coated with a protective film of varnish or paraffin. There followed a depilatory powder composition to be mixed with water prior to use and which contained "sulfurated derivatives of strontium in any of their chemical forms" (61). One such composition is still sold as a depilatory for men. Products containing "hydrosulfite of soda" were patented in 1902 (62), and as early as 1893 a soap depilatory containing sodium sulfide hydrate was claimed (63,64). In 1912 Stone disclosed a depilatory formulation containing calcium hydroxide, sodium sulfide, and calcium sulfhydrate (65).

It remained for Donner in 1921 to accomplish what has been called a milestone in depilatory formulations (66,67). The descriptive matter in Donner's patent relating to the art of depilatory making was, for its time, a masterly presentation and serves to this day as a model for preparing a patent disclosure in the pharmaceutical and cosmetic field. Prior to Donner, powder, liquid, and lotion formulas containing sulfides were, for the most part, unstable, odoriferous, and cosmetically inelegant preparations. Donner's description included sulfides, polysulfides, or hydrosulfides of the alkali and alkaline-earth metals which were suspended in emulsion-type lotion vehicles or made into semisolid vanishing creams with soaps, waxes, fats, resins, or oils. Gelatinous or mucilaginous emulsoid colloids, made with agar-agar, tragacanth, gum arabic, pectin, etc. were included in the disclosure, together with directions for the preparation of a relatively inodorous slurry, based on calcium and sodium hydrosulfides, for the above formulations.

Subsequent to Donner's disclosure, improvements in the art of making sulfide depilatories appear to have been limited to attempts to improve stability or appearance. To control the discoloration which is so characteristic of sulfide depilatory formulas on aging, the addition of nonreactive pigments of great covering power such as X-ray-grade barium sulfate (68), lithopone, zinc oxide, tin oxide, or titanium dioxide (69) to pastes and liquids has been patented. Similarly, the use of calcium carbonate, strontium carbonate, and aluminum hydroxide as fillers or amphoteric buffers was proposed (70). With the commercial availability of the sulfated fatty alcohols and alkyl aryl sulfonates, cream and lotion formulas containing the metallic sulfides were further improved, since these versatile emulsifiers are not affected by the high alkalinity. Lauryl, cetyl, and stearyl alcohols are stable at a high pH and serve to add emolliency. The irritating actions of alkali sulfides is said to be lessened by the addition of casein or its salts and by albumin degradation products (71).

Although cream formulas containing fatty acid derivatives, e.g., "Lanette wax," are commonly used abroad for strontium sulfide depilatories, mucilaginous creams seem to be preferred for those containing sodium, calcium, or lithium sulfides or hydrosulfides (57,58,72). A stable, transparent jelly composition which was designed to give better visual control over the depilating process was patented in 1936 (73). Hydrogen sulfide was passed through a 15% lithium hydroxide solution until the solution increased 8.25% in weight. This solution was added to a jelly base containing about 6% of tragacanth, karaya, or locust bean gum. The resultant clear jelly was applied and then removed when the hair assumed a curly appearance, which indicated that depilation was complete. Kamlet (74) patented a methyl cellulose gel base of a reported good stability and rapid effectiveness, containing sodium sulfide, calcium chloride, and calcium carbonate. One Japanese patent (75) describes the combination of starch paste and hydrophilic ointment in the preparation of a strontium sulfide cream depilatory; another discloses usage of 10% "hydrosulfide polymer" at pH 11 (76). The addition of a quaternary ammonium salt reportedly stabilizes sodium sulfide depilatories (77). Nonirritating compositions containing strontium sulfide with the inclusion of hydroxyquinoline sulfate to minimize skin damage are claimed (78). Examples of inorganic sulfide depilatories (29,32) are shown in Formulas 6 to 11.

Formula 6.   Powder Depilatory

Barium sulfide	31.00%
Titanium dioxide	18.00
Corn starch	50.50
Menthol	0.25
Perfume	0.25

Formula 7. Powder Depilatory

Strontium sulfide	35.0%
Corn starch	35.0
Powdered soap	5.0
Zinc oxide	23.0
Benzocaine	0.2
Perfume	1.8

Procedure: Mix with water at time of application.

Formulas for Paste Depilatories

	8	9	10	11
Sodium sulfide	4%	—	—	—
Barium sulfide	—	8%	—	—
Strontium sulfide	—	—	30.0%	35.0%
Calcium carbonate	—	32	—	—
Calcium hydroxide	4	—	—	—
Titanium dioxide	—	—	—	4.0
Powdered soap	—	4	—	—
Zinc oxide	—	—	8.0	—
Glycerol	1	2	8.0	5.0
Kaolin	32	—	—	—
Methyl cellulose	—	—	2.5	—
Menthol	—	—	1.0	2.5
Perfume	—	—	—	1.0
Water	59	54	50.5	52.5

Inorganic Nonsulfur Depilatories

Other reducing agents have been proposed for depilation in an effort to avoid the malodorous sulfide compositions. Concentrated alkaline solutions of soluble stannites have been patented as depilatories (79). Although these formulations are odorless, they have not achieved any substantial degree of popularity since they require a longer depilation time than sulfide depilatories, show a *p*H as high as 12.5 or higher, are light-sensitive, and are converted by oxidation and/or hydrolysis to stannate and stannous oxide. Various substances, including sodium potassium tartrate (72), hydroxy-aliphatic acids (such as tartaric and citric acids) (70), aliphatic amines, glycerol, vegetable gums, sugars (52,80), alkali silicates (81,82), and sodium tetraborate (83), have been proposed as stabilizers for alkali stannite depilatory compositions.

To avoid the oxidative instability of the stannites, Pacini (84) proposed nontoxic, odorless compositions containing double salts of tetravalent titanium, such as calcium sodium titanate which, upon hydrolysis, produces sufficient alkali to soften keratin.

Organic Nonsulfur Epilants

The high sulfhydryl (SH) content of hair and its rapid uptake of sulfur amino acids leave it vulnerable to agents that interfere with sulfur metabolism. People who have eaten the *coco de mono* (Paradise nut) of the Venezuelan tree *Lecythis ollaria* have lost their hair (85). Extracts from the nut have caused hair loss or prevented hair growth in mice (86). The epilating agent believed responsible for this effect is cystaselenonine, the selenium analogue of cystathionine, an intermediate in the metabolic conversion of methionine to cysteine (87,88). Mice, dehaired by plucking and then injected daily with 6 mg/kg cystaselenonine, lost hair which had grown back 16 days after plucking (89). If cystaselenonine acted as a mitotic poison, it should have interfered with growth at any time during the entire anagen phase. Instead the hair was injured in its maturity when the largest amount of sulfur was being transferred across the root sheaths. It can be speculated that the selenium compound interfered with the metabolism of sulfur compounds in the keratogenous zone of the hair follicle via enzyme or oxidative interference, by damaging the sulfur amino acid transport system or by providing an inadequate substitute for cystine.

Recent Russian publications (90–92) disclose the epilant properties of certain ketones on animal hair. Epilene [4-(dimethylaminoethoxy)phenyl phenylethyl ketone] in 4% concentration as a plaster was reputedly an effective epilant on rodents and rabbits, and was recommended for human trial for scalp epilation in dermatomycosis.

A recent patent (93) relates to the inhibition of hair growth in animals and man by applying topically certain substituted benzophenones, such as 2-amino-5-chlorobenzophenone, and other 5-halogenated 2-aminobenzophenones. In the examples cited, 2-amino-5-chlorobenzophenone was rubbed into shaven areas of the rabbit and into the shaven human forearm. After 55 days of exposure to these conditions, the areas were substantially free of hair and no irritation was noticed. In addition to examples of creams, stick, spray, powder, and cream formulas, there was included a depilatory composition embodying 1% of the benzophenone in a calcium thioglycolate cream base.

Certain unsaturated compounds, including the volatile intermediary polymers of chloroprene, oleic and linoleic acids, squalene, various allyl esters of lauric, benzoic, and diphenylacetic acids, and vitamin A have been reported to cause reversible hair loss when applied to the skin (94–98). These compounds (except vitamin A) inactivate the free sulfhydryl groups in glutathione and inhibit the sulfhydryl-containing enzyme, succinic dehydrogenase. Since epilating and sulfhydryl-inactivating effects are believed due to the alkylation of —SH groups by unsaturated bonds, human sebum (which contains unsaturated groups) produces similar reversible hair loss. Sebum production may therefore play a role in disturbances of hair growth and keratinization.

One should mention at this juncture the thallium epilants which were introduced in Europe by Sabouraud (99) and which were used extensively prior to 1930 in the treatment of ringworm of the scalp in children (100). Thallium acetate was the salt commonly used at 1% concentration in a petrolatum-lanolin base. This substance does not affect the hair structure but promotes the easy removal of the hair from the follicle, the end result being quite similar to epilation by X-ray. Papers concerned with the depilatory action of thallium acetate in rats and with the use of p-bromphenyliso-thiocyanate to increase the depilatory effect of thallium have recently appeared (101,102). The toxicity of thallium epilants (103–105) has brought their use into disrepute, and they are no longer of commercial significance.

Aliphatic Mercapto Acids

In the 1930's the literature pertaining to leather technology devoted much attention to the improvement of chemical methods for dehairing hides and skins. These investigations included studies on the mechanisms involved in the dehairing process but were directed particularly toward finding agents which would supplant or be alternative to the commonly used lime-sulfide mixtures. Probably the most significant advance which pointed the way for the use of organic sulfur compounds as depilatories was the research which led to the patent granted to Turley and Windus of Rohm and Haas in 1934 (106,107). In this disclosure, improvements in the dehairing of animal skins and hides over the lime process included the use of organic sulfur derivatives and made mention of thioglycolic acid as being effective.

Bohemen recognized the contributions to the dehairing process made by Turley and Windus and, after citing the disadvantages of inorganic sulfides, claimed invention based on aliphatic mercapto acids and salts thereof (thioglycolic, thiolactic, etc.) in depilatory formulas containing sufficient alkali to produce a pH of 10 or higher. Patents were issued to Bohemen in France through his agent Fletcher (108) and in Great Britain (109).

These patented compositions—powders, pastes, creams, jellies, or liquids—permitted the use of a wide range of perfumery materials. Practical directions for incorporating calcium thioglycolate in a cream base were given, which included (a) neutralizing thioglycolic acid with calcium hydroxide in a stearyl alcohol—"sulfonated" stearyl alcohol emulsion base—or (b) preparing a solution of thioglycolic acid by reacting monochloracetic acid with ammonium thiocyanate, neutralizing with lime, and adding to the cream base. In the United States an Evans and McDonough patent (110) claims depilatory formulations containing substituted mercaptans, particularly thioglycolic acid, in an alkaline creamy vehicle. The majority of chemical depilatories sold in the United States are alkaline thioglycolic acid compositions.

Depilatories containing thioglycolic acid are commonly employed in formulations having a pH range of 10 to 12.5. Fruton and Clark (47) have shown that the disulfide linkage in the amino acid cystine could be reduced in acid solution (e.g., with tin and hydrochloric acid), yielding the thiol acid cysteine. In neutral or alkaline solution, a thiol capable of being readily oxidized by cysteine can be used to reduce the S–S linkage. Accordingly, a departure from the high-pH thioglycolate formulations was proposed in the patent of Demuth (111), who disclosed the use of hydrazine sulfate and of di- and triisocyanates as reducing intensifiers for calcium or guanidine thioglycolate depilatory compositions. These modifications are said to permit depilation at a pH range of 8 to 10 in a suitable base containing anionic wetting agents (112). Although thioglycolate depilatories may be less disagreeable in odor than those based on inorganic sulfides, the latter appear to be more rapid in action (56). In order to increase the depilation rate of thioglycolic and thiolactic acids, such hair-swelling agents as urea (113) have been proposed to assist the hair degradation process. However, since urea is not stable at pH 12, other accelerating agents of greater stability have been proposed. Melamine, dicyandiamide, and inorganic thiocyanate salts have been claimed (114) to accelerate the action of alkali earth metal salts of thiolactic, thioglycolic, and β-mercaptopropionic acid in depilatory compositions. The addition of "pro-oxidants," including soluble manganese, iron, and copper salts, has been proposed in order to increase the efficiency and safety of thioglycolate depilatories (115,116).

Since soluble xanthates, e.g., sodium ethyl or methyl xanthates, had been proposed for use in removing hair from hides (117), a depilatory composition was patented abroad in which sodium dodecyl xanthate was used as an emulsifying agent and as a calcium thioglycolate "booster" in a cream base (118). An oil-in-water cream of pH 12 to 13 which reportedly depilated in 3 to 3.5 min contained 5% of N-(β-mercaptoethyl) aniline (119). Salts of β-mercaptopropionic acid (120,121) and the strontium salt of thiolactic acid (114) have been utilized in depilatory compositions. Lithium, strontium, and calcium salts of α-thiocarboxylic fatty acids (C_4 to C_{10}) have been reported recently as being effective for removing hair from human and animal skin (122).

Other Mercapto Compounds

Evans and McDonough proposed the use of the α or β isomers of thioglycerol in alkaline-reacting media as depilatories (123). These readily soluble and reportedly stable compounds may be formulated with natural gums, polyvinyl alcohol, and substituted cellulose ethers, since they do not react with these products as do the alkali and alkali-earth sulfides. Prior to the issuance of the patent claiming thioglycerol, these inventors obtained coverage also in Great Britain for other organic mercapto compounds which function in alkaline-reacting media as depilatories. These comprise, as a group, substituted thiols in which the hydrogen atom in hydrogen sulfide is substituted by various organic residues containing nonpolar hydroxyl, ketone, aldehyde, ether, polar amine, or sulfonic groups (124). Similar claims by Evans and McDonough are included in a French patent (125) which mentions also mercaptoethyl aniline, mercaptoethyl sulfonate, and others.

Although simple mercaptans of relatively low molecular weight and those containing a nonpolar group such as hydroxyl (e.g., ethyl mercapto- and β-mercaptoethyl alcohol) appear to function as depilatories in alkaline media (110), the odor associated with these compounds makes perfuming their compositions difficult. Moreover, mercaptans and thiols bearing nonpolar functional groups such as hydroxyl, ketone, or aldehyde are reportedly more irritating to the skin than mercaptans bearing polar groups, e.g., mercaptoacetic acid (110).

In the patent literature concerning dehairing of animal hides, there are many organic sulfur compounds which may have application in depilatory formulas (126–128). Although there is commercially available a nontoxic, relatively efficient agent, such as thioglycolic acid, this should not discourage the search for a better chemical dehairing agent.

Some Practical Aspects of Formulations with Thioglycolates

Because of the more favorable compounding characteristics permitted by their use, the salts of thioglycolic acid have largely superseded the alkali and alkali-earth sulfides as the active ingredients in depilatory formulations in the American market. Therefore this discussion of modern depilatory formulation is confined largely to those of the thioglycolate type (129).

In a previous section the objectives to be attained in depilation and the properties of an ideal depilatory were outlined. Although such an ideal product has not been formulated, the art of chemical depilation has advanced to the point where it is possible to make a reasonably elegant, effective, and safe formulation with thioglycolate—one which will remove hair within 5 to 15 min without injury to the skin of the average woman. A satisfactory depilatory of this type can be made only if careful consideration is given to certain aspects of formulation including concentration of thioglycolate, alkalinity, and composition of the vehicle (wetting properties, fillers, stability, etc.).

For cream and lotion formulas, the concentration of thioglycolic acid may vary but is usually between 2.5 and 4.0 % by weight, in the form of its alkali or alkaline-earth salts. In concentrations below 2%, depilation time is prolonged, and no practical advantage, functionally or economically, is gained by increasing the concentration beyond 4%. Evans and McDonough (110) state that the concentration of mercaptans, especially those containing acid-acting polar groups, should be between 0.1 and 1.5 M/liter, preferably at 0.5 M/liter. When the concentration of thioglycolate is low (e.g., less than 2%), it is necessary to increase the alkalinity of the preparation to produce a satisfactory rate of depilation; at a higher concentration (e.g., 5%) in the same alkali concentration, the rate of depilation is not significantly faster. Conversely, for a given concentration of thioglycolic acid, increasing the alkali concentration increases the speed of depilation but at the risk of producing skin irritation. In the hair-waving process, higher concentrations of thioglycolic acid (approximately 7%) at a pH of 9.4 will effect curling in 15 to 30 min, whereas concentrations of 3 to 3.5 % may take hours to produce the same result.

It is certainly apparent that a proper balance should be established between the concentration of thioglycolic acid and the alkalinity, in order to produce nonirritating depilation in a reasonable time. Preferably, enough alkali should be present in the formulation to give pH values between 10 and 12.5. Below a pH of 10, depilation time is too slow, and as the pH increases beyond 12.5 the risk of skin damage becomes greater. To supply the necessary alkalinity, hydroxides of the alkali and alkali-earth metals are commonly used to provide relatively soluble salts of thioglycolic acid. Barium hydroxide forms a salt with thioglycolic acid, which has limited solubility and weak depilating activity. In general, bases which have ionization constants greater than 2.0×10^{-5} are satisfactory (110).

For practical purposes, a slightly greater equivalence of alkali to thioglycolic acid (never exceeding twice the equivalent concentration) gives a depilating system without causing skin irritation (110). For neutralizing thioglycolic acid and providing the proper alkalinity, the alkali-earth hydroxides are more desirable than the alkali hydroxides because of their lower solubility since only a slight excess of unreacted alkali-earth hydroxide remains in solution in addition to that required for neutralization of the thio acid. Calcium hydroxide, because of its low solubility, is particularly useful since, even in excess in a cream or paste formulation, it provides only a slight excess of alkali in solution, but nevertheless it has a ready insoluble reservoir for keeping constant the alkalinity of the solution (110). Alternatively, stronger bases like sodium, potassium, or strontium hydroxides may be used to neutralize all or part of the thio acid, whereas a lime slurry may be used to provide a slight excess of alkalinity and an alkalinity reservoir.

Commercial calcium thioglycolate may be used, the excess alkalinity being provided by the addition of calcium and/or strontium hydroxides. Sodium thioglycolate would probably be a satisfactory depilating agent were it not for its propensity to produce skin irritation (129).

Recently, French patents (114,130–132) discussed the depilatory activity of the various salts of thioglycolic, β-mercaptopropionic acid, and thiolactic acids and advocated the use of their lithium salts because of greater solubility, hence accelerated depilatory action. Table I illustrates comparative rates of depilation with 0.4 M acid at pH 12.5. It is claimed that the

TABLE I

Salt	Thioglycolic acid (min)	β-Mercapto-propionic acid (min)
Calcium	7.0	10.0
Barium	12.0	7.0
Strontium	5.0	7.0
Sodium	4.0 (+)[a]	5.5 (+)[a]
Potassium	3.5 (+)[a]	4.0 (2+)[a]
Lithium	3.5	5.0

[a] Irritating to skin.

lithium salt of thioglycolic acid is effective at slightly lower pH than its calcium salt and, since it is nearly three times more soluble, clear liquids or gels can be made (130).

Not only does the maintenance of proper alkalinity provide a reproducible and dependable speed of depilation, but it aids in suppressing odor and to a large extent prevents undesirable coloration of the formulation by such trace metals as iron. Two well-known brands of British depilatories containing strontium sulfide are reported to have a pH of 12 or greater (133). Most of the American brands which contain thioglycolate have pH values between 12 and 12.5.

Achieving practical depilation depends also on the formulation of the base in which the alkaline thioglycolate is carried (134). The paste, cream, or lotion must be of proper consistency, capable of being localized at the site of application, easily spread, and nondrying (that is, capable of holding moisture for at least 15 min after application). It should maintain "build-up" around the hair shaft and cling to the hairy area. It should provide a wetting action, thus allowing the thioglycolate to contact the hair shaft. For the latter, surface-tension reducing agents are chosen which are compatible with alkali and alkali-earth metals and effective in alkali at pH values up to 12.5. Certain anionics, such as the alkali-metal fatty alcohol sulfates and alkyl aryl sulfonates, and several nonionics of the polyoxyalkylene alcohol or ether type, are suitable wetting agents and provide satisfactory emulsifiers for formulations of lotions and creams which contain high concentrations of electrolytes.

A creamlike consistency may be achieved by using natural or synthetic thickening agents, such as tragacanth, karaya, guar, and quince seed extractives, methyl or hydroxyethyl cellulose, polyvinyl alcohol, and chemically modified starches and sugars, or by emulsified (oil-in-water) emollient compositions containing cetyl or stearyl alcohols with the surfactants mentioned above. These thickeners have, however, various degrees of reactivity toward alkaline reducing agents, and formulas containing them should be carefully shelf-tested (134).

Since large excesses of calcium hydroxide are not necessary, and since such lime pastes are difficult to remove from the skin, the substitution of suitable fillers or "bodying agents" for part of the lime has been suggested. Agents such as precipitated chalk, magnesium oxide and carbonate, clays (kaolin, bentonite, etc.), talc and similar nonreactive insoluble powders, increase the ease of removal of the paste composition from the skin (135) and decrease the tendency for irritation. A small amount of barium sulfate or titanium dioxide may be used to "whiten" the composition. Humectants such as glycerol and propylene glycol can be used, but in certain formulas their presence appears to prolong somewhat the depilation time. Although it is desirable to retain some emolliency in the formula, such materials as petrolatum should be used sparingly since they interfere with the "wetting" of the hair. Emolliency in the proper degree can be achieved with emulsified cetyl or stearyl alcohol; less satisfactorily with the vegetable or synthetic gums. An emollient base for calcium thioglycolate designed to provide a protective layer to minimize skin irritation includes 30 to 70 % fatty acid ethanolamides, 15 to 30 % of a Kritchevsky condensate derived from fatty acid and diethanolamine, and 15 to 50 % of nonionic O/W emulsifiers, e.g., polysorbate 80 (136). The addition of an azulene compound (e.g., 0.25 % of 1,4-dimethyl-7-isopropyl azulene) is claimed to reduce irritation of thioglycolate depilatories (137).

A thioglycolate depilatory in anhydrous stick form has recently been patented (138). When applied to moist skin, it produces a cream said to depilate in 10 min. Since the composition is essentially anhydrous, good stability and freedom from objectionable odor are claimed. The composition incorporates an organic thiol, preferably calcium thioglycolate, alkaline compounds (e.g., calcium and strontium hydroxides) in a perfumed anhydrous stick base containing lanolin and ethoxylated lanolin fractions, petrolatum, paraffin, and stearyl alcohol. In use, the stick is dipped in water and applied to the wet hairy surface with a circular motion for 10 to 30 sec. The cream which forms is rubbed again with the moistened stick for 5 min.

After an additional 5 min, the cream and hair are removed with paper tissue.

Some thioglycolate-based depilatories in aerosol containers have appeared in the U.S. market and abroad. According to a French patent (139), an aerosol gel depilatory with nitrogen as propellant is effective in 4 to 5 min and is nonirritating. Examples of depilatories containing thioglycolate are given in Formulas 13 to 18.

A simple apparatus and a method for quantitatively evaluating the depilatory effect of solutions have been proposed (141). The test is independent of cuticle condition, does not rely on breaking time or stress conditions, and

Formula 12*

Strontium thiolactate	27.0 g
Propylene glycol	12.0
Strontium hydroxide	4.0
Hydroxymethyl cellulose	10.0
Urea	75.0
Methylparaben	3.0
Chlorophyll	0.3
Lemon oil, terpeneless	3.0
Water	170.0

* Ref. 139.

Formula 13

Calcium thioglycolate trihydrate	6.0%
Calcium carbonate, light, USP	21.0
Calcium hydroxide USP	1.5
Cetyl alcohol flakes NF	4.5
Sodium lauryl sulfate USP	0.5
Sodium silicate solution, 42.5°Bé	3.5
Perfume	0.5
Water, distilled or deionized	q.s. 100.0

Procedure: To make 1000 g, dissolve 4.5 g of sodium lauryl sulfate in 155 ml of hot water (heated to 65°C.). Add 35.0 g sodium silicate solution and mix. Add 45.0 g melted cetyl alcohol to this mixture while hot; then agitate while cooling, to form an emulsion. In a separate vessel, mix 210.0 g calcium carbonate with 360 ml of hot distilled water. Add the previously prepared emulsion to this slurry and agitate for 30 min at 40°C. In another vessel, make a suspension by mixing 15.0 g calcium hydroxide and calcium thioglycolate in 110 ml of distilled water containing 0.5 g of sodium lauryl sulfate. Add this suspension to the previously prepared mixture of emulsion and calcium carbonate; then agitate at 40°C. Add perfume and continue agitation for 30 min. Add water to make the proper weight. The product is then roller-milled, and should be free from gritty crystals and entrapped air. Fill the cream into wax-lined lead tubes when cooled to room temperature.

Formula 14

Evanol	6.5%
Calcium thioglycolate	5.4
Calcium hydroxide	7.0
Duponol WA paste	0.02
Sodium silicate "O"	3.43
Perfume	as desired
Water	q.s. 100.00

Procedure: Heat the water to 75°C. With stirring, add the Duponol and Evanol; continue stirring at 75°C until melted and dispersed. Remove heat, continue stirring to room temperature. Add calcium hydroxide and perfume. Finally, add calcium thioglycolate and stir until uniform. Product should assay 2.7% thioglycolic acid.

Formula 15

Calcium thioglycolate	5.4%	5.4%
Calcium hydroxide	6.8	6.6
Strontium hydroxide (8 H_2O)	3.4	3.7
Cetyl alcohol	4.3	6.0
Calcium carbonate	22.4	—
Brij 35	1.2	1.0
Perfume	0.3	0.2
Water, distilled	q.s. 100.0	q.s. 100.0

Formula 16

Calcium thioglycolate	20.0%
Calcium hydroxide	23.0
Strontium hydroxide	9.0
Sodium lauryl sulfate	1.5
Hydroxyethyl cellulose	1.0
Calcium carbonate	15.0
Magnesium carbonate	30.0
Perfume	0.5

Procedure: Place in laminated foil packets (50.0 g); for use, mix contents with 3 oz water in a plastic or ceramic bowl (avoid metal).

Formula 17*

Mercaptoacetic acid	12.0 ml
Strontium hydroxide	50.0 g
Calcium oxide	12.0 g
Colloidal clay	102.0 g
Methyl cellulose	11.0 g
Perfume	0.8 g
Water	300.0 ml

* Ref. 110.

Formula 18*

Calcium thioglycolate	15.0 g
Calcium hydroxide	5.0
Calcium carbonate	60.0
Sodium lauryl sulfate	0.5
Total	80.5 g

* Ref. 140.

Formula 19*

Thioglycolic acid	8%
Calcium oxide	8
Sodium dodecyl xanthate	15
Sperm oil, purified	5
Water	64

* Ref. 118.

Aerosols—*Formulas 20 and 21*

A. *Concentrate*	#20	#21
Evanol	2.00%	2.00%
Calcium thioglycolate (100%)	5.40	10.80
Sodium silicate "O"	3.43	3.43
Perfume	0.50	0.50
Sodium hydroxide, 40%	adjust to pH 12.4	
Water	q.s. 100.00	100.00

B. *Concentrate*	162 g.
Propellant 114	7.2
Propellant 12	10.8
	180

involves the measurement of initial cross-sectional hair diameter and the time for maximal hair swelling. Since plots of both the length and width of swelling hair vs. time are sigmoid, the slope maxima of these curves serve to define an *in vitro* index of depilatory effectiveness.

Perfuming Chemical Depilatories

The perfuming of depilatories containing inorganic sulfides is an arduous and often frustrating task. One desires not only to mask (or better, to "re-odorize") the preparation *per se*, but to maintain a pleasant fragrance during depilation. Sulfide-based formulas are particularly difficult to "reodorize" because of the liberation of hydrogen sulfide by hydrolysis on storage and when the preparation comes in contact with the hair. Perfuming depilatory formulas based on thioglycolates does not present nearly as much difficulty, although the high alkalinity of either type permits unlimited possibilities for reaction with perfume ingredients. Perfume alcohols may form evil-smelling mercaptans; esters such as benzyl acetate may be hydrolyzed to form in-odorous alcohols, and aldehydes may undergo aldol-type condensations. Discoloration may result from degradation of perfume by alkali.

The odor of depilatories containing inorganic sulfides is reported to be masked by camphor (142). Similarly, eucalyptol and diphenyl oxide may dominate the "sulfide" odor. These compounds confer, however, an un-wanted "medicinal" odor to the preparation. The ionones are capable of "reodorizing" depilatory formulas, and linalyl acetate has been suggested by Atkins (56) since it retains a floral note even after hydrolysis. He reported that nitromusks (such as musk ambrette in benzyl benzoate) oxidized sulfhydryl groups, and the depilating activity of the formulation was lost after the perfumed product was shelf-tested. In the experience of the writer, certain compositions based largely on ionones and the "rose alcohols" (such as citronellol and geraniols) have been useful in perfuming thioglycolates, and some of them have shown excellent stability. Perfume problems of this kind can best be worked out in cooperation with perfume specialists, who should base their initial selections of fragrances on the knowledge of the stability of the perfume composition under alkaline-reducing conditions. The final selection of a suitable perfume is based on extensive shelf-testing. For a general discussion of the subject, the reader is referred to several published reviews (56, 134,143–148).

The odor of hydrogen sulfide, usually produced when sulfide-based depilatories are washed from the skin, can be controlled by wiping or scraping off the cream and rinsing with dilute boric, acetic, or lactic acids, with dilute zinc sulfate or acetate solutions, or with solutions containing oxidizing agents, such as hydrogen peroxide or peracids (46,56,149,150). Sulfide odors are said to be minimized also if water-soluble ammine salts of copper, cadmium, and zinc tetraammines, cobalt hexammine, and complex nickel ammine salts are used with the depilatory composition. Insoluble metal sulfide is produced by reaction of metal ammine with hydrogen sulfide, thus rendering the depila-tion process free from unpleasant odor (151).

Packaging of Thioglycolate Depilatories

A cream formula containing thioglycolate as its active principle tends to lose a part of its depilating potency if exposed to air. Reduction of alkalinity by atmospheric carbon dioxide and oxidation of thiol retards depilation and promotes color formation in the cream. It is important, therefore, that the product be protected from atmospheric degradation. Lotion depilatories are usually packaged in jars; creams, in collapsible tubes. Air spaces between cap and product surface in the jar and between the cream and crimp seal in the tube should be reduced to a practical minimum.

The high degree of chemical reactivity of the depilatory formula makes it essential that nonreactive plastic seals and caps be used for jars and tubes, and it is good practice to use wax-lined tin or preferably lead for the bodies of the collapsible tubes. Special wax coatings have been developed for this purpose. Discoloration has been observed occasionally, however, in wax-lined lead tubes, particularly in the case of low-viscosity creams, when the thioglycolate reacts with unprotected metal in the threaded portion of the neck of the tube. For this reason, one should make sure that the cap seats firmly when tightened prior to filling.

The reactivity of thioglycolate toward metal containers leads one to consider collapsible plastic containers for packaging, particularly those made of polyethylene or polyvinylchloride. Although these containers are generally inert, container distortion has been observed, owing to partial cave-in of the side walls of round and oval polyethylene bottles. This appears to be due to consumption of oxygen in the head space within the closed container, resulting in a partial vacuum. Moreover, there are varying degrees of liquid and fragrance loss in the polyethylene containers presently available. However, the advances currently reported in plastic technology offer hope that better flexible plastic containers will be available for packaging depilatory creams and lotions.

Thioglycolic Acid and Calcium Thioglycolate

Preparation

Present manufacturing methods for thioglycolic acid enable the cosmetic chemist to use solutions of high purity, stability, and reproducibility at low cost (152). The acid was first described by Carius in 1862 (153) and made by reacting sodium hydrosulfide and chloroacetic acid (154). It may be isolated as the barium salt or extracted with an organic solvent which is in turn extracted with water. By removal of the water by distillation and fractionating the residue, nearly pure thioglycolic acid is obtained. Highly concentrated solutions of the acid undergo dehydration, forming thio- or dithioglycolide (155–158); accordingly, the commercial acid is usually marketed as a 70 to 80 % aqueous solution, which is quite stable. More dilute solutions, e.g., 40 to 50 %, may slowly oxidize in air to dithioglycolic acid, which can reach a concentration as high as 7 to 10 %, especially if the solution is contaminated with metals. The presence in a hair-waving lotion of the dithio acid in amounts over 1 % has been reported to be detrimental to the quality of the wave (159), but this is of little consequence in depilatory formulations.

Thioglycolic acid has an odor said to be due to hydrogen sulfide and other byproducts. The pure acid is miscible with water, ethyl alcohol, acetone, ethyl ether, chloroform, and benzene. A trace of iron (0.005% or less) gives a pink-to-violet coloration to a solution of alkali-metal thioglycolate.

Properties of Thioglycolic Acid and Calcium Thioglycolate

Physicochemical constants for the pure thioglycolic acid are the following:

Melting point of anhydrous acid: 16.5°C
Boiling points: 123°C/29 mm., 107°C/16 mm, 102.5 to 103°C/13 mm
Density: 1.3253/20°C
Ionization constants: K_1—4 × 10^{-4}, K_2—1 × 10^{-10} (reference 159)
$\qquad\qquad\qquad$ K_1—2.1 × 10^{-4}, K_2—2.1 × 10^{-11}
Normal oxidation potential: 0.27 V/25°C
Heat of combustion: 346.3 kilocal/M

Calcium thioglycolate, depilatory grade, is the trihydrate of the secondary salt, $\underline{Ca}$—S—CH_2—$\underline{COO}$·$3H_2O$, crystallizing in prismatic rods. It is supplied commercially as a white powder, through 100 mesh, with a faint odor of mercaptan. It assays not less than 92% as the trihydrate by iodate titration, and contains 1% dithio acid maximum and about 2 ppm iron. The approximate solubility of the hydrated salt in water at 25°C is 7 g/100 ml, and at 95°C is 27 g/100 ml. In alcohol and chloroform at 25°C, its approximate solubility is 0.015 and 0.01 g/100 ml, respectively. It is practically insoluble in ethyl ether, petroleum ether, and benzene. Heating the salt above 95°C causes loss of water of crystallization; the salt darkens at 220°C and fuses, with decomposition at around 280 to 290°C.

Assay of Calcium Thioglycolate

For assay, 0.6 to 0.7 g of calcium thioglycolate is weighed accurately and added to 100 ml of distilled water in a 250-ml Erlenmeyer flask. It should dissolve readily and show no more than a faint turbidity. Five milliliters of concentrated hydrochloric acid is added, then 3 g of potassium iodide, and the solution titrated with 0.1 N potassium iodate solution, with starch as indicator. Calcium thioglycolate reacts quantitatively with iodine in neutral solution similar to the oxidation of cystene to cystine. The trihydrate contains 21.3 to 21.7% calcium and usually assays 49 to 50% thioglycolic acid by iodometry. For further information on the assay of thioglycolates, the reader is referred to the literature (160–162).

Local Toxicity

Toxicity data on thioglycolates have been published (163,164,166–170), particularly in connection with hair waving. Although several cases of reactions to thioglycolic acid have been cited (163,171), solutions containing less than 8 % of purified thioglycolate apparently have a low sensitizing index (164–166). Spoor (172) has described predictive tests and use experiences with "reactive" cosmetics such as depilatories.

The use of proprietary thioglycolate creams for preoperative hair removal on nearly 1000 subjects has been reported (173,174).

Enzymes for Hair Removal

In the literature on leather technology, there are many references to the use of enzymes for unhairing animal skins. These include mold tryptases, bacterial proteases, proteolytic and mucolytic enzymes from animal and vegetable sources. The mechanisms by which the enzyme systems loosen hair are complex. Certain proteolytic enzymes, such as trypsin and papain, reportedly "solubilize" wool to some extent; others (e.g., mucolytic, amylytic) loosen hair by affecting nonkeratin protein skin components. Enzymes attacking keratin directly have been isolated from certain strains of actinomycetes, the moldlike organisms found in the soil.

As a rule, native keratinaceous structures such as horn, wool, or feathers are almost completely resistant to the usual proteolytic enzymes. Wool previously treated by alkaline thioglycolate is readily digested by these enzymes (175). Ghuysen (176,177), while investigating the enzymatic activity of actinomycetin, an antibiotic from *Streptomyces albus*, found a keratinase in this complex which was useful as an animal skin depilant. One strain of *Streptomyces fradiae* found by Noval and Nickerson appears to be very active in digesting keratin (178). Patents issued to Rutgers Research and Educational Foundation (179) and to the Mearl Corporation (180) concern the use of (*a*) *S. fradiae* No. 3739 to convert inexpensive keratinaceous materials into soluble products of commercial value and (*b*) the keratinase from this organism in a depilatory composition. The concentrated and purified enzyme is standardized by measuring spectrophotometrically its capacity to digest wool. One unit of keratinase activity (K-unit) is that amount of enzyme which digests wool keratin to the extent of producing an increase in optical density of 0.040 at 2800 Å. Purified powders of 200 K-units/mg have been made. Depilatory formulations based on keratinase contain 200,000 to 1,000,000 K-units/100 g of product. Although its optional *p*H range for solubilizing hair is 8.5 to 9.5, keratinase is not stable at this *p*H range and requires buffering between *p*H 7 and 8 to provide reasonable stability. The formulas below are cited in the patent; the paste depilatory can be adjusted with phosphoric acid to *p*H 7.5, or the keratinase added just prior to use to the base adjusted to *p*H 8.5 to 9.5. Alternatively, a dry mixture of enzyme, gellants, talc, bentonite, or kaolin can be prepared which needs only the addition of water prior to use.

Keratinase preparations act much more slowly than thioglycolate or sulfide depilatories. The addition of sodium sulfite or sulfoxylate reportedly increases their keratinase activity (181). They are reportedly nonirritating even if left on the skin overnight, and they can be agreeably perfumed. To the writer's knowledge, no commercial product for human use has been marketed in the United States or abroad. The use of keratinase to unhair hides has been reported (182–184).

Formulas 22 and 23

Paste Depilatory		*Dry Mixture for Depilatory Paste*	
Keratinase (200 K-units/mg)	3.3 g	Keratinase (200 K-units/mg)	3 g
Sodium stearate	1.0	Sodium lauryl sulfate	2
Calcium carbonate	27.8	Sodium alginate	6
Water	63.4	Talc	85
Sodium lauryl sulfate	0.8	Perfume	1
Potassium phosphate, dibasic		Potassium phosphate dibasic	3
0.2 M	3.7		

Adjust pH to 7.5 with phosphoric acid.

Three enzymes widely distributed in nature are known thus far to catalyze cleavage of disulfide bonds with the participation of some metabolically generated hydrogen donor (185). These are cystine reductase, glutathione reductase, and protein disulfide reductase. Cystine reductase is found in clothes-moth larvae.

The production of an exocellular proteolytic enzyme capable of causing degradation of wool and hair appears to be a unique attribute of strains of *S. fradiae*. This enzyme (keratinase) accomplishes degradation of keratin in some manner other than by cleavage of disulfide bonds (185); sulfhydryl-containing peptides are not released from wool. Unlike many microbiological and mold proteases, keratinase appears to exert proteolytic action only against those proteins from which it releases peptides (186).

The isolation of a pure keratinase from *T. mentagrophytes* has been reported (187). When incubated with guinea pig hair, the enzyme liberated water-soluble ninhydrin-positive compounds.

REFERENCES

1. *Encyclopedia Britannica*, 14th ed., Vol. 16, p. 33.

2. *The Papyrus Ebers*, trans. by B. Ebbell, Vol. 63, Copenhagen and London, 1937, p. 76; *ibid.*, Vol. 67, p. 80.

3.   Chaplet, A.: Epilatoires, *Parfum. mod.*, **24**: 429, 709, 779, 835 (1930).

4.   Downing, J. G.: Cosmetics—past and present, *J. Am. Med. Assoc.*, **102**: 2089 (1934).

5.   Friedmann, M.: Zur Geschichte der Entfernung unerwünschter Haare, *Dermatol. Wochschr.*, **90**: 451 (1930).

6.   Piesse, G.: *The art of perfumery*, 3rd ed., Longmans, Green, London, 1862, p. 313.

7.   Volk, R., and Winter, F.: *Lexikon der kosmetischen Praxis*, J. Springer, Vienna, 1936, p. 107.

8.   Niemoeller, A. F.: *Superfluous hair and its removal*, Harvest House, New York, 1939, pp. 38–43.

9.   Velez-Salas, F.: Coco de mono arbol de nuestra flora que da tanino, aceite y grasa y es un buen depilatorio, *Rev. farm.*, **87**: 161 (1945).

10.   Allen, L.: Excess Facial Hair in Women, *Am. Perf.*, **83**: 81 (October 1968).

11.   Behrman, H. T.: Diagnosis and Management of Hirsutism, Report to the AMA Committee on Cosmetics, *J. Am. Med. Assoc.*, **172**: 1924 (April 23, 1960).

12.   Kaiser, P.: Br. Pat. 195,730 (1922).

13.   Olsen, P. C.: *Drug Trade News*, January 18, 1965.

14.   Brown, C. J.: U.S. Pat. 949,925 (1910).

15.   Hiss, A. E., and Ebert, A. E.: *New standard formulary*, G. P. Engelhard, Chicago, 1910, p. 1001.

16.   Weinhold, P. M.: Br. Pat. 478,176 (1938).

17.   Cohan, F. M.: U.S. Pat. 1,474,512 (1923).

18.   Fischer, H. A.: U.S. Pat. 2,062,411 (1936).

19.   Grant, W. M.: U.S. Pat. 2,091,313 (1937).

20.   Buff, M. E.: U.S. Pat. 2,202,829 (1940).

21.   Herrmann, D., and Herrmann, E.: U.S. Pat. 2,425,696 (1947).

22.   Adams, A.: Can. Pat. 382,241 (1939).

23.   Neary, L. J.: U.S. Pat. 2,417,882 (1917).

24.   Lucas, H. V.: U.S. Pat. 2,067,909 (1937).

25.   Gernsback, H.: U.S. Pat. 1,620,539 (1927).

26.   Didier, E.: Belg. Pat. 429,447 (1938).

27.   Cimino, V.: Ital. Pat. 467,409 (1951).

28.   Gotham, M.: U.S. Pat. 2,337,774 (1944).

29.   Kalish, J.: Cosmetic manual: 34 depilatories, *DCI*, **47**: 148 (1940).

30.   Phillips, M. A.: Br. Pat. 941,081 (1963).

31.   Arion, D.: Fr. Pat. 1,396,582 (1965).

32.   Bergwein, K.: Depilation and depilatories, *Seifen-Öle-Fette-Wachse*, **94** (1): 11 (January 3, 1968).

33.   Stoves, J. L.: Histochemical studies of keratin fibres, *Proc. Royal Soc. Edinburgh*, **62B**: 132 (1945).

34.   Stoves, J. L.: The role of disulfides and mercaptans in hair chemistry, *JSCC*, **3**: 170 (1952).

35.   Schöberl, A., and Weisner, E.: Modellversuche zum oxydativen Abbau biologisch wichtiger organischer Schwefelverbindungen, *Ann.*, **507**: 111 (1933).

36.   Schöberl, A., Berninger, E., and Harren, F.: Zur Kenntnis der alkalischen

Spaltung von Disulfiden. I. Mitteil.: Verhalten der Diphenyl-dithiodiglykolsäure, *Ber.*, **67B**: 1545 (1934).

37. Schöberl, A.: Uber die Reaktionsweise tertiärer Mercaptane und ihrer Disulfide, *Ber.*, **70B**: 1186 (1937).

38. Schöberl, A., and Rambacher, P.: Die hydrolytische Aufspaltung der Disulfidbindung in Cystinderivaten, Glutathion und Insulin, *Ann.*, **538**: 84 (1939).

39. Speakman, J. B.: The reactivity of the sulphur linkage in animal fibres. Part I. The chemical mechanism of permanent set, *J. Soc. Dyers Colourists*, **52**: 335 (1936).

40. Rosenthal, N. A., and Oster, G.: Recent progress in the chemistry of disulfides, *JSCC*, **5**: 286 (1954).

41. Woliszynski, R.: Br. Pat. 26,562 (1912).

42. Moore, E. K., and Koppenhoefer, R.: Studies on the chemistry of liming. VII. The influence of various nitrogen compounds on unhairing with calcium hydroxide suspensions, *J. Am. Leather Chemists Assoc.*, **28**: 245 (1933).

43. Jayne, D. W., Jr.: U.S. Pat. 2,192,380 (1940).

44. Hill, W. H.: U.S. Pat. 2,174,497 (1939).

45. Wronski, M., and Goworek, W.: Zeszyty Nauk, *Univ. Lodz Ser. 11*, **19**: 79 (1965); through *Chem. Abstr.*, **64**: 144602 (1966).

46. Henk, H. J.: Die Veränderung des Haarkeratins durch chemische Einflüsse, *Fette Seifen*, **48**: 147 (1941).

47. Fruton, J. S., and Clarke, H. T.: Chemical reactivity of cystine and its derivatives, *J. Biol. Chem.*, **106**: 667 (1934).

48. Mercer, E. H.: Some experiments on the orientation and hardening of keratin in the hair follicle, *Biochim. Biophys. Acta*, **3**: 161 (1949).

49. Turley, H. G., and Windus, W.: *Stiasny-Festschrift*, Eduard Roether, Darmstadt (Germany), 1937, p. 396.

50. Windus, W., and Turley, H. G.: The unhairing problem. II. A proof of the reduction theory of unhairing, *J. Am. Leather Chemists' Assoc.*, **33**: 246 (1938).

51. Moore, E. K., and Koppenhoefer, R.: Studies on the chemistry of liming. VI. The influence of various sulfides on unhairing with calcium hydroxide suspensions, *J. Am. Leather Chemists' Assoc.*, **28**: 206 (1933).

52. Janistyn, H.: Modern depilatories, *SPC*, **11**: 419 (1938).

53. Gillespie, J. M.: The depilatory activity of sodium sulfide and related compounds, *Australian J. Sci. Res.*, **B4**: 187 (1951); through *Chem. Abstr.*, **45**: 8779c (1951).

54. Genot, C.: Sur l'emploi d'onguents et de lotions épilatoires à base de sulfures: leur action sur la peau et le poil, *J. Pharm. Belg.*, **4**: 925 (1922).

55. Markus, E.: Die keratolytische Wirkung des Natrium-selenids, *Deut. Parf. Ztg.*, **17**: 318 (1931).

56. Atkins, F.: Chemical depilatories, *PEOR*, **38**: 231 (1947).

57. O. E.: Zur Toxizität des Bariumsulfides im Enthaarungsmittel, *Schweiz. Apoth.-Ztg.*, **83**: 309 (1945).

58. Domenjoz, R.: Zur Toxizität von Bariumsulfid bei Verwendung als Enthaarungsmittel, *Schweiz. med. Wochschr.*, **75**: 407 (1945).

59. Freund, W. W.: Odorless depilatories with barium sulfides, *Australasian J. Pharm.* **31**: 849 (1950); through *Chem. Abstr.*, **45**: 3553h (1951).

60. Kennedy, S. R.: U.S. Pats. 330, 715–16 (1885).

61. Perl, J.: U.S. Pat. 450,032 (1891).

62. Blinn, H. S.: U.S. Pats. 707,953–55 (1902).

63. Mellinger, S.: U.S. Pat. 449,134 (1893).

64. Roth, E., and Spira, L.: U.S. Pat. 1,982,268 (1934).

65. Stone, A. H.: U.S. Pat. 1,041,897 (1912).

66. Donner, J.: Br. Pat. 163,022 (1920).

67. Donner, J.: U.S. Pat. 1,379,855 (1921); *U.S. Pat. Quarterly*, Vol. 17, p. 117.

68. Miner, C. S., and Trolander, E. W.: U.S. Pat. 1,682,181 (1928).

69. Vebbing, A. A.: U.S. Pat. 1,954,397 (1934).

70. McKee, R., and Morse, E.: Br. Pats. 405,067 and 451,600 (1934).

71. Müller, R.: Ger. Pat. 550,440 (1931).

72. Rehdern, W.: Unerwünschter Haarwuchs und seine Entfernung, *Seifensieder-Ztg.*, **68**: 305 (1941).

73. Koenigsberger, F.: U.S. Pat. 2,031,489 (1936).

74. Kamlet, J.: U.S. Pat. 2,487,558 (1949).

75. Matsukura, G.: Jap. Pat. 2999 (1953); through *Chem. Abstr.* **48**: 7855 (1954).

76. Ishizaka, O.: Jap. Pat. 4000 (1957); through *Chem. Abstr.* **52**: 16708 (1958).

77. Higginbotham, A.: Br. Pat. 941,295 (1963).

78. Ger. Pat. 965,920 (1957).

79. McKee, R., and Morse, E.: U.S. Pat. 1,899,707 (1933).

80. Stoddard, W. B., and Berlin, J.: U.S. Pat. 2,123,214 (1938).

81. Stoddard, W. B., and Berlin, J.: U.S. Pat. 2,199,249 (1940).

82. Stoddard, W. B., and Berlin, J.: Br. Pat. 516,812 (1940).

83. Buczylo, E.: Polish Pat. 41,724 (1959); through *Chem. Abstr.*, **54**: 13565h (1960).

84. Pacini, A. B.: U.S. Pat. 2,081,279 (1937).

85. Kerdal-Vegas, F.: Generalized hair loss due to the ingestion of *Coco de mono*, (*Lecythis ollaria*), *J. Invest. Dermatol*, **42**: 91 (1964).

86. Kerdal-Vegas, F., and Aronow, L.: *Dermatologia*, **1**: 57 (1966).

87. Aronow, L., and Kerdal-Vegas, F.: Seleno-cystathione, a pharmacologically active factor in the seeds of *Lecythis ollaria*: Cytotoxic and depilatory effects of extracts of *Lecythis ollaria*, *Nature*, **205**: 1185 (1965).

88. Kerdal-Vegas, F., *et al.*: Structure of the pharmacologically active factor in the seeds of *Lecythis ollaria*, *Nature*, **205**: 1186 (1965).

89. Palmer, D. D.: Epilating effect of cystoselenonine on induced hair growth in mice, *Proc. Soc. Exp. Biol. Med.*, **128**: 663 (1968).

90. Zasosow, V. A., *et al.*: Preparation of p-Hydroxyphenyl Phenylethyl ketone and some of its ethers and homologs, *Zhur. Obshchei Khim.*, **26**: 2499 (1956); through *Chem. Abstr.*, **51**: 4994b (1957).

91. Batunin, M. P., *et al.*: Effect of epilene on animal organisms, *Nauch. Zapiski Gorkovsk*, **21**: 43 (1960); through *Chem. Abstr.*, **56**: 6620e (1962).

92. Tareeva, A. I., *et al.*: Epilene toxicity study in test animals, *Farmakol. Toksikol.*, **25**: 604 (1962); through *Chem. Abstr.*, **58**: 13028g (1963).

93. Philpitt, R., and Rubacky, E.: U.S. Pat. 3,426,137 (1968).

94. Flesch, P., and Goldstone, S. B.: Depilatory action of the intermediary polymers of chloroprene, *Science*, **113**: 126 (1951).

95. Flesch, P.: Hair loss from squalene, *Proc. Soc. Exp. Biol. Med.*, **76**: 801 (1951).

96. Flesch, P.: Inhibition of keratin formation with unsaturated compounds, *J. Invest. Dermatol.*, **19**: 353 (1952).

97. Flesch, P., and Goldstone, S. B.: Local depilatory action of unsaturated compounds, *J. Invest. Dermatol.*, **18**: 267 (1952).

98. Flesch, P., and Hunt, M.: Local depilatory action of some unsaturated compounds, *Arch. Dermatol. Syphilol.*, **65**: 261 (1952).

99. Sabouraud, R.: De l'acétate de thallium et de son emploi dans le traitement de l'hypertrichose chez la femme, *La Clinique (Paris)*, **7**: 102 (1912); through *Chem. Abstr.*, **6**: 2490 (1912).

100. Bedford, G. V.: Depilation with thallium acetate in the treatment of ringworm of the scalp in children, *Can. Med. Assoc. J.*, **19**: 660 (1928).

101. Dzialek, S.: Depilatory action of thallium acetate in rats, *Ann. Acad. Med. Lodz.*, **6**: 74 (1965); through *Chem. Abstr.*, **65**: 1283h (1966).

102. Chmel, L., *et al.*: Experimental application of *p*-bromphenylisothiocyanate for increasing the depilatory action of thallium, *Cesk. Dermatol.*, **39**: 6 (1964); through *Chem. Abstr.*, **63**: 8916f (1965).

103. Ormerod, M. J.: Pharmacological and toxicological aspects of thallium, *Can. Med. Assoc. J.*, **19**: 663 (1928).

104. Thallium poisoning (editorial), *J. Am. Med. Assoc.*, **92**: 1865 (1929).

105. McCord, C. P.: Industrial thallium poisoning, *J. Am. Med. Assoc.*, **98**: 1320 (1932).

106. Turley, H. G., and Windus, W.: U.S. Pat. 1,973,130 (1934).

107. Smith, P. I.: New chemicals used in English tanneries, *Chem. Met. Eng.*, **53**: 98 (Sept. 1946).

108. Fletcher, W.: Fr. Pat. 824,804 (1938).

109. Bohemen, K.: Br. Pat. 484,467 (1938).

110. Evans, R. L., and McDonough, E. G.: U.S. Pat. 2,352,524 (1944); *U.S. Pat. Quarterly*, Vol. 72, p. 211; *U.S. Pat. Quarterly*, Vol. 75, p. 259.

111. Demuth, R. L.: Br. Pat. 636,181 (1946).

112. Jones, C. B., and Meecham, D. K.: U.S. Pat. 2,517,572 (1950).

113. Morelle, J. V.: Fr. Pat. 1,161,038 (1958).

114. Zviak, C., and Rouet, J.: Fr. Pat. 1,345,572 (1963); U.S. Pat. 3,384,548 (1968).

115. Den Beste, M., and Reed, R. E.: Can. Pat. 465,446 (1950).

116. Den Beste, M., and Reed, R. E.: U.S. Pat. 2,540,980 (1951).

117. Jaeger, A. O., and Herrlinger, R.: U.S. Pat. 2,169,147 (1939).

118. Hentrich, R., and Gilroy, G.: Ger. Pat. 719,542 (1942).

119. Nakajima, S.: Jap. Pat. 4550 (1956); through *Chem. Abstr.*, **51**: 18495g (1957).

120. Carter Products: Fr. Pat. 1,411,330 (1965).

121. Helena Rubinstein: Fr. Pat. 1,444,916 (1966).

122. Morelle, J. V., and Morelle, E.: Fr. Pat. 1,469,512 (1967).

123. Evans, R. L., and McDonough, E. G.: Br. Pat. 593,438 (1945).

124. Evans, R. L., and McDonough, E. G.: Br. Pat. 521,240 (1940).

125. Evans, R. L., and McDonough, E. G.: Fr. Pat. 844,529 (1939).

126. Moore, L. P., and Erichs, W. P.: U.S. Pat. 2,442,957 (1948); U.S. Pat. 2,453,333 (1949).

127. del Lupo, M.: U.S. Pat. 2,600,624 (1952).

128. McDonough, E. G.: U.S. Pat. 2,577,710 (1951). A listing of many thio or related compounds of cosmetic significance may be found in deNavarre, M. G.: *The chemistry and manufacture of cosmetics*, 2nd ed.. Vol. II, Van Nostrand, 1962, pp. 320–324.

129. The reader may wish to refer also to several reviews: Tobler, Leo: Chemical Depilatories, *Kosm. Parf. Drogen, Rünschau*, **11**: (3/4): 41 (1964); Walker, G. T.: Chemistry and formulation of depilatories, *Am. Perf.*, **77**: 36 (August 1962); Alexander, P.: Depilatories, *ibid.*, **83**: 115 (October 1968).

130. Zviak, C., and Rouet, J.: Fr. Pat. 1,359,832 (1964).

131. Carter Products: Fr. Pat. 1,411,330 (1965).

132. Helena Rubinstein: Fr. Pat. 1,444,916 (1966).

133. Bull, K. B.: Some principles involved in the formulation of chemical depilatories, *SPC*, **28**: 63 (1955).

134. Bull, K. B.: Some principles involved in the formulation of chemical depilatories. Part 2. Toxicity, formulation, presentation and future research, *SPC*, **28**: 655 (1955).

135. Evans, R. L.: Br. Pat. 451,611 (1936).

136. Braun, E., and Logun, J.: U.S. Pat. 3,154,470 (1964); Chemway Corp.: Fr. Pat. 1,411,308 (1965).

137. Ruemele, T. A.: Non-irritant depilatories, *Mfg. Chemist*, **30** (2): 68 (1959); Ruemele, T. A.: Br. Pat. 830,833 (1960).

138. Braun, E., and Logun, J.: U.S. Pat. 3,194,736 (1965).

139. Roy, R.: Fr. Pat. 1,405,939 (1965).

140. Hansel, H.: Depilatorien, *Parfüm. Kosmetik*, **34**: 413 (1953).

141. Yablonsky, H., and Williams, R.: A quantitative study of the effect of depilatory solutions on hair, *JSCC*, **19**: 699 (1968).

142. Lewinsohn, A.: Ger. Pat. 649,033 (1937).

143. McDonough, E. G.: Mercaptans in cosmetics, *Am. Perf.*, **50**: 445 (1947); Problems in perfuming cosmetics, *Am. Perf.*, **55**: 205 (1950).

144. Koeune, A. E.: Fundamentals of depilatory formulation and manufacture, *Mfg. Perf.*, **2**: 48 (May 1937).

145. Morel, C.: Perfuming cosmetics and toilet preparations, *SPC*, **19**: 917 (1946).

146. Morelle, J.: The problem of depilatories, *Industrie parfum.*, **11**: 176 (1956); through *Chem. Abstr.*, **50**: 13377c (1956).

147. Sagarin, E., and Balsam, M.: The behavior of perfume materials in thioglycolate hairwaving preparations, *JSCC*, **6**: 481 (1956).

148. Bergwein, K.: *Dragoco Report*, **4**: 79 (1965); also, *Givaudanian*, April 1967.

149. Knapp, M.: Fr. Pat. 764,538 (1934).

150. Henk, H. J.: Die Nachteile der Depilatorien und ihre Beheung, *Seifensieder-Ztg.*, **66**: 375 (1939).

151. Freund, W. W.: Br. Pat. 674,195 (1949).

152. Schmidt-Lamberg, H.: Manufacturing of depilatory materials, *Chem. Tech.*, **2**: 132 (1950); through *Chem. Abstr.*, **44**: 8053d (1950).

153. Carius, L.: Entstehung der Schwefelessigsäure and ihrer Analogen durch Oxydation von Sulfosäuren derselben Basicität, *Ann.*, **124**: 43 (1862).

154. Schütz, F.: Über die Herstellung reiner Thio-glykolsäure, *Angew. Chem.*, **46**: 780 (1933).

155. Whitman, R., and Eckstrom, M. G.: The mercaptan-disulfide system in permanent waving—a new mechanism and its practical implications, *Proc. Sci. Sec. TGA*, **22**: 23 (1954).

156. Andreasch, R.: Zur Kenntniss der Thiohydantoine, *Monatsh.*, **8**: 407 (1887).

157. Debray, M.: Br. Pat. 484,467 (1936).

158. Schöberl, A., and Krumey, A.: Über das Dithioglykolid und über Polythioglykolide, *Ber.*, **77B**: 371 (1944).

159. Cannan, R. K., and Knight, B. C.: Dissociation constants of cystine, cysteine, thioglycolic acid and a-thiolactic acid, *Biochem. J.*, **21**: 1384 (1927).

160. Hoshall, E. M.: Determination of sulfides in depilatories, *J. Assoc. Offic. Agr. Chemists*, **23**: 437 (1940).

161. McNall, F. J.: Report on depilatories, *J. Assoc. Offic. Agr. Chemists*, **25**: 924 (1942).

162. Fisher, H. J., Mathis, W. T., and Waldern, D. C.: *Conn. Agr. Exp. Sta. Bull.*, **460**: 448 (1942).

163. Cotter, L. H.: Thioglycolic acid poisoning in connection with the "cold wave" process, *J. Am. Med. Assoc.*, **131**: 592 (1946).

164. McCord, C. P.: Toxicity of thioglycolic acid used in cold permanent wave process, *J. Am. Med. Assoc.*, **131**: 776 (1946).

165. Harry, R. G.: *Modern cosmeticology*, 3rd rev. ed., Chemical Publishing Co., Brooklyn, N.Y., 1947, p. 172.

166. Anon.: Sensitivity of thioglycolate, *J. Am. Med. Assoc.*, July 27, 1957.

167. Draize, J. H.: *et al.*: The percutaneous toxicity of thioglycolates, *Proc. Sci. Sec. TGA*, **7**: 29 (1947).

168. Lehman, A. J.: Health aspects of common chemicals used in hair waving preparations, *J. Am. Med. Assoc.*, **141**: 84 (1949).

169. Freeman, M. V., Draize, J. H., and Smith, P. K.: Some aspects of the mechanism of toxicity of thioglycolates, *J. Pharm. Exptl. Therap.*, **118**: 296 (1956); Some aspects of the absorption, distribution, and excretion of sodium thioglycolate, *ibid.* **118**: 304 (1956).

170. Mackie, B. S.: The mechanisms of dermatitis, *Am. Perf.*, **84**: 37 (1969).

171. Masters, E.: Allergies to cosmetic products, *N.Y. State J. Med.*, **60**: 1934 (1960).

172. Spoor, H. J.: Predictive tests and use experience with "reactive" cosmetics, *Proc. Sci. Sec. TGA*, **46**: 34 (1966).

173. Vestal, P. W.: Preoperative preparation of the skin with a depilatory cream and a detergent, *Am. J. Surgery*, **83**: 398 (1952).

174. Prigot, A.: *Am. J. Surgery*, **104**: 900 (1962).

175. Geiger, W. B., Patterson, W. I., Mizell, L. R., and Harris, M.: *J. Res. Nat. Bur.*, **27**: 459 (1941).

176. Ghuysen, J. M., and Leger, F.: *Compt. Rend. Soc. Biol.*, **148**: 1691 (1954); through *Chem. Abstr.*, **49**: 8374 (1955).

177. Deltour, G., and Ghuysen, J. M.: U.S. Pat. 2,927,885 (1960).

178. Noval, J. J., and Nickerson, W. J.: Decomposition of native keratin by *Streptomyces fradiae*, *J. Bacteriol.*, **77**: 251 (1959).

179. Noval, J. J., and Nickerson, W. J.: U.S. Pat. 2,988,487 (1961); Nickerson, W. J., and Noval, J. J.: Br. Pat. 821,129 (1959).

180. Mattin, H. E., and Greenstein, L. M.: U.S. Pat. 2,988,485 (1961); Robison, R. S., and Nickerson, W. J.: U.S. Pat. 2,988,488 (1961).

181. Span. Pat. 272,921 (1962); through *Chem. Abstr.*, **59**: 7316d (1963).

182. Cordon, T. C., *et al.*: Elastase activity of some enzymes as related to their depilatory action, *J. Am. Leather Chemists Assoc.*, **56**: 68 (1961).

183. Belg. Pat. 647,059 (1964); through *Chem. Abstr.*, **63**: 11888f (1965).

184. Netherlands Application 6,413,820 (1965); through *Chem. Abstr.*, **65**: 13980d (1966).

185. Nickerson, W. J., *et al.*: Properties of keratinase conjugate, *Biochim. Biophys. Acta*, **77**: 73 (1963).

186. Nickerson, W. J., and Durand, S.: Crystalline keratinase, *Biochim. Biophys. Acta*, **77**: 87 (1963).

187. Yu, R. J., Harmon, S. R., Wachter, P. E., and Blank, F.: *Hair digestion by a keratinase of* T. metagrophytes, Archiv. Biochem., **135**: 363 (1969).

Chapter 19

SHAMPOOS

Donald H. Powers

revised by

Neil D. Stiegelmeyer and Edward W. Lang

The hair of the head has historically been associated with beauty and social distinction. Innumerable instances from all the art forms can be cited supporting the special prominence afforded the hair by people of virtually all times and cultures. Names like Samson, Godiva, and, of more recent times, Harlow call to mind how dramatically the hair can characterize the individual.

Whereas the hair has been trimmed, shaped, and even colored since the most ancient times, relatively little emphasis has been placed on the process of cleaning it. Only in this century has a real technology in the cleaning of the hair and scalp developed. First came the mass distribution of cake soap and sanitary facilities to make bodily cleanliness and personal hygiene practical. Next came the specialization of branded shampoo products for the hair and scalp, offered in a multiplicity of types and forms.

Annual sales of shampoos in the United States passed the $200 million mark several years ago (1). Sales figures demonstrating this steady growth of the shampoo market are shown in Table I.

The more spectacular rise in the total sales of all consumer hair products is due to the emergence of the hair spray and the hair coloring markets. Both of these product types have sales of the same order of magnitude as shampoos. Thus, whereas shampoo sales accounted for about half of all consumer hair products sold in 1950, current shampoo sales constitute only one-fourth of all hair products.

Originally, shampoos were made of soap or mixtures of soaps; today synthetic detergents are used in the majority of commercial products. It must not be forgotten, however, that a large, if dwindling, percentage of the male

73

TABLE I.   Annual Retail Sales of Shampoos and All
Hair Products in the United States, 1950 to 1968 (1)

Year	Shampoo sales ($ millions)	All hair products (except to professional trade) ($ millions)
1950	80.7	157.7
1955	122.7	301.6
1960	166.6	523.4
1961	172.7	559.8
1962	179.5	633.4
1963	189.7	705.1
1964	202.6	779.7
1965	220.4	880.6
1966	238.4	945.7
1967	247.6	990.7
1968	266.6	1044.5
1969	279.8	1102.2
1970	349.5	1207.9

population still uses bar soap for cleaning its hair and scalp. Synthetic detergents have been claimed to be the most important factor in the growth of the shampoo market, and certainly the greatest increase in sales has occurred since the synthetics became available. No doubt other factors have contributed to this growth. Increasing population, rising standards of living, and a growing attention to personal grooming have all helped.

The ever burgeoning number of synthetic detergents available to shampoo formulators has also played a part in providing an abundant differentiation in properties sought to attain new and improved shampoo products. McCutcheon's (2) annual listing of detergents and emulsifiers accessible to the cosmetic and toiletry formulator continues to grow.

Function of a Shampoo

Harry (3) defined a shampoo as "a preparation of a surfactant (i.e., surface-active material) in suitable form—liquid, solid, or powder—which when used under the conditions specified will remove surface grease, dirt, and skin debris from the hair shaft and scalp without affecting adversely the hair, scalp, or health of the user." He further suggested that "*only surface cleansing is required*" to make the hair "*look* and *feel clean.*" He further emphasized, "Providing the hair *looks clean* or · · · *does not* 'look dirty' and providing it has a *gloss or sheen*—and so long as it is not 'frizzy,' 'fuzzy,' or 'unmanageable' (i.e., without adversely affecting the health or *physical* properties of the hair) then the shampoo has done its job."

Wall (4) declared that a good shampoo should "cleanse hair and scalp thoroughly without stinging or irritation, and should not remove too much of the natural oil from the scalp." The success of shampoos, in replacing the cake of soap, according to Zussman (5), lies in the fact that the shampoo is not only a detergent but a cosmetic as well, and that it must impart luster, beauty, and manageability.

According to Ester, Henkin, and Longfellow (6), shampoos must do a satisfactory cleansing job without removing an excessive amount of oil. Barnett and Powers (7) suggested that the successful shampoo preparation must leave the hair fragrant, lustrous, soft, and manageable.

In summary, the primary function of a shampoo is to clean the hair and scalp. Hair soil includes the natural skin secretions, the skin debris, dirt accumulated from the environment, and the residue of hair-grooming products applied by the consumer. After accomplishing this cleansing action to the satisfaction of the user, the shampoo should leave the hair soft, lustrous, and manageable. A shampoo formulation may also be compounded to emphasize some specialized capability like minimizing eye sting, controlling dandruff, or imparting appealing fragrance to gain a more favorable acceptance from particular segments of the population.

Shampoo Types and Forms

Shampoos are available in a variety of forms and types. Any method of classification must of necessity impose an arbitrary viewpoint. A classification according to product form would consist of clear liquids, lotions, pastes, gels, and, finally, aerosols and dry products. Shampoos may be further differentiated by the specialized appeal that an unusual component or combination of components may provide, e.g., shampoos for particular hair or scalp conditions, children's or infants' shampoos, shampoos for men, etc.

Liquid Shampoos

The clear liquid type has continued to hold the position of the most popular shampoo form. According to a recent sales survey, this form accounts for somewhat more than half of all the shampoo sold.

Liquid shampoos based on soap were the products that initiated the commercial shampoo business. Cook (8) remarked, "Liquid soaps achieved almost instantaneous popularity because of ease of application, rapidity of lathering, and better rinseability." Liquid soap shampoos are frequently based on potassium soap because of its greater solubility. Usually the soap shampoos are protected with sequestering agents such as polyphosphates or salts of ethylenediamine tetraacetic acid (EDTA). These agents hinder the formation of insoluble calcium or magnesium soap curd when these ions are introduced via hard water.

Liquid shampoos are also made from sulfonated oils and promoted as oil shampoos. Mineral oils may be added to these formulations. Since the oil does not readily rinse out, it remains as a dressing to the hair.

The major liquid shampoos are based, however, on synthetics. The ethanolamine salts of these synthetics are often used for improved solubility and clarity at low temperatures. Improved odor, color, and clarity of the synthetic detergent bases have allowed formulators more latitude in choosing physical and performance capabilities. Examples of liquid shampoos are given in Formulas 1 to 9.

Lotion Shampoos

Lotion shampoos are somewhat less popular than the clear liquids, holding about 15% of recent shampoo sales. These shampoos are usually promoted

Formulas for Liquid Shampoos

	1*	2†	3†
Coconut oil	14.0%	18.0%	—
Olive oil	3.0	—	—
Castor oil	3.0	4.0	—
Potassium hydroxide, 85%	4.7	5.3	—
Glycerol	2.0	4.0	5%
Ethyl alcohol	4.0	—	10
Sodium hexametaphosphate	1.0	—	—
Perfume	0.3	0.2	q.s.
Water	68.0	68.0	40
Borax	—	0.5	—
Coconut soap potassium salt	—	—	35
Olive oil soft soap	—	—	10

   * Ref. 9.
   † Ref. 10.

Formula 4*

Water	14.0%
Sulfated castor oil, 75%	59.5
Sulfated olive oil, 75%	19.5
Mineral oil, light	3.0
Glycerol	3.5
Perfume	0.5

   * Ref. 10.

Formula 5*

Coconut diethanolamide (92%)	5.0%
Water	38.4
Sodium lauryl sulfate (30%)	30.0
Triethanolamine alkyl benzene sulfonate (60%)	13.0
Potassium soap (20%)	12.5
Phosphate buffer salt	0.5
Preservative	0.2
Perfume	q.s.

* Ref. 11.

Formula 6*

Triethanolamine lauryl sulfate (C_{10} to C_{18})	35.0%
Sodium alginate	2.5
Water	62.5

* Ref. 12.

Formulas for Liquid Shampoos

	7*	8†	9‡
Water	q.s.	56%	q.s.
Sodium lauryl sulfate	4.0%	—	—
Dimethyl lauryl amine oxide	8.0	—	—
Disodium lauryl β-imino dipropionate	5.0	—	—
Lauric diethanolamide	1.0	1	—
Sodium salt of lauryl sulfate	—	20	—
Miranol C2M	—	20	—
Hexylene glycol	—	2	—
Polyoxyethylene sorbitan monolaurate	—	1	—
GE SF-1066 silicone fluid	—	—	3.0%
Ammonium lauryl sulfate (30%)	—	—	48.0
Methocel 60 HG	—	—	0.5
Perfume	0.5	—	—

* Ref. 13.
† Ref. 14.
‡ Ref. 15.

for their conditioning action, and manufacturers often claim that lotion shampoos leave the hair in an advantageous state. Some contain lanolin, others dispersed egg powder, and many contain alkanolamines of the higher fatty acids, which are noted for their conditioning action.

The opacity of the creme lotion formulations is often achieved by "pearling" or opacifying agents deliberately added to impart the desired appearance. These components may be insoluble stearate salts, glyceryl stearates, insoluble resin latex dispersions, or even synthetic forms of mother-of-pearl. The type of opacity may vary from flat to highly reflective, depending on the size, shape, and reflective power of the opacifier, as well as its concentration in the liquid matrix. Examples of lotion shampoos are shown in Formulas 10 to 17.

Creme Paste and Gel Shampoos

Creme paste and gel shampoos are both popular forms, with recent combined sales at about 30% of total shampoo sales. Both forms have stiff

Formulas for Creme or Creme Lotion Shampoos

	10*	11†	12†	13	14‡
Ultrawet 60 L	33%	—	—	—	—
Glyceryl monostearate	2	—	—	—	—
Magnesium stearate	1	1.0%	4%	—	—
Water	64	38.0	50	q.s. 100.0%	69.25%
Sodium lauryl sulfate (C_{10} to C_{18})	—	30.0	—	40.0	—
Polyvinyl alcohol, 10%	—	20.5	6	—	—
Methyl cellulose 50 cps	—	9.0	—	—	—
Glyceryl monolaurate	—	1.0	2	—	—
Lanolin	—	0.5	1	1.0	0.50
Triethanolamine lauryl sulfate (C_{10} to C_{14})	—	—	35	—	—
Cetyl alcohol	—	—	2	—	—
Perfume	—	—	—	0.5	0.40
Ethanolamide of refined coconut fatty acids	—	—	—	5.0	—
Ethylene glycol monostearate	—	—	—	1.5	—
Hydroxyacetic acid	—	—	—	to pH 7.4	—
Triethanolamine	—	—	—	—	1.20
Behenic acid	—	—	—	—	3.50
Methyl p-hydroxybenzoate	—	—	—	—	0.15
Sodium salt of sulfated monoglyceride of hydrogenated coconut oil fatty acid	—	—	—	—	25.00

* Ref. 16.
† Ref. 12.
‡ Ref. 17.

*Formulas for Lotion Shampoos**

	15	16	17
Magnesium stearate	—	1.0	—
Water	49%	52.5%	51.5%
Cetyl alcohol	—	—	1.0
Ethylene glycol monostearate	1	—	2.0
Polyethylene glycol distearate	3	2.0	—
Propylene glycol	2	—	—
Duponol WAQ	—	40.0	—
Onyxol 336	—	3.0	—
Emcol MAS	—	1.0	—
Sodium chloride	—	0.5	—
Maprofix TLS (TEA lauryl sulfate)	—	—	40.0
Super Amide B-5	—	—	5.0
Potassium chloride	—	—	q.s.
Sodium lauryl sulfate	45	—	—

* Ref. 18.

consistencies, allowing them to be packaged in jar or tube. They must be thick enough to resist spillage by clinging to the hand, and yet be readily dispersed through the hair upon application. Both the creme paste and the gel forms usually carry higher levels of detergents and other actives than their liquid counterparts.

Creme paste shampoos are the older of the two forms. Originally they were soaps superfatted with glycerides, lanolin, and the higher fatty alcohols. Synthetic detergents are now commonly used in creme paste shampoos. Often the less soluble soaps (e.g., sodium stearate) provide both thickness and opacity. Patterson (19) has made an interesting study of the processing and environmental variables that may affect consistency and stability of a typical creme paste formula.

The gel shampoos are generally regarded as a thicker and more concentrated form of a clear liquid product. Good clarity and smoothness are consequently a necessity, since transparent plastic tubes are the most popular package for these shampoos. Consistency is often adjusted by using a greater proportion of solids relative to liquid shampoos or stiffening with electrolytes, natural gums, or cellulosic thickeners. Formulas 18 to 23 are representative of these shampoo forms.

Aerosol and Dry Shampoos

Aerosol shampoos appear periodically on the market with appeals based on convenience or novelty. These products may be either aqueous liquids that foam up as the shampoo is released from the container, or dry shampoos that

Formulas for Creme or Creme Paste Shampoos

	18*	19†	20‡
Tergitol 7	33.3%	—	—
Nacconol NRSF	16.6	—	—
Bentonite	7.4	—	—
Cellosize WSLM	5.7	—	—
Water	37.0	41%	81.00%
Sodium lauryl sulfate (C_{10} to C_{18})	—	50	—
Sodium stearate	—	8	—
Lanolin	—	1	—
Calcium alginate	—	—	2.00
Sodium citrate	—	—	1.00
Triethanolamine lauryl sulfate	—	—	10.00
Glycerol	—	—	5.00
Methyl *p*-hydroxybenzoate	—	—	0.15
Perfume	—	—	0.85

* Ref. 16.
† Ref. 12.
‡ Ref. 20.

Formulas for Creme Paste and Gel Shampoos

	21*	22†	23‡
Water	q.s.	45.75%	—
Sodium lauryl sulfate	20.550%	—	—
Lanolin	0.470	—	—
Triethanolamine lauryl sulfate	—	—	80%
Methyl *p*-hydroxybenzoate	0.140	—	—
Perfume	0.295	q.s.	q.s.
Sodium chloride	0.890	—	—
Sodium benzoate	0.235	—	—
Stearic acid	6.580	—	—
Potassium hydroxide, 34.2%	4.035	—	—
Lauric-myristic diethanolamide	1.500	—	—
Monomethylol dimethyl hydantoin	0.100	—	—
GE SF-1066 silicone fluid	—	3.00	—
Ammonium lauryl sulfate	—	49.50	—
Methocel 60 HG	—	1.75	—
Color	—	q.s.	—
Lauric diethanolamide	—	—	16
Propylene glycol	—	—	4

* Ref. 21.
† Ref. 15.
‡ Ref. 22.

are sprayed on the dry hair for a waterless cleaning operation. The foaming aerosols are applied to the wet hair and, upon application, follow the traditional pattern of lathering and rinsing.

Aerosol shampoos may be packed in either glass bottles or metal cans. Glass offers the advantage of a noncorroding container wall; however, the glass container requires the protective encasement of a plastic jacket. This checks the possibility of flying glass being projected by the released propellant pressure in the event of accidental breakage.

Corrosion is a major problem with metal cans. This is especially the case when an aqueous detergent solution is in contact with the metal and the propellant. Nonetheless, there have been many ingenious coatings and schemes for internally separating reactive components to reduce or eliminate the problem of corrosive attack on the metal.

Dry shampoos enjoy a limited use in this country. They are mixtures of dry absorbent powders and mild alkalis that pick up soil from the hair and scalp. These products may be dusted or sprayed on from an aerosol device. After a waiting period to allow the soil to be absorbed, the powder is brushed or combed out of the hair. Dry shampoos are used by those who must avoid wet hair because of illness, or when there is insufficient time for the full routine of washing, setting, drying, and styling the hair.

The added cost and effort required by aerosol packing impose a handicap on both liquid and dry products. The choice and testing of valves and fitments to insure reliable delivery and storage stability is both an art and a science in itself. However, if particular formulas must be protected from atmospheric attack, must be kept sterile, or may benefit from a propelled form of dispensing, then aerosol packaging can be attractive. Examples of aerosol and dry shampoo products are given in Formulas 24 to 28.

Formula 24*

Stearic acid	3.5%
Myristic acid	2.0
Oleic acid	3.5
Lanolin, anhydrous	0.3
Propylene glycol	20.0
Triethanolamine	5.0
PVP K-30	0.2
Cheelox BF-13	0.2
Triethanolamine lauryl sulfate	48.5
Water	16.8
Above concentrate	90%
Propellant 12/114, (40:60)	10

* Ref. 23.

Formulas for Dry Shampoos

	25*	26†	27‡
Sodium keryl benzene sulfonate	10%	—	—
Sodium sulfate	6	—	—
Sodium cetyl sulfate	20	—	—
Sodium carbonate	10	—	—
Sodium bicarbonate	5	—	83.0%
White acid clay	49	—	—
Sodium decyl benzene sulfonate	—	12%	—
Sodium dodecyl benzene sulfonate	—	42	—
Glycerol	—	42	—
Sodium alginate	—	4	—
Borax	—	—	8.3
Ammonium carbonate	—	—	8.3
Perfume	—	—	0.4

* Ref. 24.
† Ref. 25.
‡ Ref. 26.

*Formula 28**

Borax	6%
Talc	60
Sodium sesquicarbonate	17
Fuller's earth	17
Perfume	q.s.

* Ref. 3.

Strictly speaking, nonaqueous liquid shampoos may be termed "dry" shampoos. These employ organic solvents to loosen or dissolve hair soil, which is rubbed off by cotton swabs or tissues. None of the common solvents, however, such as benzene, carbon tetrachloride, ethylene dichloride, petroleum ether, or isopropyl alcohol, are safe for use on the basis of toxicity and/or flammability.

Specialty Shampoos

"Specialty shampoo" is a term referring to a product which stresses a particular component or an unusual performance property. Special components might be egg, herbs such as the saponins, lanolin and its various derivatives, protein hydrolysates, and silicone compounds. The shampoo serves as a vehicle to bring these materials into contact with the hair. Nonetheless, the main function of the product is still understood to be cleaning. The beneficial effects of the additives are ancillary and conceived of in terms of a bonus.

Specialized performance products could be described as shampoos formulated to gain differentiation by unusual or novel action or appeal. Examples of these products could be the following.

1. Shampoos of low eye sting for infants and children.

2. Shampoos that complement particular cosmetic lines by bearing matching fragrance and appearance.

3. Products intended to control or eliminate the evidence of dandruff.

4. Shampoos containing dyes intended to impose a shading or tone to hair of specified color.

It is essential for those offering specialty shampoos to determine clearly the legal responsibility they bear in reference to marketing the products. Claims that systemic changes in the body or organs are brought about may be interpreted as a drug action by the FDA. If such a product is not generally recognized by experts as both safe and effective for its intended purpose, then approval of a new drug application by the FDA would be required before the drug could be marketed. A careful review of claims is clearly necessary. Both the label and the promotional copy should be examined to ascertain proper compliance with the requirements of regulatory legislation. Formulas 9, 22, 29, 30, and 31 represent several of the wide variety of specialty products.

Raw Materials

Since the major components in shampoos are surfactants (soaps and synthetic detergents), it is appropriate to review their respective uses, differences, and advantages.

Soaps

Soaps are generally defined as the salts of fatty acids. Originally they were obtained by saponifying the natural animal and vegetable fats and oils with alkali, as sodium and potassium hydroxides. More recently alkanolamines have been used.

Over the years it has been possible to formulate soap shampoos based on mixtures of oils, so as to obtain desirable proportions of fatty acids. In such

*Formula 29**

Quillaja bark, powdered	5.0%
Ammonium carbonate	1.0
Borax	1.0
Bay leaf oil	0.1
Water	92.9

* Ref. 10.

Formulas for Antidandruff Shampoos

	30*	31†
Tween 80	4%	—
Cetyl dimethyl benzyl ammonium bromide	12	—
Perfume	q.s.	0.5%
Water	84	q.s. 100.0
Arlacel 80	—	13.0
Glyceryl monoricinoleate	—	1.0
Bentonite	—	4.0
Selenium sulfide bentonite, 1:1	—	5.0
Citric acid	—	0.4
Monosodium phosphate	—	2.0

   * Ref. 3.

   † Ref. 27.

mixtures the acids are balanced to give the desired foaming and cleaning action. Table II gives the ratios of some fatty acids as they occur in a variety of fats and oils.

In general, those oils which contain primarily the shorter-chain fatty acids (e.g., coconut oil) yield better foaming soaps. These acids contain 10 to 12 carbon atoms in a straight chain and usually work well at low temperatures. The soaps from fatty acids with somewhat longer chains (14 to 16 carbon atoms) are excellent cleaners, particularly in warm water, and those from long-chain fatty acids (16 to 18 carbon atoms) are very effective at 70°C or higher.

Table III lists the fatty acids usually found in vegetable oils with their solubilities in water at 60°C. For fatty acids with fewer than 10 carbons in the chain, the soaps are too soluble to form acceptable suds or to show appreciable detergency, whereas the soaps of fatty acids with more than 20 carbon atoms are too insoluble to function effectively at normal temperatures.

Shampoos based on coconut oil, coconut fatty acids, or a combination of oils were very popular at one time (Formulas 32 to 40). However, these shampoos do not perform well in hard-water areas. They lather poorly and have a tendency to deposit a dulling film of insoluble calcium and magnesium salts on the hair. This latter defect can be diminished somewhat by the addition of sequestrants and chelating agents, such as polyphosphates and EDTA (Formula 33).

Shampoos have also been based on fats and oils which were sulfonated or, more accurately, sulfated, without splitting the glyceride. One of the most important sulfated oils is sulfated castor oil, or turkey red oil (Formula 4).

TABLE II. Fatty Acid Constituents of Fats and Oils

Acid	Castor	Coconut	Corn	Cotton-seed	Linseed	Olive	Palm	Palm kernel	Peanut	Sesame	Soybean	Tall	Tallow
Caproic	—	0.5	—	—	—	—	—	0.5	—	—	—	—	—
Caprylic	—	8.0	—	—	—	—	—	4.0	—	—	—	—	—
Capric	—	7.0	—	—	—	—	—	5.0	—	—	—	—	—
Lauric	—	48.0	—	—	—	0.5	—	50.0	—	—	—	—	0.1
Myristic	2	17.3	—	0.5	—	9.5	2	15.0	7	8.0	0.1	—	3.0
Palmitic	1	9.0	8.0	21.0	5.5	2.0	42	7.0	7	8.0	8.0	7	29.0
Stearic	—	2.0	3.5	2.0	4.0	Trace	4	2.0	4	4.0	4.0	—	20.0
Arachidic	—	—	0.5	0.2	0.3	—	—	Trace	3	0.5	0.6	—	0.8
Behenic	—	—	—	—	—	—	—	—	2	0.1	—	—	—
Lignoceric	—	—	0.2	0.3	0.2	—	Trace	—	2	0.1	—	—	—
Myristoleic	—	—	—	—	—	—	—	—	—	—	0.1	—	0.5
Palmitoleic	—	0.2	—	—	—	—	—	0.5	Trace	—	0.2	1	2.0
Oleic	7	6.0	45.8	29.0	22.0	81.0	43	15	60	45.3	28.0	44	42.0
Ricinoleic	87	—	—	—	—	—	—	—	—	—	—	—	—
Linoleic	3	2.0	42.0	45.0	17.0	7.0	9	1	22	42.0	54.0	37	2.0
Linolenic	—	—	—	2.0	51.0	—	—	—	—	—	5.0	—	0.5
Arachidonic	—	—	—	—	—	—	—	—	—	—	—	11	0.1

Source: Armour Chemical Division (28).
Figures given are mean values, in percentages.

TABLE III.   Water Solubility of Fatty Acids

No. of carbons in acid	Solubility in H_2O at 60°C (g)	Saturated acid	Mono-unsaturated acid[a]	Diunsaturated acid[a]	Polyunsaturated acid[a]
6	1.2000	Caproic			
8	0.1100	Caprylic			
10	0.0300	Capric			
12	0.009	Lauric			
14	0.003	Myristic	Myristoleic (9–10)		
16	0.001	Palmitic	Palmitoleic (9–10)		Hiragonic (6–7, 10–11, 14–15)
18	0.0005	Stearic	Oleic (9–10)	Linoleic (9–10, 12–13)	Linolenic (9–10, 12–13, 15–16)
20	Insol.	Arachidic	Gadoleic (9–10)		Arachidonic (5–6, 8–9, 11–12, 14–15)
22	Insol.	Behenic	Erucic (13–14)		Clupanodonic (4–5, 8–9, 12–13, 15–16, 19–20)
24	Insol.	Lignoceric	Selacholeic (15–16)		Nisenic (4–5, 8–9, 12–13, 15–16, 18–19, 21–22)

[a] Position of double bond in unsaturated fatty acids.

Formulas for Coconut Oil Shampoos

	32*	33†	34†
Coconut oil	16%	21.0%	7.0%
Potassium hydroxide, 85%	4	4.1	1.1
Water	80	54.0	68.7
Perfume	—	0.5	1.0
Olive oil	—	3.0	9.9
Sodium hydroxide, 95%	—	1.9	4.4
Ethyl alcohol	—	15.0	2.8
Ethylene diamine tetraacetic acid	—	0.5	—
Palm oil	—	—	5.1

* Ref. 29.
† Ref. 94.

Formula 35*

Potassium hydroxide, 90%	14.5%
Ethyl alcohol	37.0
Glycerol	6.0
Olive oil	34.0
Corn oil	8.5

* Ref. 30.

Formula 36

Coconut oil	8.0%
Palm oil	4.0
Potassium hydroxide, 90%	5.0
Ethyl alcohol	8.0
Water	q.s. 100.0
Glycerol	6.0
Perfume	0.5
Olive oil	8.0
Oleic acid	2.5

Acid Content of Formula 36

The mixture of vegetable oils in Formula 36, when hydrolyzed by the alkali, would give potassium salts of the following acids:

Caprylic acid	0.5%	Palmitic acid	2.3%
Capric acid	0.5	Stearic acid	1.2
Lauric acid	4.0	Oleic acid	11.5
Myristic acid	1.5	Linoleic acid	1.0

Formulas for Triethanolamine Shampoos

	37*	38†	39‡
Triethanolamine	5.4%	6.0%	9.0%
Oleic acid	5.0	—	9.3
Coconut fatty acids	4.0	1.5	6.7
Propylene glycol	5.0	—	—
Versene 100	0.4	—	—
Water	80.2	q.s. 100.0	48.5
Castor oil fatty acids	—	0.5	—
Potassium hydroxide, 85%	—	3.5	—
Talc	—	0.3	—
Dodecyl benzene sulfonate, 35%	—	—	25.0
Perfume	—	—	1.5

* Ref. 94.
† Ref. 29.
‡ Ref. 20.

Formula 40*

Coconut fatty acids	21.0%
Oleic acid	28.0
Propylene glycol	25.5
Monoethanolamine	6.3
Triethanolamine	14.2
Tergitol 7	5.0

* Ref. 16.

The sulfated oils, however, are poor foaming agents. It has been claimed that excessive use of these oils tends to leave the hair dull, wiry, and difficult to manage.

Synthetic Detergents

The tendency of soap shampoos to form insoluble salts is due to the presence of the carboxylic group linked to the end of the long-chain hydrocarbon. By replacing this group, many surfactants which avoid the foaming and cleaning negatives of soap were developed.

Synthetic detergents are normally classified by the nature of their hydrophilic group. The anionics are the most widely used, with the nonionics a distant second.

Anionics

The hydrophilic portion of the anionic surfactant carries a negative charge in solution. These detergents are generally superior to other classes in terms of foaming, cleaning, and end result attributes. Some members of this class are discussed below.

Alkyl benzene sulfonates. Early studies were undertaken on the effect of replacing soap with sulfonated hydrocarbons. The results were the development of sodium alkyl benzene sulfonates and sodium alkyl naphthalene

sulfonates, R⟨ ⟩SO$_3$Na and R⟨ ⟩SO$_3$Na. By varying the size of the alkyl

(R) group, detergent properties of varying action could be obtained. Where the alkyl group contains four to eight carbon atoms, the resulting sulfonate is usually a wetting agent but not a detergent. With 10 to 14 carbons, the resulting product gives better detergency.

At one time alkyl benzene sulfonate (ABS) containing a branched alkyl chain attached to the benzene-sulfonate moiety was used as a shampoo component. It was inexpensive, but it was also resistant to the microbial enzymatic processes which degrade the surfactant to assimilable effluent. Its replacement, the analogous linear ABS, continues to be used, but rarely alone. Alkyl benzene sulfonates tend to yield an "airy" or low density foam and often are drying to the hair. They usually are combined with other anionic detergents or soaps. Shampoos in which the alkyl benzene sulfonates are used are shown in Formulas 5, 26, 37, 41, and 42.

Formulas for Alkyl Benzene Sulfonate Shampoos

	41*	*42†*
Glycerol	—	0.6%
Perfume	q.s.	—
Water	—	70.4
Oleic acid	9%	—
Coconut fatty acids	7	—
Triethanolamine	9	—
Sodium dodecyl benzene sulfonate	25	—
Ultrawet 60 L	50	22.0
Lecithin	—	0.6
Diglycol stearate	—	1.2
Isopropyl lauramide	—	4.0
Xylene sulfonate	—	0.4
Ethyl alcohol	—	0.8

* Ref. 25.
† Ref. 31.

Primary alkyl sulfates. The alkyl sulfates were developed in Germany, where vegetable oils and fats were relatively scarce and detergents resistant to hard water were badly needed. Originally these sulfates were often prepared from the corresponding primary alcohols by treatment with chlorosulfonic or sulfuric acid; the alcohols were usually prepared by hydrogenation of fatty acids. Lauric acid, for example, yields a soap which loses its activity in hard water. However, by reducing lauric acid to lauryl alcohol and subsequently sulfating, one obtains sodium lauryl sulfate, an excellent detergent and shampoo material, completely effective in the presence of hard water. More recently, newer synthetic processes for making fatty alcohols have replaced the fatty acid reduction techniques.

The alkyl sulfates soon became the backbone of the shampoo market and continue to hold that position. A wide range of alkyl sulfates is available; the most useful are those of the C_{12} to C_{18} series.

The primary alkyl sulfates, particularly those containing a mixture of C_{12}, C_{14}, and C_{16} compounds, give excellent foam and leave the hair feeling smooth and soft; such products have a good lathering effect in hard water.

A typical commercial alkyl sulfate might be made up of a mixture in the following proportions: C_{10}, 2%; C_{12}, 62%; C_{14}, 22%; C_{16}, 12%; C_{18}, 2%.

Many shampoos owe their popularity to their alkyl sulfate content. The ease with which they are perfumed, their freedom from rancidity, and the ease with which they can be rinsed out of the hair have helped establish their importance in the shampoo market.

Most of the original alkyl sulfates were the sodium salts. Now, many are neutralized as the triethanolamine, diethanolamine, and ammonium salts. In some instances, a combination of salts has been formulated most effectively.

Examples of shampoos based partly or entirely on the primary alkyl sulfates are shown in Formulas 5, 6, 9, 11, 12, 17, 19, 20, 22, 24, and 43 to 45.

Formula 43*

Behenic acid	3%
Water	71
Triethanolamine lauryl sulfate	25
Triethanolamine	1

* Ref. 17.

Formula 44*

Sodium lauryl sulfate	29.0%
Water	52.5
Glyceryl monostearate	1.5
Stearic acid	5.5
Triethanolamine stearate	6.5
Diglycol laurate	1.0
Borax	3.5
Perfume	0.5

* Ref. 32.

Formula 45

Water	32.35%
Triethanolamine lauryl sulfate	50.00
Sulfonated castor oil	12.50
Glycerol	5.00
Thymol	0.15

Secondary alcohol sulfates.   The secondary alcohols form sulfates that have proved to be quite disappointing as detergents and as components of shampoos. In going from sodium lauryl sulfate (normal lauryl group) to sodium *sec*-lauryl sulfate, the compound loses most of its foaming action and practically all of its detergent effectiveness. Although many secondary alcohols are commercially available at low cost, their sulfates are usually little more than wetting agents, with some dispersing and emulsifying action. Their structure is reported to be $RR'CHSO_4Na$, where R and R' are alkyl groups usually not the same group, and both attached to the CH.

Alkyl benzene polyoxyethylene sulfonates.   A distinct class of synthetic sulfonates which are stable in acid or alkaline solution is the alkyl benzene polyoxyethylene sulfonates, such as Triton X200,

$$R—\langle\ \bigcirc\ \rangle—OCH_2CH_2OCH_2CH_2OCH_2CH_2SO_3Na.$$

Where the alkyl group (R) contains 8 to 12 carbon atoms, the resulting sulfonate is an excellent detergent, emulsifier, and wetting agent, coming close in its effectiveness in hard water to the performance of soap in soft water. This type of sulfonate has found wide use as a detergent for doctors and in hospitals where soap is contraindicated. It is extremely stable at pH 4.0 to 5.0, comparable to the pH of the skin. It should be noted that the elimination of one of the ethoxy groups, CH_2CH_2O, gives a wetting agent with poor cleaning and emulsifying action.

Sulfated monoglycerides.   Salts of the compositions

$$NaO_3S—OCH_2—CH(OH)—CH_2OOCH_2R,$$

where R contains 11 carbons, have proved to be excellent detergents for shampoos. They must be kept neutral or slightly acid; otherwise they will hydrolyze with formation of fatty acid soaps. Properly compounded and formulated shampoos based on this type of detergent give good foam and leave the hair soft and lustrous; they also show good stability in hard water. Examples of shampoos based on a sulfated monoglyceride detergent system are shown in Formulas 46 to 48.

Formulas for Sulfated Monoglyceride Shampoos *

	46	47
Lauric monoglyceride sodium sulfate	20.00%	—
Behenic acid	3.00	2.0%
Triethanolamine	3.00	0.7
Citric acid	0.70	—
Methyl *p*-hydroxybenzoate	0.15	—
Perfume	0.40	—
Lauric monoglyceride ammonium sulfate	—	20.0
Water	72.75	77.3

* Ref. 17.

*Formula 48**

Sulfated coco monoglyceride,	
ammonium salt	21.0%
Ethyl alcohol	9.8
Polyacrylamide solution	20.0
Water	q.s.
Perfume	0.4

* Ref. 33.

Alcohol ether sulfates. Sulfates of ethoxylated fatty alcohols are relatively new. They offer the advantages of low cost and mildness, and tend to act as suds boosters. They are claimed to have the structure

$$C_{12}H_{25}(OCH_2CH_2)_n OSO_3Na,$$

where n is usually 2 or 3 (Formula 54). The product having 3 groups is reported to have excellent foam properties (34).

Sarcosines. Lauroyl and cocoyl sarcosines are playing an increasingly important role in the shampoo market. Lauroyl sarcosine is prepared by the reaction of lauroyl chloride with N-methylglycine; cocoyl sarcosine is the sarcosine derivative based on coconut fatty acids. They have excellent foaming quality and outstanding conditioning action. Eye irritation tests indicate that the sarcosines are similar in this respect to the lauryl sulfates. In combination with other anionic detergents, the sarcosines give excellent products (Formulas 49 to 51).

Sulfosuccinates. The dioctyl esters of the salts of sulfosuccinic acid are excellent wetting agents; one such product is Aerosol OT, which is made by esterifying maleic anhydride and treating it with sodium bisulfite. Sulfosuccinates are relatively low in cost, but are not voluminous foaming agents. The half-esters of longer-chain alkyl groups are more suited to shampoos and have gained attention because they are less irritating to the skin and eyes than many other anionic detergents. Whenever wetting and penetration are prime factors, these compounds should be considered. Their use in shampoos is shown in Formulas 52 to 54.

Igepon. This name is given to a group of detergents that, thus far, has played a limited role in the shampoo market. The fatty acid-isethionic acid ester (Igepon A) and the fatty acid–methyl tauride amide (Igepon T) have been used largely as textile detergents. An example of a shampoo containing Igepon TC-42, prepared by condensing a fatty acid chloride with the sodium salt of N-methyl taurine, $CH_3NHC_2H_4SO_3Na$, is shown in Formula 55.

Maypon. When proteins are carefully hydrolyzed, products are obtained which are known as protalbinic and lysalbinic acid derivates. If these hydrolysates are condensed with fatty acid chlorides in the presence of alkali,

Formulas for Sarcosinate Shampoos

	49*	50†	51‡
Sarkosyl NL-30	6%	—	—
Sodium lauryl sulfate	8	—	—
Alrosol C	1	—	—
Water	85	q.s.	q.s.
Sodium lauryl ether sulfate	—	15%	—
Triethanolamine salt of cocoyl sarcosine	—	10	10.00%
Triethanolamine lauryl ether sulfate	—	—	4.00
Monoethanolamide of coconut fatty acids	—	—	3.00
Diethanolamide of coconut fatty acids	—	—	3.00
EDTA	—	—	0.65
Ethyl alcohol	—	—	7.00
Methyl cellulose	—	—	0.75
Phenyl mercuric acetate	—	—	0.007
Perfume	—	—	0.75

* Ref. 35.
† Ref. 36.
‡ Ref. 37.

Formulas for Sulfosuccinate Shampoos

	52*	53†	54†
Dicarboxyethylsulfosuccinate	12.0%	—	—
Glyceryl monoricinoleate	1.0	—	—
Bentonite	4.0	—	—
Selenium sulfide/bentonite, 1:1	5.0	—	—
Monosodium phosphate	1.0	—	—
Citric acid to pH 4.5	0.4	—	—
Perfume	0.5	q.s.	q.s.
Water	q.s.	28%	40%
Steinapol SB-FA 30	—	62	35
Steinapol SB-Z	—	10	—
Sodium lauryl ether sulfate	—	—	22
Lauric diethanolamide	—	—	3

* Ref. 27.
† Ref. 38.

*Formula 55**

Igepon TC-42	8.0%
Triethanolamine lauryl sulfate	4.0
Lauric diethanolamide	4.0
Surfynol 82	1.0
Triethanolamine coconut/oleic soap	3.8
Water	q.s.

* Ref. 39.

they produce complex amides which have found application in shampoos. Originally developed in Germany by Landolt-Meyer and marketed under the name of Lamepon, they have been sold in this country under the trade name of Maypons. A shampoo containing Maypon 4C, the potassium salt of the coconut polypeptide condensate, is shown in Formula 56.

*Formula 56**

Maypon 4C	15.0%
Sodium lauryl sulfate	15.0
Lauric diethanolamide	5.0
Citric acid to pH 6.5	q.s.
Formalin solution	0.1
Water	q.s.
Perfume	q.s.

* Ref. 40.

Cationics

Cationic detergents are considerably less popular than anionics. With this group, the hydrophilic portion of the compound is positively charged, usually a quaternary ammonium salt. Cationics are generally poor in detergency, harsh to the skin and eyes, and rather expensive; one advantage that they may possess is bactericidal activity. Some typical cationics are distearyl dimethyl ammonium chloride, dilauryl dimethyl ammonium chloride, diisobutylphenoxyethoxyethyl dimethyl benzyl ammonium chloride, cetyl trimethyl ammonium bromide, N-cetyl pyridinium bromide, and benzethonium chloride.

When anionics and cationics are combined, often the worst of both worlds results. The anionic loses its foaming characteristics and the cationic loses whatever bactericidal activity it once may have had.

Amphoterics

In spite of their incompatibility, it is possible to combine anion-forming groups and cation-forming groups in the same detergent molecule and get a useful product. These are called amphoteric (ampholytic) or zwitterionic detergents. Compounds of this class are ionized, but the polarity of the charge depends on the solution pH. Some examples of amphoteric detergents are N-alkyl β-imino dipropionates (Formula 7), N-alkyl (10 to 20 carbons) β-amino propionates (Formula 57), and the basic quaternary

Formula 57*

Deriphat 160C	6%
Triethanolamine salt of Deriphat 170C	3
Triethanolamine lauryl sulfate	6
Ethoxylated lanolin	6
Alkanolamide	4
Water	77

* Ref. 41.

ammonium compounds derived from 2-alkyl-substituted imidazoline. One group of amphoterics of special interest for use in shampoos is the Miranols, imidazoline derivative products.

Miranol. This synthetic detergent is made by condensing aminoethyl ethanolamine with a fatty acid, such as lauric acid, to form a five-membered ring, which reacts with sodium chloracetate and alkali to form hydroxyethyl carboxymethyl alkyl imidazolinium hydroxide. The quaternary ammonium compound is reported to have a very low eye irritation potential (42) and to give excellent foaming action. The lauric derivative is known as Miranol HM, the myristic derivative as Miranol MM, the stearic derivative as Miranol DM. Miranol CM is the material derived from the mixed fatty acids of coconut oil (Formulas 8 and 58).

Formula 58*

Miranol CM	6.0%
Tetronic 908	6.0
Distearyl dimethyl ammonium chloride	4.9
Cocomethyl tauride	3.0
Water	q.s.
Concentrated H_3PO_4 to pH 5	

* Ref. 43.

Nonionics

The second most widely used class of synthetic detergents is the nonionics. Their low foam, however, has limited their use as major formula components. Nonionics have excellent resistance to hard water, even to seawater, are equally effective in alkaline or acid solutions, and are in general mild to the skin.

Most nonionic surfactants are produced by the condensation of alkylene oxide groups with an organic hydrophobic compound having an active hydrogen. The nonionic compound owes its water solubility or dispersibility to its long chain, commonly oxyethylene $(OCH_2CH_2)_x$. Different degrees of solubility can be obtained by varying the amount of ethylene oxide condensed with the base compound, but it has not yet been possible to develop high sudsing products. Some typical polyethenoxy-type nonionics are the Pluronics, the Igepals (44), and the Tweens (Formulas 30 and 59 to 61).

Formulas for Nonionic Shampoos

	59*	60†	61‡
Renex 20	22.0%	—	—
Lanolin	4.0	5%	—
Arlacel C	1.0	1	—
Sodium carboxymethyl cellulose	2.0	2	—
Perfume	0.5	q.s.	q.s.
Methyl p-hydroxybenzoate	0.2	—	—
Tween 80	—	20	—
Water	70.3	72	q.s.
Tergitol NP-35	—	—	8.0%
Pluronic F-68	—	—	8.0
Lauric diethanolamide	—	—	2.5
Lauryl dimethylamine oxide	—	—	5.0
Stearyl amine	—	—	5.0
Acetic acid to pH 8	—	—	q.s.

    * Ref. 29.
    † Ref. 3
    ‡ Ref. 45.

A second type of nonionic surfactant is the high dipoles. These nonionics are based on compounds containing high-dipole hydrophilic groups, such as teramine oxide and terphosphine oxide. Their use has been limited to date, but their interesting sudsing and mildness properties suggest that they will play an increasing role in the shampoo market.

In many cases, nonionics have been combined with anionics, chiefly for the mildness advantage of the former and the foaming advantage of the latter. However, nonionic-cationic combinations and nonionic-amphoteric combinations are also in use. Where cationics or compounds with cationic tendency are present in a formula, investigation of irritation to skin and eyes is a necessity.

Soap–Synthetic Detergent Combinations

Many shampoos are based on a combination of soap and synthetic detergent formulas (Formulas 5, 12, 19, 41, 44). The hard-water disadvantages of soap are greatly overcome and the cosmetic characteristics of the resulting shampoo are somewhat modified by the combination.

Shampoo Additives

An increasing number of compounds has been developed which contribute to the performance and acceptance of shampoos. They may affect the foam, feel, consistency, or finish imparted by the shampoo. Many are protected by patents, and the knowledge of others is confidential. Some of the better-known additives are listed here according to their chief function.

Foam Builders

Foam builders or foam stabilizers are ingredients which, when added to a formulation, increase the quality, volume, and stability of the lather. Often they also enhance viscosity and impart a slight conditioned effect to the hair. The principal foam builders are the fatty acid alkanolamides (such as lauroyl diethanolamide, lauroyl monoethanolamide, coconut monoethanolamide), the "Super" amides (46); fatty alcohols in low concentration; and, to a lesser extent, sarcosinates and phosphates (Formulas 5, 17, 21). An example of results obtained by the addition of a foam builder is shown in Table IV.

TABLE IV.   Foaming Characteristics of 10% Solution of Detergent

	Initial foam (cc)	After 15 min (cc)	After 30 min (cc)
A. Dodecyl benzene sulfonate, (37%)	120	50	25
B. A + 10% of lauroyl monoethanolamide	190	160	155
C. A + 10% of lauroyl diethanolamide	130	75	50

Dodecyl benzene sulfonate is considered a good foaming surfactant when used alone. The addition of lauroyl monoethanolamide not only increases the initial foam volume but greatly enhances the stability of the foam.

Conditioning Agents

The difference between an ordinary surfactant and a shampoo lies in the finishing or conditioning action of the shampoo. Most surfactants clean the

hair so well that it becomes light and flighty. Conditioning agents coat
the hair with a very small amount of material that either improves the handling characteristics of the hair fiber or lubricates the hair for better slip
and smoothness.

In early soap shampoos the unsaponified vegetable oils contributed to this
conditioning action. However, with the influx of synthetic detergents, a
number of agents have been incorporated to leave the hair soft, smooth, and
lustrous. The Ucons, lanolin and its derivatives, esters such as isopropyl
myristate and butyl palmitate, glycerol, and propylene glycol have been
promoted as conditioning agents. Amine oxides (13) and silicones (47) have
also been used. Lauroyl and cocoyl sarcosines are especially recommended
for the soft feel they impart to the hair (37).

Although quaternaries are often not compatible with soaps and anionic
detergents and raise questions about possible eye irritation, some show excellent conditioning action. Cationic materials condition the hair by reducing
the electrostatic charge on the hair fiber. They are adsorbed on hair and are
retained by hair after extended rinsing. (Formula 62).

Formula 62*

Sulfated cocomonoglyceride,	
ammonium salt	54.1%
Emcol E-607	2.0
Water	43.9
Perfume	q.s.

* Ref. 48.

Care must be taken in the selection and use of a conditioning agent.
Anything which is difficult to wash from the hair or leaves the hair with an
oily, sticky, or tacky feel is to be avoided.

Opacifying Agents

Since cream and lotion shampoos account for a substantial portion of total
shampoo consumption, there is a large interest in opacifying agents. The
better-known opacifiers include the higher alcohols, such as stearyl and
cetyl alcohols, and the higher acids, such as behenic acid (22 carbons).
The glycol mono- and distearates, as well as the glyceryl and propylene glycol
stearates and palmitates, are also effective opacifiers. Spermaceti has been
noted for imparting opacity, and common salt or Glauber's salt also has an
opacifying effect when the concentration is carefully controlled so as to salt
out the surfactant without causing gelation or separation. Magnesium,

calcium, and zinc stearate are opacifiers, as are magnesium silicates and some resin polymers, such as polystyrene. Finally, the amides and ethanolamides of fatty acids, in addition to building suds and conditioning, can contribute opacity as well (Formulas 10 to 23).

Clarifying Agents

The need for clarifying agents is as great as that for opacifying agents, since clear shampoos remain the most popular form. In general, coupling or solubilizing agents help maintain shampoo clarity over a wide temperature range. Care must be exercised in the selection of compounds of this type. They should be checked for possible eye irritation and toxicity. Some examples of these agents are butyl alcohol, isopropyl alcohol, terpineol, diethylene glycol, propylene glycol, and diethyl carbitol.

A sequestering agent, such as EDTA, is frequently used since it prevents the formation of calcium, magnesium, and iron soaps which may cause turbidity. Many concentrated soap solutions contain a sequestrant (citric and tartaric acids) to avoid the necessity of water purification to effect removal of calcium, magnesium, ferrous, and ferric ions. Tetrasodium pyrophosphate and tripolyphosphate have a similar lime soap-dispersing action and have an excellent buffering action. Compounds such as sodium xylene sulfonate and sodium naphthalene sulfonate have specific dispersing action on many colloidal systems. Care must be taken in using these compounds to prevent excessive drying of the hair (Formulas 1, 3, 9, 33).

Sequestering Agents

In the prevention of lime soap formation, there are two considerations, namely, the formation of insoluble calcium or magnesium soaps when the shampoo is mixed with hard water, and the precipitation of lime soap film on the hair when the shampooed hair is rinsed with hard water. In the latter case, 1 or 2 oz of shampoo may be rinsed with as much as 25 to 50 oz of water. The addition of a sequestering agent, such as citric acid, Versene 100, or Nullapon (which are salts of EDTA), prevents lime soap formation in the lathering process, provided that proportions of up to 1 % are used (Formulas 33, 37). However, large amounts of any sequestrant are required to soften the large quantities of rinse waters used. It is not usually economical to use the requisite quantities. It is preferable to add small amounts of nonionics or special ionics which disperse any lime soaps formed and prevent them from collecting and dulling the hair.

In low concentrations, nonionics, such as the Tweens, have been shown to improve the cleansing action of soaps and their lime soap dispersion.

Tetrasodium pyrophosphate and tripolyphosphate, as mentioned above, possess good clarifying action; however, these salts decrease the solubility of the soaps and detergents.

Antidandruff Agents

There are many antidandruff shampoos on the market, most of which are based on agents that are antimicrobial in nature. The shampoos contain small amounts of these actives, which are in contact with the scalp for only a short time. In order to be effective the active ingredient must work in the oil-water environment of the scalp and must be readily substantive to the scalp for continuing activity. Hence it is easy to understand why many of these antidandruff shampoos are lacking in effectiveness.

Traditional antidandruff compounds have included sulfur, salicylic acid, hexachlorophene, resorcinol, and tar. More recent additions include selenium sulfide, zinc pyrithione, the iodine-containing hydroxyquinolines, and some quaternary ammonium compounds (Formulas 30, 31). Published information on the effectiveness of these compounds is scarce; however, most authorities in this area recognize that shampoos containing selenium sulfide or zinc pyrithione are usually effective.

Thickening Agents

The problem of thickening a shampoo is not simply one of selecting the proper synthetic or natural gum, because a great many esters and amides also contribute to the viscosity of the shampoo. In general, the natural gums such as tragacanth, gum acacia, and locust bean gum have been replaced by synthetic gums such as hydroxyethyl cellulose, methyl cellulose, carboxymethyl cellulose, and Carbopol, a carboxy vinyl polymer (49), but these synthetic gums must be used with some care since they can form films on the hair.

Inorganic salts, such as sodium and potassium chloride, can be used in limited amounts; the alginates, polyvinyl alcohol, and polyvinylpyrrolidone have also found some use. However, the danger of film formation is avoided when using other thickening agents, such as the alkylolamides, the "Super" amides (46), and the glycol or glycerol stearates.

In the case of soap shampoos, with a better control of the fatty acids available, it is possible to increase viscosity by using a higher proportion of the longer-chain acids. In the case of synthetics, the use of longer-chain alcohol or amides has a similar effect. With nonionics, the length of the polyglycol chain greatly affects the viscosity in that the longer chains give greater solubility.

Preservatives

There exists a problem of protecting the shampoo from deterioration by bacterial or mold action. The solution is a matter of selecting a proper preservative from a list which may include formaldehyde, ethanol, methyl, propyl, and butyl hydroxybenzoate, phenylmercuric acetate, phenylmercuric nitrate, the alkyl anisoles, the alkyl cresols, Bronopol (50), the chlorosalicylanilides, dehydroacetic acid salts, monomethylol dimethyl hydantoin (21), and many others. In addition, sulfated detergents themselves, amide additives, and such ingredients as perfume exhibit some antibacterial activity. The best preservative for a particular shampoo can be determined only by testing the effect of that particular preservative in the shampoo formulation against all possibility of attack by microorganisms. Continual monitoring of the cleaning and sterilization of manufacturing facilities must also be part of the product preservation program.

Other Stability Additives

Sometimes it is necessary to protect the shampoo by adding stabilizers, among which are antioxidants, sunscreens, suspending agents, and pH control agents.

Reducing agents protect the product from discoloration or odor deterioration due to oxidation. Sunscreens, such as benzophenone (51) or benzotriazole derivatives, have the property of absorbing ultraviolet radiation and thus reducing product damage from sunlight exposure. Suspending agents, like Veegum and other bentonites, stabilize shampoos where solid particles are suspended in a liquid (Formula 52). Various pH control agents, which can be as simple as ordinary bases and acids, protect the product from changing color, odor, or level of irritation due to a change in pH.

Other Cosmetic Additives

All shampoos have perfume and dye in them to ensure their cosmetic acceptability, and some contain additives, such as tints and pearlescent pigments, to improve their cosmetic appeal.

Perfumes are compounded from a number of essential oils, extenders, and fixatives. Care must be taken to study the effects of oxidation, temperature, sunlight, pH, etc. when selecting a perfume. The multiplicity of reactions of perfume with shampoo components makes it difficult to predict whether a perfume will be satisfactory without actual testing of the finished formulation.

Dyes must be of the type that are certified and approved for use in cosmetics under the provisions of the Federal Food, Drug, and Cosmetic Act.

Evaluation of Shampoos

If one defines a shampoo as "a product having some cleansing and foaming action which leaves the hair soft, lustrous, and manageable," then a very wide range of compounds and chemicals will meet this definition. It becomes very difficult, however, to evaluate them because no order of relative importance can accurately be placed on foaming, cleansing, luster, softness, and manageability.

It is possible to measure a number of properties, and the most important of these properties are listed and discussed here. It should be emphasized that excellent results in these properties do not in themselves ensure the popularity of a shampoo. The question of product popularity can ultimately be answered only by the consumer in the marketplace. Nonetheless, there are useful test procedures to serve as either a screening or a comparison device for preliminary assessment of possible formulations.

In a recent study of 24 commercial shampoos, *Consumer Reports* (52) employed laboratory tests and user panels to make their evaluation. The laboratory testing included lathering and rinsing capability, pH, animal eye irritation testing, and rating of the container. The panelists answered questions on hair-handling performance, cleaning, reaction to the hands and scalp, and fragrance.

Generally speaking, *Consumer Reports* rated the great majority of the tested brands as acceptable, finding little to choose between them on the basis of the imposed criteria, thus emphasizing the difficulty of judging shampoo products.

From published literature, it is evident that much work has been done on shampoo evaluation. Longfellow (53) points out the wealth of reported laboratory evaluation techniques. A review of the literature can provide the formulator with ideas and procedural methods to make a specific and pertinent evaluation. Some of the important areas of evaluation are as follows:

1. Performance properties.
   a. Foam and foam stability.
   b. Detergency and cleaning action.
      (1) Effect of water hardness.
      (2) Surface tension and wetting.
      (3) Surfactant content and analysis.
   c. Rinsing.
   d. Conditioning action.
      (1) Softness.
      (2) Luster.
      (3) Lubricity.
      (4) Body, texture, and set retention.

e. Irritation and toxicity.
f. Dandruff control.
2. Product characteristics.
a. Fragrance.
b. Color.
c. Consistency.
d. Package.

Performance Properties

Foam and Foam Stability

Excellent lathering is one of the bellwethers of consumer shampoo acceptance. Thick rich lather can be readily seen and appreciated by the user. On the other hand, efficient cleaning action is not necessarily synonomous with foaming action. Some surfactants, particularly the nonionics, have excellent cleaning performance while developing little or no lather.

The Ross-Miles (54) foam column test is a generally accepted method for measuring foam height and stability. In this test, 200 ml of a surfactant solution is dropped into a glass column containing 50 ml of the same solution. The height of the foam generated is measured immediately and again after a specified time interval, and is considered proportional to the volume. The dimensions and detail of this procedure are all carefully specified. A later paper by Ross, Miles, and Shedlovsky (55) considered further applications of this test.

The Ross-Miles foam test is a fairly reliable index of foaming performance and is often used as a research method of reporting foaming capability. Colson (56) has added the concept of a synthetic load to make the test more realistic. Sanders (57) has given the foam height for a number of synthetic detergents and soap using this test method. The results are shown in Table V.

Laboratory evaluations have moved to making the procedure more practical and more representative of the conditions of shampoo use. Sanders developed a scheme of lathering shampoo on a standard-size hair bundle and evaluating the lather developed. Sanders, Knaggs, and Libman (58) applied a synthetic soil load to the hair. Barnett and Powers (59) developed a latherometer arrangement to measure foam speed, volume, and stability. With the latherometer it was possible to study the effect of variables such as water hardness, type of soil, and quantity of soil on lather performance. Fredell and Read (60) used a method of "titrating" actual standard oiled heads of hair with additive increments of shampoo until a persistent lather end point appeared. Powers and Fox (61) further compared ratings by a user panel to laboratory performance results. Bromley (62) described methods of evaluating shampoo foam viscosity and specific volume, using natural hair soil extracted from clippings.

TABLE V.   Foam Height (in mm) Using 0.1% of Active Material at 30°C

		Foam height	
	Activity (%)	Distilled water	Hard water (350 ppm)
Potassium coconut soap	15	160	15
Sodium lauryl sulfate (Duponol WA Paste)	31	200	125
Sodium monoglyceride sulfate (Arctic Syntex M)	32	205	205
Potassium coconut-protein condensate (Maypon 4C)	35	175	155
Sodium aralkyl sulfonate (Ultrawet K)	85	200	225
Alkyl phenol polyglycol ether (Triton X-100)	100	125	115
Sulfated castor oil (Monosulph)	68	90	20
Sodium sec-alcohol sulfate (Tergitol 4)	25	110	90
Polyoxyethylene sorbitan monolaurate (Tween 20)	100	85	80
Lauroyl imidazoline (Miranol HM)	40	220	120
Sodium dioctyl sulfosuccinate (Aerosol OT)	100	180	50

Source: Sanders (57).

The volume, creaminess, and feel of the lather doubtless are prime factors in user comparisons. Variables in the hair substrate, the soil load, the water (hardness, temperature, and relative quantity), the means of introducing the shampoo, the technique of agitation, and the time interval of use are all factors that can fluctuate among users. It would seem that any single method of mechanically duplicating shampoo lather would be subject to many compromises. Several supplementing techniques may provide a broader and better understanding of comparative lather performance.

Detergency and Cleaning Action

The cleansing or detergent action of a shampoo is a primary function. It would seem that cleansing proficiency would be a measure of a shampoo's value or popularity. Detergency has been extensively studied and reported by Perry, Schwartz, and Berch (63).

Most of the literature concentrates on laundry detergency with emphasis on detection and measurement of soil removal from cloth. "Standard soils" consisting of lipids and readily discernible components like carbon black have been devised. "Standard fabrics," spectrophotometric measurement of cloth whiteness, and even radioactive tracer measurements of activated soils have been utilized. Generally speaking, the emphasis has been on measuring removal efficiency of surfactants, with the desired end as total soil removal.

Considerable work in shampoo detergency action indicates that total removal of hair oil is not desirable. Excessive cleaning may make hair unmanageable and dry. Ester, Henkin, and Longfellow (6) measured soil removal from beauty salon hair clippings, using solvent (chloroform) removal as a basis of comparison. Their study, using popular shampoo brands, yielded wide variations in results. From 11 to 35 % of the extractable soil was actually removed by shampooing.

TABLE VI . Cleansing Power of Detergents and Soap and Hard Water and Seawater (0.25% of Active Detergent Tested)

Type of detergent	pH of 0.25% solution	Total grease removed (%)		
		In soft water	In hard water	In sea- water
Sodium alkyl sulfate	8.3	98.5	99.1	93.1
Triethanolamine alkyl sulfate	6.8	94.8	94.5	89.4
Ammonium alkyl sulfate	6.6	97.8	97.5	95.0
Alkylphenoxy polyethoxyethyl alcohol	6.3	98.1	97.5	95.7
Sodium sulfo fatty ester	5.9	97.0	93.4	85.4
Ammonium sulfated monoglyceride	4.9	96.7	94.9	92.3
Protein fatty acid condensate	6.9	90.9	80.5	58.5
Alkylphenoxy polyethoxyethyl sulfate	5.3	95.0	93.4	18.8
Polyglycol (nonionic)	7.4	93.6	79.3	62.3
Polyhydroxy amine fatty acid condensate	7.7	96.6	95.5	11.1
Sodium alkyl aryl sulfonate	7.6	95.8	93.0	53.0
Triethylanolamine alkyl aryl sulfonate	6.7	89.2	81.2	11.3
Dialkylphenoxy polyethoxyethyl sulfonate	7.7	92.7	92.8	88.0
Potassium coconut soap and sequestrant	9.6	94.2	67.8	None
Ammonium oleate	8.9	12.7	3.8	None
Alkyl dimethyl benzyl ammonium chloride	—	None	None	None

Source: Barnett and Powers (7).

Barnett and Powers (7) evaluated a number of detergents and commercial shampoos, using "wool yarn in the grease" as the test medium. These commercially available skeins of naturally soiled wool were washed under controlled conditions. Results of these investigations on detergents and soaps are shown in Table VI. Complete shampoo formulas gave results that were generally lower in grease removal than for simple surfactant solutions alone.

The rationale for using natural grease on wool (keratin) is its close approximation to the human hair and soil. Human hair, of course, does differ from wool in texture, structure, and relative proportions of amino acid components. Human lipid secretions also differ from those of other mammals (64).

It is not likely that universal agreement will be reached on what constitutes an average or representative substrate and what composes a representative soil load. Yet the need for a detergency test that closely approximates the actual use situation continues.

Effect of Water Hardness

Testing shampoos over a range of water hardness is necessary, particularly since soaps are susceptible to the impact of calcium and magnesium ions. Laboratory tests should be capable of utilizing any degree of hardness. It must be presumed that few consumers will have access to specially softened or distilled water.

It is also desirable to conduct consumer testing in areas that represent the spectrum of water hardness variations. Normally the community water plant maintains a continual record of water hardness. The U.S. Geological Survey has also provided a listing of water hardness findings throughout the nation (65). It is noteworthy that water hardness often fluctuates seasonally.

Surface Tension and Wetting

The measurements of surface (liquid to air) and interfacial (liquid to liquid) tensions are, to an extent, a guide as to how effectively a surfactant solution can surround, break up, and solubilize soil. The DuNouy ring tensiometer is a classical means of measuring this value. Normally quite low concentrations of surfactants are used, in the range of 0.1 to 0.25%.

The Draves-Clarkson (66) wetting test is a standard procedure to determine the effectiveness of wetting cotton skeins. The method is a measure of

TABLE VII.   Surface and Interfacial Tension of Soap and Detergents

Soap or detergent at 0.05% active concentration	Tension (dynes) at 30°C	
	Surface[a]	Interfacial[a]
Coconut soap–sodium salt	24.4	14.9
Duponol WA–sodium lauryl sulfate	25.2	11.8
Arctic Syntex M–sodium monoglyceride sulfate	27.6	4.0
Ultrawet K–sodium aralkyl sulfonate	26.8	1.2
Monosulph–sulfated castor oil	38.1	10.4
Triton X100–alkyl phenol polyglycol ether	29.1	1.6
Tween 20–polyoxyethylene sorbitan mono-laurate	35.8	6.6
Igepon THC–sodium oleyl taurate	28.0	7.4
Miranol HM–lauroyl imidazoline	24.0	7.0
Aerosol OT–sodium dioctyl sulfosuccinate	28.5	5.2

Source: Sanders (57).

[a] Cenco Precision DuNouy tensiometer and a Cenco Precision tensiometer gave reproducible readings.

how rapidly a surfactant solution penetrates the capillaries of thread and displaces the entrapped air which gives the skein bouyancy. Again, the concentration of the surfactant is usually of a low order.

Table VII gives surface and interfacial tension of a series of surfactants in use in some shampoos. It should be noted that the surface tension of pure water is 78 dynes/cm and that relatively low concentrations of surfactants reduce this figure by 40 to 50 dynes/cm. Increasing the concentration of the surfactant normally has little influence in further lowering the surface or interfacial tension.

The problem of converting these theoretical considerations to practical values is a real one. Normally shampoos are quite complex compositions and their use concentrations are well above that of the traditional measurement range. In a sense, these data provide some comfort in their comparison of the relative efficacy of a surfactant or a surfactant system.

Surfactant Content and Analysis

The surfactant content of shampoos frequently runs from 15 to 25 %, with the gel and creme concentrates at the higher level, whereas the liquid formulas tend to the lower side. A number of analytical techniques have been worked out for determining compositions. CTFA offers a list of standards, with appropriate analytical methods for a number of shampoo components.

Nevison (67) has offered analytical information on synthetics. Taylor (68) provided a classification of properties for the various synthetic types. The literature gives a number of techniques for analysis of surfactant solutions, both qualitatively and quantitatively.

The increasing complexity of formulations no doubt places a greater burden on the analytical chemist. The latest methods in spectroscopy and chromatography may offer some help in elucidating unknown compositions and maintaining the quality control of production.

Rinsing

The user of modern shampoos likes the rapid build-up of thick lather. But this lather must be easily rinsed from hair to continue to please. Normally a poor lathering product is easily flushed out of the hair. The evaluation of rinsing among a series of good lathering formulations presents a more difficult problem.

The *Consumer Reports* (52) shampoo evaluation, previously cited, employs the user panel rating as its method. A second technique is to employ skilled beauticians to make comparisons of the performance of several shampoos. Here rinsing can be more easily related on a comparative basis. Cryer (69) outlines some of the criteria for designing and implementing hairdresser salon testing.

The ability of shampoo lather to rinse easily from the hair depends on many factors: lather consistency, the surfactant adsorption to the hair fiber, the water conditions (temperature, hardness, rate and technique of flushing), and the quantity and length of the user's hair. It is not likely that any simple mechanical method can simulate all these ramifications.

Conditioning Action

Conditioning action is a difficult property to assess. This is because it is basically dependent on subjective appraisal. The degree of conditioning given to hair is ultimately judged by the shampoo user who is making the evaluation on the basis of past experience, present expectations, and a continuing change in the individual scalp and hair situation.

Zussman and Lennon (70), discussing sarcosines, emphasized the need to develop objective data. Cryer (69) presented techniques for appraising product trials in the salon using the expert opinions of trained beauticians. Bollert and Eckert (71) developed elaborate techniques for gravimetric determinations of the conditioning effects of various treatments.

Conditioning, nonetheless, still emerges as a composite of many facets of what is desirable, convenient, and comfortable to the user. Conditioned hair should be soft, lustrous, easily combed and coiffured, and able to retain its style. It seems that thorough, but not necessarily excessive, cleaning is desirable. Further, the shampoo should leave the hair fibers in a state of lubricity, smoothness, and control during arrangement of the coiffure.

Softness

Softness is a term that embodies a sense of resilience, smoothness of feel ("handle"), and a freedom from stiffness and stickiness; it often suggests an attendant reduction or elimination of snarls and tangling. By eliminating adhesive oils and dirt that gradually build up on the hair between washings, the shampoo restores a degree of softness. Moreover, the imparting of a capacity for reducing the development of static charge while combing and brushing the hair should be useful in reducing tangling and snarls. Henkin, Mills, and Ester (72), and Barber and Posner (73) have discussed the means of measuring electrostatic charge on hair and have described techniques for studying this phenomenon.

Luster

Hair is normally lustrous when in a clean state. Accumulation of soil and damage to the fiber surface can reduce the capability of the hair to reflect light. Thompson and Mills (74) have provided a technique for measuring luster of hair with an amplified photovolt photometer and a polarizing filter. The results indicated that soap shampoos, in hard water, deluster the hair. This effect can often be readily observed in the beauty shop.

Powers and Fox (61) compared luster ratings by panel testers against cleaning data. They could find no necessary correlation between hair cleanliness and luster. The problem of accurately measuring luster in the face of variations in hair texture, surface smoothness, and color is a formidable one.

Lubricity

The frictional aspect of hair is, to a large extent, dependent on the outer surface of the fiber, the cuticle. Since this enveloping sheath grows out in overlapping scales, it is easier to comb the hair from root to tip than in the opposite direction. This helps to explain the phenomenon of tangling. Furthermore, the swelling of the hair fiber by water uptake during shampooing makes the cuticle surface more uneven, and the thorough cleansing of the surface can increase the frictional aspects of contact as one hair passes across another, over comb teeth or brush bristle.

Added lubricity is desirable to achieve easier combing and to supplement natural oils removed by shampooing. These oils gradually work their way back onto the hair as they are secreted from the scalp.

Textile research has afforded a number of techniques that may be adapted to hair friction measurements. Schwartz and Knowles (75) have shown correlations of combing ease and subjective "handle" with frictional measurement. Waggoner and Scott (76) have devised instrumentation for measuring dry hair raspiness with an electronic comb.

Hair varies widely in texture, smoothness, previous treatment effects, and moisture content, which is dependent on the environmental humidity. All of these factors can exert a strong influence on the measurement of ease of orienting (combing or brushing) the hair and the effect of lubricating the hair.

Body, Texture, Set Retention

Hair "body" is a much desired property, especially in an era of elaborate (and bulky) coiffures. With acceptable body, the hair fibers are readily styled into the desired configuration and retain this set for long periods.

Complications arise in this evaluation by the majority of shampoo users who apply sprays, setting aids, or grooming preparations to their hair. Obviously these preparations are intended to modify the hair-fiber properties in a favorable manner. They can color the thinking of the user who may not clearly separate the effects of a multiplicity of hair products.

The proper amount of "body" depends on an individual subjective evaluation. The hair length and style have an important bearing on the rating. Hair that does not retain its set may be termed limp or "too fine," whereas strong adherence to the coiffure can cause difficulties in combing out or reworking the hair and may be described as coarse or wiry.

Irritation and Toxicity

The evaluation of the safety of a shampoo is an important step in judging its acceptability. Of major concern is the level of irritation produced when the shampoo gets into the eyes and, also, its oral toxicity potential. In addition, skin irritation, dermal toxicity, and sensitization potential are also of importance.

Eye irritation. All shampoos should be tested for eye irritation. At the present time, the most commonly used test is one which employs the rabbit's eye mucosa. Extensive work with rabbits has shown that this test leaves a great deal to be desired. For example, a rabbit frequently closes the lid over the eye and leaves it closed for a day or even longer. Obviously, when a person gets a foreign body in the eye, he blinks frequently and develops tears to free the eye of the irritant. Despite its limitations, however, the Draize (77,78) technique has been worked out as carefully as possible to obtain reliable and reproducible results, and remains the standard for eye irritation tests.

Eye irritation tests involving monkeys are becoming more popular since the monkey eye is essentially identical to the human eye anatomically. As might be expected, it has been found that monkey eyes are less subject to substances which damage the rabbit eye (93).

Hazleton (79), Meinicke (80), Cain and Markland (81), and Bonfield and Scala (82) have contributed papers on the eye irritation levels produced by various surfactants and shampoos.

Oral toxicity. The oral toxicity of a material is given in terms of its lethal dose/50 (LD/50), i.e., the number of g (fl oz) of the material per kg of body weight required to kill half of the test animals employed. Therefore the lower the LD/50, the greater is the toxicity.

In general, rats are used, but occasionally dogs are preferred since the dog is capable of emesis whereas the rat is not. The test procedure calls for fasting caged animals before dosing, which is generally accomplished with the aid of a stomach tube.

LD/50 values for a number of surfactants are given by Hopper, Hulpieu, and Cole (83). Whenever the LD/50 is 5 or higher, the surfactant is considered to have low toxicity.

Antidandruff Efficacy

There are two popular methods for evaluating the antidandruff efficacy of a shampoo. One is a visual method offered in papers by Spoor (84), Van Abbe (85), Van Abbe and Dean (86), and Botwinick and Botwinick (87), in which a trained individual grades the severity of a dandruff condition using a given numerical scale. The other method involves the use of a vacuum cleaner in an attempt to make the measurements more objective. Vander Wyk and Roia (88), Vander Wyk and Henchemy (89), and Finkelstein and Laden (90) have presented work using this technique.

Product Characteristics

Fragrance

There is no question that the fragrance of the shampoo is extremely important; indeed some surveys have indicated that it is the most important single quality at the time of purchase. The compounding of fragrances is a most complex and fascinating activity that is discussed in detail in another chapter.

Fragrance, as far as shampoos are concerned, may be evaluated in four ways:

1. The fragrance sniffed in the bottle.
2. The fragrance as the user encounters it during practical use.
3. The residual fragrance left on the hair after rinsing, drying, and coiffing.
4. The stability of the fragrance in the product during storage and the long-term effect of the fragrance on other product characteristics, such as color.

In each case the fragrance may vary in consumer appeal. It is important to test shampoo fragrance extensively in all of the above situations, if possible, to ascertain acceptance.

Furthermore, many essential oils, being natural products, can differ from season to season and year to year. It is important to monitor fragrance quality once a desirable perfume is developed.

Color

Shampoo color selection has evolved to the increasing use of light, bright, eye-catching shades. Factors supporting this trend are the disappearance of cartons for bottle and jars, the availability of lighter-colored surfactant bases, the introduction of improved color-stabilizing systems, and merchandising techniques that have made product and package mutually supporting.

Any coloring of shampoos must be done with certified colors. Although this is somewhat restrictive, a range of colors is still available. It is not likely that normal quantities of coloring could have any influence on the color of the shampooed hair; nonetheless this is an area of product performance to be checked.

Consistency

Shampoos range in consistency from waterlike fluidity to immobile cremes or gels. The relationship of form to package is important to ensure the ready dispensing of the shampoo in the desired quantity.

The manufacturer must determine whether the consistency of his product is compatible with its particular performance. Strianse (91) addressed the matter of technical research on consumer products and discussed the effect of shampoo consistency on the market. Thick products may be advantageous in reducing spillage or loss during application but may be difficult to disperse through the hair. Thinner products may hold an advantage where rapid dispersion and cleaning are desirable and where loss due to inadvertent handling is unlikely (as when a mother is applying the shampoo to a child's head).

Package

Product and package are increasingly regarded as mutually complementary parts of the whole in modern design concepts. The package, in a large degree, is the product in its presentation to the buying public. The package can determine how much of the shampoo is used, where it is stored, and how it can be promoted.

Consumer Reports (52) regards the use of nonshattering packaging as highly desirable. Increasing use of polyvinyl chloride and polyethylene plastic bottles and tubes for toiletries and cosmetics is now evident. The quality of plastic packages is improving and should continue to improve as design and marketing experience develop.

There are, no doubt, many further interesting packaging concepts awaiting reduction to practical application. We may expect the trend to tailoring the package to the needs of the public to continue. Shampoos should be contained in packages of good barrier properties. Passage of water vapor, essential oils, and air through the container poses a threat to product stability. Autian (92) presented a lucid discussion of the potential problem posed by plastic containers and offered testing programs for evaluating them. Careful and extensive testing of new packaging is always advisable for shampoos.

REFERENCES

1. This information has appeared frequently in trade publications, such as *Drug Trade News*.

2. McCutcheon, J. W.: Detergents and emulsifiers, 1968 annual, John W. McCutcheon, Inc., Morristown, N.J., 1968.

3. Harry, R. G.: *Modern cosmeticology*, 4th ed., Leonard Hill, London, 1955.

4. Wall, F. E.: *Principles and practices of beauty culture*, Keystone Publications, New York, 1946.

5. Zussman, H. W.: Shampoo formulation, *Proc. Sci. Sec. TGA*, **19**: 58 (1953).

6. Ester, V. C., Henkin, H., and Longfellow, J. M.: The use of hair clippings in the evaluation of shampoos, *Proc. Sci. Sec. TGA*, **20**: 8 (1953).

7. Barnett, G., and Powers, D. H.: The effect of tap water, hard water and sea water on the performance of shampoos and surface active agents, *Proc. Sci. Sec. TGA*, **15**: 16 (1951).

8. Cook, M. K.: Modern shampoos, *DCI*, **99**: 52 (August 1966).

9. Jannaway, S. P.: *Hairdressing preparations*, United Trade Press, London, 1946.

10. Jannaway, S. P: Modern shampoos, *PEOR*, **36**: 179 (1945).

11. Wei, L.: U.S. Pat. 3,001,944 (1961).

12. Pantaleoni, R., Shanks, J. A., and Valentine, E.: The fatty alcohol sulfates for shampoos, *Proc. Sci. Sec. TGA*, **12**: 9 (1949).

13. Lang, E. W.: U.S. Pat. 3,086,943 (1963).

14. Sipon, American Alcolac Corp., Baltimore, 1957.

15. Silicones: Cosmetics & Toiletries Handbook, General Electric Co., Waterford, N.Y.

16. Emulsions and detergents, 8th ed., Carbide and Carbon Chemicals Co. New York City.

17. Henkin, H.: U.S. Pat. 2,674,580 (1954).

18. Liquid cream shampoos, *Schimmel Briefs*, Nos. 284 and 285, 1958.

19. Patterson, R. L.: Some factors affecting the consistency of paste cream shampoos, *Proc. Sci. Sec. TGA*, **24**: (1955).

20. Morel, C.: Formulating a cream shampoo, *SPC*, **22**: 478 (1949).

21. Henkin, H.: U.S. Pat. 2,773,834 (1956).

22. Cosmetic considerations, Union Carbide Co., New York City.

23. Guide to cosmetics, General Aniline and Film Corp., New York, 1967.

24. Nabori, K., *et al.*: Jap. Pat. 7041 (1951), through *Chem. Abs.*, **47**: 5146 (1953).

25. Harris, J. C.: Shampoo formulator, *Am. Perf.*, **48**: 35 (November 1946).

26. Le Florentin, R.: *Cosmétiques et produits de beauté*, Desforges, Girandot, Paris, 1938.

27. Baldwin, M. M., and Young, A. P.: U.S. Pat. 2,694,669 (1954).

28. Composition and constants of fats and oils, Armour Chemical Division., Chicago, Ill.

29. Lesser, M. A.: Shampoos, *Soap*, **26**: 40 (December 1950) and **27**: 38 (January 1951).

30. Thomssen, E. G.: *Modern cosmetics*, 3rd ed., Drug and Cosmetic Industry, New York, 1949, p. 250.

31. Some modern cosmetic formulas, *Schimmel Briefs*, No. 225, 1953.

32. Jannaway, S. P.: Modern shampoos, *Alchemist*, **3**: 130, 151, 179, 207, 227 (1949).

33. Hansen, K. R.: U.S. Pat. 3,001,949 (1961).

34. The lauryl ether sulfates, *Schimmel Briefs*, No. 299, 1960.

35. *Sarcosyl n-acyl sarcosine surfactants*, Geigy Industrial Chemicals, Ardsley, N.Y., 1963.

36. Seidenfaden, M.: Fatty acid condensates, *Parfum., Cosmét., Savons.*, **6** (2): 55 (1963).

37. Anderson, J. A.: U.S. Pat. 3,085,067 (1963).

38. Steinapol, SBFA 30, Data sheet, REWO, Steinau.

39. Laiderman, D. C.: U.S. Pat. 3,072,580 (1963).

40. The Maypons, data sheet, Stepan Chemical Company, Maywood, N.J.

41. Deriphates, amphoteric surfactants, General Mills Chemical Division, Kankakee, Ill., 1961.

42. Mannheimer, H. S.: Baby shampoos, *Am. Perf.*, **76**: 115 (1961).

43. Goff, S. R.: U.S. Pat. 2,950,255 (1960).

44. Mixtures of soaps and detergents in the form of solid gels, *Schimmel Briefs*, No. 296, 1959.

45. Gillette Co.: Br. Pat. 1,051,461 (1966).

46. Liquid cream shampoos, *Schimmel Briefs*, Nos. 284 and 285, 1958.

47. Geen, H. C.: U.S. Pat. 2,826,551 (1958).

48. Anderson, J. A.: U.S. Pat. 2,928,772 (1960).

49. Schwarz, T. W., and Levy, G.: A report on the oxidative degradation of neutralized Carbopol, *J. Am. Pharm. Assoc., Sci. Ed.*, **47**: 442 (1958).

50. Bryce, D. M., and Smart, R.: The preservation of shampoos, *JSCC*, **16**: 187 (1965).

51. Signore, A., and Woodward, F. E.: Ultraviolet light absorbers in cosmetics, *JSCC*, **9**: 358 (1958).

52. Shampoos, *Consumer Reports*, **33**: 529 (1968).

53. Longfellow, J. M.: Shampoos, *DCI*, **84**: 444 (April 1959).

54. Ross, J., and Miles, G. D.: An application for comparison of foaming properties of soaps and detergents, *Oil and Soap*, **18**: 99 (1941).

55. Ross, J., Miles, G. D., and Shedlovsky, L.: Film drainage: a study of the flow properties of films of solutions of detergents, and the effects of added materials, *J. Am. Oil Chem. Soc.*, **27**: 268 (1950).

56. Colson, R.: *Industrie parfum.*, **9**: 53 (1954).

57. Sanders, H. L.: Surfactant performance, *Soap*, **27**: 39 (1951).

58. Sanders, H. L., Knaggs, E. A., and Libman, O. E.: Alkylolamides in shampoos, *JSCC*, **5**: 29 (1954).

59. Barnett, G., and Powers, D. H.: Factors attributing to the performance of shampoos and to consumer acceptance, *Proc. Sci. Sec. TGA*, **24**: 24 (1955).

60. Fredell, W. G., and Read, R. R.: Shampoos: practical method of evaluation, *Proc. Sci. Sec TGA*, **25**: 30 (1956).

61. Powers, D. H., and Fox., C.: The role of detergents in shampoos, *JSCC*, **10**: 116 (1959).

62. Bromley, J. M.: Modern trends in the assessment of shampoos, *JSCC*, **15**: 631 (1964).

63. Schwartz, A. M., Perry, J. W., and Berch, J.: *Surface active agents and detergents*, Interscience, New York, 1958.

64. Nicolaides, N., Fu, H. C., and Rice, G. R.: The skin surface lipids of man compared with those of 18 species of animals, *J. Invest. Derm.*, **51**: 83 (1968).

65. Industrial utility of public water supplies in the United States, U.S. Geological Survey, Water Supply Paper 1299 and 1300, Government Printing Office, Washington, D.C., 1954.

66. Draves, C. Z., and Clarkson, R. G.: A new method for the evaluation of wetting agents, *Am. Dyestuff Rep.*, **20**: 201 (1931).

67. Nevison, J. A.: Analysis of syndets, *J. Am. Oil Chem. Soc.*, **29**: 576 (1952).

68. Taylor, A.: Synthetic detergents: types and applications, *JSCC*, **4**: 201 (1953).

69. Cryer, P. H.: Design and analysis of product performance in the hairdressing salon, in A. W. Middleton (Ed.), *Cosmetic Science*, Macmillan, New York, 1963.

70. Zussman, H. W., and Lennon, W.: Acylated amino acids in cosmetics, *JSCC*, **6**: 407 (1955).

71. Bollert, V., and Eckert, L.: Quantitative Verfolgung Haarkosmetischer Prozesse, *JSCC*, **18**: 273 (1967).

72. Henkin, H., Mills, C. M., and Ester, V. C.: Measurement of static charge on hair, *JSCC*, **7**: 466 (1956).

73. Barber, R. G., and Posner, A. M.: A method for studying the static electricity produced on hair by combing, *JSCC*, **10**: 236 (1959).

74. Thompson, W. E., and Mills, C. M.: An instrument for measuring the luster of hair, *Proc. Sci. Sec. TGA*, **15**: 12 (1959).

75. Schwartz, A. M., and Knowles, D. C., Jr.: Frictional effects in human hair, *JSCC*, **14**: 455 (1963).

76. Waggoner, W. C., and Scott, G. V.: Instrumental method for the determination of hair raspiness, *JSCC*, **17**: 171 (1966).

77. Draize, J. H., and Kelley, E. A.: Toxicity to eye mucosa of certain cosmetic preparations containing surface-active agents, *Proc. Sci. Sec. TGA*, **17**: 1 (1952).

78. Draize, J. H., Woodard, G., and Calvery, H. O.: Methods for the study of irritation and toxicity of substances applied topically to the skin and mucous membranes, *J. Pharm. Exptl. Therap.*, **82**: 277 (1944).

79. Hazleton, L. W.: Relation of surface active properties to irritation of the rabbit eye, *Proc. Sci. Sec. TGA*, **17**: 5 (1952).

80. Meinicke, K.: Irritating effect of surface active substances on the eyes of rabbits, *Riechst. Parfüm. Seifen*, **62** (2): 107 (1960).

81. Cain, R. A., and Markland, W. R.: Some eye irritation studies of shampoos and other hair preparations, *Proc. Sci. Sec. TGA*, **17**: 10 (1952).

82. Bonfield, C. T., and Scala, R. A.: The paradox in testing for eye irritation: a report on thirteen shampoos, *Proc. Sci. Sec. TGA*, **43**: 34 (1965).

83. Hopper, S. H., Hulpieu, H. R., and Cole, V. V.: Some toxicological properties of surface-active agents, *J. Am. Pharm. Assoc.*, *Sci. Ed.*, **38**: 428 (1949).

84. Spoor, H. J.: Clinical evaluation of antidandruff formulations, *JSCC*, **14**: 135 (1963).

85. Van Abbe, N. J.: The investigation of dandruff, *JSCC*, **15**: 609 (1964).

86. Van Abbe, N. J., and Dean, P. M.: The clinical evaluation of antidandruff shampoos, *JSCC*, **18**: 439 (1967).

87. Botwinick, I. S., and Botwinick, C. G.: Methods for evaluating antidandruff agents, Joint meeting of the TGA and Committee of Cutaneous Health and Cosmetics of AMA, New York, May 1967.

88. Vander Wyk, R. W., and Roia, F. C.: The relationship between dandruff and microbial flora of the human scalp, *JSCC*, **15**: 761 (1964).

89. Vander Wyk, R. W., and Henchemy, K. E.: A comparison of the bacterial and yeast flora of the human scalp and their effect upon dandruff production, *JSCC*, **18**: 629 (1967).

90. Finkelstein, P., and Laden, K.: An objective method for evaluation of dandruff severity, *JSCC*, **19**: 669 (1968).

91. Strianse, S. J.: Technical research on consumer products, *DCI*, **95**: 504 (1964).

92. Autian, J.: Plastic packaging and cosmetic products: potential problems, *Proc. Sci. Sec. TGA*, **43**: 14 (1965).

93. Buehler, E. V., and Newmann, E. A.: A comparison of eye irritation in monkeys and rabbits, *Toxicol. Appl. Pharm.*, **6**: 701 (1964).

94. Keithler, W.: The clear shampoo, *DCI*, **75**: 610 (1954).

Chapter 20

HAIR-GROOMING PREPARATIONS

RICHARD K. LEHNE

Hairdressings of one type or another have had a popular appeal for men and women throughout the ages. Today they are used in the most exclusive beauty salons and by the primitive tribes of Africa. The tombs of Egyptian kings dating as far back as 3500 B.C. contain evidence of perfumed hair oils, and it is known that the Greeks and Romans made extensive use of available fats and oils for personal grooming (1–3).

Most of the products formerly used were homemade, being derived from special or secret recipes, although during the last few centuries many shops and specialty stores sold their own brands. People used whatever was at hand—wines, herbs, animal and plant by-products. The American frontiersmen, for example, simply used bear and goose grease to keep the hair out of the eyes.

The end of the nineteenth century saw the establishment of some successful custom brands, with ever-increasing attention paid to uniformity of product and cosmetic elegance. The contributions of technical men became apparent as much of the hit-and-miss of formulation was removed. Newer raw materials became available, and quality was greatly improved. The chemist now has a much better understanding of the factors involved in making a hair dressing and a knowledge of how to keep the product from spoilage, although by no means has he reached the ultimate in hair dressings. Improvements are constantly being made on existing types, and newer products are being introduced frequently. For example, the currently popular clear ringing gels are less than ten years old, but have already become a major factor in hairdressings.

The need for hairdressings appears to be constantly increasing. Hair not only acts as a protective covering for the head but is also an attractive feature of men and women. Both of these factors—protection and attraction—are

117

best accomplished by healthy hair, and the beauty of the hair is a measure of its health and of the efficiency with which it accomplishes its task. The scalp normally secretes more oils (sebum) than do other skin surfaces. This oil coats the hair fibers and prevents loss of moisture; it keeps the hair in place and provides some luster. The oil film also directly protects the hair and scalp from wind, rain, heat, and sharp temperature changes (4).

If the flow of oil is below normal, or if the hair has experienced chemical or thermal damage, dehydration can occur. This makes the hair brittle, allowing it to split and break. Bleaching, dyeing, and permanent waving of the hair have been studied intensively, and safer, more efficient products and methods have been evolved. However, excessive use or unskilled handling of such products can cause considerable damage to the hair. In addition, the synthetic detergents used in shampoos are quite "defatting" in action, despite some fairly successful attempts to modify this effect. When used regularly, particularly on dry hair, the protective oil film is removed and evaporation of moisture occurs (5).

Animal, vegetable, or mineral oils are used to supplement some of the natural oils removed from or lacking on the hair and scalp. They provide adequate protective properties as well as a desirable sheen, but they can do nothing for the brittleness of the hair. To overcome this, moisture must be added, but the direct application of water is of little benefit. Although the hair swells and softens readily in the presence of water, the evaporation is equally rapid and an equilibrium is soon reached. A means must be provided to keep the absorbed water from evaporating, and emulsions have proved of great value in this respect. Some of the oil-in-water emulsions are of particular benefit, because the external aqueous phase is absorbed, the emulsion breaks, and the remaining oil forms a protective film on the hair shaft.

Properties of a Good Hairdressing

The consumer is interested in a product which will give the hair a natural, healthy appearance. He expects good grooming, luster without greasiness, protection from the elements, and some degree of hair conditioning.

The chemist must employ raw materials which will combine these properties. For the proper luster, one needs a glossy material such as an oil or fat, or a solubilized wax which is adsorbed to the hair shaft. The material must give good wetting action and be left in a uniformly thin film.

To achieve good grooming, a balance of cohesion and adhesion is required. The product must adhere to the hair fiber and provide good cohesion as well; otherwise the hair becomes unruly. There are many materials which fit this description, but they must possess another property—lubrication. This is important for allowing the hairs to slide past each other when combed, and then to remain put.

The major property required for hair conditioning—in addition to freedom from toxic and sensitizing effects—is "moisturizing." The water must be offered in such a form as to be available to the hair and scalp. In the case of gum-base and alcoholic lotions, most of the water evaporates, and the resulting film grooms and provides luster without relieving brittleness. The oil-in-water emulsions and, to a lesser extent, the water-in-oil emulsions can give the best moisturizing action. Lanolin, fatty acid amides, and some long-chain quaternary ammonium compounds are substantive to the hair and act as emollients. These, along with the fats and oils, give a protective film which prevents the subsequent loss of moisture and greatly minimizes damage by the elements.

Nothing enhances the beauty of the hair more than the effective control of loose dandruff. The regular application of a hairdressing should aid in keeping the sloughed-off skin soft rather than dry and flaky, and help to "anchor" it to the scalp and base of the hair shafts without packing in hard scales.

The introduction of new cosmetic ingredients, and of new combinations of old ones, presents a potential danger. What appears to act as a wonderful new conditioning or grooming agent may cause primary irritation or sensitization. It behooves every chemist to be constantly aware of this possibility and always to demand toxicity tests of any new product.

A bibliography of modern scientific literature on hair dressings is presented for further study (6–28).

Types of Hairdressing

The hairdressings in vogue today can be divided into nine major groups. There is some overlapping, as will be shown later, since the inventiveness of the cosmetic chemist has cut across borderlines. These groups, which will be considered individually, are as follows:

1. Brilliantines, liquid and solid.
2. Alcoholic lotions.
3. Two-layer systems.
4. Hair tonics.
5. Gum-base hair dressings.
6. Oil-in-water emulsions.
7. Water-in-oil emulsions.
8. Gels.
9. Aerosols.

Brilliantines

The main purpose of a brilliantine is to add a measure of grooming and to afford a sheen to the hair. If the natural oils of the hair are deficient or have been removed, the hair has a dull appearance. This can be overcome by adding oil to it. Historically, the vegetable and animal oils were first used for this purpose. Since these materials are subject to spoilage, they were largely replaced by mineral oil after V. Chapin Daggett, shortly before 1900, used white mineral oil in his cold cream to avoid rancidity. Mineral oil is not absorbed by the hair but provides a thin protective film owing to its low viscosity and penetrating power. The heavier the viscosity, the better its grooming properties, but the more difficult it becomes to spread thinly and uniformly. Deodorized kerosene can be used to dilute the heavy oil, making it easier to spread and penetrate. After evaporation, a thin uniform layer of viscous oil is deposited, imparting good grooming and high gloss to the hair (19,20).

Liquid Brilliantines

The simplest liquid brilliantines are described in Formulas 1 and 2. The

Formulas for Liquid Brilliantines

	1	2
Mineral oil, light	100%	75%
Deodorized kerosene	—	25
Color and perfume	q.s.	q.s.

Procedure: After mixing thoroughly, filter the solution to ensure brilliance.

viscosity of the mineral oil will vary with the final effect desired and the method of application. If a product is to be sprayed, a low-viscosity oil is desired. If it is to be combed in or applied by massage, heavier grades can be used. In all cases, a good-quality oil should be used, preferably one containing an antioxidant. Even the best grades can develop off-odors on aging. The pour point should be low enough to ensure free flow at low temperatures, depending of course on the climatic conditions to which the brilliantine is to be exposed (21).

Although mineral oil and deodorized kerosene are relatively odor-free, a problem of perfuming does arise because of the low solubility of many per-fume ingredients. Resinous and crystalline materials will not dissolve, or may precipitate on standing. Since perfect brilliance is paramount, the effect of aging, sunlight, and heat on the final product should be determined. The addition of "coupling" agents often aids in keeping the brilliantine clear. Even a small percent of a vegetable oil, a fatty alcohol, a fatty ester, or some nonionic surfactant will usually solubilize a perfume oil.

Some authorities claim that mineral oil, being foreign to the human body, should not be used in cosmetics. They suggest vegetable or animal oils as being more similar to the human sebum. Both are said to be absorbed to some extent, whereas mineral oil is not (13). However, only non- or semidrying oils with an iodine number less than 105 should be used on the hair. Most commonly employed oils are olive, castor, sesame, peanut, avocado, almond, apricot and peach kernel. Two or more of these are blended to obtain the optimum balance of lubrication, cohesion, and adhesion, as shown in Formulas 3 to 5.

Formulas for Liquid Brilliantines

	3	4	5
Castor oil	80%	6%	85%
Almond oil	20	49	—
Olive oil	—	45	—
Deodorized kerosene	—	—	15
Color and perfume	q.s.	q.s.	q.s.

Procedure: Thoroughly mix the ingredients to dissolve the color and the perfume, and then filter bright.

Vegetable oils are subject to rancidity and therefore require the addition of some antioxidants. The naturally occurring antioxidants are usually removed during the refining and bleaching operations, but the oils can be reconstituted by the addition of tocopherols, which are fairly effective as antioxidants. Some materials give good preservation but are themselves oxidized to colored substances, e.g., hydroquinone. The lower alkyl gallates, e.g. propyl, have proved of interest, particularly when combined with other compounds such as butylated hydroxy anisole and sequestering agents. Octyl, dodecyl, cetyl, and stearyl gallates have been found effective owing to their greater stability. Propyl p-hydroxybenzoate, nordihydroguaiaretic acid (NDGA), and di-*tert*-butyl-p-cresol are also effective in preventing rancidity. Quantities as low as 0.01 to 0.10% are usually adequate (12).

The phenol derivatives are also active preservatives; thymol, pyrogallol, or p-chloro-m-cresol may prove effective. It has been observed that the addition of mineral oil to the extent of 10% will retard rancidity of vegetable oils.

Vegetable oils are added to mineral oil to impart to brilliantines certain "plus" features long associated in the minds of the consumer with a given oil. They also serve to modify the feel of the mineral oil. Examples of such products are given in Formulas 6 to 9. The relative quantities and even types of oils can be varied by the experienced chemist within wide limits, depending on the properties sought.

Lanolin has been added to brilliantines, although its low solubility in mineral oil is a disadvantage. Partially esterified lanolin and lanolin fatty acids, of which many derivatives are now available, have better mineral oil solubility, and should be employed rather than lanolin itself. Liquid lanolin,

Formulas for Liquid Brilliantines

	6	7	8	9
Mineral oil	99 to 80%	66%	90%	80%
Olive oil	1 to 20	—	—	—
Peanut oil	—	34	—	—
Almond oil	—	—	—	20
Sesame oil	—	—	10	—
Color and perfume	q.s.	q.s.	q.s.	q.s.

Procedure: Prepare simply by mixing and then filtering bright.

a dewaxed fraction of regular lanolin, also has excellent oil solubility. Increased solubility can be effected also by using certain fatty acid esters, such as isopropyl myristate or palmitate, as disclosed in the Verblen patent (22). The liquid acetylated monoglycerides have a coupling effect, and a higher percentage of lanolin can be added to mineral oil in their presence.

The addition of these fatty esters themselves makes an agreeable modification of the mineral and vegetable oils, as do fatty alcohols such as oleyl alcohol and partially ethoxylated fatty alcohols. They are completely miscible, resistant to rancidity, absorbed by the hair fiber, and give emolliency as well as gloss. Formulas 10 to 15 illustrate such preparations.

Formulas for Liquid Brilliantines

	10	11	12	13	14	15
Mineral oil	75%	33%	—	—	60%	92.5%
Ethyl myristate	25	—	—	—	—	—
Castor oil	—	33	60%	—	—	—
Isopropyl myristate	—	34	—	—	20	—
Ethyl oleate	—	—	40	—	—	—
Methyl oleate	—	—	—	25%	—	—
Olive oil	—	—	—	75	—	—
Liquid lanolin	—	—	—	—	20	—
Lanolin	—	—	—	—	—	2.5
Oleyl alcohol	—	—	—	—	—	5.0
Color and perfume	q.s.	q.s.	q.s.	q.s.	q.s.	q.s.

Other emollient or grooming agents can be incorporated if they are compatible and if, of course, the final product is appealing as far as appearance, feel, and odor are concerned.

A large number of synthetic nonionic materials are being offered as replacements of naturally occurring oils. A partial list of such potential additives includes polyglycol esters, fatty alcohol lactates, dialkyl sebacates and adipates, and hexadecyl alcohol. They usually have one great advantage—that of not being subject to rancidity. Since many of these materials are polyoxyethylene condensation products of glycols and polyols, their solubilities in polar and nonpolar solvents can be varied at will. Unctuous grooming materials soluble in water or in hydroalcoholic solutions are available.

Occasionally one finds a brilliantine for which an advantage is claimed by including such substances as an insect repellent or a sunscreen. At present, the market for these items is rather specialized; but since prolonged exposure to strong sunlight does affect the hair adversely, the use of such protective additions appears logical. Mineral oil itself will contribute somewhat to this protection by reflecting a portion of the ultraviolet light of the sun spectrum.

Solid Brilliantines

Solid brilliantines and pomades are almost traditional in composition, and little has been done to present a basically new item in this category. The terms "pomade" and "solid brilliantine" are almost synonymous at present, although the former usually refers to a product of softer consistency, such as a perfumed petrolatum. Originally a pomade was the residual fatty material left from the enfleurage process of extracting floral odors. The solid brilliantines, on the other hand, are usually vegetable or mineral oils which have been stiffened to the desired consistency with the aid of certain waxes.

The consistency of solid brilliantines varies from that of a soft petrolatum to a definite wax, and from a greasy, stringy material to products which crumble to the touch. They are therefore packaged in wide-mouthed jars or in collapsible tubes.

Solid brilliantines are best suited for unruly or kinky hair which has to be forcibly held in place, and are very popular for use on closely cropped hair. The lubricity is low—actually just enough to permit uniform spreading—and much greater grooming is therefore obtained.

The products are invariably opaque, and the degree of opacity increases as the wax content increases. This is often carried to an extreme to give a definite crystallized effect. A typical simple product is shown in Formula 16 (29).

Formula 16

Stearic acid	23%
Mineral oil	77
Color and perfume	q.s.

Procedure: Melt the two ingredients together at 60 to 70°C and add color and perfume. Fill the hot melt into warm jars and cool slowly in a warm room for 12 to 18 hr. This gives a marbled product which crumbles at the touch.

Aluminum stearate, 7 to 10%, has been used to impart a gel structure to heavy mineral oil in a manner similar to its use in the grease industry (30). Strict exclusion of all moisture is necessary, and high processing temperatures are needed, up to 170°C. Great care is required in selecting the proper grade and type of metal salt, since mono, di, and tri aluminum derivatives are possible, and since the composition of the stearic acid itself varies considerably. Additives such as castor oil or lanolin sometimes help in preparing a stable gel. Cooling rates are important; too fast gives a friable gel, and too slow, a liquid. Oil seepage, or syneresis, occurs on aging. Aluminum lanolate can also be used for gelling, as can triethanolamine stearate, although the latter affords a very soft ointment. Typical formulas are shown in Formulas 17 and 18.

Formulas for Gelled Mineral Oil

	17	18
Mineral oil, heavy	89%	85%
Aluminum stearate	8	—
Castor oil	3	—
Aluminum lanolate	—	7
Palmitic acid	—	3
Lanolin, anhydrous	—	5
Color and perfume	q.s.	q.s.

Procedure: Blend the oils and, while heating, sift in the solid materials under good agitation. Heat to about 150°C, stir until all solids are melted and dissolved, cool to 90 to 100°C, and pack into tubes. Allow to cool to room temperature without disturbing to permit gelling.

Mineral oil "bodied" with paraffin wax would appear to be the most obvious formulation. However, paraffin wax is not the best material to use, since it tends to crystallize and to permit oil exudation. It also shrinks on cooling, thereby drawing the brilliantine mass away from the sides of the jar (1). The simplest means of overcoming this is to add some petrolatum, shown in Formulas 19 and 20.

Better control of the sweating-out of mineral oil is obtained with ozokerite, spermaceti, or ceresin wax. Some of the products shown in Formulas 21 to 26, suggested as a starting point for formulation, indicate the widest variation of consistency possible.

Formulas for Solid Brilliantines

	19	20
Paraffin wax	20%	15%
Mineral oil	50	25
Petrolatum	30	60
Color and perfume	q.s.	q.s.

Vegetable oils and other additives, such as lanolin derivatives, have been added to solid brilliantines, just as they have been to liquid brilliantines. Formulas 27 to 31 indicate the general proportions of oils to waxes.

An ingredient widely used in Europe in many cosmetic formulations is rosin. This is a very economical material which can improve gloss and has a good fixative action on the hair. Although it is a hard solid at room temperature, it will not give a solid product when incorporated in hot mineral oil and cooled (21). It is necessary to add a wax, such as paraffin wax, to stiffen it. The amount of rosin used can vary from 1 to 25 %, depending on the tackiness desired. Examples of preparations in which rosin is used are shown in Formulas 32 to 34.

Formulas for Solid Brilliantines

	21	22	23	24	25	26
Mineral oil	67%	42%	86%	56%	70%	75%
Ceresin	11	8	—	—	—	—
Lanolin	22	—	—	—	—	—
Petrolatum	—	50	—	—	—	8
Spermaceti	—	—	10	22	5	—
Beeswax	—	—	4	22	—	—
Paraffin wax	—	—	—	—	15	—
Stearic acid	—	—	—	—	10	—
Ozokerite	—	—	—	—	—	17
Color and perfume	q.s.	q.s.	q.s.	q.s.	q.s.	q.s.

Procedure: Melt together all ingredients at the lowest possible temperature, add perfume as cooling occurs, and run the material into warmed jars. Allow to cool slowly. Care must be taken not to work the materials when partially solidified; otherwise air bubbles will be permanently incorporated.

Formulas for Solid Brilliantines

	27	28	29	30	31
Castor oil	50%	—	—	80%	—
Almond oil	30	85%	—	—	—
Spermaceti	20	10	—	—	—
Cocoa butter	—	5	—	—	—
Coconut oil	—	—	75%	—	—
Ceresin	—	—	25	—	—
Beeswax	—	—	—	20	—
Petrolatum	—	—	—	—	50%
Acetylated lanolin	—	—	—	—	10
Isopropyl lanolate	—	—	—	—	5
PEG 400 monostearate	—	—	—	—	5
Mineral oil 180	—	—	—	—	30
Color and perfume	q.s.	q.s.	q.s.	q.s.	q.s.

Formulas for Solid Brilliantines

	32	33	34
Rosin	48%	1.5%	25%
White oil	12	16.5	20
Paraffin wax	24	15.0	25
Petrolatum	16	67.0	—
Ceresin wax	—	—	30
Color and perfume	q.s.	q.s.	q.s.

The synthetic waxes and some of the high-molecular-weight polyoxyethyl-ene derivatives could be used to advantage in the formulation of solid brilliantines. Compatibilities must be carefully watched, since many of these newer materials, although giving the appearance and feel of waxes, are very lipophobic.

The perfuming of solid brilliantines is not quite so difficult as that of clear liquid preparations. The amount of perfume used is usually higher than in a comparable liquid brilliantine, since it is not quite so volatile in the more viscous medium, and since a much smaller quantity of the grooming material is applied to the hair.

Articles and patents of special interest dealing with brilliantine and po-mades are cited in the bibliography (29–41).

Alcoholic Lotions

Alcoholic lotions have been popular as hairdressings since before the turn of the century. By diluting viscous oils with alcohol, it is possible to get good wetting action, and after evaporation of the alcohol, deposition of a uniformly thin layer of oil. Furthermore, many people like the temporary stimulation the scalp receives from the alcohol.

Castor oil is unique among the fixed oils, in being alcohol-soluble in all proportions; it is used frequently as the oil in an alcoholic lotion. Another material often used is glycerol.

The habitual use of concentrated alcohol on the scalp is not recommended, since it may act as a dehydrating agent (11). If the relative humidity is low, moisture will be drawn from the hair and scalp, leaving them dry and brittle. If the alcohol is diluted with water, the degree of dehydration is reduced proportionately. Some authorities claim that the addition of oils to the alcohol prevents it from having a drying effect, since emollients are being added. As has been pointed out, brittleness is not due to loss of oil, and there-fore is not subject to correction by adding oil, but is due to loss of moisture. Alcohol tends to dissolve or extract oil from the skin, and the addition of some oil to it will retard this extraction.

Anhydrous alcohol is used extensively in aerosol hair sprays without problem, since the fine mist is deposited almost exclusively on the surface of the hair (thus not reaching the scalp as a large volume of liquid) and evaporates rapidly to leave the holding resin.

A material often used, either by itself or diluted with alcohol, is glycerol. This cannot be recommended since it is extremely hygroscopic and, rather than "superfatting" the product, may intensify the dehydration of the hair and scalp, sometimes leading to dandruff formation (21).

Formulations will vary with the viscosity and degree of grooming desired. Several examples are presented in Formulas 35 to 38.

Formulas for Alcoholic Lotions

	35	36	37	38
Ethyl alcohol	50%	80%	60%	83%
Castor oil	50	20	10	15
Glycerol	—	—	30	—
Tincture of benzoin	—	—	—	2
Color and perfume	q.s.	q.s.	q.s.	q.s.

Tincture of benzoin is said to reduce the stickiness of castor oil. The materials are mixed well at room temperature and filtered bright. It is essential to use a good grade of deodorized castor oil; nevertheless, the matter of perfuming is a difficult one, since upon aging the castor oil may develop a strong, characteristic odor. Antioxidants have not been used to any great extent in this type of product, but surely are worth investigating.

Fatty alcohols, fatty acids, and ethoxylated and propoxylated derivatives of lanolin, lanolin alcohols, and fatty alcohols have been used in alcoholic and aqueous alcoholic solutions. Examples are shown in Formulas 39 to 43.

Popular materials for the formulation of hydroalcoholic hairdressings are the Ucons. They are nonrancidifying polyalkylene glycols which can be

Formulas for Alcoholic Lanolin Hairdressings

	39	40	41	42	43
Isopropyl myristate	5%	—	—	—	—
Glycerol	5	—	2.5%	—	—
Ethyl alcohol	90	96%	45.0	57%	90%
Water	—	—	47.5	36	—
Oleic acid	—	4	—	—	—
Ethoxylated lanolin	—	—	5.0	—	—
Propoxylated lanolin alcohol	—	—	—	7	—
Isopropyl lanolate	—	—	—	—	10
Color and perfume	q.s.	q.s.	q.s.	q.s.	q.s.

obtained with any degree of solubility in alcohol or water. Almost any desired viscosity and tack are available, and less greasy products can be formulated than with some of the other oils. Since larger amounts of water can be added than with castor oil, the resulting hairdressing is much less dehydrating.

Lotions representing the use of various Ucons are shown in Formulas 44 to 46. Attempts have been made to solubilize other oils in alcohol. Caimi

Formulas for Alcoholic Hair Lotions

	44	45	46
Ucon 50HB660	15%	—	—
Ucon LB 1715	—	20%	—
Ucon LB 3000	—	—	5%
Ethyl alcohol	20	50	60
Water	63	26	32
Ethoxylated lanolin	2	—	—
Diisopropyl adipate	—	—	3
Ethoxylated oleyl alcohol	—	4	—
Color and perfume	q.s.	q.s.	q.s.

and Caimi (42) described in a patent the solubilizing of olive oil with the aid of deodorized kerosene. U.S. Patents 2,865,859 and 2,942,008, issued to Lubowe, describe the solubilization of mineral vegetable and animal oils in ethanol by the use of low-molecular-weight esters of saturated and unsaturated fatty acids and by fatty alcohols. Examples are presented in Formulas 47 and 48.

Formulas for Alcoholic Hair Lotions

	47	48
Ethyl alcohol	54%	35%
Olive oil	19	—
Kerosene	27	—
Mineral oil	—	30
Isopropyl myristate	—	35
Color and perfume	q.s.	q.s.

Several of the most interesting patents and an article on alcoholic hairdressings are cited in the bibliography (42–45).

Hair Tonics

Hair tonics have long been presented for the specific purposes of curing baldness, relieving oily or dry scalp, and preventing or curing dandruff. For all of their long use, none of these preparations has ever been known to grow a new crop of hair once it has been lost.

Some cases of falling hair can be checked if the cause is due to mistreatment or carelessness leading to a clogging of the sebaceous glands and hair follicles dehydration of the scalp, poor circulation, or infection. Proper care, stimulation, and prophylaxis offer the best means of promoting a healthy hair growth, but they will not induce new growth on a bald head.

Most of the hair tonics described in Formulas 49 to 66 present a combination of at least two of the three ingredients of a hair tonic—a sebaceous-gland stimulant, a rubefacient, and an antiseptic. The list of rubefacients commonly used (21) includes chloral hydrate (2 to 4%), formic acid spirits (10 to 12%), quinine and its salts (0.1 to 1.0%), tincture of cantharides (1 to 5%), tincture of capsicum (1 to 5%), tincture of cinchona (1 to 10%), and such tars as cade, pine, and birch (0.1 to 1.0%).

When used in low concentration, most of these materials will cause a reddening of the skin and a warm feeling owing to an increased flow of blood in the skin capillaries. Stronger concentrations should be avoided to prevent irritation or even necrosis. Since individual sensitivity varies greatly, all of these materials should be used with care, especially in newer formulations made with wetting agents. Here the possible synergistic effect should not be overlooked; all such new products embodying any of the above stimulants should have a record of patch and use tests.

The usual antiseptics are either phenolics or quaternary ammonium compounds, although quinine, formaldehyde, and ethyl or isopropyl alcohol will also contribute an antiseptic quality. Phenol itself is not used, because of its toxic and irritant action; however, the halogen, alkyl, and aryl derivatives which show a higher antiseptic coefficient combined with lower toxicity are preferred. Some useful phenolic derivatives are salicylic acid, *p*-chloro-*m*-cresol, *p*-chloro-*m*-xylenol, *o*-phenyl phenol, *o*-chloro-*o*-phenyl phenol, *p*-amyl phenol, chlorothymol, resorcinol, β-naphthol, and hexachlorophene.

Quaternary ammonium compounds are especially interesting because they are usually more substantive than phenolics and because their spectrum of activity extends to the fungi as well as the bacteria. The most useful "quats" are alkyl dimethyl benzyl ammonium chloride, cetyl trimethyl ammonium bromide, lauryl isoquinolinium bromide, *N*-soya-*N*-ethyl morpholinium ethosulfate, and cetyl pyridinium chloride.

Less than 1% of these antiseptics is normally added to a hair tonic, since in higher concentrations some of them may become irritating. Resorcinol is the major exception; it is being used in concentrations up to 5%. In common with all phenolic materials, it will discolor when exposed to sunlight, and will turn fair hair dark, especially in the presence of traces of alkali or soap. The monoacetate is much less of an offender in this respect, but under hydrolyzing conditions will, of course, revert to resorcinol.

Among the various materials claimed to stimulate the sebaceous glands are the following: quinine and its salts, tincture of jaborandi leaves or pilocarpine, resorcinol and resorcinol monoacetate, cholesterol, salicylic acid, ethyl alcohol, methyl linoleate, sulfur, and lecithin. Quinine and pilocarpine, when administered internally, were thought to be able to stimulate hair growth. Along with resorcinol, they are traditional in hair tonic formulas, and have been well received by the consumer. Potassium arsenite and the mercurials were once used to a great extent, but the use of such materials is now strictly controlled by federal and state laws, and their usage must comply with official regulations.

Other additives which have been used to impart soothing, healing, or conditioning action are panthenol, polypeptides, polyunsaturates, camomile, comfrey, and allantoin which is the active ingredient in comfrey. Panthenol is the alcohol analogue of pantothenic acid, to which it is converted in the skin. Pantothenic acid is a component of the vitamin B complex and has been found essential for growth and maintenance of skin and hair (46). It is normally used at a level of about 1 %, and must be adjusted to a pH of 4 to 7 to prevent undue hydrolysis. It is recommended as an aid to healing and to promote epithelization. The unsaturated fatty acids and esters were once called "vitamin F" because they seemed to be essential basic ingredients for general skin care. Although no longer officially recognized as a vitamin, these emollients have been shown to be of value in scalp and skin preparations (11).

Camomile, in addition to being a surface dye enhancing blond highlights, has been used for centuries as an antiinflammatory and is a well-received constituent of hair lotions. Azulene has been identified as the effective substance in camomile and can be used at levels of about 0.01 to 0.02 %. Comfrey is another plant whose extracts have been used since the Middle Ages to heal wounds and reduce inflammation, both topically and internally. Allantoin has been isolated from comfrey and appears to have the same soothing effect, in addition to being capable of dispersing dandruff scales. Levels of about 0.2 % of allantoin are normally recommended.

Proteins from various animal sources have been used in cosmetic preparations for many years (47), but the newer, shorter-chain polypeptides derived from them by hydrolysis are significantly more useful. Enzyme hydrolysates are preferred over acid or alkali hydrolysates because of fewer by-products. These polypeptides are substantive to the hair shaft when used in high-enough concentrations and add to the body and feel of the hair (48). They are an important component of hair-conditioning lotions.

Hair tonics intended for oily scalps are invariably alkaline and often astringent. Tannin has merit in this respect; possibly the efficacy of tincture of cinchona is due to its tannin content.

The formulas presented are roughly divided into those intended to retard loss of hair, those to control dandruff and alleviate oily scalp, and those designed for dry scalp. Bay rum was once a very popular item and still has its adherents; since it is often used as the vehicle for some of the drugs and antiseptics, a brief description of it will be given first.

Bay rum is said to have originated in the West Indies; the original product is made with Jamaica rum and oil of bay. Denatured alcohol can be used in less expensive products, with ethyl acetate or heptaldehyde added to suggest the odor of rum. Oils of bay, pimento, cloves, and cinnamon leaf can be used to produce the typical bouquet. Formulas 49 to 52 are suggested for such products.

Formulas for Bay Rum Hair Tonics

	49	50
Jamaica rum	12%	10.0%
Ethyl alcohol	45	50.0
Oil of bay	2	2.0
Glycerol	5	—
Water	36	37.5
Oil of pimento	—	0.5

Procedure: Dissolve the oils in the alcohol, add the remainder of the ingredients, and mix well. Filter bright and bottle.

Formulas for Bay Rum Hair Tonics

	51	52
Oil of bay	0.15%	0.25%
Oil of cloves	0.15	—
Ethyl alcohol	60.00	65.00
Tincture of quillaia	10.00	—
Water	29.55	34.59
Ethyl acetate	0.15	—
Oil of cinnamon leaf	—	0.05
Quassia extract, solid	—	0.10
Heptaldehyde	—	0.01

Procedure: Dissolve the essential oils in the alcohol, add the ethyl acetate or heptaldehyde, dissolve the quassia extract in the water with heat, and mix with the rest of the materials. Stir well and filter bright.

Formulations using resorcinol or its monoacetate normally contain a rubefacient, as demonstrated in Formulas 53 to 57.

Pilocarpine-containing lotions can be prepared from either the alkaloid or the tincture of jaborandi, as indicated in Formulas 58 and 59.

Formula 60 to 62 are said to stimulate the sebaceous glands and to relieve a dry scalp (49).

Some degree of relief can be obtained from an oily scalp by frequent shampooing and application of tonics designed to regulate the secretions of the sebaceous glands. Formulas 63 to 66 contain astringents, stimulants, and cleansing agents. Very little change has taken place in this type of product, and its popularity has decreased. The severe cases of seborrhea and pityriasis should be left to the dermatologist, but modern hair tonics can be formulated to give aid and some control to milder dandruff cases.

Formulas for Resorcinol Hair Tonics

	53	54	55	56	57
Resorcinol	5%	0.8%	0.3%	—	—
Resorcinol monoacetate	—	—	—	3%	2.5%
Tincture of capsicum	5	—	—	—	—
Chloral hydrate	—	1.5	—	—	—
Spirits of formic acid	—	—	—	20	—
Pine tar oil	—	—	2.7	—	—
Ethyl alcohol	85	80.0	—	70	93.0
Castor oil	5	—	—	7	—
β-Naphthol	—	0.8	—	—	—
Sulfonated castor oil	—	16.9	—	—	—
Soft soap	—	—	0.5	—	—
Potassium sulfate	—	—	3.0	—	—
Water	—	—	93.5	—	—
Methyl linoleate	—	—	—	—	2.5
Cinnamein	—	—	—	—	2.0
Perfume and color	q.s.	q.s.	q.s.	q.s.	q.s.

Procedure: Add the tinctures and oils to the alcohol in which the resorcinol or resorcinol monoacetate has been dissolved; then add perfume and color. After stirring well, filter the batch to clarify.

Formulas for Jaborandi Hair Tonics

	58	59
Tincture of jaborandi	5.0%	—
Tartaric acid	0.5	—
Ethyl alcohol	5.0	9.00%
Triple rose water	82.5	—
Glycerol	7.0	—
Pilocarpine nitrate	—	0.05
Tincture of cantharidine	—	0.95
Water	—	85.00
Glyceryl borate	—	5.00
Perfume	q.s.	q.s.

Procedure: Prepare same as previous formulations, great care being taken to obtain a brilliantly clear product on filtration.

Formulas for Dry Scalp Tonics

	60	61	62
Ammonia water	1.5%	—	—
Sulfonated castor oil	9.5	—	—
Tincture of capsicum	0.8	—	—
Ethyl alcohol	88.2	87%	2.500%
Chloral hydrate	—	3	—
Castor oil	—	10	—
Potassium sulfate	—	—	5.500
Water	—	—	91.690
Hydrochloric acid	—	—	0.004
Glacial acetic acid	—	—	0.006
Pine tar oil	—	—	0.300
Perfume	q.s.	q.s.	q.s.

Procedure: Mix the ingredients well and filter to obtain a brilliant product.

Formulas for Oily Scalp Tonics

	63	64	65	66
Tannin	5.00%	—	—	—
Formaldehyde	0.75	—	—	—
Water	83.75	32.5%	—	56.5%
Ethyl alcohol	10.50	—	70%	40.0
Bay rum	—	30.0	—	—
Rose water	—	24.5	6	—
Ammonia 26°	—	5.0	—	—
Glyceryl borate	—	7.0	—	—
Tincture of capsicum	—	1.0	—	3.0
Eau de cologne essence	—	—	10	—
Glycerol	—	—	4	—
Tincture of cinchona	—	—	4	—
Tincture of quillaia	—	—	6	—
Chlorothymol	—	—	—	0.1
Quinine sulfate	—	—	—	0.1
Benzoic acid	—	—	—	0.3
Perfume	q.s.	q.s.	q.s.	q.s.

Procedure: Dissolve the water-soluble materials in the water, and the oils, tinctures, and perfume in the alcohol. Then mix, stir well, and filter bright.

The exact etiology of dandruff still has not been elucidated, although it is undoubtedly the sum total of many different factors working simultaneously. Some external relief can be given through regular massage, coupled with a mild rubefacient, to help stimulate the blood circulation in the scalp as well as the sebaceous flow. Antiseptics, as mentioned above, can reduce the bacterial and fungal population, whether this be cause or effect, and frequent shampooing can remove the loose matted scales. Zinc pyridine thione has been found to be of value in controlling dandruff from both shampoos and hair dressings. Since it is substantive to the scalp, only low concentrations are required.

Several formulas for antidandruff preparations are given in Formulas 67 to 69. Care must be exercised in formulating with phenolics or quaternary

Formulas for Antidandruff Tonics

	67	68	69
Salicylic acid	1.0%	—	—
Polysorbate 80	3.0	—	—
Isopropyl palmitate	3.0	—	—
Isopropyl alcohol	70.0	—	—
Water	23.0	84.40%	77.5%
Cetyl trimethyl benzyl ammonium chloride	—	1.00	—
Stearyl dimethyl benzyl ammonium chloride	—	0.10	—
Polyethylene glycol 600	—	4.50	—
Ethyl alcohol	—	10.00	—
Protein hydrolysate, 55%	—	—	20.0
Tween 20	—	—	1.5
Panthenol	—	—	1.0
Perfume	q.s.	q.s.	q.s.

ammonium compounds, lest the surfactants or nonionic additive inactivate the germicides.

A few patents of special interest dealing with hair tonics are mentioned in the bibliography (49–54).

Two-Layer Lotions

In an attempt to overcome the inherent greasiness of the brilliantines, diluents such as water and alcohol have been added, resulting in two-layer products. Before application, it is necessary to shake the container vigorously in order to form a temporary emulsion, thereby distributing the two phases proportionately. In most preparations, the breaking of the emulsion and the agglomeration of oil droplets are so rapid that a disproportionate amount of one phase is removed, and often before the bottle is emptied, one phase or the other has been completely used up. This shortcoming can be minimized by adding a small amount of an emulsifier to either or both phases, so that a temporary emulsion is formed with slight shaking. This emulsion will cream or break within a few minutes and revert to a two-layer system upon standing.

XX. HAIR-GROOMING PREPARATIONS

The specific gravities of each phase must be sufficiently different to permit a clear-cut, well-defined separation over a wide range of temperatures, without having inversion occur. Each phase should be tinted separately to ensure complete solution of the dye, and when the two phases are mixed no diffusion of color should take place. The perfume must be selected with care to prevent turbidity or precipitation at the interface.

The simplest and least expensive formulation is a mixture of mineral oil and water, suitably colored and perfumed. The water can be replaced in part or completely by alcohol to give a faster-drying product and to prevent bottle breakage on freezing. Formulas 70 to 72 describe such products.

Formulas for Two-Layer Lotions

	70	71	72
Mineral oil	50%	65%	50%
Water	50	—	32
Ethyl alcohol	—	35	18
Perfume and color	q.s.	q.s.	q.s.

Modifications can be made by replacing part or all of the mineral oil with a vegetable oil in order to obtain the kind and degree of grooming desired. When vegetable oils are used, it is wise to add an efficient antioxidant. Rancidity is promoted by traces of metals; therefore it is advisable to use distilled water or to add sufficient sequestering agent to remove the metal traces. Only the best grades of refined vegetable oils should be employed, since the possibility exists of precipitating phosphoproteins and gummy material upon contact with water or alcohol. The two phases should be mixed intimately, and then allowed to stand with only intermittent stirring and finally filtering to give a crystal clear product. Some suggested preparations can be made from Formulas 73 to 80.

Formulas for Two-Layer Lotions

	73	74	75	76	77	78	79	80
Mineral oil	—	50%	32%	—	80%	28%	—	—
Castor oil	—	16	—	—	2	—	3%	—
Olive oil	8%	—	—	10%	—	28	—	—
Deodorized kerosene	—	—	—	—	—	—	—	5%
Sesame oil	—	—	10	32	—	—	—	—
Almond oil	—	—	—	—	—	—	40	5
Ethyl alcohol	45	—	58	58	18	44	57	—
Water	47	34	—	—	—	—	—	90
Perfume and color	q.s.	q.s.	q.s.	q.s.	q.s.	q.s.	q.s.	q.s.

The two-layer hairdressings lend themselves to the incorporation of any of the materials which have been used to give a "tonic" effect, be they soluble in oil, alcohol, or water. A number of such products are found on the market. Very little work has been done on this type of hairdressing, since the demand is not too great. It is possible to use the fatty alcohols and fatty esters to produce more emollient effects as well as sheen. The duration of the temporary emulsion should certainly be prolonged by the judicious use of the newer emulsifiers. Rider and Gershon patented such an improvement (55). The product in question has an 8% oil phase containing an oil-soluble emulsifier and an aqueous alcohol phase with a water-soluble emulsifier. Slight shaking will give a good temporary emulsion which is easily applied to the hair.

Other patents and articles of interest on the two-layer hairdressings are listed in the bibliography (56–58).

Gum-Base Hairdressings

A hairdressing almost approaching a wave set in its action can be based on any of the available gums. Such preparations do not impart much sheen to the hair, but they hold unruly strands in place for long periods of time; they are excellent for training the hair of children. Light-colored and gray hair is often darkened by the application of oils. For that reason, mucilaginous hair creams are often preferred, although they do not add much luster to the hair (10).

A variety of natural and synthetic gums is available. Gums tragacanth and karaya are used most often, although quince seed mucilage is preferred by some. Use has also been made of the pectins, Irish moss, gum arabic, gelatin, flax seed mucilage, and water-soluble shellac. The greatest drawback to these natural materials is the tremendous variation in quality and properties, from source to source and from year to year.

Hairdressings based on gums vary from 0.5 to about 2% in gum content, depending on the type of gum and, of course, on the viscosity desired. If increased to 6 to 8%, jar or tube products known as "gominas" result. Upon aging, the products sometimes change in viscosity, usually thinning out. This may be due either to microbial decomposition or to an enzymatic reaction. It is imperative to add a good fungicide or germicide. It is often necessary to destroy extracted enzymes by heat in order to achieve a product with the desired shelf life.

Gum tragacanth, the most popular of the gums, is available in ribbon and powdered form. Dispersing it in water is a slow and often a tedious task. Air bubbles are invariably trapped, as agglomerates of powder are superficially wetted by the water, thereby forming a gelatinous capsule about the air bubble. Even prolonged stirring does not remove them. It is best first to wet the powder thoroughly with some inert water-soluble solvent, such as alcohol or glycerol, thus expelling all the air. When this is added to water, a more rapid dispersion without occluded air results. The flake form will disperse more slowly, sometimes requiring up to a week with frequent stirring.

Karaya gum, at the same concentration, gives a much less viscous mucilage than tragacanth, but does have the advantage that it disperses more rapidly in water, giving a whiter product. Again, it is advantageous to wet the powder first, with alcohol, for example. Karaya sometimes has a characteristic odor of acetic acid which may be removed by adding borax or other mild alkali.

Arbitrarily dispersing a given weight of the gum, or extracting a prescribed weight of quince or flax seed, will give erratic results. Each shipment should be standardized for viscosity before use, if final specifications are to be met.

This variation in viscosity can be largely overcome if one uses the newer modified gums, of which the alginates are the best example (59). Sodium, ammonium, and triethanolamine alginates are available in powder form. These are produced synthetically and are available with uniform properties, such as swelling rate and resultant viscosity. The polyvalent metals all have greater gelling power and are prepared *in situ* by adding calcium citrate or another calcium salt to the solution. The viscosity of the final product can be carefully controlled by the amount of calcium salt added. For a given calcium ion concentration, the viscosity also increases with decreasing pH. Distilled or deionized water should always be used with the alginates, since tap water can cause great variations in viscosity because of the presence of polyvalent metals.

When applied to the hair, the products based on the gums will gradually dry out, leaving a continuous and more or less flexible film, which holds the hair in place. Drying can be accelerated by adding some alcohol. Too great a proportion of alcohol will precipitate the gums, however; it is usually not possible to go above 10%.

Polyvinylpyrrolidone (PVP) is alcohol- as well as water-soluble and will dry to a continuous film. It may be plasticized with glycols, sorbitol, the water-soluble lanolin derivatives, or a copolymer of methyl vinyl ether and maleic anhydride, to give quick-drying, nontacky products. These have been especially well adapted to the aerosol package.

Methyl cellulose and sodium carboxymethyl cellulose both afford films on drying. They flake considerably, however, and most attempts to plasticize them have failed thus far. These materials are therefore not recommended for hair dressings.

Polyvinyl alcohols, polyethylene glycols (such as the Carbowaxes), soluble starches, and abietic acid esters are other basic ingredients which can be and have been used to prepare mucilaginous hair dressings.

Many preparations form dull or brittle films which crumble and flake off upon brushing. Both of these shortcomings can be overcome to some extent by using plasticizers. Castor oil, mineral oil, and some of the polyols have been used. Some of the newer nonionics are compatible with mucilaginous materials and should lend themselves for this purpose. The water-soluble lanolin derivatives, for example, afford excellent plasticity as well as emollience and should be considered. To keep the film flexible, it is necessary to have moisture present; the inclusion of a humectant is therefore indicated.

Perfumes for gum-base hairdressings should be chosen with care, since some of them will cause the product to discolor with age.

In Formulas 81 to 86 several preparations made with gums are presented. These are excellent products in themselves; they will serve also as a starting point for further experimentation with newer plasticizers, emollients, and gloss-producing agents.

Formulas for Gum-Base Hairdressings

	81	82	83
Gum karaya	—	—	2%
Gum tragacanth, powdered	1.2%	1%	—
Ethyl alcohol	15.0	6	5
Glycerol	2.0	1	—
Castor oil	—	2	—
Water	81.8	90	93
Preservative	q.s.	q.s.	q.s.
Perfume and color	q.s.	q.s.	q.s.

Procedure: First wet the gum with alcohol and stir slowly to expel the air. Add the glycerol, preservative, castor oil, and perfume, and add all of the water at once. Continue to stir until all of the gum is dispersed uniformly. Filter or strain and allow to reach its maximum viscosity upon standing for a few hr. The addition of castor oil aids in plasticizing the gum film and in preventing dullness.

Formula 84

Gum tragacanth, powdered	1.0%
Isopropyl alcohol	2.0
Glycerol	4.0
Mineral oil, heavy	1.8
Water	91.0
Formalin	0.2
Perfume and color	q.s.

Procedure: Mix the powdered tragacanth with the alcohol, add the glycerol, perfume, and the mineral oil, and then all of the water. Stir until the dispersion is complete and allow to stand for several hr to thicken up before straining. The appearance of the product can be improved by running it through a homogenizer.

Formulas for Gum-Base Hairdressings

	85	86
Sodium alginate	1.25%	1.5%
Glycerol	2.50	3.0
Calcium citrate	0.10	0.3
Water, distilled	96.15	86.2
Tincture of benzoin	—	4.0
Balsam Peru, 25%	—	5.0
Preservative	q.s.	q.s.
Color and perfume	q.s.	q.s.

Procedure: Add the sodium alginate to half of the water, then add the glycerol and perfume, and stir well to dissolve. In the meantime, dissolve the calcium citrate in the remainder of the water. When the alginate solution is smooth, pour in the calcium solution. Agitate well and allow to "body up" by standing a few hr. The glycerol can be replaced by alcohol, if desired. The viscosity can be increased by adding more calcium citrate or citric acid. Resins have been added to Formula 86 to modify the film and to give a more opaque product. Add the tinctures directly to the alginate solution.

A more creamlike product can be prepared by adding mineral oil, which is emulsified by the gum. More sheen is apparent on the hair, but the film strength will be decreased by the admixture of mineral oil. Such a product is described in Formula 84.

Two new resins, Carbopol and EMA, have found increasing use in the formulation of clear gumlike preparations. The former is a high-molecular-weight polyvinyl carboxylic acid formed by cross-linking acrylic acid; the latter, a lower olefin maleic anhydride-vinyl crotonate copolymer. They have great uniformity of viscosity and are easier to disperse than the natural gums, lending themselves to clear viscous preparations for tube or jar. Neutralization is necessary for gel formation once the polymers are dispersed in water. Although sodium or potassium hydroxide may be used, various amines are preferred—triethanolamine, diisopropanolamine, and 2-amino-2-methyl-propanol being the main ones. The degree of neutralization determines the viscosity and stability of the gel. Both materials will tolerate large amounts of alcohol, unlike the natural gums; hence many medicaments and additives can readily be incorporated for their emollient, lusterizing, plasticizing, or conditioning action. Much greater grooming and gloss are possible in Carbopol-based formulations because the resin tolerates additives without crumbling. Tackiness during drying can be overcome to a great extent by combining several resins, such as Carbopol, PVP, PVP/VA, and EMA, and by adding plasticizers such as ethoxylated alcohols, lanolins and lanolin alcohols, as well as fatty acid polyglycol esters.

Several grades of Carbopol are available and the best one for the purpose must be determined. On prolonged storage, the viscosity sometimes drops markedly, and this has been attributed to the action of metals and ultraviolet light. It is wise therefore to incorporate a sequestrant and an UV absorber to minimize the viscosity decrease. Although thorough agitation is necessary to blend all ingredients intimately and to promote uniform gelation during neutralization, care must be taken not to incorporate unwanted air, since the air bubbles will remain trapped for the life of the product. For some products this is desired, because it adds life and sparkle to the preparation.

Formulas 87 to 90 show how Carbopol and EMA can be used to prepare clear gellike hairdressings.

Formulas for Gum-Base Hairdressings

	87	88	89	90
Carbopol 934	1.0%	—	—	—
Carbopol 940	—	2.0%	0.4 parts	—
EMA DX840-91	—	—	—	1.5%
PVP K30	—	—	2.5	—
Coco fatty acid diethanolamide	3.0	—	—	—
Ethyl alcohol, 95%	54.0	—	5.0	55.0
Triethanolamine	—	3.0	—	—
Ethoxylated (4) fatty acid polyglycol ester	—	10.0	—	—
Water	42.0	85.0	89.6	34.6
Diisopropylamine	—	—	0.3	—
UV absorber	—	—	0.1	—
Ethoxylated (21) oleyl alcohol	—	—	0.5	—
2-Amino-2-methyl-1-propanol	—	—	—	0.9
Polyethylene glycol 600	—	—	—	8.0
Color and perfume	q.s.	q.s.	q.s.	q.s.

In the bibliography the author refers the reader to several articles and patents covering this type of preparation (60–71).

Oil-in-Water Emulsions

Although known in Europe for a number of years, oil-in-water emulsions for hairdressings were introduced in America only during World War II (72). Scarcity and allocation of alcohol and oils at that time forced manufacturers to try other materials. Emulsions proved a way of preparing an esthetically acceptable product with the then available materials. The acceptance of these hairdressings was immediate, for a number of reasons other than the novelty. Emulsions present a smooth, attractive, creamlike product. Good pouring and spreading consistency can be obtained, resulting in a thin, uniform film of oils on the hair. The greasy feel is greatly reduced, and the residue on the hands can be easily removed. Because of their oil content, emulsion hairdressings may still stain, but the stains are usually more readily removed, since the emulsifiers present will aid in reemulsifying the oils. By the same token, the hairdressing can be readily removed by shampooing. From the manufacturer's point of view, too, these products offer the advantage of a reduction in cost by replacing a portion of the raw materials with water.

For good shelf life, emulsions have to be well balanced and stable. The question has been raised whether they are "too stable," in the sense that the inner phase may not be reaching the hair. There may be some validity to this question when water-in-oil emulsions are involved, but certainly not in the case of oil-in-water emulsions. Owing to evaporation of water and absorption of it by the hair, the emulsion is soon broken. A continuous film of residual oils and fats is then present to protect the hair shaft, to help seal in the absorbed moisture, and to afford grooming and luster.

Since, in an emulsion, the interfacial area between oil and water is increased considerably, more rapid deterioration of vegetable oils and fatty materials is possible. Most of the prooxidants are water-soluble, and rancidity can develop much faster. The risk of bacterial and mold attack is also enhanced, and every cosmetic chemist soon realizes that he is dealing with a perishable product.

Efficient preservatives and antioxidants are necessary in the formulation of emulsions. Many organisms can develop resistance to a given germicide, or can consume or alter it, and after an interval of apparent antisepsis will suddenly contaminate an entire plant.

Since each emulsion is different, it therefore presents a different culture medium; and since the sources of contamination vary tremendously, no blanket statement can be made as to which preservative would be most effective. This agent often must be found by trial and error. It is essential that the preservative be water-soluble, and if also oil-soluble, that it establish an equilibrium so that some of it will be present in the aqueous phase.

To avoid bacterial contamination of emulsions, cleanliness during manufacture, storage, and filling cannot be overemphasized. Periodic bacteriological testing will catch any upswing in bacterial count and allow the necessary preventive measures to be taken. It might be desirable to use two equally effective, mutually compatible preservatives, alternating their use periodically to prevent the build-up of resistant strains.

All properties mentioned previously as being important to the formulation of a good hairdressing naturally apply to oil-in-water emulsions. A fine emulsion must be cosmetically elegant in itself and should supply the gloss-producing fats and oils in an easily applied, nongreasy manner. Beeswax or other waxes will enhance the gloss; however, one must not use too high a content of solids, since these tend to leave a white deposit on the hair. Furthermore, using a number of ingredients with a wide difference in melting points will often result in unstable emulsions, particularly upon freezing (73).

Uniform wetting of the hair usually is obtained with oil-in-water emulsions, since the emulsifiers lower the surface tension of the aqueous phase. Even when viscous oils are used, spreading is easy since the particle size is small and lubrication is therefore good.

A degree of hair conditioning is obtained, as described above, by furnishing water in an absorbable form. Further conditioning can be achieved by adding to the oils any of the emollients mentioned so far, namely, lanolin and its many derivatives, fatty alcohols, fatty esters, ethoxylated derivatives, silicone oils, polyglycerols.

Proper grooming is arrived at by the careful choice of materials for the oily phase. Mineral oil of light or medium viscosity is used most often. Petrolatum and various waxes are added to increase the body of the emulsion and to enhance the grooming effect. Vegetable oils can be added, or even used to replace the mineral oil, in order to alter the feel of the residual oils.

Gums have been added to increase the fixative properties on the hair, and to control the viscosity and stability of the emulsion. Synthetic materials, such as magnesium aluminum silicate or polyvinylpyrrolidone, have been used, and Carbopol is one of the most useful ingredients for this purpose.

The main factors to be considered in evolving a formula for an emulsified hairdressing are (a) selection of an appropriate emulsifying agent, (b) proper balance between the oil and water phases, and (c) correct viscosity of both phases.

Thin lotions with a low oil content tend to foam when applied to the hair; they leave a white film because of the occluded air bubbles. White deposits are also caused by the inclusion of too large an amount of stearic acid or of other solids. Creaming of an emulsion, as opposed to separation owing to failure of the emulsifiers, can often be corrected by balancing the density and viscosity of the two phases. Thus a different grade of mineral oil may overcome some of the creaming.

Greater stability is usually conferred on an emulsion by adding a small amount of a so-called antagonistic emulsifier. Thus, to stabilize an oil-in-water emulsion, it is often helpful to add a minor amount of a water-in-oil emulsifier, such as lanolin, an absorption base, or a short-chain ethoxylated fatty alcohol. A combination of this sort is used frequently when it is desired to prevent an emulsion from breaking suddenly during application. The watery feel which accompanies the application of many lotions, as they are worked into the hair and scalp, can be overcome to some extent by using an antagonistic emulsifier. As evaporation occurs, the oil-in-water emulsion can invert to a multiphase one rather than separate. The smooth feel of the emulsion is thus retained to the end.

The simplest formula consists of oil and water, with a sufficient quantity of emulsifier to give a stable emulsion. The number and variety of emulsifiers are growing every day, and any and all can be used in hairdressings. It would be impossible to give a detailed account of all possible emulsifiers and their application, but the major ones will be considered.

The oldest known emulsifiers, still widely used because of their good stability and ease of use, are the soaps.

Triethanolamine stearate is the preferred soap, although other fatty acid salts can be and have been used. The sodium salt (often prepared with borax), the potassium salt, and such amino derivatives as 2-amino-2-methyl-1,3-propanediol are all being used. Morpholine has the "ideal" property of volatilizing from the surface, leaving an insoluble film which cannot re-emulsify by itself; yet it should never be used in hair preparations, since it is irritating and toxic.

Stearic acid, triple pressed, is used most often to produce the soap by direct neutralization, although other acids can also be used. Many oil-in-water emulsions will become more viscous when frozen or subjected to low temperatures for any length of time, often to the point where they will no longer flow. This can sometimes be remedied by adding a small amount of oleic acid or of another unsaturated fatty acid to the stearic acid.

Formula 91 will give a heavy product. Thinner lotions can be prepared by altering the oil-to-water ratio, as shown in Formula 92. It is usually better to prepare the soap *in situ*, as was done in the case of Formula 91. Greater

Formula 91

Mineral oil	44.0%
Stearic acid	6.0
Water	48.5
Triethanolamine	1.5
Perfume and color	q.s.

Procedure: Heat the mineral oil and stearic acid to about 70°C until solution is complete. Heat the water and triethanolamine about 2°C higher, and slowly run into the heated oil, under moderate agitation. It is the usual practice to have the phase which is being added a few degrees warmer than the other to compensate for the loss of heat while pouring or pumping. Continue agitation while the emulsion is cooled, and add the perfume at around 45 to 50°C. It should not be necessary to cool below 40°C, since the emulsion should be stable at that temperature.

Formula 92

Mineral oil	25%
Triethanolamine stearate	7
Water	65
Beeswax	3
Color and perfume	q.s.

Procedure: Dissolve the triethanolamine stearate in the water and heat to 70°C. Heat the mineral oil and beeswax to 72°C and slowly run into the aqueous phase. Continue agitation until cool, adding the perfume at 45 to 50°C. It may be found necessary to pass this lotion (as well as Formula 91) through a colloid mill or a homogenizer to ensure stability.

activity seems apparent and a finer emulsion results, owing to the continuous formation of the emulsifier at the oil–water interface where it is needed. The degree and amount of agitation required to form the emulsion is also decreased when the soap is formed during the mixing.

The differences between Formulas 91 and 92 serve to demonstrate another fact: The greater the concentration of the dispersed phase, the greater is the viscosity of the resultant emulsion; hence the ease of thinning an oil-in-water emulsion by adding water (7).

Formula 93, with good grooming qualities and an excellent stability, has been suggested by Jannaway (74). The addition of 0.2 % of glyceryl monostearate confers added stability and smoothness of texture. Formula 94 is a rather viscous hairdressing with good gloss properties, because of the increased wax content.

A great variety of materials can be added to the rather fundamental Formulas 93 to 95. These include glycerol, propylene glycol or other polyol as a humectant to protect the emulsion from drying out upon exposure to air (and to act as an antifreeze), and gums to increase the stability of the emulsion and to add fixative properties (75). Emollience and manageability can also be achieved with vegetable oils, fatty acids and esters, alkanolamides, lanolin esters, and lanolin oils. Formula 96 is a preparation to which such emollients and humectants have been added. Soaps other than those of triethanolamine can be used either alone or in conjunction with the latter. Formula 96 also includes this variation.

In Formula 96 the mineral oil is completely replaced by Carbowax 1500, and the resultant emulsion is much less greasy. Its grooming properties are good. The operating procedure is similar to that applied in the previous examples; the heated aqueous phase is added to the heated oils.

A combination of emulsifiers is commonly used and often gives better results than when either one is employed singly, as demonstrated in Formula 97.

The partially sulfated fatty alcohols are efficient emulsifiers in their own right, and can be used as such to form the emulsion, or the fatty alcohol can be

Formulas for Emulsified Hairdressings

	93	94	95
Mineral oil	43.0%	40.0%	5%
Beeswax, bleached	3.0	1.5	2
Stearic acid	2.4	3.5	14
Glyceryl monostearate	0.2	—	—
Triethanolamine	1.2	1.5	1
Lanolin	—	—	1
Cetyl oleate	—	—	10
Isopropyl myristate	—	—	3
Carnauba wax	—	1.0	—
Stearamide	—	1.0	—
Water	50.2	51.5	59
Sorbitol (70%)	—	—	5
Perfume, color and preservative	q.s.	q.s.	q.s.

Procedure for Formula 93: Add the heated aqueous triethanolamine solution to the melted oils and fats at 75 to 80°C, stirring until cool, and perfuming below 45°C.

Procedure for Formulas 94 and 95: Pour the melted oils into the aqueous triethanolamine at 75 to 80°C, with vigorous agitation. Upon cooling, agitate more slowly.

Formula 96

Carbowax 1500	12.0%
Propylene glycol	3.0
Carbitol	5.0
Stearic acid	5.0
Lanolin	1.0
Triethanolamine	0.2
Potassium hydroxide, 85%	0.1
Sodium alginate, 2%	4.0
Water	69.7
Perfume and preservative	q.s.

Formula 97

Stearic acid	2%
Cetyl and stearyl alcohols, 10% sulfated	2
White paraffin, soft	8
Cocoa butter	4
Mineral oil	50
Water	33
Borax	1
Color and perfume	q.s.

Procedure: Heat the borax and water and add to the hot fats at 75 to 80°C, with moderate stirring. Continue slow stirring until the emulsion is cool, adding the perfume at 45°C.

dissolved in the oil phase and the sulfated alcohol in the water phase to cause emulsification during the mixing process.

Such emulsions can be varied in viscosity by varying the amount of partially sulfated alcohol. Stearyl alcohol affords the most body, whereas oleyl alcohol gives a softer cream. A softer cream or a thinner lotion can be obtained also by using triethanolamine lauryl sulfate in place of the sodium salt (19). Although they are anionic emulsifiers, partially sulfated fatty alcohols can be used at a slightly acid pH; the trend at present seems to be toward neutral and acid formulations to protect the acid mantle of the scalp. Citric or tartaric acid may be added safely to emulsions made with the sulfated fatty alcohols (76).

Basically the preparation contains oil and water, with sufficient emulsifier to give the desired viscosity and to prepare a stable emulsion, as shown in Formula 98; a preparation with better grooming properties can be based on Formula 99.

Formula 98

Cetyl–stearyl alcohol, 10% sulfated	3%
Mineral oil	20
Water	77
Perfume and preservative	q.s.

Procedure: Add the heated water to the melt of mineral oil and sulfated alcohols at 75°C and stir the mixture until it is cool. This gives a fluid lotion which, owing to its low oil content, would be inadequate for controlling unruly hair.

Formula 99

Stearyl alcohol	5%
Mineral oil	33
Petrolatum	10
Sodium lauryl sulfate, 15% solution	52
Color and perfume	q.s.

Procedure: Melt the stearyl alcohol, mineral oil, and petrolatum together to 70°C and run this mixture into the warmed sodium lauryl sulfate solution under agitation. Add the perfume at 45°C. Cetyl alcohol or the softer myristyl alcohol may be used to replace the stearyl alcohol.

Vegetable and animal fats can be used with the partially sulfated fatty alcohols to give less greasy hair conditioners, as shown in Formulas 100 and 101. The latter gives a stiff cream which can be offered as a lanolin-enriched hair conditioner.

The polyhydric alcohol esters of the fatty acids have been widely used to prepare stable, attractive emulsions. Of these, glyceryl monostearate is no doubt the best known, and is presented in a number of different forms. The self-emulsifying (s.e.) grade is probably the most widely used, since it requires

Formulas for Hair Conditioners

	100	101
Cera emulsificans	2%	15%
Peach kernel oil	18	—
Beeswax	1	—
Castor oil	3	—
Lanolin, anhydrous	—	3
Citric or tartaric acid	—	1
Water	76	81
Perfume and preservative	q.s.	q.s.

Procedure: Dissolve the water-soluble materials in the water, heat to 75°C and pour into the hot oil phase. Stir until cool and add the perfume at 45 to 50°C.

no other surfactant to form the emulsion, in contrast to the pure material which requires an auxiliary emulsifier, such as soap, to yield a stable emulsion. A form which is stable at acid pH is also available.

Other things being equal, the viscosity of the emulsion is directly proportional to the amount of glyceryl monostearate present. Up to 3% of the emulsifier will give a lotion, while 10% will give a cream which can be packaged in tubes or wide-mouthed jars. The emulsifier has good emolliency and will tend to cut the greasiness of the oils used. It is compatible with other fats and oils, but is subject to mold attack and must be preserved. The esters of *p*-hydroxybenzoic acid have been found rather effective in this respect.

The ease of use of glyceryl monostearate is another factor in its popularity. Usually all of the ingredients, water and oils, can be mixed at once, heated until fluid, and stirred until cool. Propylene glycol monostearate can be used in place of the glyceryl ester in most formulations to give a softer cream.

The number of esters of this type is quite large, and only a few of the most useful ones will be mentioned. In the development of a new product, other esters should not be overlooked, since they may be particularly effective for specific mixtures.

Diglycol stearate and laurate are, next to glyceryl monostearate, most frequently used. Diglycol laurate, a liquid, can emulsify in the cold, provided both phases are fluid and homogeneous.

Some rather interesting and useful esters have been prepared by reacting the fatty acids with polyethylene glycols of various molecular weights. This changes the solubility of the products, and almost any hydrophilic or lipophilic characteristic can be obtained. The mono- and dipolyethylene glycol (PEG) esters of the fatty acids have become very important and afford elegant emulsions.

Several lotions and creams are suggested in Formulas 102 to 116. The lotions are usually designed for hair-grooming purposes; they are enjoying

Formulas for Hair Conditioners

	102	103
Mineral oil	30%	32%
Tegin	6	10
Water	64	50
Beeswax	—	3
Castor oil	—	5
Perfume and preservative	q.s.	q.s.

Procedure: Melt together all of the ingredients and agitate until cool. Add the perfume at about 45°C. At first, the emulsion will have a very characteristic gelled appearance, but upon cooling, this will thin out to a smooth emulsion.

an ever-growing appeal to women. The creams, packaged either in tubes or jars, are—because of their consistency—suggestive of a higher content of solids and are used as hair conditioners or treatments. Greater amounts of emollient, of penetrating material, such as the emulsifiers themselves, or of lanolin and the fatty alcohols, can be incorporated in the more viscous products.

Formula 102 is a soft cream which cannot be poured. Variations can be made by adding beeswax for luster, and vegetable oils to cut or replace the mineral oil, as shown in Formula 103.

Humectants are frequently added to prevent or to retard the drying out of a cream. Fatty alcohols and lanolin can of course be added, to afford emollient action as well as to act as auxiliary emulsifiers, and thus to produce a

Formulas for Hair Creams

	104	105	106	107	108
Glyceryl mono- stearate	12%	9.0%	13.5%	12%	3.0%
Mineral oil	2	25.0	8.5	2	8.0
Lanolin	4	—	3.5	10	—
Cetyl alcohol	—	1.5	—	—	—
Beeswax	—	1.0	1.5	—	1.5
Triethanolamine stear- ate	7	—	—	—	—
Water	75	59.0	59.5	68	78.5
Glycerol	—	4.5	4.5	3	5.0
Cholesterol esters	—	—	9.0	—	—
Spermaceti	—	—	—	5	—
Petrolatum	—	—	—	—	3.0
Stearic acid	—	—	—	—	1.0
Perfume and preserv- ative	q.s.	q.s.	q.s.	q.s.	q.s.

Formulas for Hair Creams

	109	110	111
Diglycol laurate	—	—	14%
Diglycol stearate	7%	8%	—
Mineral oil	20	—	36
Water	73	60	50
Castor oil	—	16	—
Almond oil	—	16	—
Perfume and preservative	q.s.	q.s.	q.s.

Formulas for Emulsified Hair Creams

	112	113	114
PEG 300 monostearate	—	—	10.0%
PEG 400 monostearate	6.0%	4%	—
PEG 400 monolaurate	0.5	1	—
Lanolin	1.0	1	—
Polyethylene glycol 400	2.5	—	—
Water	90.0	92	54.5
Propylene glycol	—	2	—
Beeswax	—	—	8.0
Stearic acid	—	—	8.0
Mineral oil	—	—	18.0
Triethanolamine	—	—	1.5
Perfume and preservative	q.s.	q.s.	q.s.

Procedure for Formulas 112 and 113:   Heat all of the ingredients to 70 to 75°C, stir well, and cool with moderate agitation.

Procedure for Formula 114:   Pour the triethanolamine–water solution into the heated oils at 70 to 75°C. Add the perfume at about 45°C, as usual.

treatment product as well as one for hair grooming. Formulas 104 to 108 incorporate one or more of these features.

Diglycol stearate of laurate can be employed as readily as glyceryl mono-stearate to give stable, smooth hairdressing creams, as shown in Formulas 109 to 111.

These basic formulations lend themselves to the same modifications as does self-emulsifying glyceryl monostearate, and the procedure for preparation is also the same.

The use of the polyethylene glycol (PEG) derivatives is exemplified in Formulas 112 to 114.

One can prepare stable oil-in-water emulsions, with much less than 50% water, using the fatty acid condensates of sorbitol and manitol, and their polyoxyethylene derivatives, known as the Spans and Tweens respectively. A large number of emulsifiers is available, and the choice depends on the nature of the fatty materials used, each emulsifier or combination being somewhat specific in regard to the type of oils it can best emulsify. Lotions as well as creams can be prepared with such products, as Formulas 115 and 116 illustrate. Formula 115 is a free-flowing lotion hairdressing, whereas Formula 116 represents a cream suitable for filling into tubes or jars.

Formulas for Emulsified Hairdressings

	115	116
Petrolatum	6.0%	15%
Mineral oil	37.5	10
Lanolin	3.0	20
Beeswax	12.0	12
Arlacel 83	3.0	—
Arlacel 20	1.0	—
Span 60	—	5
Tween 20	2.0	—
Tween 60	—	5
Borax	0.5	1
Water	35.0	32
Perfume and preservative	q.s.	q.s.

Procedure: Heat the oil phase plus the emulsifiers to 70°C and the aqueous phase to 72°C, and slowly add the latter, with agitation. Add the perfume at 45°C and stir continuously until cool.

Polyvinyl carboxylic acid (Carbopol), when neutralized with long-chain fatty amines such as stearyl amine, will give oil-in-water emulsions in addition to gelling the system. Excellent creams can be made, utilizing many lipophilic materials in addition to mineral oil. Formulas 117 and 118 demonstrate typical formulations.

Ethoxylated phosphate ether esters are very stable emulsifying agents resistant to hydrolysis and oxidative degeneration. When neutralized with diethanolamine, for example, either alone or in combination with other oil-in-water emulsifiers, high ratios of mineral oil can be smoothly incorporated to produce stable emulsions, as shown in Formulas 119 and 120.

A nonionic emulsifying system of good versatility, capable of giving both cream and lotions, is available in the blend of higher fatty alcohols and ethoxylated derivatives known as Emulsifying Wax. Since they are ethers, rather than esters, they have greater resistance to acid or alkaline hydrolysis and can be used over a wide pH range. Since they are nonionic, additives and medicaments of all kinds can be added without destroying the emulsion or inactivating the ingredient. Although no other emulsifiers need be added to produce stable emulsions, a combination can be made with these materials,

Formulas for Emulsified Hair Creams

	117	118
Carbopol 934	0.25%	0.5%
Sodium hydroxide, 10%	0.75	2.0
Ethomeen C25, 10%	1.25	—
Water	41.75	49.6
Propylene glycol	—	8.0
Polypropylene glycol 5000	—	2.0
Liquid lanolin alcohols	2.50	—
Acetylated lanolin	2.50	—
Ethoxylated lanolin alcohols	1.00	—
Petrolatum	25.00	5.7
Paraffin wax MP 151	2.50	—
Mineral oil 70	17.50	30.6
Ucon 50 HB 660	5.00	—
Lanolin	—	0.5
PEG 400 monococate	—	1.0
Stearyl amine	—	0.1
Perfume, preservative	q.s.	q.s.

Procedure: Disperse the Carbopol in water with high speed, heat to 50°C, then add the sodium hydroxide solution. Slowly add the "oils" heated to 60°C and blend until smooth. Cool to 45°C and add perfume.

Formulas for Emulsified Hair Creams

	119	120
Neutralized phosphate ester (Crodafos N-10)	10%	2.0%
Mineral oil 70	60	60.0
Water	30	10.2
Myristic acid	—	2.6
Triethanolamine (TEA)	—	1.2
Carbopol 940, 1.0%	—	24.0
Color and perfume	q.s.	q.s.

Procedure for Formula 119: Dissolve Crodafos N-10 neutral in the mineral oil and heat to 60°C. Add the hot water, also at 60°C with good agitation, cool to 30°C, and color and perfume.

Procedure for Formula 120: Combine oils and heat to 75°C. Add TEA and the water also at 75°C with agitation. When smooth, start cooling with slow agitation, adding the Carbopol solution at 70°C, and continue stirring cold. Add color and perfume at 45°C.

especially for lotions where low levels of emulsifier are desirable. Typical formulations are illustrated in Formulas 121 to 123.

Germicidal agents may be incorporated into emulsion hair-grooming preparations as well as into alcoholic lotions, and many of the same ingredients mentioned before are used. These are first dissolved in the water or

Formulas with Low Emulsifier Level

	121	122	123
Ethoxylated fatty alcohols	5%	5.0%	9.0%
Ethoxylated (65) lanolin	—	3.0	—
Mineral oil 70	25	24.0	—
Lanolin	—	2.0	2.0
Lanolin alcohols	—	—	1.5
Oleyl alcohol	—	—	4.5
Cetyl alcohol	—	—	5.0
Benzethonium chloride	—	0.4	—
Water	70	65.6	78.0
Color and perfume	q.s.	q.s.	q.s.

Procedure: Dissolve the ethoxylated fatty alcohols and ethoxylated lanolin in the oily phase and heat to 70 to 75°C. Add the hot water (containing the benzethonium chloride in Formula 118) at the same temperature with good agitation and stir till cold. Add color and perfume at 45°C.

oil phase, if at all possible, to ensure uniform dispersion. Solids such as the zinc salt of 1-hydroxy-2-pyridinethione (Zinc Omadine) or selenium polysulfide are added as micropulverized materials, usually thoroughly wetted out with one of the ingredients of the formulation, and then distributed evenly by passing the completed emulsion through a colloid mill.

It is usually not possible to add these active chemicals to a finished emulsion and obtain the same stability as before. Each formulation must be built around the active ingredients to ensure adequate shelf life. Great care must also be exercised to maintain the expected antimicrobial activity, because some emulsifiers can inactivate them partially or completely.

Water-in-Oil Emulsions

Water-in-oil emulsions have been used on and off as hair-grooming preparations since Eugene Rimmel first introduced his lime cream about 1864. This emulsion, as reported by Jannaway (77), consisted of almond oil and lime water, which had to be shaken well before being used. Although improvements have been constantly made, water-in-oil emulsions are more difficult to prepare; as a matter of fact, it may be said that a water-in-oil emulsion (liquid) whose stability approaches that of an oil-in-water emulsion has yet to be made. Despite the tendency of the water-in-oil emulsions to separate and to show a layer of oil or water, a number of such hairdressings have achieved worldwide popularity. In general appearance, they present a smooth, lustrous cream which gives excellent grooming and a high sheen to the hair.

The film deposited on the hair is greatly water-resistant and is therefore the ideal type of hairdressing in rainy climates and for those who swim a great deal.

As a rule, humectants as such are not required in the formulation of water-in-oil emulsions, since the external, continuous phase consists of the oils. For the same reason, however, water-in-oil emulsions have a greasier feel, although not so pronounced as that of brilliantines. They will not rinse away as oil-in-water emulsions do, but they can be removed readily from the hands or scalp by soaping or shampooing.

The water-in-oil dressings seem to be more popular with men than with women, probably because of the greater grooming action, coupled with the slightly oily feel. Thus far, the products have not lent themselves to compounding preparations of the hair-conditioning type, although it would appear possible to develop such formulations.

Several types of emulsifiers and of combinations of emulsifiers have been used to prepare the water-in-oil hairdressings. The polyvalent soaps are probably the oldest of these, although beeswax and borax have long been used.

More stable emulsions of finer particle size, and hence good luster and feel, can be prepared with absorption bases, with polyglycerol esters and with ethoxylated alcohols as primary emulsifier. Auxiliary water-in-oil emulsifiers are lanolin, lanolin alcohols, cetyl alcohol, and sorbitan sesquioleate, which can be judiciously added to the primary ones to ensure better shelf stability, to incorporate special emollients, or to modify the feel and rub-out. These will all be discussed separately. Homogenization is extremely helpful for many of these water-in-oil emulsions, especially the creams. An empirical approach is usually necessary to find the proper temperature, pressure, and speed of through-put to ensure the desired consistency and stability, and once this has been determined for a given formulation, very little leeway is possible in changing the manufacturing procedure.

Calcium oleate and stearate, prepared from lime water or from saccharated lime water containing a higher concentration of calcium ions, were the original emulsifiers used; but it has been found subsequently that the magnesium salt is much more stable. Zinc and aluminum stearates have found extensive use as water-in-oil emulsifiers, although they are not usually employed alone. A combination of water-in-oil emulsifiers has proved much more effective than any single emulsifier. The simplest preparation, a modern version of Rimmel's lime preparation, is shown in Formula 124 (19). This is not a very stable emulsion, but it represents the basis for work along these lines. Beeswax or other waxes can be added for gloss; vegetable oils can be used in place of mineral oil, and lanolin or the like can be added for emolliency as well as to increase the emulsion stability. Formulas 125 to 128 are suggested to illustrate these modifications.

Formula 124

Mineral oil	49%
Stearic acid	1
Lime water	50
Perfume and preservative	q.s.

Procedure: Melt the stearic acid in the mineral oil and add the lime water after heating to about the same temperature. Stir until cool and add the perfume at about 45°C.

Formulas for Emulsified Hair Groom

	125	126	127	128
Mineral oil	45%	—	41.7%	48.0%
Oleic acid	12	20.0%	1.0	—
Beeswax	2	1.0	1.5	2.5
Lanolin	2	0.5	—	—
Lime water	19	33.5	53.8	—
Saccharated lime water	20	5.0	—	—
Olive oil	—	40.0	—	—
Magnesium sulfate, 25%	—	—	2.0	—
Stearic acid	—	—	—	1.5
Absorption base	—	—	—	5.0
Petrolatum	—	—	—	12.5
Magnesium oleate	—	—	—	2.5
Water	—	—	—	28.0
Perfume and preservative	q.s.	q.s.	q.s.	q.s.

Procedure: Add the aqueous phase at 70 to 75°C to the heated oils, with moderate agitation, and stir until cool. In Formula 127 add the magnesium sulfate solution after the lime emulsion has been prepared, and then stir the emulsion until cool.

A water-in-oil emulsion can be built around beeswax–borax as the emulsifying system, and one such product is illustrated by Formula 129.

A combination of emulsifying agents is often effective, as shown by Formulas 130 to 132.

Absorption bases have also been used to prepare hairdressings. These bases, formulated from lanolin, lanolin alcohols, cholesterol, or cholesterol esters combined with mineral oil, petrolatum, and waxes, can retain a large volume of water, forming water-in-oil emulsions; usually creams rather

Formula 129

Beeswax, white	2.50%
Mineral oil	62.55
Water	34.80
Borax	0.15
Perfume	q.s.

Procedure: Melt the wax in half of the mineral oil and add the hot borax water. Then add the remainder of the oil. Homogenization is of great value with this formula.

Formulas for Emulsified Hair-Grooming Preparation

	130	131	132
Mineral oil	40.0%	27.0%	23.4%
Petrolatum	19.4	6.0	9.0
Beeswax, white	17.6	1.8	2.0
Oleic acid	0.4	—	—
Absorption base	0.8	—	—
Cetyl alcohol	—	—	2.0
Lanolin alcohols	—	0.2	2.0
Lanolin	—	—	25.0
Paraffin wax	—	—	3.0
Borax	0.4	0.1	0.1
Magnesium sulfate	1.0	—	—
Sodium hydroxide	0.4	—	—
Water	20.0	65.0	33.5
Perfume and preservatives	q.s.	q.s.	q.s.

Procedure: Add the aqueous phase at 70 to 75°C to the heated oils with good agitation, and stir cool. Add perfume at 45°C and pass through a homogenizer.

than lotions are produced. These absorption bases give water-in-oil emulsions whose stability is much greater than that of the lime creams. The tendency to bleed oil or water on standing is greatly reduced, and the texture of the finished cream is smooth. Good grooming as well as a high gloss is obtained with these creams.

The basic formula is simply a mixture of the absorption base and water in the proper ratio to ensure emulsion stability and to give the desired cream consistency. A much thinner emulsion, with higher gloss, can be made by the incorporation of beeswax and mineral oil. These two approaches are illustrated in Formulas 133 and 134.

One of the newer emulsifiers for water-in-oil emulsions, and one which can be used in conjunction with any of the above described systems, is Arlacel 83 (sorbitan sesquioleate). Its use permits the preparation of stable lotions of low viscosity with 40 to 50% water content. Formulas 135 to 138 are given to indicate the scope of its use with these other emulsifiers.

Formulas for High-Gloss Hair Emulsion

	133	134
Absorption base	63%	6.00%
Beeswax	—	3.00
Mineral oil	—	60.25
Glycerol	—	1.50
Triethanolamine	1	0.25
Water	36	29.00
Perfume	q.s.	q.s.

Procedure: Melt the oil phase and heat to 70 to 75°C and slowly pour the hot triethanolamine solution into the oil phase, with good agitation. A polyphase emulsion is formed first, but upon continued agitation and cooling, this changes substantially to a water-in-oil emulsion, and no further water can be added. Homogenization will increase the stability of the emulsion.

Formulas for Low-Viscosity Hair Emulsions

	135	136	137	138
Mineral oil	45.00%	37.5%	36.5%	33.0%
Petrolatum	8.00	7.5	8.0	—
Beeswax	3.00	2.0	2.0	3.5
Absorption base	7.00	—	—	—
Arlacel 83	4.00	3.0	3.0	4.0
Lanolin	—	3.0	0.5	—
Lanolin esters	—	—	—	10.0
Ceralan	—	—	—	5.0
Zinc stearate	—	1.0	—	—
Water	32.25	45.5	49.7	44.0
Borax	0.75	0.5	0.1	0.5
Magnesium sulfate	—	—	0.2	—
Perfume	q.s.	q.s.	q.s.	q.s.

Procedure: Add the aqueous phase slowly to the oil phase at about 75°C, with moderate agitation. Continue to stir while cooling, and add the perfume below 45°C. Homogenization, although not necessary, will greatly add to the shelf stability of these water-in-oil emulsions. Care must be taken to carry this out at a given constant temperature, or temperature range, lest the viscosity vary from batch to batch.

In these formulas, the sorbitan sesquioleate is used with the beeswax–borax system, to which an absorption base or a polyvalent soap is sometimes added. The judicious use of an oil-in-water emulsifier, a so-called antagonistic emulsifier, is often said to increase emulsion stability, although the writer has found in many cases that such additions have markedly decreased it.

Two other emulsifier systems currently finding more use are the polyglycerol esters like decaglycerol decaoleate, and a blend of ethoxylated fatty alcohol and ethoxylated lanolin alcohol such as the 3-M adduct of oleyl alcohol and the 5-M adduct of refined lanolin alcohols. The first-mentioned polyglycerol esters exhibit excellent emollient and conditioning properties and therefore contribute greatly to the lipophilic phase. Typical formulas are Formulas 139 and 140.

Formulas for Emollient Hair Emulsions

	139	*140*
Mineral oil 70	35.50 parts	12%
Petrolatum, white	5.00	—
Decaglycerol decaoleate	3.00	—
Decaglycerol decalinoleate	3.00	—
Liquid lanolin alcohols	5.00	—
Beeswax	2.00	—
Ethoxylated (3) oleyl alcohol	—	3
Ethoxylated (5) lanolin alcohol	—	1
Glycerol	—	4
Borax	0.50	—
Veegum HV	0.25	—
Water	47.75	80
Perfume and preservative	q.s.	q.s.

Procedure for Formula 139: Add the aqueous phase at 70°C to the heated oils, cool to 45°C, add perfume, and homogenize.

Procedure for Formula 140: Blend the oils, add the aqueous glycerol solution at room temperature or slightly warmed with vigorous agitation. When smooth, add perfume and blend.

The literature of emulsified hair dressings is rather vast; several patents and articles of special interest are cited in the bibliography (71–95).

Scarcely ten years old, the transparent microemulsion gels, with great consumer appeal, have created a whole new product category. Properly formulated hairdressings have a characteristic "ring" to them when struck, are easy to apply because they spread well without tack, white streaks, or mess, and leave the hair lustrous and well groomed. These crystal-clear gels are usually packaged in metal tubes and are extruded as rigid, nonflowing ribbons.

The ethoxylated lanolin alcohols were first employed to formulate micro-emulsions. However, large amounts of the surfactant were required to produce a stable system with mineral oil, and the gels were tacky, defatting to the skin, and often very irritating. Only low levels of water could be incorporated, and the ratio of surfactant to mineral oil was often 3:1, thus making this a very expensive product. Typical formulas are given in Formulas 141 to 144. Note that isopropyl myristate can replace mineral oil and often affords a softer, more stable gel.

Formulas for Gels

	141	142	143	144
Ethoxylated (20) lanolin alcohol	60%	—	—	—
Mineral oil	20	—	—	14%
Ethoxylated (15) lanolin alcohol	—	35.0%	—	30
Ethoxylated (24) cholesterol	—	—	15%	—
Lanolin alcohols	—	6.5	—	5
Isopropyl myristate	—	28.5	25	7
Oleyl alcohol	—	—	5	—
Ethoxylated (25) hydrogenated castor oil	—	—	15	—
Propylene glycol	—	—	5	—
Polyethylene glycol 200	—	—	—	5
Water	20	30.0	35	39
Perfume, preservative, color	q.s.	q.s.	q.s.	q.s.

Procedure: The hot aqueous phase is added with good agitation to the hot blend of oils and emulsifier, between 60 and 70°C. On cooling, the perfume is added and the product packaged.

The polyethoxylated ethers of oleyl alcohol were subsequently found to permit markedly lower ratios of emulsifier to oil, and lower total lipophilic phase. Gels thus produced were less tacky and defatting. The phosphate esters of the above ethers, when combined with either alkylolamides or the ethoxylated oleyl alcohols, were found to give highly desirable gels with higher water levels and greater stability. To this emulsifier list can be added the polyethylene glycol esters of fatty acids like lauric and oleic. All of these emulsifiers can be used either alone or in various combinations depending on the other ingredients to be included in the formulation. Blends are usually best, to ensure clarity and stability. Detailed information is covered in U.S. patents 3,101,300, 3,101,301, and 3,175,949 issued to B. Siegal *et al.*, and in British patent 1,002,466 (96–99).

Formulas illustrative of each of the aforementioned emulsifiers and blends are given in Formulas 145 to 153.

The transparent gels often become hazy on aging, liquefy with time, or markedly change viscosity at very low or very high temperatures. Couplers, such as 2-ethyl-1,3-hexanediol, various polyhydric alcohols, and polyethylene glycols can enhance the clarity and stability of the gels. Flexibility in formulation is possible by varying the length of the polyethylene oxide chain of the ether or phosphate ester, in the nature of the fatty alcohol, in the kind

Formulas For Gels

	145	146	147
Ethoxylated (20) lanolin alcohol	8.6%	—	—
Ethoxylated (15) lanolin alcohol	2.1	—	—
Lanolin alcohols	2.7	—	—
Mineral oil 70	20.0	—	13.7%
Lauric myristic diethanolamide	6.7	6%	—
Isopropyl myristate	—	22	—
Ethoxylated (10) oleyl alcohol	—	5	15.5
Ethoxylated (15) cetyl stearyl alcohol	—	5	—
Decaglycerol tetraoleate	—	4	—
Ethoxylated (25) hydrogenated castor oil	—	—	15.5
Propylene glycol	—	—	8.6
Sorbitol solution, 70%	—	—	6.9
Water	59.9	58	39.8
Color, perfume, preservative	q.s.	q.s.	q.s.

Formulas for Gels

	148	149
Mineral oil	20.0%	—
Lanolin alcohols	2.7	—
Lauric myristic diethanolamide	3.0	6%
TEA bis(lauryl tetraethylene glycol) monoether phosphate	9.0	—
TEA bis(oleyl octoglycol ether) phosphate	—	10
Ethoxylated (3) lauryl alcohol	—	4
Ethoxylated (23) lauryl alcohol	—	6
Hexadecyl alcohol	—	15
Water	65.3	59
Color, perfume, preservative	q.s.	q.s.

of alkylolamide, the choice of coupling agent, and the type of lipophile used in addition to, or in place of, mineral oil.

The total emulsifier level can be varied between 15 and 30% with a major part of it an ethylene oxide adduct or mixture of two or more. Increasing the mols of ethylene oxide in the ether chain will usually afford a stiffer gel, as will increasing the ratio of long-chain ethers to short-chain ethers.

Although the formulation of crystal-clear, stable gels is largely an empirical matter, considerable theoretical work has been done to better understand the microemulsion, and the work of L. I. Osipow (107) and P. Becher (105) should be consulted. Transparency is due to the extremely small particle size of the dispersed phase, which is less than one-fourth the wavelength of light. Greatest emulsion stability is obtained when the mineral oil, or other lipophile, has a shorter hydrocarbon chain length than the emulsifier, and

Formulas for Gels

	150	151	152	153
Ethoxylated (10) oleyl ether phosphate diethanolamide	4%	4%	—	—
Ethoxylated (3) oleyl ether phosphate diethanolamide	—	3	6.8%	6%
Lanolin fatty acids	—	4	—	—
Ethoxylated (10) lanolin	—	—	—	6
Ethoxylated (3) oleyl alcohol	4	—	4.1	—
Ethoxylated (10) oleyl alcohol	12	9	—	—
Polyethylene glycol 200	10	—	—	—
Glycerol	—	10	—	15
Mineral oil 70	20	20	13.6	15
Ethoxylated (5) oleyl alcohol	—	—	2.7	8
2-Ethyl-1,3-hexanediol	—	—	3.4	—
Propylene glycol	—	—	1.4	—
Water	50	50	68.0	50
Color, perfume, preservative	q.s.	q.s.	q.s.	q.s.

Procedure for Formulas 145 to 153: The hot aqueous phase is added to the hot blend of oils and emulsifiers, between 60 and 70°C. On cooling, the perfume and color are added, and the product packaged.

when a large amount of emulsifier is present to form the proper interfacial film. Light-viscosity mineral oil is preferred over the more viscous grades for product clarity, because of the chain length of the hydrocarbons.

Preservatives are required for these oil-in-water emulsions, but because the ethylene oxide adducts often inactivate them, every formulation must be carefully tested for adequacy of preservation.

The manufacturing of transparent gels is not difficult if a properly balanced formula has been developed. Normal agitation is all that is required, since the emulsion readily forms when hot water is added slowly to the oil and emulsifier phase. The mass is often quite cloudy while hot, but gradually clears and becomes brilliant on cooling.

The rapidity of gel formation depends on the emulsifiers and ingredients used; sometimes a day or more is required before a stable state is obtained. This permits easy filling of the tubes, with subsequent gelling right in the package. Viscosities up to 1,500,000 centipoise at 20°C are optimum for proper application to the hair, though of course this property will depend on the nature of the components.

The fast-growing literature in this category is exemplified in references 96 to 108.

Aerosol Hairdressings

Aerosol hair sprays were originally designed for women, primarily for holding the coiffure in place after styling. The aerosol spray is efficient in uniformly dispensing a film-forming resin on the outside of the hair mass to form a confining shell without disturbing the hairset. It should not be brittle or easily relaxed by moisture, and should be readily combed and shampooed out. Aerosol sprays have also been used for grooming women's hair and in setting curls, because of the quick-drying nature of the spray. Men too have recently found hair sprays to be excellent for holding heavy, long, or unruly hair in place, much as do the gum preparations discussed earlier. The advantages of aerosol spray-dispensing are many. The product is applied directly and need only be combed or brushed through to groom the hair, thereby avoiding soiled hands. The product can also be applied after grooming to help hold the set in place. The danger of tipping over a bottle of hairdressing is eliminated, as is the dripping which often stains the surface on which the bottle rests.

Two types of sprays are popularly used, the film former and an oily type which serves more as a dressing and to give gloss. These are illustrated in Formulas 154 to 159. For a comprehensive discussion on aerosol hair products, see the chapter on aerosols; we merely wish to point out here that products of this type are also used for hair grooming.

Formulas for Aerosol Hairdressings

	154	155	156
Dicrylan 325-50 (acrylic copolymer resin)	1.60%	—	—
Isopropyl myristate	0.07	—	—
Ethyl alcohol	38.28	24.5 parts	28.5 parts
Polyvinyl pyrrolidone–vinyl acetate (50%)	—	3.0	—
Ethoxylated (65) lanolin	—	0.5	0.5
Isopropyl lanolate	0.05	1.5	—
Ucon 50 HB 660	—	1.5	—
Propellant 12	24.00	24.5	—
Propellant 11	36.00	45.5	42.0
Propellant 114	—	—	28.0
dl-Panthenol	—	—	1.0
Zelec NK (fatty alcohol phosphate salt)	—	—	0.1
Perfume	q.s.	q.s.	q.s.

Procedure: All ingredients, including the fragrance, are dissolved in the alcohol, with gentle warming if necessary. This so called concentrate is filtered bright and filled into aerosol cans, and the propellants added.

Formula 157

Decaglycerol monolaurate	2.0%
Polypropylene (200) monooleate	3.0
Ethoxylated (10) lanolin alcohols	1.0
Propylene glycol	2.0
Ethyl alcohol, anhydrous	39.5
Protein polypeptide (20% alcoholic)	1.2
Isopropyl myristate	1.3
Propellant 11	15.0
Propellant 12	35.0
Perfume	q.s.

Mineral oils should never be used in aerosol sprays, since the atomized hydrocarbons can be inhaled and cause lipid pneumonia. There are many emollient oils available which can both be metabolized by the body and give good luster to the hair.

An interesting aerosol form is the quick-breaking foam, based primarily on emulsifying waxes such as Polawax, a preparation of higher fatty alcohols and ethylene oxide reaction products. The emulsifying wax, used at a 1 to 4% level, is dissolved with slight warming in a hydroalcoholic medium with the other active ingredients to give a clear single-phase solution, and packaged with about 10% propellant. When dispersed through a foam actuator, the expanding propellant produces a small-bubbled foam which readily collapses on touching or with body heat to afford a thin liquid. The next two formulas presented here are illustrative of hairdressings of the collapsible-foam type.

Formulas for Aerosol Foam Hairdressings

	158	159
Polawax A31	1.40 parts	4.0%
Ethyl alcohol	54.00	52.2
Ucon 50 HB 5100	2.75	—
Polyvinylpyrrolidone	—	2.8
Polyvinylpyrrolidone–vinyl acetate	1.80	—
Glycerol	—	1.0
Water	32.55	30.0
Propellant 12/114 (20/80)	8.00	—
Propellant 12/114 (40/60)	—	10.0
Perfume	q.s.	q.s.

Procedure for Formulas 158 and 159: Dissolve all ingredients in slightly warmed ethyl alcohol, avoiding loss of the alcohol, add the water, and agitate well to disperse any haze. Filter the concentrate and fill into aerosol containers. Add propellants.

Water-based hairdressings, propelled with hydrocarbons, are effective as hairdressings because they supply moisture to the hair, permitting better control and grooming, and evaporate more slowly than alcohol-based products, thus affording time to comb or brush. Formulations 160 to 162 describe such aerosol sprays. These formulations, because of the cheaper hydrocarbon and water, are also less expensive and can be expected to gain a large share of the male aerosol hair-grooming market.

Formulas for Aerosol Hairdressings

	160	*161*	*162*
PVP K30	1.50%	—	—
PVP/VA E735	—	3.00%	—
Resyn 28-1310	—	—	2.00%
2-Amino-2-methyl-1,3-propanediol	—	—	0.20
Acetyl tributyl citrate	0.15	—	—
Ethoxylated (65) lanolin	—	0.10	0.10
Ethyl alcohol	25.00	41.75	44.55
Water, distilled	38.20	30.00	28.00
n-Butane	12.50	—	—
Isobutane	—	25.00	25.00
Propellant 114	12.50	—	—
Propellant 12	10.00	—	—
Perfume	0.15	0.15	0.15

Procedure: Dissolve the resins in the alcohol (in Formula 162, first dissolve the 2-amino-2-methyl-1,3-propanediol), add the plasticizers and perfume, then water and agitate till clear. Filter the concentrate and fill into aerosol containers. Add propellants.

REFERENCES

General

1.   Lourie, D.: Hair fixatives for home and export, *SPC*, **13**: 238 (1940).
2.   Hilfer, H.: Hair dressing, *DCI*, **61**: 605 (1947).
3.   Wall, F. E.: "Cosmetics," in *Encyclopedia of chemical technology*, Vol. 4, Interscience, New York-London, 1947, pp. 529–562.
4.   Schwarz, H.: Haar Öle, *Seifensieder-Ztg.*, **62**: 910 (1935).
5.   Powers, D. H.: Polyols in cosmetics, *DCI*, **71**: 610 (1952).
6.   Jannaway, S. P.: Hair fixatives, *Alchemist*, **2**: 226 (1948).
7.   Redgrove, H. S.: Hair creams, *PEOR*, **31**: 225 (1940).
8.   Sedgwick, F. H.: Cosmetic pot-pourri, *SPC*, **13**: 770 (1940).
9.   Jannaway, S. P.: Specialties for export, *PEOR*, **31**: 371 (1940).
10.   Jannaway, S. P.: Hair fixatives, *Alchemist*, **3**: 8 (1949).
11.   Harry, R. G.: *Modern cosmeticology*, 5th ed., Leonard Hill, London, 1962.
12.   Hilfer, H.: Preservatives in cosmetic preparations, *DCI*, **71**: 38 (1952).
13.   Kalish, J.: Cosmetic manual, *DCI*, **47**: 398 (1940).

14. Chilson, F.: *Modern cosmetics*, 1st ed., Drug & Cosmetic Industry, New York, 1934, pp. 247–264.

15. deNavarre, M. G.: *The chemistry and manufacture of cosmetics*, Van Nostrand, New York, 1962.

16. Poucher, W. A.: *Perfumes, cosmetics, and soaps*, 7th ed., Vols. I–III, Van Nostrand, New York, 1960.

17. Goodman, H.: *Cosmetic dermatology*, McGraw-Hill, New York, 1936, pp. 263–266.

18. Atlas Powder Co.: *Guide to cosmetic and pharmaceutical formulation*, 1970, Wilmington, Del.

19. Janowitz, H.: Moderne Haarcrems, *Seifen-Öle-Fette-Wachse*, **76**: 166 (1950).

20. Jannaway, S. P.: Hair fixatives, *Alchemist*, **2**: 175 (1948).

21. Jannaway, S. P.: Hair preparations for export, *PEOR*, **32**: 74 (1941).

22. Verblen, J.: U.S. Pat. 2,498,727 (1950).

23. Chester, J. F. L.: Hair preparations, a review, *PEOR*, **58**: 539 (1967).

24. Bergwein, K.: Haarcrems, ihre Geschechte und ihr Aufbau, *Seifen-Öle-Fette-Wachse*, **92**: 811 (1966).

25. Wells, F. V., and Lubowe, I.: *Cosmetics and the skin*, Reinhold, New York, 1964.

26. Amerchol laboratory handbook for cosmetics and pharmaceuticals, American Cholesterol Products, Edison, N.J.

27. Keithler, W. R.: *The formulation of cosmetics and cosmetic specialties*, Drug & Cosmetic Industry, New York, 1956.

28. Croda cosmetic and pharmaceutical formulary, 1967, Croda, Inc., New York.

Brilliantines and Pomades

29. Jannaway, S. P.: Hair fixatives, *Alchemist*, **2**: 200 (1948).

30. Singiser, R. E., and Beal, H. M.: Metallic soap–petrolatum ointment pases, *J. Am. Pharm. Assoc.*, **47**: 6 (1958).

31. Ekmann: Haaröle, Brillantinen, Pomaden, Fixative, *Riechstoff Ind.*, **14**: 1 (1939).

32. Schwarz, H.: Haar Glanz, *Seifensieder-Ztg.*, **65**: 265 (1938).

33. Cordero, T.: U.S. Pat. 2,382,398 (1945).

34. Cordero, T.: U.S. Pat. 2,402,473 (1946).

35. Lougavay, B.: U.S. Pat. 1,884,015 (1933).

36. Schwarz, H.: Haarpomaden, *Seifensieder-Ztg.*, **68**: 176 (1941).

37. Matuura, K., and Takazi, K.: Jap. Pat. 91,885 (1931).

38. Perret, J.: Brit. Pat. 298,167 (1927).

39. Kerasin S. A.: Fr. Pat. 765,018 (1934).

40. Belzarelli, F., and Baldini, C.: Fr. Pat. 778,236 (1935).

41. Novavita A. G.: Swiss Pat. 253,837 (1948).

Alcoholic Hairdressings

42. Caimi, D., and Caimi, A.: U.S. Pat. 2,158,791 (1939).

43. Stetson, R. A.: Glycerine in hair preparations, *Am. Perf.*, **61**: 285 (1953).

44. Boleau, E. A.: Fr. Pat. 939,931 (1948).

45. Foldes, M.: Hung. Pat. 103,586 (1931).

Hair Tonics

46. Rubin, S. H., Magid, L., and Scheiner, J.: Panthenol in cosmetics, *Proc. Sci. Sec. TGA*, **32**: 6 (1959).

47. Burnett, R. S.: Proteins in cosmetics, *Am. Perf.*, **78**: 69 (October 1963).

48. Karjala, S. A., Johnsen, V. L., and Chiostri, R. F.: Substantive proteins in cosmetics, *Am. Perf.*, **82**: 53 (October 1967).

49. Chilson, F.: *op. cit.* (ref. 14), pp. 253–259.

50. Borreca, W.: U.S. Pat. 1,612,255 (1927).

51. Shiseido, K. K.: Jap. Pat. 102,053 (1933).

52. Wüste, F.: Ger. Pat. 656,384 (1938).

53. Roia, F. C.: The use of plants in hair and scalp preparations, *Economic Botany*, **20**: 17 (1964).

54. Lehne, R. K.: Hair and scalp lotions, *Am. Perf.*, **83**: 97 (October 1968).

Two-Layer Hairdressings

55. Rider, T., and Gershon, S. D.: U.S. Pat. 2,543,061 (1951).

56. Adinolfi, J.: Br. Pat. 439,011 (1935).

57. Jannaway, S. P.: Hair fixatives, *Alchemist*, **2**: 199 (1948).

58. Poucher, W. A.: *op. cit.* (ref 16), pp. 430*ff.*

Gum-Base Hairdressings

59. Jannaway, S. P.: Alginates in cosmetics, *SPC*, **21**: 1003 (1948).

60. Harry, R. G.: *op. cit.* (ref. 11), pp. 476–477.

61. Jannaway, S. P.: Hair fixatives, *Alchemist*, **3**: 33 (1949).

62. Lesser, M. A.: Pectins, *DCI*, **45**: 549 (1938).

63. Kern, J. C., and Stetson, R. A.: Glycerine in new cosmetics, *Am. Perf.*, **56**: 139 (1950).

64. Luckenbach, W. F.: U.S. Pat. 2,305,356 (1948).

65. Omohundro, A. L., and Fauto, E. C.: U.S. Pat. 2,440,555 (1948).

66. Klug, E. D.: U.S. Pat. 3,210,251 (1965).

67. Bodensiek, A.: Ger. Pat. 1,150,179 (1963).

68. B. F. Goodrich formulary, 1968.

69. Gross, R. I., and Lehne, R. K.: U.S. Pat. 3,215,603 (1965).

70. Bodensiek, A., and Scheller, K.: Ger. Pat. 1,165,204 (1964).

71. Dragoco Report, **9**: 212 (1967).

Emulsion Hairdressings

72. deNavarre, M. G.: Recent developments in cosmetics, *Am. Perf.*, **58**: 105 (1951).

73. Karas, S.: Luster in cosmetics, *Am. Perf.*, **60**: 353 (1952).

74. Jannaway, S. P.: Hair fixatives, *Alchemist*, **2**: 252 (1948).

75. Stetson, R. A., and Kern, J. C.: Use of glycerine in simple lotions, *Am. Perf.*, **54**: 467 (1949).

76. Harry, R. G.: *op. cit.* (ref. 11), pp. 477–481.

77. Jannaway, S. P.: Hair fixatives, *Alchemist*, **2**: 277 (1948).

78. Peel, N. S.: Brilliantine creams, *SPC*, **26**: 275 (1953).

79. Dicken, W. H.: Developments in aliphatic chemistry and their influence on the cosmetic industry, *JSCC*, **2**: 126 (1951).

80. Stetson, R. A.: Glycerine in hair preparations, *Am. Perf.*, **61**: 287 (1953).

81. Lesser, M. A.: Readers' questions, *DCI*, **71**: 801 (1952).

82. Lesser, M. A.: Readers' questions, *DCI*, **71**: 531 (1952).

83. Lesser, M. A.: Readers' questions, *DCI*, **72**: 237 (1953).

84. deNavarre, M. G.: Desiderata, *Am. Perf.*, **56**: 453 (1950).

85. deNavarre, M. G.: Desiderata, *Am. Perf.*, **61**: 439 (1953).

86. Bennett, H.: *Practical emulsions*, 2nd ed., Chemical Publishing Co., New York, 1947, pp. 351–353.

87. Wells, F. V.: Cosmetic uses of Cera Emulsificans, Part III, *Am. Perf.*, **59**: 183 (1952).

88. Sinigalia, C.: It. Pat. 461,105 (1951).

89. Clauzel, M. A.: Fr. Pat. 941,372 (1949).

90. Bergwein, K.: Haarcrems: Ihre Geschichte und ihr Aufbau, *Seifen-Öle-Fette-Wachse*, **92**: 811 (1966).

91. Clark, E. W.: Lanolin derivatives in hair dressings, *Chem. Drug.*, July 3 1967.

92. Henkin, H., and Lehne, R. K.: Belg. Pat. 644,855 (1964).

93. Jass, H.: Belg. Pat. 671,120 (1966).

94. Kalish, J.: Hair dressing for men, *DCI*, **11**: 621 (1962).

95. Parnassum, G. A.: Some formulations based on mineral oil, *SPC*, **41**: 35 (1968).

96. Siegal, B., Petgrave, R., and Thau, P.: U.S. Pat. 3,101,300 (1963).

97. Siegal, B., and Petgrave, R.: U.S. Pat. 3,101,301 (1963).

98. Siegal, B.: U.S. Pat. 3,175,949 (1965).

99. Chesebrough-Ponds: Br. Pat. 1,002,466 (1965).

100. Kaufmann, T., and Tkuczuk, R.: U.S. Pat. 3,341,465 (1967).

101. Shiseido Co.: Br. Pat. 1,042,499 (1966).

102. Cosmetic and pharmaceutical formulary, Croda Co., Inc., 1967.

103. Martin, J. R. L.: Clear gels, *SPC*, **39**: 894 (1966).

104. Bergwein, K.: Transparent creme hair dress, *Seifen-Öle-Fette-Wachse*, **92**: 812 (1966).

105. Becher, P.: Nonionic gel systems, *Am. Perf.*, **82**: 41 (1967).

106. Janowitz, H.: Glasklare "Haarcremes und Brilliantinen," *Seifen-Öle-Fette-Wachse*, **11**: 352 (1963).

107. Osipow, L. I.: Transparent emulsions, *JSCC*, **14**: 227 (1963).

108. Kaufman, T., and Blaser, R.: Clear gel cosmetics, *Am. Perf.*, **80**: 37 (December 1965).

Chapter 21

PERMANENT WAVING

S. D. Gershon, M. A. Goldberg, and M. M. Rieger

Nessler, a London hairdresser, produced permanent waves by heating hair after it had been wound on a mandrel and moistened with an aqueous solution of alkali. His method combined the wigmaker's art of waving hair, using hot alkalies, with Marcel's technique of heating hair on the head. It is generally accepted that modern permanent waving began with Nessler's method for permanently waving human hair on the head (1,2).

Initially, modern hair waving was a skill practiced by artisans using secret formulas and cumbersome apparatus. The recipient of the wave was subjected to hours of discomfort and occasionally even to physical harm. Hillier's (3) account of this aspect of permanent waving is rather humorous and is quoted here in part:

> It took all day to give a permanent wave . . . . Women bragged about their burns to other women. And it was torture. The heaters were so heavy that only a few curls could be baked at a time. The pads . . . would stick to the heaters. It was necessary to tap the heaters with a hammer to release them . . . . The hair dresser would use a nut cracker . . . to release the pads . . . that had been baked on the wound hair. Sometimes the hair would come off with the pad . . . .

In spite of its disadvantages, Nessler's system was adopted widely. Improvements on his original process involved the development of various methods for supplying heat to the hair. Sartory (4) utilized the exothermic reaction between calcium oxide and water to generate steam which was conducted to the hair by means of a chandelier. Winkel (5) conceived the idea of placing the chemicals in a small envelope (heat pad) which could be applied directly to the hair, thus eliminating the need for the cumbersome and expensive Sartory equipment. Following this, a number of patents issued covering additional calcium oxide (6) and exothermic oxidation-reaction compositions for use in heat pads (7,8). During the course of these developments, the waving fluid was changed to contain more volatile alkali and to include a sulfite to accelerate waving.

Cold waving, i.e., the method of waving hair without externally applied heat, began in the 1930's with the so-called "overnight" wave, a process which required 6 to 8 hr for completion at room temperature. This was followed (in 1940) by a fast but short-lived cold-wave process based on bisulfides. Almost simultaneously, cold-wave lotions based on thioglycolates became available, and today the use of such lotions is almost universal. In contrast to early methods of permanent waving, present-day hair waving has become a routine, safe procedure, requiring no special apparatus, and performed daily by thousands of women (9).

THE PRACTICE OF HAIR WAVING

General Considerations

In principle, hair waving is accomplished by altering the configuration of hair while it is maintained in the curled position. In so-called "water waving" of hair, a wet strand is wound in a circular fashion and allowed to dry. Such a process curls the strand of hair, and the hair will remain curled in the absence of moisture. This type of curl is commonly called a "temporary or cohesive set" (10) and depends on physicochemical forces in the hair that are readily affected by water. In contrast, permanent waving produces a "permanent set" by altering additional chemical forces in the hair that are stable to water at temperatures normally encountered by the human body. Permanent waving, therefore, requires that some of the water-stable forces that normally hold the hair in a straight configuration be broken. The disruption of the water-stable forces is the so-called "softening" of the hair. It is accomplished by chemical means and is followed by what may be called the "rearrangement." During the "rearrangement," the polypeptide chains, or parts thereof, in the "softened" hair are forced to move relative to each other under an externally applied strain. This physical reaction appears to take place in both hot and cold waving. It is the reaction that changes the configuration of the hair.

"Rearrangement" can occur only after the fiber has been "softened" and with the fiber under strain. The time required for this physical rearrangement to take place depends on the extent of "softening" and the amount of strain.

The final step in permanent waving may be referred to as "hardening." The purpose of this step is to reverse the "softening," i.e., to reestablish the water-stable forces existing in the original fiber, thus making the rearrangement permanent.

A satisfactory permanent wave as it is known today will not change the chemical and physical properties of hair significantly. Hair color should remain unchanged and foreign odor should be absent. A good permanent wave should last until the hair has grown so long that the waved hair has to be cut off.

All methods of permanent waving require that the hair be held in a curled or waved position while the various chemical changes that produce the wave take place (11). It has not been possible to wave hair that is free to assume any configuration by the simple application of chemicals or heat. Therefore the hair is held in the desired position with the aid of a curling device and then forced to conform—at least partially—to its shape.

All curling devices have one thing in common. The hair is wound on or around a mandrel and held in place while the essential chemical or physical operations take place. Two types of curlers have been used in hair waving. The first, the spiral rod, is used in spiral winding. For this purpose, the hair on the head is sectioned, i.e., divided into 4 to 6 sections. These sections are then blocked, i.e., subdivided into squares about 2.5 by 2.5 cm. The hair in such a block is then grasped near the proximal end and wound around a mandrel in a helical fashion. The distal ends are held in place with the aid of crepe wool, string, or similar device. The second or modern type, the croquignole curler, is used somewhat differently. The hair is sectioned into blocks 1 to 2.5 cm by 5 to 7.5 cm large. The hair in such a block is combed, and the distal ends are placed centrally on the mandrel. The hair is then wound by rotating the mandrel and is finally secured in place by a rubber band or other suitable means. Mayer (12), Squerso (13), Szlanyi (14), Bishinger (15), and Grant (16,17) received patents for curlers intended to be used in this manner.

A series of more recently granted patents includes devices which curl the hair by suction without the need for the more conventional, tedious winding procedures (18–20).

In order to facilitate collection of the distal ends of hair of different lengths, most modern permanent waving methods employ so-called "end papers." These end papers are made from high wet-strength paper or nonwoven fabric and are placed over the hair-ends before winding is begun. End papers have also been used as carriers for the waving agent (cf. Dry Thioglycolate Preparations) or for suitable hair-protective emollients (21).

Although curler rods of the described types are used widely for permanent waving, a few modern home-waving processes utilize the well-known bobby pins or similar clamps to hold the hair in circles (22).

The Effect of Curler Diameter

The quality of the wave produced is dependent on the dimensions of the curler used to wind the hair. Generally, the diameter of the curler will determine the tightness of the permanent wave.

Obviously, the number of turns that a hair can describe around a curler depends on the circumference of the curler. Any given length of fiber will yield more turns when wound on a curler of smaller diameter than on one

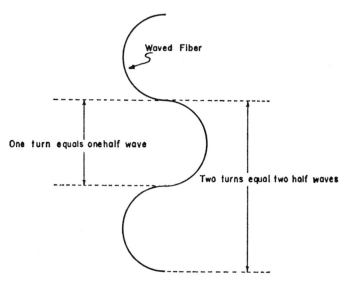

Fig. 1. Schematic drawing of waved hair fiber showing relationship between wave formation and turns around curler. [After Hillier (23).]

of larger diameter. One complete turn of the hair around the curler will yield one half-wave (Figure 1) at the completion of the process (23). Therefore the thinner mandrel will produce more half-waves than the larger one.

The character of the permanent wave is determined by the number of half-waves per unit length of fiber. For example, if there are too few half-waves, the overall appearance of the permanent wave will be too loose. On the other hand, too many half-waves will produce a tight, frizzy wave. In common practice, the diameter of the curler varies from about 0.3 to 1.5 cm. Selection of the proper curler diameter makes possible the formation of waves having almost any desired curl tightness; for a tight permanent wave, use of a small-diameter curler is indicated; for the more modern casual wave (body wave), larger-diameter curlers are preferred.

The Effect of Hair Diameter

Another factor which influences curl tightness is the diameter of the hair. If hair is wound around a circle, there is tension (γ_+) on the outer part of the hair shaft. Conversely, the inner part of the hair is under compression (γ_-) whereas in the center of the hair (the so-called neutral plane) neither compression nor tension exists (Figure 2).

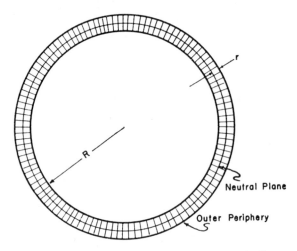

Fig. 2. Schematic drawing of hair fiber placed concentrically around curler.

When r equals the radius of the hair and R the radius of the curler, the strain (γ_+) at the outer periphery of the hair can be calculated from the following equation:

$$\gamma_+ = \frac{r}{R + r}.$$

Since R is considerably greater than r,

$$\gamma_+ = \frac{r}{R}.$$

This equation indicates that coarse hair (r large) is under greater strain than fine hair (r small) at the constant curler radius R. Moreover, at constant hair radius r, a large curler radius R will produce less strain than a small curler radius.

As indicated above, the "rearrangement" step cannot take place unless the individual fibers are under strain. Hair wound on a curler is under a differential strain, and the tendency for segments of the hair fiber to move relative to each other depends on the extent of this differential strain. Therefore all conditions that increase the differential strain also increase the rate of waving or the amount of wave produced in a given time. In addition to initiating physical "rearrangement" of the hair, it is possible that differential strains in hair affect the rate of waving by facilitating rupture of strained chemical (disulfide) linkages in preference to unstrained ones, as suggested by Speakman and Whewell (24) and by Astbury and Woods (25). The differential strain due to curvature is small and usually less than the longitudinal strain which can result from stretching the hair. Stretching is known to speed cold waving (26) but must be employed with caution in order to avoid serious damage or even breakage.

On the basis of the effects exerted by both curler and hair diameter, a fairly adequate assessment can be made of the merits of the various curlers used in permanent waving. A satisfactory croquignole curler should have a fairly small diameter. Otherwise the proximal part of the wound hair will be under too little strain for waving because this hair is wound after the curler diameter has been increased by previously wound hair. For the same reason, even with a curler of proper diameter, it is important that not too much hair be wound on a curler. As might be expected, the croquignole curler frequently yields tight "end" curls; i.e., the distal portion of the fiber is curled tightly.

Although the geometrical relationships are such that croquignole winding yields slightly looser scalp curls than spiral winding, the former method is generally preferred because it is easier to use.

Pin curls made without special winding devices have a large diameter, and permanent waves given to hair wound in this fashion are generally very loose.

The diameter of the hair may affect the quality of the wave not only by way of the indicated geometrical relationships but also in a more subtle manner. Stoves (27) attributed some of the differences in ease of waving between fine and coarse fibers to the cuticle/cortex ratio. Since the cuticle/cortex ratio is large for fine fibers, and since the cuticle, because of its scale structure, is unable to support a wave, such fine fibers would be expected to be waved more difficultly and to relax, i.e., straighten, more rapidly after waving.

Lotion Penetration

The speed of waving and the quality of the wave depend not only on geometrical relationships involving hair and curler diameter but also on the chemical reaction of the lotion with the hair. The rate of this reaction depends on the temperature, the composition of the lotion, and the speed with which the lotion penetrates the hair. Not surprisingly, the speed of penetration into hair depends on the lotion's chemical composition (28) and on the so-called "porosity" of the hair, which in turn depends on its morphology and on its previous chemical or physical treatment. "Porosity" varies not only from head to head and hair to hair but also along the hair shaft (29). Thus Freytag (30) has demonstrated that, because of weathering alone, the distal portion of a fiber is more readily influenced by chemicals than the proximal portion. This fact, though well known to beauticians, is difficult to quantify and is frequently neglected in scientific papers on permanent waving (28). Hair or that portion of hair that has been bleached (chemically or by UV light), dyed, previously waved, etc., is more "porous" than virgin hair. Generally, "porous" hair is waved more rapidly than "nonporous" hair (31).

In addition to the porosity of the hair, thorough saturation of the wound tress with lotion is important to the success of the wave. Whereas porosity is concerned with the intrafiber penetration of lotion, saturation of the tress affects the interfiber penetration. The importance of thorough saturation of the wound tress with lotion—especially in cold waving—has been treated theoretically by Reed and his co-workers (11). Assuming that all air is expelled, these authors show that in a close-packed system of circular fibers, the ratio of lotion in the interstices to fiber is constant in any given volume, regardless of the diameter of the fiber. This rather fortunate circumstance indicates that it is impossible to use too much of a properly formulated waving lotion because the tress cannot hold an excess. Obviously, too little waving lotion because of incomplete saturation is a common cause of wave failure.

Reed indicates further the necessity for freeing the hair from dirt and sebum prior to waving—with a shampoo of the detergent type—in order to facilitate wetting of hair and penetration of the lotion between the fibers. Recently Freytag showed how geometrical considerations determine the optimum size for strands during hair waving (31a). He also pointed out that the skill of the operator is an important factor in determining the quality of the resulting wave.

Processing Time

The time that a given waving lotion remains in contact with the rolled hair—also called the processing time—determines the tightness of the curl. During this time, the "softening" reaction and the "rearrangement" take place. (Menkart et al. (32) use the term "creep" for this step.) Actually, these two processes occur almost simultaneously in heat waving. In cold waving, the "rearrangement" requires appreciably more time than the "softening" and cannot proceed until some "softening" has been effected. Once the "softening" has taken place, the lotion itself may be removed by a water rinse. Wave development, or "rearrangement," continues until the hair is "hardened" (33). Since the extent of "rearrangement" determines the tightness of the wave, proper timing is extremely important to the success of the wave. The recently introduced method of tepid waving (at temperatures between 35 and 50°C) utilizes the accelerating effect of temperature on the rearrangement step to decrease the processing time.

Some methods of permanent waving take advantage of the fact that hard-to-wave hair can be curled satisfactorily by an increase in the processing time (28). Similarly, damaged hair can sometimes be waved safely if the processing time is reduced sharply. In practice, it is possible to decrease the speed of waving by use of a weak lotion or to accelerate wave formation with the aid of a stronger lotion (34).

In the case of heat waving, the processing or steaming time is a function of the temperature. An increase in the steaming time or temperature is accompanied by an increase in disulfide bond breakdown and other chemical reactions. "Softening" continues until the temperature drops.

In contrast, during cold waving, the extent of chemical reaction ordinarily reaches its maximum during the first few minutes following application of the lotion. No further "softening" takes place, although additional processing time is required for the "rearrangement."

The effect of temperature on cold waving was studied by Böss (35), who pointed out that climatic temperature differences affect the processing time.

STRUCTURAL, CHEMICAL AND PHYSICAL PROPERTIES OF HAIR PERTINENT TO WAVING

The structural, chemical, and physical properties of hair constitute a very complex subject. For information beyond this summary, which is designed to facilitate understanding of the chemistry and physics of permanent waving, the reader is referred to recent reviews by Mercer (36), Crewther et al. (37), and Lundgren and Ward (38).

Human hair is composed almost entirely of keratin, a protein which is also the major constituent of horn and feathers and other mammalian hair, particularly wool. Outstanding properties of native keratin include insolubility in acids and alkalies and in solvents that dissolve other proteins, resistance to enzymatic digestion, and high mechanical strength. Keratin, unlike other proteins, is characterized by a comparatively high percentage of combined cystine, an amino acid which is responsible for many of the specific physical and chemical properties of this protein.

Structure of Hair

Morphologically, one can distinguish three major components in wool, hair, and other mammalian fibers—namely, a cuticle, a cortex, and a medulla (36,39). The cuticle, which accounts for about 10% by weight of the hair fiber, consists of overlapping flat scales which surround the remainder of the hair shaft (40). There has been considerable debate as to whether the cuticular layer is coated by a continuous film. On the basis of work of Mercer and co-workers (41), it would appear that the cuticle of wool consists of several continuous layers among which is the epicuticle, which is chemically extremely resistant. On the other hand, the photomicrographs of human hair by Orfanos and Ruska (40,40a) show no evidence for a continuous layer similar to the epicuticle. However, these authors find good evidence for the presence of an intercellular lamella which cements the cuticle cells to each other. Still more recently Leeder and Bradbury (42) present specific evidence that the epicuticle cannot be regarded as continuous over the whole surface of the fiber but, instead, covers each cuticle cell separately. Chemical analyses by King and Bradbury (42a) and by Lofts and Truter (42b) indicate that the epicuticle is composed primarily of proteinaceous materials.

The remainder (about 90 %) of the hair fiber is made up of the cortex. The cortex consists of spindle-shaped keratin cells which are oriented along the axis of the fiber. In the case of wool, a highly crimped fiber, the so-called ortho and para cortex portions can be differentiated. These portions lie side by side and are wound around each other in phase with the fiber crimp (cf. e.g. 43). In contrast, hair fibers from Caucasians probably consist only of the so-called para cortex, which is chemically more resistant than the ortho cortex (44). It is not known whether the crimped fibers of virgin Negroid hair are exclusively para cortex or, in addition, contain some ortho component.

In the center of human hair fibers, one can from time to time distinguish the so-called medulla which is not believed to make a major contribution to either the physical or the chemical properties of the fiber. The medulla possesses a relatively open structure, may contain air pockets, and probably consists of proteins which differ chemically from those present in the cortex (39).

Electron microscopy and techniques for the disintegration and staining of keratin fibers have yielded further insight into the various components of the cortical cells, which are of primary importance for permanent waving (45,45a,46). In our discussion of the fine details of the structure of keratin, the terminology employed by Mercer and co-workers (47) will be used exclusively. Each cortical cell is believed to be surrounded by a fairly resistant cell membrane. Within the cells are found the so-called fibrils which are also referred to as tonofibrils; they have a diameter of approximately 0.2 μ (Figure 3). Each fibril consists of a group of filaments which are also frequently referred to as microfibrils; they have a diameter of approximately 60 Å (48). Each filament is built up by an assembly of protofilaments (49) which have also been called protofibrils or α-filaments. They are believed (50) to have a diameter of approximately 20 Å and are embedded in a continuous amorphous interfilamentary matrix which is commonly referred to as γ-matrix. The final component of this system is the so-called intra-fibrillar matrix, the material which may be presumed to "glue" filaments together to form the fibril.

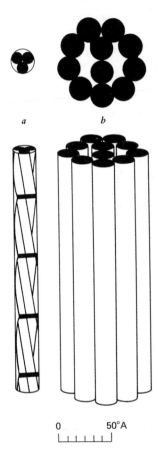

a

b

0   50°A

Fig. 3. Model of the molecular structure of α-keratin. (*a*) Protofibril consisting of three α-helices coiled into a rope. Within the axial repeat of 200 Å three similar, but not identical, subunits are shown. Between each subunit there is a major interruption in electron density. (*b*) Microfibril containing 11 protofibrils, each consisting of a three-strand rope. [After Fraser, Macrae, and Rogers (49).]

There is still considerable debate concerning the chemical composition of the various matrices and of the arrangement of the so-called protofilaments. Fraser and co-workers (49,51) suggest that the protofibril consists of three α-helices coiled into a rope, as shown in Figure 3. Dobb (52) has recently suggested that a protofibril may contain either two or three α-helices. The arrangement of the protofilament in the (9 + 2) system shown in Figure 3 originated with Filshie and Rogers (53). The existence of this system has been questioned by Sikorski and co-workers (cf. e.g., 54,55), although sup-. port for the (9 + 2) arrangement has been provided by Wilson (56). Very recently Parry has suggested several alternate means of packing α-helices into the protofilament (56a,56b). The exact structure of the microfibril is still being debated, and the interested reader should consult current literature.

The structural features of keratin fibers discussed to this point are discerned either by the naked eye or by means of the microscope or electron microscope. Still finer details of keratin structure are yielded by X-ray analysis (57). Work that was pioneered by Astbury (58) has been further refined in more recent years (38,59). For the sake of this discussion, it may be concluded that the α-keratin pattern of Astbury has been fitted by Pauling (60) into aggregates of 3.7 residue helices, the α-helix. It has already been noted that two or three α-helices combine to make up the protofilament. It is generally accepted that the fibrillary component of keratin is crystalline and responsible for the characteristic α-pattern. On the other hand, the matrix is believed to be amorphous.

The next finer level of protein structure, the primary structure, is, in effect, a description of the order and/or sequence of amino acids which make up the polypeptide chain. Despite much effort to elucidate details of this structure, information to date is still limited (37,61). Detailed knowledge concerning the sequence of amino acids in the chain which forms the α-helix and in the matrices would be required for a full understanding of the lateral aggregation of the various elements which combine to form a keratin fiber.

Chemistry of Hair

Fairly complete amino acid analyses of various keratin fibers have been reported in the literature (37,62–67). The outstanding feature of these analyses is the high cystine content which is considered specific for keratin. The cystine content of human hair is higher than that of lamb's wool and ranges from about 15 to 17 %. According to Stary (68), and this is generally accepted by most workers in keratin chemistry, cystine disulfide bonds are responsible for cross-linking adjacent chains of protein molecules.

In addition to cystine disulfide bonds, a variety of other bonds in keratin fibers helps to account for the tightness of the structure and the mechanical properties of the fiber. Such bonds (cf. Figure 4) not only are responsible for holding the α-helix in the coiled position but are also important for the lateral aggregation of the various structural elements discussed above. These bonds include salt linkages, hydrogen bonds, van der Waals forces, and covalent peptide and ester linkages.

Salt linkages, also called Coulomb forces, are believed to exist as a result of the electrostatic interaction between positively charged ammonium ions and negatively charged carboxylate groups (69). Using load elongation curves (which will be discussed later), Speakman (70) has been able to show that these bonds contribute approximately 35% of the strength of keratin fibers and are readily ruptured by acids and probably also by alkalies.

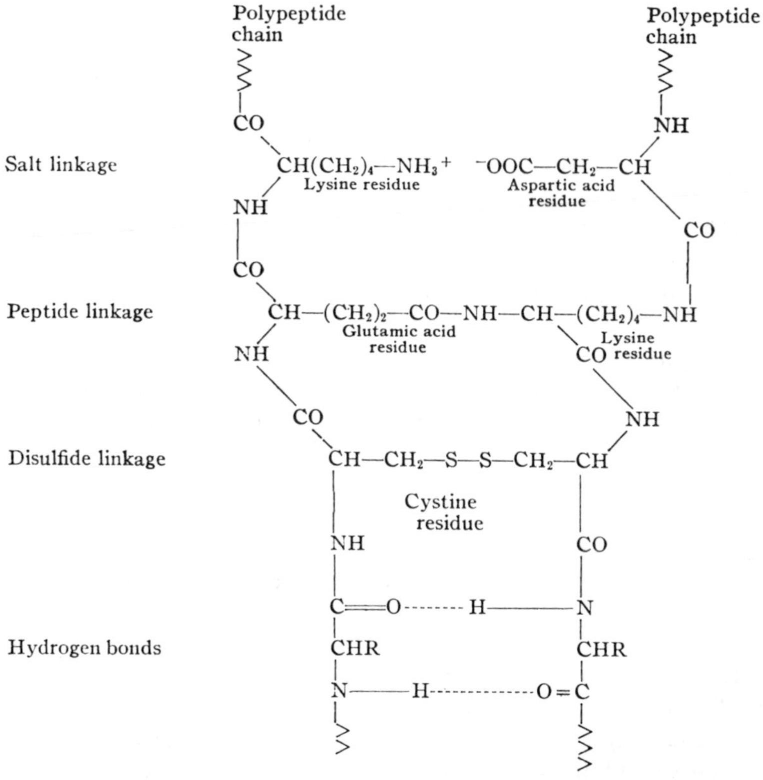

Fig. 4.   Schematic representation of various types of cross-linkages between polypeptide chains in hair.

The disulfide linkage is of particular importance to the chemistry of hair waving. One of the earliest references to the existence of this linkage in keratin was made by Stary (68), who suggested that cystine disulfide links connect neighboring polypeptide chains like the rungs of a ladder. It is generally accepted that disulfide cross-links are broken during waving, permitting rearrangement of the polypeptide chains. After completion of this process, disulfide bonds and other cross-linkages are reestablished, thus making the wave permanent (11,71,72).

Rupture of disulfide bonds by thioglycolates reduces the strength of keratin fibers significantly, the extent of this reduction depending on the length of treatment and pH. There is, however, some question regarding the specific contribution of disulfide bonds to the strength of keratin fibers. Elöd and his co-workers (73) believe that disulfide bonds contribute relatively little to the strength of the fiber. Speakman (70) and Bogaty (74) indicate that disulfide bond rupture is accompanied by hydrogen bond and salt linkage breakdown. As a result, the specific contribution of the disulfide bond to fiber strength has not been ascertained.

It is generally believed that the γ-matrix is particularly rich in cystine; it has, therefore, the ability to cross-link to cystine residues of the protofilaments. Details of this structure have not been established, although working models have been suggested for hair (75) and wool fibers (76–78). These models are based on physical properties of fibers, including those in which disulfide bonds have been broken.

Hydrogen bonds, especially those between an amide nitrogen and an adjacent carboxyl oxygen, make a major contribution to the strength of the fiber. Hamburger and Morgan (79) showed that dry fibers are more difficult to elongate than fibers immersed in water. These authors explain the observed difference by assuming that hydrogen bonds are weakened in the presence of water and therefore offer less resistance to the unfolding of polypeptide chains.

The magnitude of the contribution of hydrogen bonds to the strength of wet keratin fibers is about the same as that of the salt linkage, i.e., about 35 %. This estimate is based on results òf fibers stretched in monochloroacetic acid (11), which breaks both salt links and hydrogen bonds, and in lithium bromide, which breaks only hydrogen bonds (80–82). Bogaty (74) has presented persuasive evidence that hydrogen bonds—not disulfide bonds alone—are involved in present-day hair-waving practices.

It is also believed (74,82a) that the so-called temporary or cohesive set, i.e., the water wave, depends primarily on the strength of various types of hydrogen bonds formed during drying at or near ambient temperatures of wet hair. Treatment of hair with water at or near the boiling point of water causes more significant chemical changes and can be used for permanent waving.

Although the existence of peptide cross-linkages has never been established or disproved, it seems reasonable that such bonds exist in hair, especially in view of the high content of trifunctional amino acids in keratins. Similarly, the presence of ester cross-linkages seems likely in view of the serine content of hair.

Attractive forces of the van der Waals type may exist between any two atoms in close proximity. They probably contribute to the mechanical strength of the fiber, not by virtue of their individual strength but by virtue of their number. In the past, it was generally concluded that nonpolar groups interact with each other or with the solvent only by means of van der Waals forces (83). More recently it has been recognized that changes in the structure of water surrounding nonpolar groups may have a stabilizing effect even on the structure of insoluble proteins (84,85). The interaction between nonpolar groups has also been attributed to the so-called hydrophobic bonds described by Kauzmann (86). His approach is based on the thermodynamic tendency of nonpolar groups to contact each other in preference to contacting water. It has not been possible to separate the effect of these secondary valence forces from that of other linkages because chemical treatments of keratin fibers usually break more than one type of linkage.

Generally speaking, keratin fibers are considered relatively inert to chemical attack and chemical reaction. Whewell (87) has tabulated those reactions which keratin can undergo under suitable conditions. He has also pointed out (88) that keratins can be attacked at the peptide links of main chains, at the tyrosine and serine residues attached to the main chains, at the disulfide and salt links adjoining adjacent peptide chains, and at hydrogen and other secondary valence bonds between adjacent peptide chains.

Keratin fibers under mechanical strain are considerably more reactive toward chemicals than unstrained fibers (32,89,90). Since hair is under some strain during waving, this fact has a direct bearing on permanent waving.

Physical Properties of Hair

Hair is a combination of various morphological elements formed under relatively mild conditions by living organisms. It is nevertheless an extremely durable fiber and possesses high mechanical strength. The load to rupture a wet fiber, i.e., its tensile strength, is high (19 kg/mm²) and exceeds the load necessary to pull the hair from the scalp (91).

One of the most fruitful techniques for examining keratin fibers has been the so-called stress-strain technique or the determination of the load elongation curve.

This important mechanical testing method will be described here briefly because it has contributed more to the study of hair and wool than any other physical or chemical procedure. The "load elongation" curve or "stress-strain" diagram (Figure 5) is determined by stretching a single fiber and plotting the load necessary to extend the fiber versus the amount of elongation. A complete stress-strain diagram (Figure 5) actually consists of two branches, an upper one for extension and a lower one for contraction of the fiber.

The usefulness of such curves would be limited if wet keratin fibers did not exhibit perfect recovery, as evidenced by the fact that a hair stretched approximately 20% of its original length will give the same stress-strain diagram if the stretching is repeated after a 24-hr rest period. In other words, it is possible to stretch a hair frequently and obtain the identical stress-strain

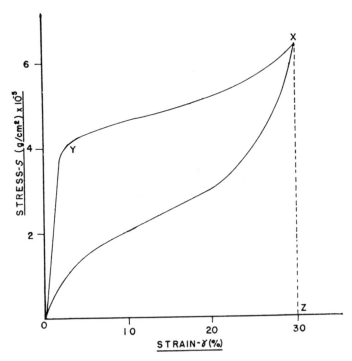

Fig. 5.   Typical stress-strain diagram of human hair fiber in water.

diagram, provided that the hair is allowed to rest after each stretching, that the stretching is done with wet hair, and that the hair is not injured mechanically or chemically. It is this fact that permits comparison of the stress-strain diagrams of the same hair before and after the fiber has been subjected to any given treatment. Extensions up to 30% have also been used (70,92–94) by workers in the field.

It can be noted that the initial slope of the load elongation curve changes markedly at point Y (Figure 5). This point is referred to as the yield point. Following the yield point, comparatively large changes in elongation take place with small additional loads until another point, X, after which a marked change in slope again occurs. The slope of the curve after point X is referred to as the "post-yield" slope. The work necessary to stretch the fiber a given length can be determined from the area $OYXZ$ under the elongative portion of the load elongation curve. This amount of work is called the "toughness."

The frequently employed "30% (or 20%) index" refers to the ratio of the toughness after the fiber has been treated to that of the untreated fiber when the fiber is stretched 30% (or 20%) in both instances. Briefly, damage to the fiber is indicated by a lowering of the toughness and a decrease in the 30% (or 20%) index. By using reagents that break only a specific type of fiber linkage, its specific effect on the strength of the fiber may be ascertained.

The 20 or 30% index of chemically treated human hair fibers has been widely used to assess hair damage and other permanent changes which may occur during various cosmetic treatments. These applications will be described in some detail later.

Load elongation curves of dry or wet fibers can be used directly to help interpret chemical and physical changes in keratin fibers. For example, Ciferri (95) uses such diagrams to support the concept that the $\alpha \rightleftharpoons \beta$ transformation, which takes place on stretching wet wool fibers, is a first-order phase transition between crystalline modifications. Feughelman (96,97) correlates mechanical measurements with the sulfhydryl-disulfide interchange in reduced fibers.

Typical load extension curves of human hair treated with an excess of waving lotion have been recorded by Hamburger and Morgan (79) (Figure 6). These curves include the "post-yield slope" which Hamburger and Morgan consider directly related to the number of disulfide bonds, the post-yield slope decreasing with disulfide bond rupture.

Other mechanical methods for studying human hair fibers have been reported. Thus Hirsch (98) studied load rotation curves which can be determined because human hair rotates around its own axis when it is stretched. Bogaty (98a) determined the torsional modulus of waved and unwaved hair and described the influence of relative humidity on this parameter. From these measurements he concluded that torsional properties of single fibers and of arrays are important in permanent waving and in styling of hair. Wall and co-workers (99) have used stress relaxation at various temperatures following small tensile strains on hair to elucidate differences between bleached and unbleached hair.

The permanent waving of hair is accomplished generally by three different methods. One of these, the heat wave, depends on the action of steam in the presence of alkalies and/or other chemicals to soften the fiber. A second method, the cold wave, employs in part reducing chemicals at room temperature to produce a permanent wave. The third and final one, the tepid wave, operates at intermediate temperatures and utilizes chemicals used in hot and/or cold waving. These three waving methods actually depend on the same principle: Physicochemical linkages in hair are broken and later re-formed to produce the desired wave. As a corollary and extension of this principle, it can be said that hair cannot be curled permanently without some change in the attractive forces that hold the keratin fiber in a straight,

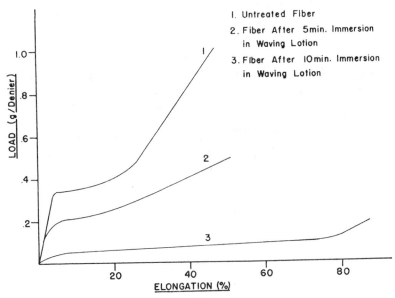

1. Untreated Fiber
2. Fiber After 5min. Immersion in Waving Lotion
3. Fiber After 10min. Immersion in Waving Lotion

Fig. 6.   Load elongation curves of wet human hair. [After Hamburger and Morgan 79.]

uncurled position. After or during rupture of these attractive forces, the "softened" fibers can be bent or curled into a given conformation. Finally, in order to make the new conformation permanent and to avoid damage to the fiber, the fiber must be "rehardened" to complete the waving process.

The various reactions that are used singly or in combination for the purpose of waving hair are discussed in some detail below.

Reaction of Hair with Steam

For purposes of hair waving, the reaction between keratin and steam is generally carried out in the presence of alkalies. Although the reactions taking place during steaming have been studied for some time, no general agreement exists regarding the chemistry of this process. Speakman (92,100), among the first to study the steaming of wool in detail, suggested that the preliminary chemical reaction involves breakdown of disulfide links. Parallel with this breakdown reaction occurs linkage rebuilding. Speakman reasoned that the new linkages are formed between nitrogen and sulfur atoms because linkage rebuilding does not take place with deaminated fibers.

$$\text{protein—S—S—protein} \xrightarrow[\text{OH}^-]{\text{steam}} \text{protein—SH} + \text{protein—SOH}$$

$$\text{protein—SOH} + \text{protein—NH}_2 \longrightarrow \text{protein—S—NH—protein} + \text{H}_2\text{O}$$

The existence of such S–N bonds in steamed fibers could not be substantiated by Phillips (101).

Schöberl and co-workers (102,103) suggested the accompanying reaction scheme in which the fate of the sulfenic acid residue (I) plays a dominant part. In the presence of alkali, the disulfide bond is hydrolyzed to form a mercaptan (II) and sulfenic acid (I). Each of these fragments is then involved in a series of reactions, the mercaptan (II) giving rise to an olefin (V), the sulfenic acid (I) to an aldehyde (III), a primary alcohol (IV), and sulfinic (VI) or sulfonic acid (VII). The mercaptan and olefin react to form lanthionine (VIII), while the aldehyde and protein, the latter reacting through a primary amino group, form an aldimine (IX).

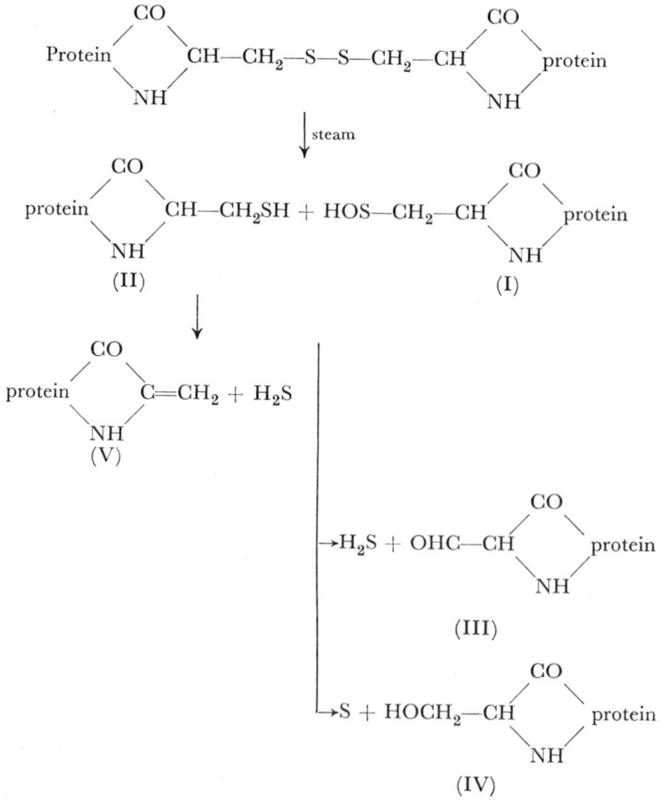

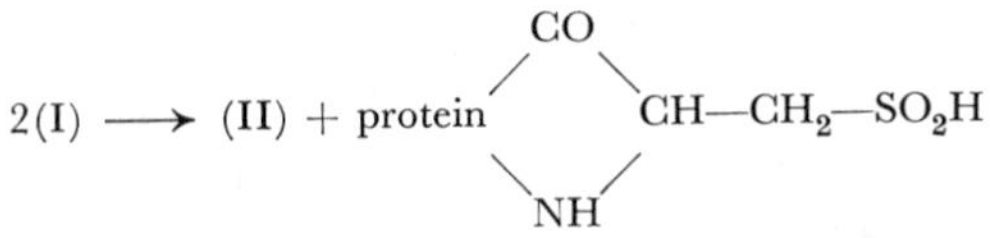

$$2(I) \longrightarrow (II) + \text{protein} \underset{\diagdown_{NH}\diagup}{\overset{\diagup^{CO}\diagdown}{}} CH-CH_2-SO_2H$$

$$(VI)$$

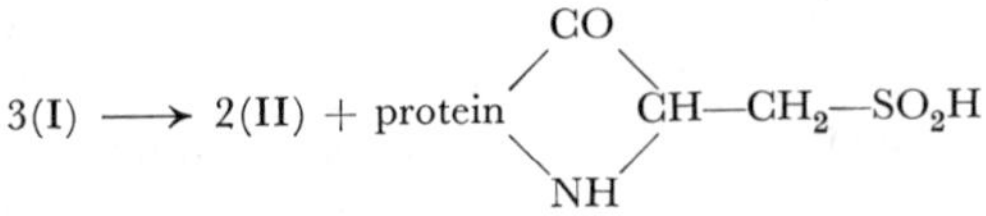

$$3(I) \longrightarrow 2(II) + \text{protein} \underset{\diagdown_{NH}\diagup}{\overset{\diagup^{CO}\diagdown}{}} CH-CH_2-SO_2H$$

$$(VII)$$

$$(II) + (V) \longrightarrow \text{protein} \underset{\diagdown_{NH}\diagup}{\overset{\diagup^{CO}\diagdown}{}} CH-CH_2-S-CH_2-CH \underset{\diagdown_{NH}}{\overset{\diagup^{CO}\diagdown}{}} \text{protein}$$

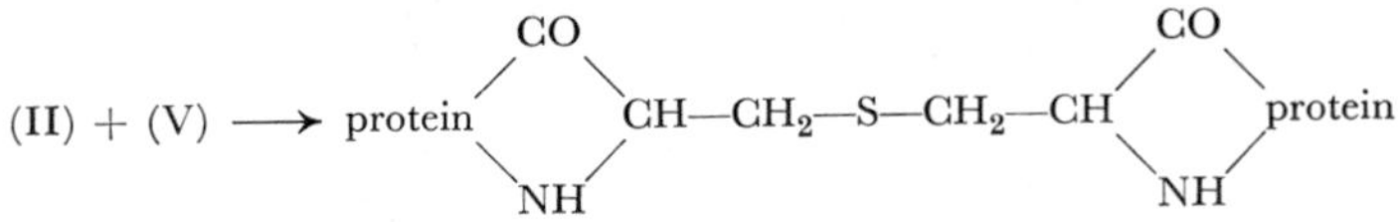

lanthionine residue
$$(VIII)$$

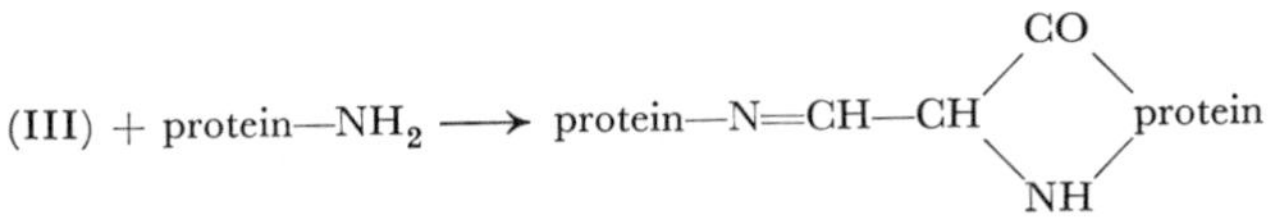

$$(III) + \text{protein}-NH_2 \longrightarrow \text{protein}-N{=}CH-CH \underset{\diagdown_{NH}}{\overset{\diagup^{CO}\diagdown}{}} \text{protein}$$

$$(IX)$$

Cuthbertson and Phillips (104) were not able to detect the presence of a
—C=N bond in steamed wool. Mizell and Harris (105) suggested that the
sulfenic acid (I) is not formed and that combined cystine is decomposed by
steam in accordance with the reaction, as shown here.

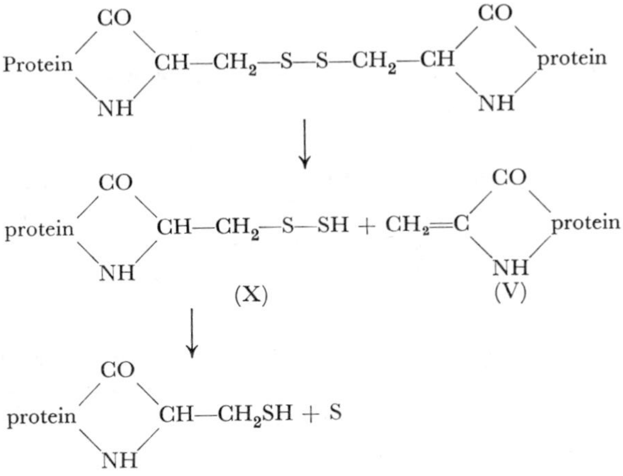

$$\text{Protein} \underset{\diagdown_{NH}\diagup}{\overset{\diagup^{CO}\diagdown}{}} CH-CH_2-S-S-CH_2-CH \underset{\diagdown_{NH}}{\overset{\diagup^{CO}\diagdown}{}} \text{protein}$$

$$\downarrow$$

$$\text{protein} \underset{\diagdown_{NH}\diagup}{\overset{\diagup^{CO}\diagdown}{}} CH-CH_2-S-SH + CH_2{=}C \underset{\diagdown_{NH}}{\overset{\diagup^{CO}\diagdown}{}} \text{protein}$$

$$(X) \qquad\qquad (V)$$

$$\downarrow$$

$$\text{protein} \underset{\diagdown_{NH}\diagup}{\overset{\diagup^{CO}\diagdown}{}} CH-CH_2SH + S$$

$$(II)$$

The mechanism suggested by Schöberl and that indicated by Mizell receive support from the fact that lanthionine has been isolated from hydrolysates of steamed or alkali-treated wool or hair fibers (106), although it has been claimed that lanthionine is an artifact in protein hydrolysis (107).

The extensive studies by Zahn and his co-workers (108) demonstrate that lanthionine is generally formed when keratin is treated with an alkali at elevated temperatures. The mechanism by which these various reactions occur during treatment has been the subject of much debate. Schöberl's scheme is a hydrolytic one depending evidently on nucleophilic substitution of the OH⁻ ion on a disulfide sulfur atom. The Mizell and Harris scheme basically involves a nucleophilic elimination of a β-carbon atom, as shown in the accompanying reaction.

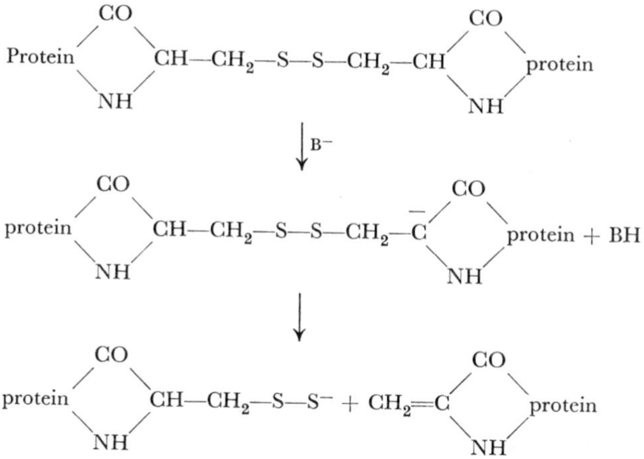

$$
\text{Protein}\!\!\underset{NH}{\overset{CO}{<}}\!\!CH—CH_2—S—S—CH_2—CH\!\!\underset{NH}{\overset{CO}{>}}\!\!\text{protein}
$$

$$\downarrow B^-$$

$$
\text{protein}\!\!\underset{NH}{\overset{CO}{<}}\!\!CH—CH_2—S—S—CH_2—\overset{-}{C}\!\!\underset{NH}{\overset{CO}{>}}\!\!\text{protein} + BH
$$

$$\downarrow$$

$$
\text{protein}\!\!\underset{NH}{\overset{CO}{<}}\!\!CH—CH_2—S—S^- + CH_2{=}C\!\!\underset{NH}{\overset{CO}{>}}\!\!\text{protein}
$$

(XI)

The formation of the persulfide, thiocysteine, analogous to compounds (X) and (XI), in the reaction between dilute alkali and cysteine in the presence of copper has recently been demonstrated by DeMarco and co-workers (109). They further suggest that this compound may give rise to a host of other S-containing substances. On the other hand, Zahn and Golsch (110) considered the so-called β-elimination unlikely because they were unable to find spectroscopic evidence for the presence of the thiocysteine anion.

Regardless of the initial reaction, the interaction between combined α-aminoacrylic acid and cysteine can account for the formation of combined lanthionine. These intermediates are also required in the mechanism proposed by Elliott and co-workers (cf. e.g. 111), who suggest initial formation of trace quantities of cysteine or combined cysteine from the interaction of alkali with keratin. Cysteine then attacks the α-carbon of cystine to yield lanthionine.

According to Parker and Kharasch (112) and Swan (113), with some support from Zahn (108), lanthionine formation is best explained without recourse to unusual intermediates but involves two successive nucleophilic attacks, as illustrated.

$$\underset{\substack{\diagup \\ NH}}{\overset{\substack{CO \\ \diagup \quad \diagdown}}{\text{Protein}}}\!\!\diagdown\, CH\!-\!CH_2\!-\!S\!-\!S\!-\!CH_2\!-\!CH \overset{CO}{\underset{NH}{\diagup\diagdown}} \text{protein} + B^-$$

$$\updownarrow$$

$$\underset{\substack{\diagup \\ NH}}{\overset{\substack{CO \\ \diagup \quad \diagdown}}{\text{protein}}}\!\!\diagdown\, CH\!-\!CH_2\!-\!SB + {}^-S\!-\!CH_2\!-\!CH \overset{CO}{\underset{NH}{\diagup\diagdown}} \text{protein}$$

$$\downarrow$$

$$\underset{\substack{\diagup \\ NH}}{\overset{\substack{CO \\ \diagup \quad \diagdown}}{\text{protein}}}\!\!\diagdown\, CH\!-\!CH_2\!-\!S\!-\!CH_2\!-\!CH \overset{CO}{\underset{NH}{\diagup\diagdown}} \text{protein} + BS^-$$

B^- can be OH^- or any one of the commonly used disulfide-splitting nucleophilic reagents, such as CN^-, HSO_3^-, etc.

Rosenthal and Oster (114) questioned any mechanism which postulates the formation of a sulfenic acid, a compound which has not been isolated from the reaction. In its place they suggest, supported by spectrographic evidence, a mechanism involving the elimination of a proton from the α-carbon atom (adjacent to a sulfur atom) in accordance with the following equation:

$$OH^- + R\!-\!CH_2\!-\!S\!-\!S\!-\!CH_2\!-\!R \rightleftharpoons R\!-\!\overline{C}H\!-\!S\!-\!S\!-\!CH_2\!-\!R + H_2O$$

$$R\!-\!\overline{C}H\!-\!S\!-\!S\!-\!CH_2R \rightarrow R\!-\!CH\!\!=\!\!S + \overline{S}\!-\!CH_2R$$

$$R\!-\!CH\!\!=\!\!S + H_2O \rightarrow RCHO + H_2S$$

In a recent study, Danehy and Hunter (115) concluded that disulfides may be decomposed by aqueous alkali by several (and perhaps simultaneous) mechanisms. Whether nucleophilic attack on the sulfur atom or proton abstraction from an α- or β-carbon takes place depends on the charge of the disulfide. Their conclusions suggest that reactions of aqueous solutions of cystine and related compounds may take place by mechanisms which are entirely different from those occurring with combined cystine in keratin.

Although this discussion has been limited to describing reactions of disulfide bonds under the influence of alkali and/or steam, there can be no doubt that other functional groups in keratin can be attacked under such drastic conditions. Peptide hydrolysis can undoubtedly take place, but more interesting are the observations of Ziegler (116) and of Asquith and Garcia-Dominguez (117) pointing out that new peptide cross-links can be formed by alkali treatment of wool.

Reaction of Hair with Sulfites

In 1932 Clarke (118) pointed out that the reaction between cystine and sulfite is a double decomposition analogous to the reaction between cystine and cyanides described by Mauthner (119).

$$^-OOC—CH(NH_2)—CH_2—S—S—CH_2—CH(NH_2)—COO^-$$

$$\swarrow X_2SO_3 \qquad\qquad \searrow XCN$$

$$^-OOC—CH(NH_2)—CH_2—SX + \qquad ^-OOC—CH(NH_2)—CH_2—SX +$$

$$^-OOC—CH(NH_2)—CH_2—S—SO_3X \qquad ^-OOC—CH(NH_2)—CH_2—S—CN$$

Goddard and Michaelis (120) reported that sulfites react fastest with keratin at neutral or slightly acid pH. Elsworth and Phillips (121) obtained similar results in the reaction between wool and sulfite, reporting an optimum pH between 4.0 and 6.0. Evidence was presented that the rate of reaction between wool and sulfites increases with temperature and is complicated by side reactions at temperatures above 60°C. Since only about half of the cystine in wool reacts with sulfites, Phillips *et al.* (122,123) suggested that wool contains fractions of cystine exhibiting different degrees of reactivity.

The combined cysteine-S-sulfonate formed in the reaction between hair and sulfite is an organic thiosulfate or "Bunte salt." Such compounds are decomposed by heat, acids, or alkalies, leading to mercaptans or sulfenic acids (124).

The existence of an equilibrium between disulfides and sulfites has been demonstrated by Stricks and Kolthoff (125).

$$R—S—S—R + X_2SO_3 \rightleftharpoons RSX + R—S—SO_3X$$

The re-formation of disulfide bonds during the washing of bisulfited wool, observed by Stoves (126), can be explained by the foregoing equilibrium. This equilibrium has been studied by Speakman (127) and by Wolfram and Underwood (128) with the conclusion that cleavage of disulfide bonds at equilibrium reaches a maximum between pH 3 and 7, with a peak at pH 4.7 at 35°C. This is in agreement with the results of other investigators and is very close to the pH at which the HSO_3^- concentration is at a maximum.

The foregoing characteristics of the disulfide–sulfite reaction make the SN_2 mechanism postulated by Zahn and co-workers (cf., e.g., 108) appear reasonable:

$$\text{protein---S---S---CH}_2\text{---protein} + HSO_3^-$$

$$\updownarrow$$

$$\text{protein---S}^- + \text{protein---CH}_2\text{---S---SO}_3\text{H}$$

This reaction is reversible by nucleophilic attack of the mercaptide or the divalent sulfur atom of the Bunte salt. Zahn also suggested that nucleophilic attack of the mercaptide on the α-carbon of the Bunte salt leads—as a side reaction—to lanthionine and thiosulfate. An alternate mechanism has been suggested by Rosenthal and Oster (114), who consider the reaction between sulfites and disulfides to be of the radical type.

Quite recently Blankenburg (129) reported that the supercontraction of wool in aqueous 5% bisulfite containing 50% n-propanol occurs at a lower temperature than in the absence of the alcohol. He attributes this effect to the influence of the n-propanol on hydrophobic bonds. The original discoverer of this effect, Speakman, applied this finding to permanent waving (130,131); more recently this finding has found some application in hair-straightening products. Speakman and Blankenburg's observations are probably related to the finding by Zahn and Osterloh (132) that keratin disulfide bonds are more readily broken in hydroalcoholic media than in aqueous solutions (cf. also 133,134).

Reaction of Hair with Mercaptans

The "softening" of hair, which is a required step in permanent waving, is accomplished in cold waving by the reaction of hair with a mercaptan, such as thioglycolic acid. The stoichiometry of the reaction is indicated by the following equation:

$$\text{protein}\begin{array}{c}CO\\NH\end{array}\!\!CH\text{---CH}_2\text{---S---S---CH}_2\text{---CH}\begin{array}{c}CO\\NH\end{array}\!\!\text{protein}$$

cystine residue

$$+ 2HS\text{---CH}_2\text{---COOH} \rightleftharpoons$$
thioglycolic acid

$$\text{protein}\begin{array}{c}CO\\NH\end{array}\!\!CH\text{---CH}_2\text{---SH} + HS\text{---CH}_2\text{---CH}\begin{array}{c}CO\\NH\end{array}\!\!\text{protein}$$

cysteine residues

$$+ HOOC\text{---CH}_2\text{---S---S---CH}_2\text{---COOH}$$
dithiodiglycolic acid

In 1920 Lecher (135) demonstrated that mercaptides (RS^-) can reduce organic disulfides. Several years elapsed before this general reaction (disulfide interchange) was applied to the disulfide bonds in keratin.

The reaction between keratin and thioglycolic acid was studied by Goddard and Michaelis (120), who showed that this reaction takes place rapidly and appreciably only in alkaline solutions. Bersin and Steudel (136) demonstrated that the simpler reaction between cystine and thioglycolic acid is reversible and has an equilibrium constant approximating unity. The velocity of the reaction increases with pH and temperature and is reported to depend on the extent of ionization of the mercaptan group of thioglycolic acid. These authors suggest, therefore, formation of the thioglycolide anion,

$$(COO-CH_2-S)^=,$$

and indicate the following mechanism which includes a mixed disulfide (I) as a key unstable intermediate.

$$HS-CH_2-COOH + 2OH^- \rightleftharpoons (S-CH_2-COO)^= + 2H_2O$$

$$(S-CH_2-COO)^= + (OOC-CH(NH_2)-CH_2-S)_2^= \rightleftharpoons$$

$$(OOC-CH(NH_2)-CH_2-S-S-CH_2-COO)^= + (OOC-CH(NH_2)-CH_2-S)^=$$
$$(I)$$

$$(OOC-CH(NH_2)-CH_2-S-S-CH_2-COO)^= + (S-CH_2-COO)^= \rightleftharpoons$$
$$(1)$$

$$(OOC-CH_2-S-S-CH_2-COO)^= + (OOC-CH(NH_2)-CH_2-S)^=$$

The kinetics of the thiol-disulfide interchange have been studied by Fava and his co-workers (137), who conclude that the reacting species is the mercaptide ion. This identification of the reactant is also supported by the detailed equilibrium analysis by Wolfram and Underwood (128) and is generally accepted (138). On the other hand, the free radical mechanism suggested by Rosenthal and Oster (114) has received no additional support in recent years.

In 1953 Schöberl claimed (139) that the mixed disulfide (I) between cysteine and thioglycolic acid may exist in hair treated "by a cold wave process." Schöberl's evidence for the existence of the mixed disulfide was based on the observation that treated hair contains more sulfur than unwaved hair. Actually, the observation of increased sulfur content in reduced and reoxidized wool was first made by Patterson and co-workers in 1941 (140). They also indicated that removal of residual thioglycolic acid is possible by washing with ethanol and that the reason for the increased sulfur content may be formation of the mixed disulfide.

Subsequently, Schöberl and Gräfje (141,142) supported this supposition by the synthesis of several mixed disulfides. Schöberl and Gräfje (143) also suggested that the mixed disulfide in permanently waved hair is in part responsible for hair damage in permanently waved hair. Actually, these authors provided no evidence for the existence of the mixed disulfide in *waved* hair. Some support for the existence of the mixed disulfide can be found in the analyses by Gerthsen and Gohlke (144) and the radioactive studies by Schulte and co-workers (145). In this connection, it is noted that Whitman and Eckstrom (146) claimed that the mixed disulfide can be formed in hair, especially if some disulfide is present in the reducing agent. These authors also claimed that the mixed disulfide is undesirable and accounts for hair damage.

Though it is widely assumed today that a mixed disulfide forms during hair waving and that it is present in fairly large amounts, this interpretation of published data may not be correct. Schulte and co-workers (145) clearly show that only very little radioactive thioglycolate (9 to 10 mg/g hair) is found in hair if hair *soaked* in an excess of labeled thioglycolate waving lotion is rinsed for 1 min before oxidation. Whether the quantity of retained thioglycolic acid is altered by conditions which do not include soaking in waving lotion is not discussed. On the other hand, these authors demonstrate the presence of radioactive mixed disulfides after 4 hr (!) of soaking in a large excess of waving lotion and subsequent oxidation *without* rinsing. Most other demonstrations of the presence of the mixed disulfide routinely include acid hydrolysis of the hair. According to Zahn (147), the mixed disulfide can be formed during hydrolysis of keratin in the presence of thioglycolic acid, and he concludes that the amount of mixed disulfide formed under practical hair-waving conditions is so small as to be essentially insignificant. In this connection, it is important to note that acid hydrolysis of keratin can lead to a variety of sulfur-containing compounds which evidently are not present in the original keratin (148,149).

The reaction between hair and alkaline thioglycolate rapidly reaches an equilibrium, the extent of reaction depending on the ratio of hair to thioglycolate. The equilibrium position of the reaction can be altered in hair-waving practice by the incorporation of dithiodiglycolic acid into the waving lotion (150).

Prolonged immersion of hair fibers at room temperature in an excess of alkaline thioglycolate produces extensive swelling (31,151–156) of the fiber and reduces the toughness almost to zero. Sanford and Humoller (67) reported that approximately 85 % of the cystine present in hair is reduced in 8 min during immersion of hair fibers in an excess of waving lotion.

Experimental results obtained by immersion of hair into cold-waving products and other chemical solutions must be examined with care. As

Reed (11) has demonstrated, a wound tress of hair can hold only a limited quantity of waving liquid. Depending on the tightness of winding, the liquor/hair ratio rarely exceeds 2:1 in practice. Any significant experimental deviation from this ratio makes the practical value of such experiments questionable. Thus, during normal hair waving, approximately 25 % of the cystine present in hair is reduced in less than 10 min, lowering the 20 % index of reduced hair to no less than about 0.80 (Tables I and II). This level of reduction is not exceeded during prolonged processing unless additional waving lotion is employed. This fact does not require that only 25 % of the cystine in hair be involved in the permanent waving process. Many more disulfide linkages in hair may be broken by the thiol–disulfide interchange mechanism, but at any given moment, no more than about 25 % of the total disulfide bonds is in the reduced form.

Although this discussion of the chemical reaction between hair and mercaptans has been concerned exclusively with thioglycolic acid, the reactions with other mercaptans are undoubtedly analogous. Differences in the degree of ionization of the mercaptan group to —S⁻ and the charge on the mercaptan probably alter only the rate of reduction and/or the equilibrium position (157).

Reactions of Hair That Establish Cross-Linkages

The various reactions discussed above are used in permanent waving to "soften" the hair fiber primarily by rupture of disulfide linkages. The second chemical step required for permanent waving of hair re-forms the ruptured linkages or creates novel cross-linkages. Several of these reactions have already been discussed in connection with the steaming and alkali treatment of hair. Among these may be mentioned the possible formation of S–NH linkages and of lanthionine.

The most satisfactory method of "hardening" hair during cold waving is an oxidation process which restores the fiber to its original condition by re-formation of disulfide linkages. In addition, a few nonoxidative processes will be mentioned briefly, because they have been employed for hair waving.

Oxidative Cross-Linking

Since oxidation of sulfhydryl groups in hair may be accomplished either by means of a chemical oxidizing agent or by air alone, it will be convenient to discuss these two methods separately.

1. Chemical oxidation. A number of the usual chemical oxidizing agents are suitable for the oxidation of sulfhydryl groups to the corresponding disulfide. Hydrogen peroxide, perborate, and bromate are widely used in hair waving. Sanford and Humoller (67) have shown that the cysteine

TABLE I. Chemical Oxidation

History of hair sample	Cysteine (%)	20% Index
Untreated	0.40	1.00
Reduced[a] for 10 min	4.00	0.85
Reduced[a] for 20 min	4.20	0.86
Reduced[a] for 30 min	4.20	0.83
Reduced[a] for 30 min, oxidized with 0.2 M bromate (wet)	0.70	0.91
Reduced[a] for 30 min, oxidized with 0.1 M perborate (wet)	0.57	0.91
Reduced[a] for 30 min, oxidized with 0.2 M bromate, fibers dried for 24 hr at room temperature	0.44	0.93

[a] The reduction was carried out under actual waving conditions with a lotion containing 0.65 M of sodium thioglycolate per liter adjusted to pH 9.5 with ammonium hydroxide.

present in extensively reduced fibers disappears rapidly and almost completely after treatment with bromate. Similarly, during actual waving, the disappearance of cysteine from hair during the "neutralization" of hair previously reduced by thioglycolate proceeds rapidly and is accompanied by an increase in the 20% index of hair fibers. This is indicated in Table I (based on unpublished results by the authors).

Similar conclusions have been reached by Whitman (10) on the basis of results obtained under somewhat more drastic conditions.

Although ample evidence has been presented demonstrating that sulfhydryl groups disappear during chemical oxidation (10,67), the nature of all the oxidation products is by no means established. Formation of the already mentioned mixed disulfide (139,144,145) is only one of many possible reactions. The problem is further complicated by the fact that the results of Sanford and Humoller (67) indicate that the cystine content of thioglycolate-reduced and bromate-oxidized hair is lower than that of untreated hair. This may be due to oxidation of combined cysteine beyond the desired cystine to combined cysteic acid (158).

$$\text{protein} \underset{\diagdown NH \diagup}{\overset{\diagup CO \diagdown}{}} \text{CH}-\text{CH}_2-\text{SO}_3\text{H}$$

This and similar side reactions during the "neutralization" of waved hair appear plausible in view of the work of Rutherford and Harris (159), who showed that the action of hydrogen peroxide on wool may produce a variety of sulfur-oxygen compounds. Zahn (147) confirmed the formation of combined cysteic acid during the oxidation of thioglycolate-reduced hair by dilute acidic hydrogen peroxide.

The mechanism for the oxidation of mercaptans by hydrogen peroxide in homogeneous aqueous systems has been studied extensively. Heavy metal catalysis of peroxide decomposition or heavy metal complexation of mercaptan has been suspected (160) in the rate-determining step (cf. 161). Nevertheless, the details of the mechanism in solution are not fully established, and the problems in the heterogeneous system, reduced hair and hydrogen peroxide, are obscure.*

Stoves (162) studied the effect of a 6.25% solution of 100-vol hydrogen peroxide on human hair and showed that damage to hair increases with pH and becomes more severe if decomposition of hydrogen peroxide is catalyzed by metal ions, such as Cu^{2+}, Ni^{2+}, Co^{2+}, or Mn^{2+}.

2. Air oxidation. In 1913 Thunberg (163) showed that mercaptans are oxidized readily by air in the presence of small amounts of certain heavy metal ions. Kharasch and his co-workers (164) reported that copper is the most effective catalyst for the oxidation of thioglycolic acid to its disulfide and of cysteine to cystine. Manganese is almost as effective as copper, whereas iron is a relatively weak catalyst for this air oxidation. Manganese has been incorporated into waving lotions where it serves a twofold function, to oxidize excess thioglycolate and to effect re-formation of hair-disulfide bonds (165).

The mechanism of the air oxidation of mercaptans probably proceeds through a free radical mechanism (160,166). In the presence of a metal catalyst, such as iron, the following series of reactions takes place (167): H_2O_2 oxidizes the Fe^{2+}–mercaptan complex to the Fe^{3+} stage which is unstable and decomposes to Fe^{3+} ion and the disulfide.

On the basis of load elongation studies and cysteine analyses, it can be shown that catalyzed air oxidation restores reduced hair as well as does chemical oxidation. The results presented in Table II (from unpublished results of the authors) were obtained using hair that was waved with a 0.63-N solution of ammonium thioglycolate at pH 9.2 and containing 32 ppm of manganese. The wound hair was wet with lotion, blotted after 20 min, and then allowed to dry in air.

Whitman (10) presented similar results and showed further that, in order to ensure maximum strengthening of the fiber, air oxidation should proceed slowly and should not be accelerated by heat.

---

* Unpublished results by one of the authors (M. M. R.) show that the disappearance of combined cysteine from hair under the influence of hydrogen peroxide proceeds more rapidly and more completely if the remains of the permanent-waving lotion are not completely removed from the hair.

TABLE II. Air Oxidation

History of hair sample	Cysteine (%)	20% Index
Untreated	0.25	1.00
Reduced for 20 min	4.50	0.83
Reduced for 20 min, blotted, air-dried 40 min	0.65	—
Reduced for 20 min, blotted, air-dried 4 hr	0.20	—
Reduced for 20 min, blotted, air-dried 24 hr	0.25	0.95

Load elongation curves presented by Harris and his co-workers (93,94) indicate that oxidation of wet wool fibers with oxygen accomplishes re-formation of cross-links.

Nonoxidative Cross-Linking

Nonoxidative methods for the cross-linking of reduced hair have been described but have found only limited use in hair waving. Speakman (168,169) claims that bivalent metal ions, such as barium, form stable cross-linkages between two adjacent sulfhydryl groups in hair, and that this reaction is useful for hair waving:

$$2 \text{ protein—SH} + Ba^{2+} \rightarrow \text{protein—S—Ba—S—protein} + 2H^+$$

Patterson and his co-workers (170) consider ionic bonds of this type unstable.

The reaction of reduced wool and alkylene dihalides has been studied by Harris and his group (cf. e.g. 171) and has been suggested for hair-waving purposes (172).

Several other nonoxidative cross-linking agents have been suggested for neutralization. These are generally bifunctional compounds, such as dialdehydes, or even simple aldehydes such as formaldehyde (173). None of these has achieved prominence in hair waving.

The Permanent Effects of Hair Waving on Hair

It was already noted that treatment of keratin with an excess of alkaline mercaptan can cause severe damage, especially if the fiber is under tension during reduction (90). In normal hair-waving practice, no significant permanent damage occurs, and most of the weakening of fibers caused by reduction is repaired by the oxidizing neutralization step. Nevertheless, certain permanent changes in hair fibers take place during the waving step.

First of these is, of course, the permanent wave itself. Its quality and permanence are of prime importance, and laboratory methods for studying waving performance are important for development and control purposes. Such methods have been described by Kirby (174) and by Stavrakas and co-workers (175). The problem of permanency of waves has been studied more recently by Freytag (175a), who described means for evaluating the mechanical effects of combing, brushing, humidity, and temperature on permanently waved hair. Permanent waving usually causes some additional hair changes even under ideal conditions. The changes can be assessed mechanically (79,176,177), by swelling measurements (31,178), and by chemical procedures (179–181). Robbins and Kelly (181a) have recently established that permanent waving causes minor changes in the amino acid composition of hair. Thus they have found that the cystine content is slightly reduced after completion of a normal permanent wave and that this change is accompanied by a corresponding increase in the cysteic acid level in the hair fiber. Although these modifications of hair during permanent waving are of scientific and forensic interest, they are of no concern to the consumer under normal circumstances.

HEAT WAVING

Heat waving, as the name implies, is a method for permanent waving of hair by the utilization of heat. Wound hair, moistened with a suitable waving fluid before, during, and/or after winding, is heated to about 93 to 104°C (182) electrically or with the aid of so-called heat pads.* Heat waving at such very high temperatures often requires heavy and expensive equipment or very carefully controlled processes to generate steam on the head. It is not surprising, therefore, that heat waving has lost its popularity, and much of this discussion of heat waving is only of historical interest. Information derived from studies of heat waving is now applied to a new technique, called tepid waving (see pp. 223–224). This latter method of permanent waving utilizes moderate temperatures which can be obtained under the hoods of conventional hair dryers.

In order to avoid scalding or burning of the scalp during heat waving, so-called "spacers," made generally of felt, are placed between the rolled tress and the scalp. After the heating is completed, the hair may be rinsed with water or dilute acid to remove residual alkali or with solutions of oxidizing agents. Originally, "neutralizers" were used to remove excess alkali left in the hair and to return the hair to its isoelectric point. "Neutralizers" are generally oxidizing rinses but have retained the name which described their original function.

---

* It is beyond the scope of this review to discuss the various methods used to supply heat to the hair. A number of good reviews on this subject are available (183–186).

Formulation of Heat-Waving Preparations

Compositions Containing Alkalies and Sulfites

Early heat-waving preparations contained nonvolatile alkalies as active ingredients. As early as 1913 Grosert (187) received a patent for a waving composition combining borax with sodium hyposulfite (thiosulfate). A similar combination of active ingredients with oily materials was patented by Ingrassia (188) in 1926.

Compositions containing nonvolatile alkalies exclusively are of historical interest only. They were generally aqueous solutions of one or more alkaline salts or alkalies in water. Typical ingredients include alkali carbonates and borax (sodium tetraborate). Borax, which has been a favorite ingredient of such preparations since the days of Nessler, has a tendency to crystallize or even "fuse" on the hair (189). Allen (189) cites a formulation, shown as Formula 1, which is considered typical.

Formula 1

Potassium carbonate	40 g
Borax	10 g
Tragacanth mucilage	100 cc
Coumarin	5 g
Methyl acetophenone	1 cc
Ethyl alcohol	100 cc
Rose water, to make	1000 ml

Kietz and his co-workers (190) were granted one of the first patents covering the use of sulfites in hair waving. The composition of a heat-waving lotion containing sulfite and nonvolatile alkalies is presented in Formula 2 (191).

Formula 2

Sodium sulfite	35 g
Sodium bicarbonate	20
Sodium borate	7
Powdered soap	4
Water, to make	1 liter

Compositions Containing Volatile Alkalies and Sulfites

Since about 1930, workers in the field of heat-wave preparations have been concerned with practical ways and means to protect hair against the damaging and drying effects of heat and alkalies of the nonvolatile type (189). One of the important milestones in the attack on this problem was the substitution of volatile bases for the permanent alkalies used in older preparations (192). The use of volatile bases such as ammonia, ammonium carbonate, morpholine, etc. provides a strongly alkaline medium (*p*H of 9 to 11) during the initial period of heating. As the hair is softened and weakened, the alkali is volatilized; the *p*H begins to drop, thus reducing danger of overwaving or of partial hydrolysis of the protein by the alkali.

In addition to the indicated advantage of mildness, lotions containing volatile bases require less nonvolatile alkalies and, therefore, have less tendency to deposit or concentrate hair-harshening chemicals on the hair. The disadvantages of the lotions containing volatile alkalies are the objectionable odors and premature loss of the alkali from the solution. To minimize this loss of alkali, modern formulations often contain one or more of the alkanolamines in addition to ammonia.

It appears that the introduction of sulfites into practical heat-wave lotions occurred about the same time as the use of volatile bases. The combination of sulfites with volatile alkalies was patented by Steinbach (193) in 1935.

Rüttgers (194) suggests that sulfite-containing waving lotions are preferred over those containing only alkalies because the former require shorter heating periods, do not yellow hair, and, overall, give better waves.

Heat waving may be performed over a wide range of temperatures. The composition of the lotion to be used depends on the temperature to which the hair is heated, the duration of heating, and the length of time the product remains on the hair. Generally useful typical formulations of heat-wave compositions of the liquid and creme type have been published (195) and are shown in Formulas 3 and 4.

Formula 3.   Liquid Heat-Wave Preparation

Monoethanolamine	6.0%
Potassium sulfite	1.5
Potassium carbonate	1.5
Borax	0.5
Ammonium carbonate	2.5
Sulfonated castor oil	1.0
Water, distilled	87.0

Formula 4.   Creme Heat-Wave Preparation

Ammonia water, 28%	8.0%
Ammonium carbonate	5.0
Potassium carbonate	3.0
Sodium sulfite	4.5
Sodium hydroxide	4.0
Water	55.5
Oxycholesterol absorption base	20.0

Formula 5

"Furmament Emulsifier"	7 to 12%
Kerosene, purified	4 to 40 parts
Hydrous wool fat, purified	1 to 20
Commercial wool wax	1 to 20
Paraffin wax	0.00 to 0.50
Mineral oil, refined	1.00 to 5.00
Castor oil	0.00 to 1.00
Sodium sulfite	3.00 to 10.00
Sodium hexametaphosphate	1.00 to 4.00
Glycerol	0.25 to 1.00
Triethylamine	0 to 2.00
Ethylenediamine	0 to 1.00
Water	q.s. 100.00

Procedure: Melt together the fatty ingredients at 90°C and dissolve the water-soluble materials in warm water. Then add the aqueous phase to the oil phase with continuous stirring until a thick white creme emulsion is obtained. The stable emulsion is conveniently packaged in collapsible tubes from which it can be applied directly to the hair.

Furman (196) presented detailed instructions for the preparation of thick hair-waving cremes, with the compositions shown in Formula 5.

For the waving of dyed hair, Auch (197) recommends a preparation, shown as Formula 6, that contains very little volatile alkali in the form of ammonium carbonate.

For hard-to-wave hair, Sarensen (198) suggests Formula 7.

Formula 6

Potassium sulfite	5.50%
Borax	2.65
Sodium carbonate	0.45
Ammonium carbonate	0.25
Perfume	0.50
Water, distilled	q.s. 100.00

Formula 7

Sodium sulfite	10.0 parts
Ammonium carbonate	8.0
Ammonia, 26%	2.0
Glycerol	0.5
Powdered acacia	0.5
Water	107.0
Total	*128.0 parts*

Most heat-wave preparations contain, in addition to the active ingredients (alkalies and sulfites), one or more surface-active agents to aid the wetting and penetration of the hair (199).

In order to protect sulfite-containing waving lotions against air oxidation, DeMytt and Reed (200) patented the use of a sequestering agent which stabilizes the sulfite solutions by removal of metallic ions.

To minimize possible hair damage during waving, the addition of dissolved hair to waving lotions has been recommended (201). The incorporation of "oily" substances into heat-waving compositions is considered beneficial to the hair. Virtually every emollient available to the cosmetic chemist may serve in this capacity. Glycerol, fatty alcohols, lecithin, lanolin, mineral oil (202), cholesterol, and sulfonated castor oil are some of the favored ingredients. Quaternaries (203) have also been suggested as components of heat-waving compositions. Friedman and Goldfarb (204) described the preparation of a starch paste that is claimed to provide a protective coating on the hair. Jannaway (184) indicated that the use of emollients in permanent waving solutions is of debatable merit. Benk (205), who reviewed this subject, pointed out that nothing is known regarding the mode of action of these materials.

A U.S. patent protecting a *dry* composition (cf. Formula 8) useful for heat waving was issued to Steinbach (206). The mixture is intended for use in aqueous solution.

Formula 8

Potassium sulfite	20.5%
Potassium carbonate	43.3
Ammonium sulfite	36.2

Patents have been issued to Coriolan G.m.b.H. (207) protecting the use of ketone or aldehyde addition products of bisulfite for hair-waving purposes. These addition compounds may be useful in preparations of this type because they serve as a source of sulfite that is unaffected by oxygen until they are hydrolyzed during heating. The use of a 10% sodium formaldehyde sulfoxylate solution for heat waving was patented by Ströher (208) and by Amica (172). In contrast to bisulfite addition compounds, this substance is subject to air oxidation and reacts with keratin disulfide bonds without prior decomposition by heat (209).

As pointed out above, the elevated temperatures required to wave hair by means of sulfite–alkali waving lotions may promote excessive amounts of undesirable side reactions, particularly between the alkalies and hair, tending to produce harsh, dull, weak fibers. Speakman (100), in efforts to minimize these undesirable reactions, suggested formula changes intended to permit hair waving at lower temperatures than were required by conventional lotions of the time. Speakman improved upon his sulfite (210,211) and bisulfite (212,213) waving lotions by incorporating a water-soluble alcohol or glycol (130,131) to accelerate waving and to permit waving at moderate temperatures (cf. tepid waving, pp. 223–224).

Compositions Containing Reducing Agents Other then Sulfites

A number of heat-waving compositions containing reducing agents other than sulfites have been suggested. These materials have not gained wide acceptance. Sulfides have been used, but their use in a heat-wave solution (214) is of no interest because of odor and toxicity.

Reed and his co-workers (215) described a method of hair waving that combines the use of a thioglycolate-containing shampoo with the technique of heat waving. The hair is shampooed with a preparation of the composition indicated in Formula 9.

Formula 9

Sulfated lauryl alcohol	25 g
Thioglycolic acid	3 g
Ammonium hydroxide to adjust alkalinity to pH 9.0	
Water	q.s. 100 ml

After about 15 min, the hair is rinsed, wound on curlers, and then subjected to a temperature of about 82°C. It is claimed that the heating temperature may be lowered if the mercaptan content is raised to about 5%.

It has also been suggested that thioglycolates may be used as the active components in a heat-wave preparation. Grant (216) patented a process in which the hair is moistened with a 1 to 5% solution of thioglycolic acid, adjusted to a pH of 7.0 to 8.5. After a suitable heating period, the hair is saturated with a dilute solution of acetic acid containing hydrogen peroxide and then reheated.

Heat-Waving Processes Based on Deposition of Polymers

The idea of depositing a polymeric material within or on the fiber to maintain it in a permanently curled position has been studied by a number of investigators. Such a treatment should leave the hair soft and supple, and yet the polymer must be rigid enough to support a wave. The polymer should be deposited in a manner that would avoid subsequent adhesion of the hair fibers. In view of the difficulties of these requirements, it is not surprising that this method of permanent waving has not been perfected.

In 1928, Brown (217) received a patent for a waving process in which dissolved keratin is deposited from an ammoniacal solution at a temperature of 100°C to produce a permanent wave. Maeder and Sims (218) patented a similar process using an aqueous alcoholic solution of partially hydrolyzed hair.

Calva (219) patented a method for changing the configuration of hair, involving the reaction between a "reacting agent" and an "activating agent." Formaldehyde was cited as a suitable "reacting agent" with a sulfonated phenol (e.g., resorcinol disulfonic acid) or an amine (e.g., triethanolamine) as "activating agents." In addition to the reaction between the aldehyde and the hair, condensation products (such as those resulting from the reaction of the formaldehyde and the phenol) were deposited. Nowak (220) reported that urea-formaldehyde polymers have been suggested for hair waving.

In 1942 Luckenbach (221) patented the incorporation of a water-soluble polyacrylate into a creme-type heat-waving preparation based on ammonium and monoethanolammonium carbonate. It was claimed that the polymer was deposited as a thin hygroscopic coating which held the hair in the form or shape it had been given.

In 1953 Gant (222) patented a waving process in which a polysiloxane is deposited on the fiber by evaporation of solvent or by continued polymerization of a lower-molecular-weight polysiloxane under the influence of high heat (150 to 260°C) in the presence of a catalyst. Preferably a nonpolymerizable silicone fluid is included in the waving composition to act as a lubricant and softening agent. Gant (223) also patented a process which utilizes an aqueous dispersion of the polysiloxane. The use of diepoxides (224) and diepoxides containing a silicone substituent (225) for hair waving has been patented in Germany.

The wool industry has been concerned for years with upgrading low-quality wools and with improvements which include "setting." Setting in the textile industry refers to the stabilization of a textile structure (fiber or yarn) by a physical or chemical process (225a). None of these developments has been successfully applied to permanent waving of hair.

COLD WAVING

In contrast to heat and tepid waving, cold waving requires no externally applied heat to achieve the desired result. Thus a typical cold-waving process utilizes waving preparations at room temperature. Accordingly, the chemical reactions normally take place at temperatures between ambient and body temperature, usually about 30°C. Generally, a detergent shampoo precedes waving to ensure subsequent thorough saturation of the hair with the lotion. The hair is sectioned, blocked, and wound on the curler. Winding may be done with damp hair, dry hair, or hair moistened with the waving fluid. Wherever suitable for the desired hair style, the hair is wound as closely to the scalp as possible to ensure a close scalp curl. In contrast to heat waving, stretching of the hair is carefully avoided during winding. Unless a very weak reducing solution is employed, stretching of hair can result in extensive hair damage and even breakage. Usually the wound hair is saturated thoroughly with waving fluid after winding. Depending on the strength of the lotion and the desired curl tightness, the lotion may be allowed to remain in contact with the hair for 2 min up to 2 hr. In modern practice, the processing time ranges from about 10 to 30 min. Next, the hair is rinsed with water and/or towel-blotted to remove as much remaining waving lotion as possible. The hair is finally oxidized either by exposure to air or by treatment with a "neutralizer." Directions supplied with most modern waving products suggest that the hair remain wound on the curlers until oxidation is almost complete. If a chemical oxidant is employed, a final water rinse is generally required before the hair is set or styled. If air is relied upon as the oxidizing agent, the hair is permitted to remain on the curlers until essentially dry.

Cold waving began during the 1930's. In 1936 Malone (226) described the use of an alkaline solution containing a water-soluble neutral salt and a small amount of trypsin for hair waving in a process requiring no external heat. This method did not require spacers to protect the scalp against burns, and it therefore permitted waving of the proximal portion of hair. Shortly thereafter, Malone (227,228) patented a second method for cold waving involving an amphoteric metal salt (such as stannite) in the presence of auxiliaries (such as sulfites, sulfides, and enzymes).

In 1939 Brown (229) received a patent for a waving process which involved the use of a 5% solution of ammonium sulfite for 3 to 6 hr at *substantially* normal room temperatures. To complete the permanent wave, the hair was finally subjected to treatment with an oxidizing agent.

Later that year, Willat (230) patented the use of sulfides or bisulfides at room temperature for hair waving. Willat specified the use of croquignole curlers for winding because they permit waving close to the scalp. For beauty salon use, Willat designed a system that allowed circulation of the sulfide waving fluid through the curlers and the hair, thus reducing the attendant offensive odor (231).

Also in 1939, Pye (232) received a patent for the use of sulfides or bisulfides as cold-waving agents. His disclosures included as waving agents all compounds containing sulfur having a valence of −2 in the pH range of 8 to 11. Pye's process differs from Willat's in that the former employs a neutralizing step utilizing a 4% solution of acetic acid saturated with cane sugar. Sulfides were also included in Speakman's list of reducing agents for permanent waving (212,233). The use of alkanolammonium sulfides (234) and magnesium sulfide (235) for cold waving has also been patented.

In March 1941 a death was attributed to hydrogen sulfide poisoning resulting from a wave given with an ammonium bisulfide lotion. The Food and Drug Administration confiscated lotions of this type and precluded their use for hair-waving purposes (235). In view of their short and unfortunate history, waving lotions containing sulfides are of no importance today.

Thioglycolates have been the active waving agents in almost all successful cold waving lotions. Such lotions were demonstrated and marketed as early as 1941 (236). The early history of cold waving with alkaline thioglycolates is obscured by patent claims by various inventors and by the fact that some of the major patents covering thioglycolate cold-wave lotions have been declared invalid in the United States (236,237).

In 1936 Coustolle (238) disclosed that monoethanolammonium thioglycolate is "beneficial" for scalp and hair. Martin (239), in 1944, mentioned the use of thioglycolic acid as an "accelerator" in cold-waving lotions containing hydroxylamine sulfate and an alkaline substance, such as ammonium or sodium hydroxide.

The next year, Baker (240) described cold-waving lotions containing thioglycolic acid or monochloroacetic acid as one of the ingredients. His preferred compositions contained, in addition, ammonium hydroxide, ammonium thiocyanate, and ammonium sulfite.

A fairly complete disclosure of a variety of mercaptans, including thioglycolic acid, has been made by McDonough in a series of patents (241–247) concerning active ingredients for hair-waving compositions. McDonough (245) covers a permanent waving composition containing a mercaptan having a molecular weight less than 121 in a solution containing an alkaline material having a dissociation constant less than 5×10^{-3}. The mercaptan, RSH, carries a vapor-pressure-depressant substituent of the acidic (thioglycolic acid), basic (mercaptoethylamine), or nonpolar (glyceryl monomercaptan) type. Solutions of pH from 7.0 to 9.5 are included in the claims of the early McDonough patents (244,245). The more recent McDonough patents are not as broad and identify the pH and the choice of mercaptan more specifically (246,248). The U.S. patents issued to McDonough have been challenged in the courts and have been held invalid (236,237,249).

The Formulation of Cold-Waving Preparations

The preparation of a cold-waving product requires not only the formulation of ingredients that will wave hair but also their further modification by virtue of perfumes, opacifiers, color, and hair-conditioning additives. The composition of opacifiers and the nature of conditioning additives will be discussed later.

It is convenient to divide the discussion of the formulation of cold-waving preparations into several sections concerned with (a) products based almost exclusively on thioglycolic acid, (b) products based on mercaptans other than thioglycolic acid, and (c) products based on nonmercaptan reducing agents.

Products Based on Thioglycolic Acid

Cold-waving lotions may be used either by professional beauticians in beauty salons or by unskilled consumers in their own homes. Products designed for beauty salon use, generally, are stronger and faster-acting than those intended for home use. This is accomplished in practice by one or more of the following: an increase in pH, mercaptan content, choice of alkali, or total alkalinity of the lotion (31). In the case of preparations for professional use, the beautician judges the time of contact of the product with the hair. As a result, a fast-acting preparation may be employed without danger of overwaving. Formulations intended for home use are slower-acting and designed to be effective with less critical timing.

In its simplest form, a typical cold-waving formulation may contain only thioglycolic acid which has been adjusted to the desired pH with a suitable alkali (250). Generally, the mercaptan content of home permanent waving formulations does not exceed 0.65 M, corresponding to about 6.0% thioglycolic acid, whereas waving products for professional use may contain as much as 8.5% thioglycolic acid.

The pH of thioglycolate-containing waving preparations without accelerating additives is generally maintained between 8.5 and 9.5, with a preferred range of 9.1 to 9.3. The choice of alkali used to reach this pH is wide and includes mixtures of various amines and other alkalies. Products containing at least some ammonia or a low-molecular-weight amine generally wave hair faster than those containing only an alkali metal hydroxide (251,252). Ammonia is preferred over nonvolatile amines because it escapes during processing. The resultant drop in pH of the waving product reduces its activity and minimizes the possibility of overprocessing.

According to McDonough (244,245), the preferred alkali should have a dissociation constant of less than 5×10^{-3}. However, for practical purposes, alkalies with a dissociation constant ranging from 1×10^{-4} to about 1×10^{-6} are required for formulating a reasonably effective cold-waving preparation containing thioglycolic acid.

Heilingötter (253) concluded that the waving action of thioglycolates depends on pH and requires free ammonia, i.e., ammonia in excess of that

TABLE III.   Effect of Ammonia and pH on Waving Power of Thioglycolic Acid

Waving agent	pH	Free ammonia (%)	Results
8.8% Thioglycolic acid	2.0	—	Weak or no wave
9.2% Ammonium thioglycolate	7.0	0.05	Weak or no wave
9.2% Ammonium thioglycolate	8.8	0.24	Weak wave
9.2% Ammonium thioglycolate	9.2	0.75	Adequate wave
9.2% Ammonium thioglycolate	>9.2	>0.75	Wave becomes stronger

required to neutralize the carboxyl group of the acid. Some of his hair-waving results at constant waving time are summarized in Table III.

Obviously, the addition of ammonia to ammonium thioglycolate increases the free ammonia, free alkalinity, and the pH. It is, however, possible to increase the free alkalinity of a waving product at constant pH, thereby increasing the effectiveness of the waving product. Practical procedures for this purpose will be discussed later, but the theoretical considerations deserve some comment: The addition of an ammonium salt (254) to ammoniacal ammonium thioglycolate lowers the pH (law of mass action); in order to keep the pH constant, additional ammonium hydroxide is required in the case of a product with added ammonium salt, e.g., NH_4Cl. For this reason, too, a waving preparation in which the carboxyl groups of thioglycolic acid have been neutralized with, e.g., sodium hydroxide to a pH of approximately 5 to 6 will require much less ammonium hydroxide for adjustment to pH 9.5 than a comparable product in which neutralization is effected with ammonium hydroxide. From practical experience, it is known that the effectiveness of waving preparations at a given pH is dependent on the amount of ammonium hydroxide utilized beyond that required for neutralization of the carboxyl group.

Modern cold-wave products for home and professional use have been specialized to fit the needs of users with various types of hair. Products have been marketed in three strengths, for hard-to-wave, normal, and easy-to-wave hair. Special preparations utilizing either higher free alkalinity, higher pH, or higher thioglycolate concentrations are also available for waving the resistant hair of children. The thioglycolic acid and ammonia contents of some typical commercial products are presented in Table IV.

Shansky (255) analyzed a variety of commercial products designed for normal hair, and some of his results are presented in Table V.

The effect of pH and free alkalinity on hair waving with thioglycolate-containing products can be interpreted as follows: An increase in pH is probably related to an increase in the formation of mercaptide ion by dissociation of —SH groups and a corresponding increase in reaction with

TABLE IV. Composition of Commercial Cold-Waving Lotions for Various Types of Hair

Lotion	Thioglycolic acid (%)	Total ammonia as NH_3(%)	pH
Children "A"	5.23	2.01	9.52
Children "B"	6.59	2.14	9.22
For "hard-to-wave" hair ("A")	6.62	2.11	9.24
For "hard-to-wave" hair ("B")	5.64	0.935[a]	9.65
For "normal" hair	5.64	2.04	9.27
For "easy-to-wave" hair	4.63	1.41	9.20
For professional use	8.50	2.70	9.25

[a] This lotion contains approximately 2.5% sodium hydroxide.

TABLE V. Analyses of Commercial Cold-Waving Lotions for Normal Hair

Type	Thioglycolic acid (%)	Alkali (%) Free	Alkali (%) Total	pH
Home	5.0	MEA[a] 2.1	MEA 5.8	9.2
Home	6.7	NH_3 1.0	NH_3 2.3	9.2
Professional	6.9	NH_3 0.9	NH_3 0.9 + MEA 4.6	9.3
Professional	8.1	NH_3 1.3	NH_3 2.8	9.3

[a] MEA = monoethanolamine.

protein disulfide bonds. Such an effect has been observed by Bersin and Steudel (136) in the reaction between cystine and thioglycolic acid. A high concentration of mercaptide ion is probably involved in the use of thioglycolates as depilating agents (256) at pHs above 10.0. The effect of free alkalinity on the rate of waving is more difficult to explain, but an attempt has been made by Hermann (156), who reported that the diffusion rates of mercaptans and fiber swelling are highly dependent on alkalinity in the heterogeneous reaction between hair keratin and mercaptan solutions.

Most of the formulations shown in Table III, IV, and V contain fairly large quantities of ammonia. Although ammonia-containing thioglycolate waving lotions are excellent waving agents, the odor of ammonia is considered objectionable. Complete or partial replacement by nonodorous and non-toxic alkanolamines is widely practiced throughout the industry. Brant (257) considers secondary alkanolamines well suited to form the reservoir of free alkalinity in waving lotions and prefers diisopropanolamine. He further recommends that a primary amine be used to neutralize the acidity of thio-glycolic acid and that the secondary amine be used to raise the pH to the required level. Some typical compositions from his patent are shown in Table VI.

TABLE VI.   Recommended Compositions of Cold-Waving Products (257)

Ingredient	Concentrations (M)			
Thioglycolic acid	0.80	—	0.55	0.55
Ammonium thioglycolate	—	0.50	—	—
Diisopropanolamine	0.61	0.55	0.65	0.60
Ammonia	0.69	—	—	—
Monoethanolamine	—	—	0.50	—
Monoisopropanolamine	—	—	—	0.55

The use of alkanolamines, either by themselves or in mixtures with various other alkalies, has been the subject of several patents (258–260). In a brief note published in 1964, Saphir (261) stated that these patents do not represent original invention and that alkanolamines have been widely used in cold-wave lotions in Europe prior to the issuance of several of the patents. Before leaving the subject of alkanolamines, it should be mentioned that alkanolamines are claimed to yield waving lotions which are significantly less toxic than comparable ammonium hydroxide-containing preparations (262).

A British patent provides an alternate solution to the odor problem owing to ammonia by suggesting that lithium thioglycolate solutions containing some free lithium hydroxide are capable of waving hair as well as conventional ammonia-containing preparations (263).

Another means of supplying free alkali in the form of ammonium hydroxide is the subject of a number of patents (264–267). Briefly, the disclosures of these patents concern the incorporation of urea into essentially neutral solutions of thioglycolic acid. Urease is added to the solution just prior to use or is incorporated into a suitable end paper. Depending on temperature and other conditions, ammonia is slowly liberated by the action of urease on urea. It is claimed that hair can be permanently waved more effectively by this process in the presence of a low concentration of thioglycolic acid and that such preparations damage hair relatively little and cause no irritation to the hands of the operator.

Other means of controlling the pH of the waving lotion, either to reduce the speed or extent of action of the lotion, to protect the hands against irritation, or to overcome the danger of overprocessing damaged ends of hair, have also been patented (254,268–271b).

An interesting method for increasing the free alkalinity of thioglycolate-containing waving lotions has been patented by DeMytt and Hannigan (272,273). It involves the addition of ammonium bicarbonate or ammonium carbonate to ammoniacal solutions of ammonium thioglycolate. Advantages claimed for such products include the following: the pH of waving lotions

TABLE VII.   Recommended Compositions of Cold-Waving Products (272,273)

Ingredient	Concentrations		
Ammonium thioglycolate	0.6M	0.65M	0.65M
Ammonium bicarbonate	0.6M	0.10M	0.70M
Ammonium carbonate	—	0.25	—
Ammonium hydroxide	0.5M	—	0.55M
pH	8.8	8.5	8.7

can be lowered; despite the lower pH, the presence of the carbonates increases the speed of waving; the resulting lotions reduce the tendency to overwave or damage hair; hair condition at the end of the permanent waving process is improved; the solutions are less irritating to the skin and reduce the irritating odor of ammonia; finally, the solutions have less tendency to yellow gray hair. Some typical compositions disclosed by DeMytt are shown in Table VII.

A related method for increasing the free alkalinity of thioglycolate waving lotions has been patented by Zviak and Rouet, who utilize the addition of carbon dioxide to a thioglycolate preparation which has been made alkaline with the aid of alkanolamines or other amines (274).

Although, as shown above, the speed of thioglycolate-containing waving products can be controlled by pH, alkalinity, choice of alkali, and concentration of mercaptans, scientists continue to work on other means of accelerating or retarding the waving process.

For example, Brown (275) discloses the addition of urea or of a low-molecular-weight alkyl urea and an ammonium salt of a mineral acid to a thioglycolate waving lotion. It is claimed that urea, by accelerating the swelling of hair, permits a decrease in the mercaptan concentration of the lotion. The inorganic ammonium salt is said to limit the swelling of hair by alkaline mercaptans and urea without materially interfering with the waving effectiveness. An example cited is as follows:

Urea	2.6 M
Ammonium thioglycolate	0.45 M
Ammonium sulfate	0.5 M
Ammonium hydroxide to reach pH 9.4	

This lotion is claimed to produce a very good wave without danger of overprocessing. Other swelling agents, such as diamides of dicarboxylic acids (276) or simple amides, e.g., dimethylformamide (277), have also been patented.

Grein (278) has claimed that the inclusion of isopropanol in waving lotions increases the speed of waving. In connection with Grein's disclosure, it is interesting that Speakman and his co-workers (279) have shown that, contrary to expectations, wool is more easily extended in mixtures of water and isopropyl alcohol than in water alone, as long as the concentration of isopropyl alcohol is less than about 75 % (cf. also 129–134).

Means of retarding the action of thioglycolate waving lotions are desired in order to protect normal hair against damage or to permit waving of damaged hair with reasonable confidence (177). Chemical means towards this end have been studied and patented. For example, compositions have been patented containing (a) a mercaptan with a Bunte salt* and (b) a disulfide with a sulfite (280). Regardless of the ingredients of the lotions, an equilibrium of the following type is established in an alkaline medium:

$$RSH + R'—S—SO_3^- \rightleftharpoons R'—S—S—R + HSO_3^-$$

These waving lotions contain appreciable amounts of all four compounds. They are reported to soften hair rapidly and efficiently, to be gentle, and to protect against hair damage.

Waving lotions have been marketed that contain from 2 to 3 % of dithiodiglycolic acid in addition to 4 to 7 % of thioglycolic acid. These formulations have a pH between 9.0 and 9.4 and are claimed to protect hair against damage during waving and to be safe for use on bleached and dyed hair (150,281). In contrast, Whitman and Eckstrom (146)—probably erroneously —indicated that hair waved with thioglycolate-based waving lotion containing dithiodiglycolic acid may be damaged through extensive formation of the previously mentioned mixed disulfide in hair.

Protection of hair against damage has also been claimed for a wide variety of different additives. Emollients and lubricating additives are included in the discussion of emulsified hair-waving lotions and creams on pp. 215–218. In addition, such materials as dissolved hair (282), protein hydrolysates (201,283,284), gelatin (285), various amino acids (205,286), milk (287), plant extracts (288), substituted amino acids (289), and silicones (289a) have been patented in various countries.

Leberl (290) has discussed the protective effects of colloids, such as gums, cellulose derivatives, proteins, polyvinyl alcohol, and polyglycols, and attributes these benefits to the formation of a protective film on the hair.

Products Based on Mercaptans Other than Thioglycolic Acid

The patent literature concerned with cold waving describes a variety of mercaptans with reportedly desirable qualities. The fact that today almost all commercially available cold-waving products are based on thioglycolic or

---

* Salts of the general formula $R—S—SO_3^-\ X^+$.

thiolactic acid attests to the fact that other mercaptans have not been able to achieve prominence for reasons of safety, odor, or cost. In spite of this, mercaptans other than thioglycolic acid suggested for cold-waving purposes will be reviewed and, for this purpose, divided into three arbitrary groups: (*a*) mercaptans unrelated to thioglycolic acid, (*b*) derivatives of thioglycolic acid, and (*c*) waving agents intended for use at low *p*H.

Mercaptans unrelated to thioglycolic acid. α-Thioglycerol (CH₂SH— CHOH—CH₂OH) is one of the most prominent examples of a mercaptan mentioned in the early patent literature (241,244,245). Home permanent waves based on this mercaptan were marketed in France (291), the United States, and Switzerland (292). An example (245) of a waving lotion based on thioglycerol is given in Formula 10.

Formula 10

Glyceryl monomercaptan	5.0 g
Ammonium hydroxide	2.5 cc
Water	g.s. 100.0 cc

Thyroid hyperplasia following administration of thioglycerol was reported by Draize and his co-workers (293), Kensler and Elsner (294), and Lehman (295). In addition, it has been reported that thioglycerol is a sensitizer (295,296). Despite this adverse publicity, claims in a recent German patent (297) include favorable comments about the safety of thioglycerol. As a matter of interest, this patent also describes the use of a neutral or slightly acidic solution of thioglycerol for waving under tepid conditions at 40°C.

Among the mercaptans in this group should be mentioned β-aminoethyl mercaptan and mercaptoethyl alcohol. The use of essentially odorless β-mercaptoethane sulfonic acid salts for cold waving is described in a British patent (241). Since this acid is reported to be difficult to prepare, Raecke (298) recommends instead the use of several more readily accessible isomeric mercaptobutane sulfonates for cold waving.

Morelle (291,299–303) has studied the usefulness of selected mercaptans, especially thiolactic acid (301–303), for waving purposes. A comprehensive study of a variety of mercaptans was conducted by Haefele and Broge (304). These workers described the synthesis and chemical properties of many types of mercaptans. Later the authors followed up with an excellent study of the waving characteristics of this series of compounds (157). From their data, it would appear that, for practical purposes, thioglycolic acid, β-mercaptopropionamide, mercaptoethylacetamide, thioglycolamide, thio-lactic acid, methyl mercaptoethyl sulfone, mercaptoethyl nitrile, and mer-captoethyl trifluoroacetamide are fairly strong waving agents in the *p*H range of interest. The toxicity of these compounds was studied by Voss (305).

The use of thioparaconic acid and its derivatives has been patented by DeMytt (306). Another patent by this worker is concerned with the utility of α,α'-dimercaptoadipic acid (307). This substance is reportedly relatively free of mercaptan odor, is nontoxic, and can be formulated into effective waving lotions by addition of alkali to about pH 9.2 to 9.3. The results of chemical studies and hair-waving experiments with this compound have been reported by Finkelstein $et\ al.$ (308). It is also of interest to note that dimercaptoadipic acid can be used in the formulation of dry-waving products or impregnated end papers (307).

Cysteine has been reported (308a) to be an effective waving agent. Its use, however, is complicated by the fact that the cystine formed during waving is deposited on hair as a dulling insoluble film or as crystals. Instead, it has been suggested to utilize N-acetyl cysteine, the oxidation product of which is soluble in water (308b).

The use of 1,4-dimercapto-2,3-butanediol for permanent waving is the subject of a recent U.S. patent (308c). It is claimed that this dithiol is an effective waving agent at relatively low concentration (0.2 M) at pH 9.1 (with ammonia) and that it is still moderately effective at pHs as low as 7.9 in the presence of 0.2 M urea.

Derivatives of thioglycolic acid. This group of chemicals recommended for cold waving includes compounds which (a) liberate thioglycolic acid by decomposition or reaction in the presence of water just before or during actual use or (b) are derivatives which are active waving agents by virtue of their free SH groups.

A compound which liberates thioglycolic acid has been named "masked" thioglycolic acid by Heilingötter (309). Typical among these are carbaminyl thioglycolic acid (310), glycolic acid esters of di- and trithiocarbonic acids (311,312), dithiocarbamyl derivatives (313,314), and heterocyclic compounds, such as 2,4-thiazoledione (315). A most unusual approach has been used by Charle and co-workers (316). Their patent describes the preparation of the thioglycolic acid esters of monomethyl, dimethyl, or trimethyl silanol. Addition of one of these esters to a hydroalcoholic solution of ammonia causes immediate hydrolysis to thioglycolate. The simultaneously formed silanol polymerizes to a polysiloxane, which protects the hair during waving.

A more conventional approach to the use of a thioglycolic acid derivative involves dithiodiglycolic acid salts and their $in\ situ$ reduction to thioglycolates. As pointed out earlier, Haefele (280) has patented cold-waving lotions based on alkaline mixtures of sulfites and dithiodiglycolates. Kalopissis and Viout (317) have suggested that a polyacrylic acid polymer containing sulfhydryl groups (which incidentally has been recommended (318) for cold waving $per\ se$) be mixed with a dithiodiglycolate just prior to use to

yield an effective waving lotion. It is not apparent why such fairly complicated systems should be preferred over the direct use of a thioglycolate.
Improved odor has been claimed as a major advantage, but the only practical reason for their use appears to be the fact that dry mixtures can be
prepared with the aid of dithiodiglycolates. For this reason such products
will be discussed in more detail in the section on "dry" waving agents.

The most important derivatives of thioglycolic acid useful for waving,
per se, are the simple amides and esters. Thioglycolamide is the most carefully
studied member of this group since this compound has been shown to wave
hair well at *p*Hs below those required for good waving by thioglycolic acid
salts (304,157). At the same time, thioglycloamide does not cause as much
hair damage as most other mercaptans. This substance nevertheless has not
achieved commercial importance because of its intrinsic instability and potential toxicity (305). In order to avoid repetition, this and related compounds
are discussed in greater detail in the section on waving agents intended for
use at low *p*H.

Waving agents intended for use at "low" pH. Permanent waving with a
mercaptan, such as thioglycolic acid, requires an alkaline medium. Aside
from the fact that the odor of ammonia or of other amines is considered
objectionable, it has been implied that the alkalinity of cold-wave preparations is responsible to some degree for possible irritation. It is not surprising,
therefore, that a concerted effort has been made to find mercaptans which are
able to wave hair at *p*Hs below 7. As a result, numerous patents have appeared, especially in Europe, which describe various waving agents which are
claimed to be active in acid solution. Among the compounds of major interest
are esters and amides of thioglycolic acid. Many of these have been described
in considerable detail by Haefele and co-workers (304,157). The rather
definitive studies of Voss (305) have clearly demonstrated that amides and
esters of thioglycolic acid are potential sensitizers (cf. 319).

In view of the potential danger, "acidic" waving agents publicized to date
are of no practical importance today. For the sake of completeness, patents
describing the activities of cosmetic chemists in this field will be mentioned
briefly. Although these chemicals are discussed here as "acidic" waving
agents, most of them will yield permanent waves when used in an alkaline
medium (cf. section on derivatives of thioglycolic acid). Patents for the use of
monothiopropylene glycol, α-thioglycerol, various esters of monothioglycol
or thioglycolic acid, and substituted thioglycolamides have been issued
(320–322).

Amides of thioglycolic acid generally suffer from the fact that they split off
hydrogen sulfide and hydrolyze upon prolonged storage, especially in neutral
or acidic media (323). Nevertheless, several patents have been issued describing the use of various thioglycolamides (as solids or as solutions) for hair

waving. These patents include thioglycolic acid hydrazide (324), β-amino-ethyl thioglycol amide (325), thioglycolic acid amides of amino acids (326), and bis-thioglycolic acid imide (327). Kalopissis (328) recognized the intrinsic instability of thioglycolamide. He described, therefore, means of dispensing a weakly alkaline solution of thioglycolamide which is acidified just before use with an organic acid, such as citric acid.

Products Based on Nonmercaptan Reducing Agents

Among cold-waving lotions based on nonmercaptan reducing agents, preparations utilizing organic and inorganic salts of sulfurous acid have gained most prominence. For the purpose of cold waving, bisulfites are preferred over sulfites (212,213,329). Michaels and Lustig (330) indicate that the commonly used inorganic cations, more specifically Na^+, inhibit the swelling of hair. They suggest, however, that the bisulfites of organic nitrogen bases are useful for effective swelling and relaxation of keratinous fibers, and recommend solutions of monoethanolamine and guanidine bisulfites at pH of 5 to 6. A suggested lotion is prepared as follows:

Ten grams of monoethanolamine is dissolved in 10 g of water, and sulfur dioxide is passed into the solution until a dilution of the solution containing 10% by weight of sulfur dioxide will have a pH of 5.5.

Guanidinium or ammonium bisulfite was recommended by Whitman and DenBeste (331), who also suggest the use of hexamethylene tetramine for neutralization. More recently, alkyl or aryl substituted guanidinium bisulfites have been suggested for cold waving (332).

Although such lotions are free from mercaptan odor, they wave relatively slowly and have a tendency to bleach hair. As a result, many attempts have been made to accelerate the rate of waving by bisulfites through chemical additives. As early as 1950, Mora (333) indicated that the action of waving lotions containing sulfites can be accelerated by the inclusion of quinones, polyphenols, etc. (cf. 334). Another means of accelerating the action of bisulfites is the inclusion of a protein denaturant or "swelling" agent such as urea or formamide (335). Finally, it is noted that the combination of a denaturant and a polyhydric phenol was patented in the United States (336).

Formamidine sulfinic acid (337) and its potassium salt (338) were recommended for odorless safe cold waving; better results are claimed (339) if a disulfide (such as dithiodiglycolic acid, cystine, etc.) is included in the waving lotion.

Cold waving with bisulfites has not been widely used commercially. Sulfite-containing preparations can be used in tepid waving (q.v.) and find some use in hair straightening (in the presence of large amounts of urea and *i*-propanol).

Bogaty and Brown (340) patented the use of a stabilized solution of a borohydride, especially potassium borohydride, in 1956. In order to improve its long-term stability, it has been recommended to disperse the borohydride in a paste of inert but water-soluble materials, such as polyoxyethylene or fatty alcohol ethers thereof (341). A dry mixture comprising sodium borohydride and anhydrous sulfite is the subject of a German patent (342).

Another group of strong reducing agents, the phosphines, has been recommended to be odor-free and nondamaging to hair (343). Both borohydrides (344,345) and phosphine derivatives, such as tetrakishydroxymethylphosphonium chloride (346), can reduce disulfide bonds in keratin but apparently have not been used commercially for hair waving.

Forms of Cold-Waving Preparations

Cold-waving preparations based on the actives described above can be prepared in various physical forms. Included among these are transparent and opaque (lotions) solutions, creams, gels, aerosols, and dry forms.

Lotions

Aqueous solutions of most of the chemicals used to "soften" the hair are colorless and transparent. Initially, cold-waving lotions were marketed as transparent liquids. Today, transparent cold-waving lotions are used mainly by professional operators. In the absence of a fragrance, their formulation requires only addition of a soluble wetting agent to assist in lotion penetration. Since almost all mercaptan-containing compositions require a masking perfume, formulation of a clear transparent (and colorless) waving lotion is difficult. Careful selection of perfume ingredients and of wetting agents which yield stable solutions in the presence of electrolytes is required.

Since consumers perceive opaque waving lotions as being richer and gentler than transparent lotions, most marketed preparations are opacified. The lotions can be clouded by emulsifying oily materials or by the addition of dispersed water-insoluble materials.

Waving lotions containing suspensions of polyacrylates to impart opacity have been very popular. Such a lotion has been described in a patent by Peterson (347) and is shown in Formula 11.

The use of a polystyrene latex for preparing an opaque hair-waving lotion has also been patented (348). The described creme base required to make 1 liter of opaque lotion has the composition shown in Formula 12.

Today, opacifiers based on synthetic polymers have been replaced by emulsions of oily materials. This change was a natural one because the latter,

Formula 11

Thioglycolic acid	0.90 fl. oz.
Ammonium hydroxide	1.66
Water	13.44
Creaming agent	0.70
Acrylic acid derivative	
(copolymer)	Rhoplex and Triton
Dispersing agent	720 in equal parts
Water	

Formula 12

Amorphous polystyrene latex	
(40% solids)	5.15 g
Lauryl ether of polyethylene	
glycol (mol. wt. 900)	10.00
Water, distilled	16.18
Perfume (optional)	2.00
Total	*33.53 g*

in addition to conferring opacity, contribute an emollient effect (349). The composition of stable oily emulsions is a closely guarded trade secret, and few if any satisfactory formulas have been published.

The use of silicones in emulsions for cold-waving lotions has been patented by Freytag (350) and Deutsch and Temblett (351). An example from the latter disclosure is shown in Formula 13.

Mineral oil is probably the most commonly used raw material for preparing emulsified cold-waving lotions. Arnold (352) has described a variety of typical "creaming agents," such as the one shown in Formula 14.

Formula 13

Dimethyl polysiloxane (50 cps)	2.0 parts
Polyethylene glycol 400	
monostearate	1.5
Thioglycolic acid	7.5
Ammonium hydroxide (SG 0.88)	1.7
Perfume and color	q.s.
Water, to make	100.0

Formula 14

Mineral oil, light, 50/60	74%
Sulfonated castor oil (Triol)	5
Diglycol oleate (containing 2 to	
5% of oleic acid soap)	11
Lanogene	10

Two to 5 % of this creaming agent is incorporated into the finished lotion, which in addition to the active ingredients contains from 1 to 2 % of a fatty acid condensate of a protein hydrolysate (Maypon C, Lamepon). It is claimed that a lotion of this type separates very slowly and can be reconstituted readily by gentle shaking in case of slight separation.

The preparation of stable oil-in-water emulsions for use in permanent waving lotions has been patented by Mace (353). Highly chlorinated hydrocarbons are incorporated into the oil phase to increase its specific gravity to approximate that of the aqueous phase, thus reducing the tendency of the preparation to separate. A representative composition from the patent is given in Formula 15. Similar compositions are covered by a British patent (354).

Formula 15

Chlorinated paraffin (26 C atoms, 43% Cl)	1.0%
Sorbitan monooleate	1.0
Sorbitan monooleate polyoxyalkylene derivative	2.0
Mineral oil	0.4
Lanolin	0.2
Gelatin (optional)	0.1
Borax (optional)	0.1
Ammonia (as NH_3)	1.6
Ammonium thioglycolate	8.0
Water	85.6

Procedure: Melt together the oily materials, including the wetting or emulsifying agents, at about 50°C. Introduce this mixture into the heated water (45 to 50°C), which may contain a small amount of gelatin and borax as dispersing agents. Stir the resulting mixture until it reaches room temperature, at which time introduce the ammonium thioglycolate and ammonia to complete the formulation. If desired, perfume, dyes, or even keratin solutions may be added.

Creams and Gels

Creams containing thioglycolates have been used for hair straightening and, to a lesser extent, for waving. Such products, packaged into aluminum tubes, achieved some prominence in the United States and Europe, but their popularity has decreased significantly in recent years. A cream composition suggested for either use was patented by Ramsey (355) in 1947. Generally, creams for hair straightening should be stiff to hold the hair in place; creams for cold waving should be soft to permit ready penetration into wound curls and to allow removal from curls by simple rinsing. Properly formulated creams or gels are claimed to be "dripless" and to impart a desirable feel and luster to the hair (356). A self-neutralizing cold-waving cream is described in a French patent (357), and a typical composition is shown in Formula 16.

Formula 16

Ammonium thioglycolate	7.00 g
Ammonium hydroxide, to reach	
pH of 9.5	
Manganese sulfate	0.05
Stearyl alcohol	15.00
Sulfonated cetyl alcohol*	5.00
Water, to make	100.00

* It seems likely that the patent was meant to refer to "sulfated" cetyl alcohol.

The inclusion of cationics, castor oil, lanolin, cholesterol, or lecithin for their beneficial effect on hair is further recommended. Similar compositions based on anionic surfactants have been described (358,359).

Deadman (356) describes a number of polyoxyethylene ethers and esters to prepare creams and gels. Gums, such as acacia and tragacanth (360), and polycarboxylic acids, such as polyacrylates and alginates (361), have been suggested for the formulation of thickened cold-waving products. Salts of long-chain S-alkylmercaptoethane sulfonic acid also have been recommended for thickening cold-waving products (362).

The preparation of a stable waving cream by use of a combination of nonionic emulsifiers has been reported (363) and is shown in Formula 17.

Formula 17

Cetyl oleyl alcohol ether of polyoxyethylene glycol (mol ratio 1:20)	15 to 30 g
Cetyl oleyl alcohol ether of polyoxyethylene glycol (mol ratio 1:10)	20 to 40
Stearyl alcohol	10 to 20
Perfume	q.s.
Ammonium thioglycolate (50% thioglycolic acid)	100 to 150
Ammonium hydroxide, 28%	70 to 100
Water, distilled, to make	1000

Procedure: The ethers, the stearyl alcohol, and the perfume are emulsified in about 500 g of water at 60 to 70°C. After emulsification and cooling, the ammonia and ammonium thioglycolate are added.

Aerosols

Under most circumstances, cold-wave lotions are packaged in glass containers; the consumer is generally instructed to discard any portion of the lotion which remains unused. The reason for this precaution is to avoid use of substandard lotion. The air space in a partially filled container of waving lotion will cause loss of reducing agent owing to air oxidation. It is not surprising, therefore, that manufacturers have searched for means of dispensing waving lotions which would be less wasteful and permit storage of partially used containers. One such approach is the previously discussed use of waving creams packaged in tubes. Another is the use of pressurized packages. A further advantage which arises from the use of an aerosol is that it permits application of the waving product directly to the hair without the need for constant exposure of the hands to the alkaline waving product.

Although it is possible to dispense alkaline thioglycolates as sprays, a more aesthetic, safer, and less wasteful approach is an aerosol which dispenses the product in the form of a quick-breaking foam. Products of this type have been patented by Banker and co-workers (364), who describe an aerosol product which comprises, in addition to the alkaline thioglycolate, a nonionic fatty ether as the wetting agent, propellants, and a so-called vapor pressure depressant. The vapor pressure depressant is said to be desirable to lower the pressure of the aerosol composition. The materials recommended include a variety of fatty esters, such as isopropyl myristate, dioctyl phthalate, etc. The selection of the propellant must be made judiciously to ensure stability of the propellant in the presence of alkaline thioglycolate and to provide the proper pressure and the proper rate of foam collapse. The latter can also be controlled by means of the amount and the type of wetting agent and vapor pressure depressant added.

A typical example from the disclosures is cited below in Formula 18. The example includes, in addition to monoethanolammonium thioglycolate, a sufficient amount of dithiodiglycolate to reduce potential hair damage and a small quantity of sodium silicate to minimize corrosion of the aluminum container.

The described compositions separate into two separate liquid phases, but the compositions are designed to form reasonably stable dispersions of the propellant in the aqueous phase so that only infrequent shaking of the aerosol container is necessary during application of the product to the hair. In order to avoid the need for shaking, Shepard and co-workers (365,366) patented compositions which, in addition to the propellant and actives, include a sufficient amount of ethyl alcohol to yield a homogeneous single liquid phase. Brechner (366a) and co-workers recommend the addition of various glycols, such as propylene glycol or 1,3-butylene glycol, to aerosolized

Formula 18

Monoethanolammonium thioglycolate	7.60%
Monoethanolammonium dithio-diglycolate	3.05
Monoethanolamine (to adjust *p*H to 9.3)	2.10
Isopropyl myristate	4.78
Difluoroethane	4.35
Polyoxyethylene lauryl ether with an average of 23 oxyethylene groups per lauryl group	0.95
Perfume	0.95
SiO$_2$ (added as sodium silicate having an SiO$_2$/Na$_2$O ratio of 2.5:1)	0.01
Water	q.s.

waving lotions. It is claimed that the addition of polyhydric alcohols improves the waving efficacy of aerosolized thioglycolate-containing waving lotions.

A self-pressurized thioglycolate waving preparation has been patented by Wajaroff (367), who recommends mixing of an acidic formulation of thioglycolic acid with a solid or liquid alkaline composition containing a large quantity of carbonates. The carbon dioxide liberated during the neutralization is sufficient to propel the material from the container. Wajaroff describes several means of packaging such ingredients in containers to keep them separate until they are mixed for use.

Dry Preparations

In order to avoid the need for shipping dilute aqueous solutions of thioglycolates, efforts have been made to provide users with dry thioglycolate waving compositions. This presentation of a waving product not only avoids the danger of breakage of glass bottles in shipment but, at the same time, reduces the shipping weight.

One of the first solid cold-waving compositions was patented by Schnell (368), who used calcium thioglycolate, which is produced commercially in dry form. A typical composition which must be dissolved in water before use is shown in Formula 19.

Formula 19

Calcium thioglycolate	80 parts
Ammonium chloride and/or	47 to 65
Ammonium formate	55 to 77

A dry composition of thioglycolic acid, ammonium carbonate, and sodium carbonate has been patented by Deadman (369). The preparation of various dry amine or ammonium salts of thioglycolic acid has been described by Martin (370), whose disclosure includes such salts as the dithioglycolate and the monothioglycolate of ethylenediamine, ammonium thioglycolate, and guanidinium thioglycolate. A somewhat more complicated dry-waving compound has been suggested by Schweizer (371). He describes the preparation of the dipotassium salt of zinc bis-thioglycolic acid and suggests that it be dispensed dry together with a large quantity of a sequestering agent and a suitable alkalizing agent such as a solid amine. A typical composition is shown in Formula 20.

Formula 20

Dipotassium salt of zinc bis-thioglycolic acid	32.7%
Trishydroxymethyl aminomethane	30.0
Sodium hydroxide	2.7
Tetrasodium ethylenediamine-tetraacetate	32.7
Perfume	1.9

The resulting powder mixture is reportedly stable, and an aqueous solution—to yield the equivalent of 5 % thioglycolic acid—is reportedly a good waving agent. Presumably, the sequestering agent liberates the free mercaptan from the zinc mercaptide by complexing the zinc to provide free thioglycolate as the waving agent.

Dry mixtures of thioglycolic acid salts, such as those described above, require dissolution in the proper quantity of water prior to use. In order to simplify the process still further, recommendations have been made to impregnate end wraps with dried thioglycolates. These end wraps are generally referred to as end papers but usually are manufactured from a nonwoven fabric. Typical among these are the disclosures by Hsiung (372), who recommends treatment of end papers with a mixture of solid amines and a magnesium salt of thioglycolic acid. A related approach has been recommended by Moore (373), who suggests the use of a mixture of sodium or preferably potassium thioglycolate with a dry amine, such as 2-amino-2-methyl-1,3-propanediol or trishydroxymethyl aminomethane. It is apparent that the impregnation and drying of end papers must be performed under conditions which reduce air oxidation to a minimum and which deposit the precise amount of active waving agents on each end paper. A more complicated system for combining not only the reducing agent but also the neutralizing agent in a single end paper has been recommended by Kochenrath in a German patent (374).

Impregnated end papers are used as follows: Clean damp hair is wound with the aid of the paper; the wound tresses are wet with water and then processed as usual. Impregnated end papers achieved some commercial acceptance in the United States, but their success was apparently limited by the fact that waves produced with their aid yielded primarily end curls but rather loose scalp curls. As a result, the lasting qualities of permanent waves produced by treated end papers were drastically reduced and frequently unsatisfactory.

Salts of dithiodiglycolic acid are relatively stable in air and can be crystallized readily. As a result, efforts have been made to package dry salts of dithiodiglycolic acid with materials capable of reducing them to thioglycolates upon dissolution in water. Brown and Bogaty (375) recommended mixtures of diammonium dithiodiglycolate and thiourea dioxide (as the reducing agent) for this purpose. These workers also disclosed the use of other salts of dithiodiglycolic acid as well as other reducing agents, such as diammonium thiomalate and diammonium α,α'-dimercaptoadipate.

Miscellaneous Considerations

Thioglycolate cold-wave preparations and products based on other mercaptans exhibit objectionable odors unless they are properly perfumed. During cold waving, three types of odors must be covered by the fragrance: the odor of the mercaptan, the odor of the free ammonia or amine, and the odor which develops by the interaction between the hair and the waving product. In addition, one may wish to make provision for the odor contributed to the waving preparation by traces of hydrogen sulfide, which is a common impurity. The problems involved in perfuming thioglycolate hair-waving preparations have been discussed by Sagarin and Balsam (376). A more recent review by Cook (376a) includes a series of starting perfume compositions for mercaptan-containing waving products. Cyclohexyl-cyclohexanone has been recommended (377) as a covering odorant for thioglycolate-containing products. Finally, Freund (378) suggests the addition to cold-waving products of zinc tetrammine sulfate (0.1 to 3.0%), with the thought that any hydrogen sulfide would react to form insoluble zinc sulfide.

A second problem in formulating waving lotions arises from the tendency of alkaline mercaptans to yield an intense purple coloration with iron in the presence of air. The addition of chelating agents to overcome this has been the subject of two patents (379,380).

Before leaving the subject of cold-waving preparations, it is worth noting that hair dyeing with direct dyes can be combined with permanent waving merely by incorporating suitable dyestuffs into the cold-wave preparation

(381,382). Products have also been marketed in which oxidation dye intermediates have been incorporated into the waving lotion. After completion of the reduction step, the neutralizer (hydrogen peroxide) not only reoxidizes protein disulfide bonds but also produces the typical colored pigments from the intermediates.

TEPID WAVING

Since the days of heat waving, it has been well known that the reaction between sulfites and hair is temperature-dependent and can be accelerated greatly by a relatively moderate increase in temperature. It is not surprising, therefore, that during recent years attempts have been made to modify (a) cold-waving processes by slightly increasing the temperature during processing (383) and (b) heat-waving processes by reducing the temperature slightly. These variations of conventional procedures are called tepid waving. Although of limited value to the home user, a tepid-waving process is easily adapted to salon practice and, under normal circumstances, can be performed safely and at temperatures which can be well tolerated by the consumer (under a hair dryer).

For the purpose of tepid waving, the strength of conventional heat-waving preparations is increased, whereas the active content of conventional cold-wave preparations is reduced. On the other hand, the relatively weak waving agent thiourea peroxide (imino amino methane sulfinic acid or formamidine sulfinic acid) has been recommended for tepid waving (384). Unneutralized solutions of mercaptans, such as thioglycerol (385,386) or thioglycolic acid (387), have also been recommended. The latter is rather interesting and deserves additional comments. In this process, the straight hair is treated with an unneutralized aqueous solution of thioglycolic acid (approximately 8%) for a period of approximately 15 min, reportedly to form a mixed disulfide. The hair is then rinsed, subjected to treatment with a dilute solution of ammonium hydroxide, and wound. The hair is next exposed to a temperature of about 60 to 80°C to generate a permanent wave. The inventors claim that the mixed disulfide is broken during the second step to yield first a sulfinic acid and then a mercaptan.

A patent (388) for generating the temperature required for tepid waving just before use of the lotion has been issued to Böss and co-workers, who utilize the heat of neutralization of an acid mixture (citric acid) of reducing agents with monoethanolamine to raise the temperature of the waving product to about 20°C above ambient.

Unilever received a series of patents (389) for preparations which contain, in addition to ammonium thioglycolate, some ammonium sulfite, urea, and ammonia and which are to be used at a temperature at or near 40 to 60°C. The combination of a sulfite and a phenolic substance in a hydroalcoholic medium has also been patented (390) for use in tepid waving. A modification of a typical bisulfite formulation with a water-soluble polyamide (prepared from diethylene triamine and adipic acid) has been recommended for use in tepid waving (391). Several polyfunctional thiophenols, such as 2,6-dihydroxythiophenol and 2,4,6-trihydroxythiophenol, have been recommended for tepid waving (392). The so-called acidic mercaptan waving agents are reportedly useful in tepid processes (325). An interesting approach has been patented by Kalopissis and Viout (393), who describe the preparation of S-acetyl mercapto acetamide. This is made alkaline with ammonium hydroxide to prepare the waving lotion which is to be used at about body temperature. This compound is, of course, an amide of thioglycolic acid which cannot split off H_2S but apparently yields thioglycolamide rapidly on exposure to dilute alkalies at moderate temperatures.

Goble (393a) recently described a novel process for imparting a permanent wave to hair. The method involves treating wound wet hair with heated air modulated at an ultrasonic frequency.

NEUTRALIZATION

As pointed out above, neutralization serves two basic functions, removal of excess waving agent from the hair and restoration of the hair to its original condition, thus giving the wave the desired permanence.

Two methods of neutralization, both involving oxidation, have been used widely in cold permanent waving. In one method (chemical neutralization), the hair, after contact with the waving lotion, is treated with a solution of a suitable oxidizing agent. In the second method (self-neutralization), oxidation by air is relied upon to make the wave permanent. A third method of neutralization relies on cross-linking of sulfhydryl groups to "harden" reduced hair. Methods based on this principle have been patented but apparently are not widely used for commercial purposes.

Chemical Neutralization

Chemical neutralization is utilized by almost all modern waving preparations because of its speed. It is performed after sufficient time has elapsed for the waving lotion to alter the configuration of the hair. The neutralizer may be applied to the hair after a water rinse or after the waving lotion has been removed partially by blotting. If desired, the hair may be treated several times with the neutralizer. The time required for complete neutralization varies from about 5 to 20 min. Before the hair is set in the desired style, the neutralizer is removed from the hair by a water rinse.

Complete removal of the reducing agent from the hair is desired. Semco (394) suggested that the color reaction between nitroferricyanide and mercaptans may be used to determine when this has been accomplished.

Solutions of hydrogen peroxide in the presence of organic acids, such as acetic, tartaric, or citric, were the first chemical oxidizing agents used for permanent waving.

Solid oxidizing agents were once considered more suitable for packaging and, at one time, were used almost exclusively in home permanent kits. The two outstanding examples of this type are potassium bromate and sodium perborate. Although simple aqueous solutions of oxidizing agents are satisfactory neutralizers, lotion- and cream-type neutralizers have been widely used in order to achieve special effects (395). Today neutralizers based on hydrogen peroxide are used almost universally.

Regardless of the type of oxidizing agent employed for neutralization, means have been described to speed up the reaction between reduced hair and the oxidant. The addition of large amounts of urea has been disclosed by McGoldrick and McDonough (396). Reiss and Lichtin (397) recommend the addition to the oxidant of large quantities of a polyvalent metal salt, such as magnesium sulfate, calcium acetate, etc. Interestingly, one of these processes makes use of a swelling agent whereas the other's effectiveness probably depends on osmotic effects to deswell the hair.

Hydrogen Peroxide

One of the first cold-waving preparations marketed employed a 3 % solution of hydrogen peroxide, diluted 2 oz to 1 pint, for neutralization. Hydrogen peroxide solutions for neutralizing can be prepared by diluting the more concentrated commercially available solutions to the desired strength. Suppliers of the concentrated solutions recommend necessary stabilizers for the dilute solutions (cf. e.g. refs. 398–400a). Extreme care in the manufacture and packaging of hydrogen peroxide solutions is necessary to assure stability. Before use, a solution or crystals of an acid-reacting material, such as tartaric acid, may be added (239). Incorporation of a wetting agent (401) or of a thickening agent (402) has been recommended.

Heilingötter (403) reported that hydrogen peroxide oxidizes thioglycolic acid faster at a pH of 9.4 than at a pH of 4.0. At the lower pH, this oxidation, reportedly, is faster in the presence of hydrochloric acid than citric acid. It is not made clear whether the same considerations apply to the re-formation of disulfide linkages in hair. It is reasonable that the use of acidic or neutral solutions of hydrogen peroxide should be preferred over alkaline preparations since the latter have a tendency to bleach hair. In order to control the oxidizing action of hydrogen peroxide, the use of ascorbic acid has been suggested (404).

The usual concentration of modern hydrogen peroxide-containing neutralizers is 1.5 to 2.0%, but, depending on use, other levels have been employed. By judicious selection of the emulsifiers and oils, it is possible to develop stable (with regard to loss of oxygen) emulsions or suspensions in hydrogen peroxide. Although such compositions have been marketed, no information regarding their composition has been published.

Bromates

In 1951, Reed and his co-workers (405) patented the use of alkali-metal bromates and iodates as cold-waving neutralizers. The wide acceptance of potassium and sodium bromate for the purpose of re-forming disulfide bonds from adjacent mercaptan groups in hair is due to the fact that these salts can be packaged conveniently in dry form.

Originally, potassium or sodium bromate was supplied in the waving kit as a dry solid, packed in a hermetically sealed envelope. The quantity varied from about 10 to 30 g. The solid was dissolved in approximately 500 to 1000 ml of water before use. In dry form and in the absence of organic materials, the bromates are stable. They may be used in slightly acidic media if care is taken to avoid formation of bromine. Monosodium orthophosphate appears to be suitable for this purpose, and several modern cold-wave kits include an envelope containing 5 to 10 g of this substance or of a similar acidic material, in addition to the bromate.

It is claimed that a stable, dry neutralizing mixture can be prepared from a bromate and an acid phosphate in the presence of a carbonate (406), as shown in Formula 21. The carbonate is claimed to act as a dehydrating agent, thus avoiding formation of bromine.

Formula 21

Sodium bromate	98 parts
Monosodium orthophosphate	75
Sodium carbonate	2
Total	*175 parts*

In order to improve the luster and softness of permanently waved hair, neutralizers have been made available that contain hair-conditioning agents. The reagents preferred for this use generally belong to the group of quaternary ammonium compounds carrying one long-chain aliphatic group. Liquid emulsions and a cream-type composition containing both a quaternary and bromate appeared on the American market several years ago. The quaternary has also been packaged in a separate emulsion, which is added to the bromate at the time of dissolution in water. To avoid dangerous fires, bromates generally should not be packaged with organic materials in the dry form.

The formulation of the so-called "instant neutralizers," which have achieved wide popularity for both professional and home use, was made possible by the commercial availability of sodium bromate at reasonable cost. Because concentrated solutions are used, such preparations neutralize reduced hair rapidly. Generally, 100 to 125 ml of 10 to 15% solutions of sodium bromate is utilized. A variety of hair conditioners and related materials can be added to impart the desired characteristics.

The addition of urea to bromate neutralizers was mentioned above (396). Urea reportedly retards combustion, acts as a buffer, and increases the speed of neutralization. Finished neutralizers of this type may be solids, liquids, or emulsions, as shown in Formulas 22 to 24.

Formulas 22 to 24

Solid (22)
Urea	150 g
Sodium bromate	150
Duponol	9
Ammonium chloride	12

Liquid (23)
Urea	150 g
Sodium bromate	150
Duponol	30
Ammonium chloride	40
Water	q.s. about 1300

Emulsion (24)
Atlas G2135	5.5 g
Cetyl alcohol	7.0
Lauryl alcohol	2.0
Water	1000.0
Urea	172.0
Sodium bromate	150.0

Finally, neutralizers with lathering qualities have been marketed for home use as "shampoo neutralizers."

Despite the fact that the "instant bromate neutralizers" reoxidize hair rapidly, several inventors evidently believe that this reaction is still too slow. Patents have been issued which describe the catalysis of the reaction between mercaptans and bromates by heavy-metal ions. Typical are the patents issued to Saphir and co-workers (407) from whom an example is cited (Formula 25).

Ammonium salts are included in the disclosed compositions to reduce flammability. Saphir *et al.* also describe the use of nickel or cobalt salts as

Formula 25

Sodium bromate	4.0%
Lauryl alcohol sulfate	1.5
Diammonium phosphate	4.0
Ferric ammonium citrate	0.3
Water	q.s. 100.0

catalysts. A similar patent discloses the use of metavanadates (408). In the United States the use of a sequestered iron salt in sodium bromate has been patented by Grant (409).

Perborates

Despite their desirable stability and use characteristics, bromates are being replaced by the more economical and less toxic sodium perborate (410).

A British patent (411) covers the use of a sodium perborate monohydrate. This perborate is preferred over the tetrahydrate because of its greater stability. Sodium or potassium hexametaphosphates or tetraphosphates have been recommended to improve the solubility of the perborate in hard water. A typical dry mixture consists of 4.2 g of the perborate and 1.0 g of sodium hexametaphosphate. A solution of this composition in about 450 ml of water yields a satisfactory neutralizer. A combination of bromate and perborate stabilized against loss of oxygen by sodium tripolyphosphate has been suggested (412) as a neutralizer and is shown in Formula 26.

Formula 26

Potassium bromate	10 parts
Sodium perborate, monohydrate*	2
Sodium tripolyphosphate	5
Total	17 *parts*

* Percarbonates, perpyrophosphates, and perborosilicates may be substituted as neutralizers.

Seventeen grams of this mixture may be dissolved in 1 liter of water to form a neutralizer solution. It is claimed that the high pH of this formulation is desirable, because it does not reduce the swelling of hair prior to diffusion of the oxidizing agent. It is also indicated that the perborate activates the bromate and that the polyphosphate acts as a calcium-sequestering agent and promotes solution of the neutralizer in water.

It is generally advisable to package perborate carefully because, like other "per" salts, it is subject to decomposition by traces of heavy metals.

Other Neutralizers

A number of other chemical oxidizing agents have been suggested and used from time to time. Sodium chlorite has been used in the United States and was patented in England (413) and Switzerland (414). Urea peroxide has been used for some time in Germany (383). The use of melamine perhydrate for neutralization is the subject of a series of patents (415). Another patent recommends use of the perhydrates of α-aminoalkylphosphonic acids (416).

Several agents which are not usually considered oxidizing agents are reportedly able to effect the formation of disulfide bonds in reduced hair without undesirable side reactions. Among these are the *S*-oxides of *N*-dialkyl-substituted thioalkanoic acid amides* (417), the related *S*-oxides of sulfinamides† (418), and alkaline polythionates‡ (419).

Self-Neutralizing Waving Lotions

The second method of neutralization depends on air oxidation to destroy any reducing agent on the hair and to re-form disulfide linkages. Whitman (10) has pointed out that air oxidation requires considerably more time than chemical oxidation. In view of the time factor, this method of neutralization is not suitable for professional use. In the case of home waving, a long oxidation step, during which the hair remains on the curlers, is not objectionable. This method of neutralization has been used successfully in home permanent waving, both in the presence and absence of a catalyst.

The first self-neutralizing waving lotion to be marketed was based on a conventional thioglycolate lotion to which was added a heavy-metal catalyst (165). The preferred catalyst is manganous ion in a concentration of 20 to 200 ppm. As would be expected, the higher the concentration of catalyst, the faster is the oxidation. Directions for use of lotions containing a heavy-metal catalyst suggest towel-blotting of the hair to remove excess waving lotion, followed by air-drying on the curlers.

$$* \ R-\underset{\substack{\| \\ SO}}{C}-N\underset{R''}{\overset{R'}{<}}$$

$$\dagger \ R-CH_2-\underset{\substack{\| \\ O}}{S}-N\underset{R''}{\overset{R'}{<}}, \text{ where } R- \text{ may be } R'\underset{R''}{\overset{}{>}}N-CO-$$

‡ E.g., K_2S_4O

Freytag (420) claims that addition of 0.1 % of 4-amino-2-methyl naphthol as an oxidation catalyst to a conventional thioglycolate waving solution yields a satisfactory self-neutralizing waving lotion. Another self-neutralizing thioglycolate waving composition based on hexamethylene tetramine has been patented (421) in France. Finally, a French patent discloses that the incorporation of a peroxide-forming terpene into the waving lotion yields a satisfactory self-neutralizing waving lotion (422).

In 1956 Reed (423) described details of the evolution of a self-neutralizing waving process which required a water rinse but no oxidation catalyst. Reportedly this process yields tighter, longer-lasting waves than those from a more complicated process which includes chemical oxidation. Reed also implied that the self-neutralizing process causes less hair damage, as shown by 10 % indices and cystine analyses.

Despite promotional effort, consumer acceptance of self-neutralizing processes (in which the hair is allowed to air-dry on the curling rods) is steadily decreasing. This fact may be due to the frizziness of the water wave created when hair is allowed to dry on small-diameter curlers. A process similar to that described by Reed has been patented in England by Bunford (424), who discloses drying the hair with the aid of a heating pad after rinsing.

Neutralization by Cross-Linking

Although cross-linking of sulfhydryl groups by substances which can react with two —SH moieties appears attractive, this method of neutralization has achieved no prominence. One of the difficulties is the fact that these reactions do not reform disulfides but yield thioethers. The latter are not attacked by reducing agents, and their formation in the hair may create irreversible changes. A second problem is the likelihood that reagents and conditions favoring cross-linking of sulfhydryl groups may also show side reactions with the skin. Under the circumstances, methods for cross-linking sulfhydryl groups for the purpose of neutralization will be only listed here:

Simple aldehydes (425,426).
Bifunctional aldehydes (e.g., glyoxal) (172).
Alkylene dihalides (e.g., dibromoethane) (172, cf. 209).
Diolefines (e.g., ethylene glycol dimethacrylate) (427).
Substituted mercaptides (427a).

Safety Studies

The toxicology of alkaline ammonium thioglycolate solutions was studied by Draize *et al.* (428). The results of this work established the general safety of this chemical. The percutaneous toxicity of thioglycolate solutions appears to be low despite the fact that they can penetrate intact skin as shown by sulfur excretion (428). Schulte (429) confirmed Draize's penetration studies by means of ^{35}S-labeled thioglycolate using the rat as the test animal. Some additional data have been recently reported by Norris (430) attesting to the fact that this chemical is relatively safe and that toxic and irritational effects are associated with the alkalizing agent and the pH of thioglycolate formulations, not the thioglycolate moiety. For example, Shansky (431) reported that guanidinium thioglycolate lotions produce faster waves than conventional waving lotions, with less potential irritation from "free" alkalinity. Shortly thereafter, Bogaty and Giovacchini (432) indicated that, regardless of alkalinity, guanidine thioglycolate is probably not a safe waving agent for general use in view of adverse percutaneous toxicity tests on animals and that guanidine salts *per se* are not recommended for human topical use.

A similar claim for greater safety of one thioglycolate salt over another was made by Whitman and Brookins (262) in 1956, who claimed that a monoethanolammonium thioglycolate waving lotion was less toxic to animals and humans than a "comparable" ammonium thioglycolate preparation. Actually, their preparations differed significantly in pH, and the observed differences in safety are probably related to this fact, not to the choice of alkalizing agent.

Brunner (433) has reviewed available toxicity and irritation data up to 1952 and concluded that ammonium thioglycolate cold-wave formulations normally sold for human use are not apt to cause primary irritation or sensitization. On the other hand, Borelli (296) in an extensive series of tests concluded that beauticians who routinely handle thioglycolate waving lotions may show evidence of primary irritation owing to the continuous corrosive action of these products. His extensive series of patch tests confirms Brunner's conclusions that properly formulated thioglycolate formulations are safe for general use and especially that sensitization to thioglycolates is rare. He was unable to demonstrate any undue effects from normally found trace impurities (434) in thioglycolate formulations, but a large percentage of his subjects reacted to an unidentified thioglycolic acid ester (434,435). In 1958 Voss (305) studied the sensitization potential of more than 40 mercaptans in man and in guinea pigs. Sensitization in guinea pigs was determined via intradermal injection whereas sensitization in man was based on the results of a repeated insult patch test with (0.14 M) solutions of the mercaptans. His results again show that thioglycolic acid and 3-mercaptopropionic acid cause no sensitization.

During the late 1950's a variety of formulations containing amides and esters of thioglycolic acid appeared on the European market. These compounds reportedly are capable of waving hair at relatively low pH. The safety of waving preparations based on these substances is suspect since they can elicit sensitization (305,319,434–436).

Toxicity information on the commonly used oxidants for neutralizers is less readily available; however, some data on various bromates, perborates, and hydrogen peroxide have been recorded by Norris (430).

The safety of the active constituents used for permanent waving is not sufficient to attest to the safety of the finished preparations. Excipients (such as emulsifiers, perfumes), misuse (such as accidental ingestion), etc., can cause effects which are difficult to predict and can normally only be established on the basis of specific safety tests of the final formulation. In this connection, it is noted that permanent waving preparations and neutralizers can accidentally enter the human eye during use. Under these circumstances, those nonionic detergents and amides which can cause eye anesthesia are particularly insidious because the user may not sense that a potential irritant has been introduced into the eye (437).

THE MANUFACTURE OF COLD-WAVING LOTIONS

A general discussion of the more important factors involved in the manufacture of cold-waving lotions would appear advisable prior to the conclusion of this chapter. A suggested manufacturing procedure for such lotions has been outlined by Hollenberg (71), and some of its salient features are presented below.

Cleanliness and absence of contaminants during every step in the manufacture of mercaptan waving lotions are of utmost importance. As mentioned previously, thioglycolic acid is subject to air oxidation, particularly in the presence of certain heavy-metal salts. It is advisable, therefore, to avoid contamination of the lotion by either copper or manganese and to limit air exposure or aeration of the waving fluid during manufacture. Contamination by iron is especially objectionable because in the presence of air even small quantities of this metal (0.25 ppm) produce a purple coloration that is considered undesirable in a cosmetic product.*

In order to avoid contamination, glass-lined equipment should be used wherever possible. Contact of free thioglycolic acid with metals should be avoided. Salts of thioglycolic acid—especially when made alkaline—can be handled safely in aluminum or stainless steel equipment (302,304, or 316 type). A number of plastic materials, suitable for use in manufacture of cold-waving lotions, have become available recently. Polyethylene and Teflon are good examples.

Most other plastic materials should be evaluated before they are allowed to come in contact with waving lotions, especially if oils, perfumes, and wetting agents are incorporated into the lotion.

* Although this color is discharged slowly in the absence of air, it reappears when the lotion contacts oxygen.

In the preparation of the waving lotion, the thioglycolic acid may be neutralized (by cooling) with a suitable alkali and diluted, or the ammonium salt may be diluted directly with distilled or deionized water. Further alkali is added until the required quantity of free alkali or the desired pH is reached. All other ingredients (opacifier, perfume, dye, etc.) are then incorporated into the lotion, which is then brought to final volume by dilution with water. Suitable control analyses are performed and appropriate adjustments made at this point. The lotion may be filtered prior to filling into bottles.

In addition to a mercaptan analysis, determination of free alkali and pH should be considered minimum control tests for each batch of waving lotion.

Detailed procedures for the iodometric determination of thioglycolic acid have been published (438). The pH determination of waving lotions is conveniently carried out with a pH meter, using a glass electrode. The use of indicators or pH papers appears unsatisfactory.

Total ammonia is best determined by a Kjeldahl distillation after excess sodium hydroxide has been added to the sample of the lotion.

Free alkali, i.e., alkali above a specified pH, depending on the cation or cations present, can be determined by potentiometric titration or by titration using a suitable indicator.

On the basis of the authors' experience, properly formulated waving lotions should be chemically stable indefinitely, although some physical changes owing to emulsion instability, etc., may take place. In order to make certain of the desired long-term stability, mercaptan waving lotions must be properly packaged to prevent loss of water or ammonia and to protect the mercaptan against contact with air. Normally this is achieved by packaging waving products into glass bottles sealed with plastic or aluminum caps with liners that make a hermetic seal and can withstand the action of the alkali and of the mercaptan. Plastic containers, especially polyethylene and polypropylene bottles, should not be used unless rigorous testing has shown them to be suitable for the intended purpose. A U.S. patent (439) discloses that polyethylene bottles lined with a special protective cured epoxy resin yielded "commercially satisfactory" results with a thioglycolate waving lotion containing a mixed alkali of ammonium hydroxide and monoethanolamine.

REFERENCES

1. Nessler, C.: Br. Pat. 2931 (1910).

2. Wall, F. E.: *The principles and practice of beauty culture*, 2nd ed., Keystone Publications, New York, 1946, p. 26.

3. Hillier, N. G.: *Profitable permanent waving*, published by the author, New York, 1948, pp. 8–9.

4. Sartory, P.: U. S. Pats. 1,565,509–10 (1925).

5. Winkel, F.: U. S. Pat. 2,051,063 (1936).

6. Evans, R.: U. S. Pats. 1,892,389–91 (1932).

7.   Reed, R. E.: U. S. Pats. 2,040,406-7 (1936).

8.   Evans, R. L., and McDonough, E. G.: U. S. Pat. 2,111,558 (1938); reissue 22,600 (1945).

9.   FDA Publication No. 26 (1965): Cosmetics—facts for consumers.

10.   Whitman, R.: The role of the neutralizer in cold waving, *Proc. Sci. Sec. TGA*, **18**: 27 (1952).

11.   Reed, R. E., Den Beste, M., and Humoller, F. L.: The permanent waving of human hair, *JSCC*, **1**: 109 (1949).

12.   Mayer, J.: U. S. Pat. 1,622,957 (1927); reissue 18,852 (1933).

13.   Squerso, J.: U. S. Pat. 1,884,891 (1932).

14.   Szlanyi, W.: U. S. Pat. 1,400,637 (1921).

15.   Bishinger, R.: U. S. Pat. 1,718,025 (1921).

16.   Grant, S.: U. S. Pat. 2,395,965 (1946).

17.   Grant, S.: U. S. Pat. 2,396,782 (1946).

18.   Mizell, L. R., Tewksbury, C. G., and Vitello, J. P.: U. S. Pat. 3,213,860 (1965).

19.   Underwood, D. L.: U. S. Pat. 3,213,863 (1965).

20.   Vitello, J. P.: U. S. Pat. 3,213,861 (1965). See also Br. Pat. 992,274 (1965).

21.   Ashe Laboratories, Ltd.: Br. Pat. 726,788 (1955).

22.   Société d'Etudes et de Recherches: Fr. Pat. 921,545 (1947).

23.   Hillier, N. G.: *op. cit.* (ref. 3), p. 100.

24.   Speakman, J. B., and Whewell, C. S.: The reactivity of the sulphur linkage in animal fibres. Part II. The action of Baryta and caustic soda on human hair, *J. Soc. Dyers Colourists*, **52**: 380 (1936).

25.   Astbury, W. T., and Woods, H. J.: X-ray studies of hair, wool, and related fibers. II. The molecular structure and elastic properties of hair keratin, *Philos. Trans. Roy. Soc. London*, **232A**: 333 (1933).

26.   L'Oréal: Br. Pat. 988,386 (1965).

27.   Stoves, J. L.: Hair structure and the formulation of permanent waving solutions, *PEOR*, **43**: 232 (1952).

28.   Becker, G.: Essai de controle des liquides frisants de permanente froide, *Parfum., Cosmet. Savons*, **7**: 403 (1964).

29.   Robbins, C.: Weathering in human hair, *Text. Res. J.*, **37**: 337 (1967).

30.   Freytag, H.: Beiträge zur Kenntnis des physikalischen und chemischen Verhaltens von Humanhaaren, *JSCC*, **11**: 555 (1960).

31.   Freytag, H.: Untersuchungen über das Phänomen der Dauerverformung Menschlichen Haares—IV. *JSCC*, **15**: 667 (1964).

31*a*.   Freytag, H.: Zu Fragen der Dosierung, Geschicklichkeit und Reproduzierbarkeit bei der Haardauerverformung, *Parfüm. Kosmetik*, **50**: 377 (1969).

32.   Menkhart, J., Brown, A. E., and Bogaty, H.: Study of keratin setting reactions: Hair waving and wool setting, *Third Int. Wool Tex . Res. Conf.*, Section **2**: 329 (1965).

33.   Brown, A. E.: U. S. Pat. 2,688,972 (1954).

34.   Société d'Etudes et de Recherches: Fr. Pat. 932,009 (1948).

35.   Böss, J.: Kaltwelle und Raumtemperatur, *Seifen-Öle-Fette-Wachse*, **78**: 633 (1952).

36. Mercer, E. H.: *Keratin and keratinization*, Pergamon Press, New York (1961).

37. Crewther, W. G., Fraser, R. D. B., Lennox, F. G., and Lindley, H.: "The chemistry of keratins," in Anfinsen, Jr., C. B., Edsall, J. T., Anson, M. L., Richards, F. M. (Eds.): *Advances in protein chemistry*, Vol. 20, Academic Press, New York, 1965, p. 191.

38. Lundgren, H. P., and Ward, W. H.: "The keratins," in Borasky, R. (Ed.): *Ultrastructure of protein fibers*, Academic Press, New York, 1963, p. 39.

39. Robson, A.: Two-phase materials and wool, *J. Text. Inst.* (Text. Inst. & Industry), **4**: 37 (1966).

40. Orfanos, C., and Ruska, H.: Die Feinstruktur des menschlichen Haares—I. Die Haar-Cuticula, *Arch. Klin. Exptl. Dermatol.*, **231**: 97 (1968).

40a. Mahrle, G., *et al.*: Haar und Haarcuticula im Raster-Elektronenmikroskop, *Arch. Klin. Exptl. Dermatol.*, **235**: 295 (1969).

41. Mercer, E. H., Lindberg, J., and Philip, B.: The "subcutis" and other cuticular preparations from wool and hair, *Text. Res. J.*, **19**: 678 (1949).

42. Leeder, J. D., and Bradbury, J. H.: Conformation of epicuticle on keratin fibres, *Nature*, **218**: 694 (1968).

42a. King, N. L. R., and Bradbury, J. H.: The chemical composition of wool—V. The epicuticle, *Australian J. Biol. Sci.*, **21**: 375 (1968).

42b. Lofts, P. F., and Truter, E. V.: The constitution of the epicuticle of wool, *J. Text. Inst.*, **60**: 46 (1969).

43. Menkhart, J., Wolfram, L. J., and Mao, I.: Caucasian hair, Negro hair, and wool: Similarities and differences, *JSCC*, **17**: 769 (1966).

44. Dusenbury, J. H., and Jeffries, E. G.: The effect of bilateral structure on the chemistry of keratin fibers, *JSCC*, **6**: 355 (1955).

45. Randebrock, R.: Neue Erkenntnisse über den morphologischen Aufbau des menschlichen Haares, *JSCC*, **15**: 691 (1964).

45a. Orfanos, C., and Ruska, H.: Die Feinstruktur des menschlichen Haares—II. Der Haar-Cortex, *Arch. Klin. Exptl. Dermatol.*, **231**: 264 (1968).

46. Rogers, G. E.: Electron microscope studies of hair and wool, *Ann. N.Y. Acad. Sci.*, **83**: 378 (1959).

47. Mercer, E. H., Munger, B. L., Rogers, G. E., and Roth, S. I.: A suggested nomenclature for fine-structural components of keratin and keratin-like products of cells, *Nature*, **201**: 367 (1963).

48. Rogers, G. E., Electron microscopy of wool, *J. Ultrastruct. Res.*, **2**: 309 (1959).

49. Fraser, R. D. B., MacRae, T. P., and Rogers, G. E.: Molecular organization in alpha-keratin, *Nature*, **193**: 1052 (1962).

50. Dobb, M. G., Fraser, R. D. B., and MacRae, T. P.: The structure of the keratin filament, *Third Int. Wool Text. Res. Conf.*, Section **1**: 95 (1965).

51. Fraser, R. D. B., MacRae, T. P., Miller, A., Stewart, F. H. C., and Suzuki, E.: The molecular structure of α-keratin, *Third Int. Wool Text. Res. Conf.*, Section **1**: 85 (1965).

52. Dobb, M. G.: The structure of keratin protofibrils, *J. Ultrastruct. Res.*, **14**: 294 (1966).

53. Filshie, B. K., and Rogers, G. E.: The fine structure of α-keratin, *J. Mol. Biol.*, **3**: 784 (1961).

54.   Johnson, D. J., and Sikorski, J.: Alpha-keratin, *Nature*, **205**: 266 (1965).

55.   Johnson, D. J., and Sikorski, J.: Fine and ultrafine structure of keratin (V), *Third Int. Wool Text. Res. Conf.*, Section **1**: 147 (1965).

56.   Wilson, H. R.: The molecular arrangement in α-keratin, *J. Mol. Biol.*, **6**: 474 (1963).

56a.   Parry, D. A. D.: An alternative to the coiled-coil for α-fibrous proteins, *J. Theor. Biol.*, **24**: 73 (1969).

56b.   Parry, D. A. D.: A proposed conformation for α-fibrous proteins, *J. Theor. Biol.*, **26**: 429 (1970).

57.   Dickerson, R. E.: X-Ray analysis and protein structure, in H. Neurath (Ed.): *The proteins—composition, structure and function*, Vol. 2, 2nd ed., Academic Press, New York, 1964, p. 603.

58.   Astbury, W. T.: The molecular structure and elastic properties of hair, in Savill, A., and Warren, C.: *The hair and scalp*, 5th ed., Williams and Wilkins, Baltimore, 1962, pp. 66–79.

59.   Lundgren, H. P., and Ward, W. H.: "Levels of molecular organization in α-keratins," *Arch. Biochem. Biophys. Suppl.*, **1**: 78 (1962).

60.   Pauling, L., and Corey, R. B.: The structure of hair, muscle, and related proteins, *Proc. Nat. Acad. Sci. U. S.*, **37**: 261 (1951).

61.   Lindley, H., and Haylett, T.: Occurrence of the cys-cys-sequence in keratins, *J. Mol. Biol.*, **30**: 63 (1967).

62.   Clay, R. C., Cook, K., and Routh, J. I.: Studies in the composition of human hair, *J. Am. Chem. Soc.*, **62**: 2709 (1940).

63.   Blackburn, S.: Changes produced in wool by processing, *Chem. Ind.*, 1950, p. 718.

64.   Block, R. J.: Chemical classification of keratins, *JSCC*, **2**: 235 (1951).

65.   Block, R. J., and Bolling, D.: *The amino acid composition of proteins and foods*, 2nd ed., Charles C Thomas, Springfield, Ill., 1951, pp. 236–240.

66.   Simmonds, D. H.: The amino-acid composition of keratins. V. A comparison of the chemical composition of Merino wools of differing crimp with that of other animal fibers, *Text. Res. J.*, **28**: 314 (1958).

67.   Sanford, D., and Humoller, F. L.: Determination of cystine and cysteine in altered human hair fibers, *Anal. Chem.*, **19**: 404 (1947).

68.   Stary, Z.: Aufschliessung des Keratins für die Trypsinverdauung, *Z. physiol. Chem.* **175**: 178 (1928).

69.   Stoves, J. L.: Some physico-chemical properties of keratin fibres and their significance in cosmetology, *JSCC*, **2**: 158 (1951).

70.   Speakman, J. B.: Mechano-chemical methods for use with animal fibres, *J. Text. Inst.*, **38**: T102 (1947).

71.   Hollenberg, I. R.: Formulation of cold wave preparations, *Proc. Sci. Sec. TGA*, **13**: 9 (1950).

72.   Freytag, H.: Beitrag zur Kenntnis des physikalischen und chemischen Verhaltens von Humanhaaren, *Aesthet. Med.*, **11**: 278 (1962).

73.   Elöd, E., Nowotny, H., and Zahn, H.: Struktur und Reaktionsfähigkeit der Wollfaser. X. Über den Zusammenhang zwischen Schädigung und Cystingehalt der Wolle, *Klepzigs Textil-Z.*, **45**: 663 (1942).

74. Bogaty, H.: Molecular forces in permanent waving, *JSCC*, **11**: 333 (1960).

75. Richter, R., and Fuhrmann, W.: Theoretische und praktische Folgerungen aus Studien über die Einwirkung von Thioglykolsäure und Alphamonothioglycerin auf menschliche Haare, *Arch. Dermatol. Syphilol.*, **198**: 274 (1954).

76. Skertchly, A. R. B.: A unified hypothesis for the physical structure and deformation behaviour of wool keratin, *J. Text. Inst.*, **55**: T324 (1964).

77. Crewther, W. G.: The stress–strain characteristics of animal fibers after reduction and alkylation, *Text. Res. J.*, **35**: 867 (1965).

78. Skertchly, A. R. B.: Fundamentals of keratin structure, *Third Int. Wool Text. Res. Conf.*, Section **1**: 161 (1965).

79. Hamburger, W. J., and Morgan, H. M.: Some effects of waving lotions on the mechanical properties of hair, *Proc. Sci. Sec. TGA*, **18**: 44 (1952).

80. Steele, R.: Recent developments in the structure of keratin fibers, *JSCC*, **3**: 99 (1952).

81. Alexander, P.: Über die Einwirkung von Lithiumbromidlösungen auf Wolle und die Rolle der Wasserstoffbindungen, *Melliand Textilber.*, **31**: 550 (1950).

82. Klotz, I. M., and Franzen, J. S.: The stability of interpeptide hydrogen bonds in aqueous solution, *J. Am. Chem. Soc.*, **82**: 5241 (1960).

82a. Speakman, J. B.: The science of permanent waving, *SPC*, **22**: 269 (1949).

83. Scheraga, H. A.: Intramolecular bonds in proteins—II. Noncovalent bonds, in Neurath, H. (Ed.): *The Proteins—Composition, Structure and Function*, Vol. 1, 2nd ed., Academic Press, New York, 1963, p. 477.

84. Berendsen, H. J. C., and Migchelsen, C.: Hydration structure of fibrous macromolecules, in *Forms of water in biologic systems*, *Ann. N.Y. Acad. Sci.*, **125**: (Art. 2): 365 (1965).

85. Zahn, H.: Hydrophobe Wechselwirkungen in Faserproteinen, *Kolloid Z.*, **197**: 14 (1964).

86. Kauzmann, W: Some factors in the interpretation of protein denaturation, in Anfinsen, C. B., Bailey, K., Anson, M. L., and Edsall, J. T. (Eds.): *Advances in Protein Chemistry*, Vol. XIV, Academic Press, New York, 1959, p. 1.

87. Whewell, C S: The chemistry of hair, *JSCC*, **15**: 423 (1964).

88. Whewell, C. S.: The chemistry of hair, *JSCC*, **12**: 207 (1961).

89. Hoare, J. L., Ripa, O, and Speakman, J B: Action of iodine on strained keratin, *Nature*, **196**: 268 (1962).

90. Wolfram, L. J.: Reactivity of disulphide bonds in strained keratin, *Nature*, **206**: 304 (1965).

91. Hamburger, W. J., Morgan, H. M, and Platt, M. M.: Some aspects of the mechanical behavior of hair, *Proc. Sci. Sec. TGA*, **14**: 10 (1950).

92. Speakman, J. B.: The intracellular structure of the wool fibre, *J. Text. Inst.*, **18**: T431 (1927).

93. Harris, M., Mizell, L. R., and Fourt, L.: Elasticity of wool: Relation to chemical structure, *Ind. Eng. Chem.*, **34**: 833 (1942).

94. Harris, M.: Chemistry of keratin, *JSCC*, **1**: 223 (1949).

95. Ciferri, A.: The α-β-transformation in keratin, *Trans. Faraday Soc.*, **59**: 562 (1963).

96. Feughelman, M.: The mechanical properties of permanently set and cystine

reduced wool fibers at various relative humidities and the structure of wool, *Text. Res. J.*, **33**: 1013 (1963).

97. Feughelman, M.: Sulfhydryl–disulfide interchange in extended wool fibers, *Textile Res. J.*, **36**: 293 (1966).

98. Hirsch, F.: Structure and synchronized stretch-rotation of hair keratin fibres, *JSCC*, **11**: 26 (1960).

98a. Bogaty, H.: Torsional properties of hair in relation to permanent waving and setting, *JSCC*, **18**: 515 (1967).

99. Wall, R. A., Morgan, D. A., and Dasher, G. F.: Multiple mechanical relaxation phenomena in human hair, *J. Polymer Sci.*, Part C, **14**: 299 (1966).

100. Speakman, J. B.: The reactivity of the sulphur linkage in animal fibres. Part III. Methods for realizing a permanent set at low temperatures, *J. Soc. Dyers Colourists*, **52**: 423 (1936).

101. Phillips, H.: *Soc. Dyers Colourists, Symp. Fibrous Proteins*, 1946, pp. 38–49.

102. Schöberl, A., and Rambacher, P.: Über die Reaktionsfahigkeit des Keratins der Schafwolle, *Biochem. Z.*, **306**: 269 (1960).

103. Schöberl, A., Rambacher, P., and Wagner, A.: Über die Einwirkung von Natronlauge auf menschliche Haare, *Biochem. Z.*, **317**: 171 (1944).

104. Cuthbertson, W. R., and Phillips, H.: The action of alkalis on wool. I. The subdivision of the combined cystine into fractions differing in their rate and mode of reaction with alkalis, *Biochem. J.*, **39**: 7 (1945).

105. Mizell, L. R., and Harris, M.: Nature of the reaction of wool with alkali, *J. Res. Nat. Bur. Stand.*, **30**: 47 (1943).

106. Schöberl, A.: Über die Isolierung von Lanthionin aus vorbehandelter Schafwolle, *Ber.*, **76**: 970 (1943).

107. Dowling, L. M., and Maclaren, J. A.: The formation of lanthionine as an artifact in protein hydrolysis, *Biochim. Biophys. Acta*, **100**: 293 (1965).

108. Zahn, H.: *N,O*-Peptidylverschiebung, Disulfidaustausch und Lanthioninbildung in Wolle und anderen cystinhaltigen Proteinen, *Chimia*, **15**: 378 (1961).

109. DeMarco, C., Coletta, M., and Cavallini, D.: Cystine cleavage in alkaline medium, *Arch. Biochem. Biophys.*, **100**: 51 (1963).

110. Zahn, H., and Golsch, E.: Über Reaktionen von schwefelhaltigen Aminosäuren, I. Zersetzung von Cystin, Cystindihydantoin, Lanthionin und Lanthionindihydantoin in wässrigen Lösungen, *Z. physiol. Chem.*, **330**: 38 (1962).

111. Elliott, R. L., Asquith, R. S., and Hobson, M. A.: Some observations on the degradation of cystine in ethylamine solution, *J. Text. Inst.*, **51**: T692 (1960).

112. Parker, A. J., and Kharasch, N.: The scission of the sulfur–sulfur bond, *Chem. Rev.*, **59**: 583 (1959).

113. Swan, J. M.: The possible formation of thiazoline and thiazolidine rings in peptides and proteins, in Albert, A., Badger, G. M., and Shoppee, C. W. (Eds.): *Current Trends in Heterocyclic Chemistry*, Academic Press, New York, 1958, p. 65.

114. Rosenthal, N. A., and Oster, G.: Recent progress in the chemistry of disulfides, *JSCC*, **5**: 286 (1954).

115. Danehy, J. P., and Hunter, W. E.: The alkaline decomposition of organic disulfides. II. Alternative pathways as determined by structure, *J. Org. Chem.*, **32**: 2047 (1967).

116. Ziegler, K.: New cross-links in alkali-treated wool, *J. Biol. Chem.*, **239**: PC2713 (1964).

117. Asquith, R. S., and Garcia-Dominguez, J. J.: New amino acids in alkali-treated wool, *J. Soc. Dyers Colourists*, **84**: 155 (1968); and Crosslinking reactions occurring in keratin under alkaline conditions, *ibid.*, **84**: 211 (1968).

118. Clarke, H. T.: The action of sulfite upon cystine, *J. Biol. Chem.*, **97**: 235 (1932).

119. Mauthner, J.: Über Cystin, *Z. physiol. Chem.*, **78**: 28 (1912).

120. Goddard, D. R., and Michaelis, L.: A study of keratin, *J. Biol. Chem.*, **106**: 605 (1934).

121. Elsworth, F. F., and Phillips, H.: The action of sulphites on the cystine disulphide linkages in wool. I. The influence of *p*H value on the reaction, *Biochem. J.*, **32**: 837 (1938): II. The influence of temperature, time and concentration on the reaction, *ibid.*, **35**: 135 (1941).

122. Carter, E. G. H., Middlebrook, W. R., and Phillips, H.: Chemical constitution and physical properties of bisulfited wool, *J. Soc. Dyers Colourists*, **62**:203 (1946).

123. Middlebrook, W. R., and Phillips, H.: The action of sulfites on the cystine disulfide linkage of wool—III. Subdivision of the combined cystine into four fractions differing in their reactivity towards sodium bisulfite, *Biochem. J.*, **36**: 428 (1942).

124. Gutmann, A.: Über die Einwirkung von Laugen auf Äthylnatriumthiosulfat, *Ber.*, **41**: 1650 (1908); Über die Einwirkung von Säuren auf Natrium-äthylthiosulfat, *Ber.*, **42**: 228 (1909).

125. Stricks, W., and Kolthoff, I. M.: Equilibrium constants of the reactions of sulfite with cystine and with dithioglycolic acid, *J. Am. Chem. Soc.*, **73**: 4569 (1951).

126. Stoves, J. L.: The reactivity of the cystine linkage in keratin fibres. Part II. The action of reducing agents, *Trans. Faraday Soc.*, **38**: 261 (1942).

127. Speakman, P. T.: The mechanism of the reaction between cystine in keratin and sulphite/bisulphite solutions at 50°C—Part I, *Biochim. Biophys. Acta*, **25**: 347 (1957); Part II, *ibid.*, **28**: 284 (1958).

128. Wolfram, L. J., and Underwood, D. L.: The equilibrium between the disulfide linkage in hair keratin and sulfite or mercaptan, *Text. Res. J.*, **36**: 947 (1966).

129. Blankenburg, G.: Kraft-Dehnungs-Eigenschaften und Superkontraktion von Wollkeratin in wässrig-alkoholischen Lösungen, *Melliand Textilber.*, **48**: 686 (1967).

130. Speakman, J. B.: Br. Pat. 591,932 (1947).

131. Speakman, J. B.: U. S. Pat. 2,400,377 (1946).

132. Zahn, H., and Osterloh, F.: New methods of lanthionine formation in wool, in *Proc. Int. Wool Text. Res. Conf.*, C (Part 1),: C-19 (1955).

133. Miro, P., and Garcia-Dominguez, J. J.: Action of nucleophilic reagents on wool, *J. Soc. Dyers Colourists*, **83**: 91 (1967).

134. Gerthsen, T., and Meichelbeck, H.: The reactivity of wool cystine in water-solvent mixtures, *Third Int. Wool Text. Res. Conf.*, Section **2**: 65 (1965).

135. Lecher, H.: Über eine Reduktion organischer Disulfide durch Alkalimercaptide, *Ber.*, **53B**: 591 (1920).

136. Bersin, T., and Steudel, J.: Polarimetrische Untersuchungen über das Thiol-Disulfid-System, *Ber.*, **71**: 1015 (1938).

137. Fava, A., Iliceto, A., and Camera, E.: Kinetics of the thiol–disulfide exchange, *J. Am. Chem. Soc.*, **79**: 833 (1957).

138. Foss, O.: Ionic scission of the sulfur-sulfur bond, in Norman Kharasch, (Ed.): *Organic Sulfur Compounds*, Vol. I, Pergamon Press, New York, 1961, p. 83.

139. Schöberl, A.: Über die Thioglykolsäurebehandlung von Haaren und die Wiedererzeugung von Disulfidgruppen, *Naturwiss.*, **40**: 390 (1953).

140. Patterson, W. I., Geiger, W. B., Mizell, L. R., and Harris, M.: The role of cystine in the structure of the fibrous protein, wool, *Am. Dyestuff Rep.* **30**: 425 (1941).

141. Schöberl, A., and Gräfje, H.: Aminocarbonsäuren und Haarkeratin mit unsymmetrisch eingebauten Disulfidbindungen, ein Beitrag zu dem Problem des Disulfidaustausches, *Ann. Chem.*, **617**: 71 (1958).

142. Gräfje, H.: Recherches et observations sur la cinétique des reactions se produisant au cours de l'ondulation permanente à froid, *Parfum., Cosmet., Savons*, **2**: 564 (1959).

143. Schöberl, A., and Gräfje, H.: Über Disulfid-Austauschreaktionen bei nieder- und hochmolekularen Verbindungen, *Fette-Seifen-Anstrichmittel*, **60**: 1057 (1958).

144. Gerthsen, T., and Gohlke, C.: Chemische Analysen an dauergewellten Haaren, *Parfüm. Kosmetik*, **45**: 277 (1964).

145. Schulte, K. E., Mleinek, I., and Hobl, H. D.: Beitrag zur Frage der Thioglykolsäure-Retention im Haar, *Parfüm. Kosmetik*, **45**: 87 (1964).

146. Whitman, R., and Eckstrom, M. G.: The mercaptan-disulfide system in permanent waving—a new mechanism and its practical implications, *Proc. Sci. Sec. TGA*, **22**: 23 (1954).

147. Zahn, H., Gerthsen, T., and Kehren, M.-L.: Anwendung schwefelchemischer Analysenmethoden auf dauergewelltes Haar, *JSCC*, **14**: 529 (1963).

148. Lewis, B., Robson, A., and Tiler, E. M.: An investigation of the sulphur compounds in acid hydrolysates of wool, *J. Text. Inst.*, **51**: T653 (1960).

149. Fletcher, J. C., and Robson, A.: The occurrence of bis-(2-amino-2-carboxyethyl) trisulphide in hydrolysates of wool and other proteins, *Biochem. J.*, **87**: 553 (1963).

150. Sanders, J. H.: U. S. Pat. 2,719,815 (1955).

151. Eckstrom, M. G.: Swelling studies of single human hair fibers, *JSCC*, **2**: 244 (1951).

152. Valko, E. I., and Barnett, G.: A study of the swelling of hair in mixed aqueous solvents (I), *JSCC*, **3**: 108 (1952).

153. Powers, D. H., and Barnett, G.: A study of the swelling of hair in thioglycolate solutions and its reswelling, *JSCC*, **4**: 92 (1953).

154. Shansky, A.: The osmotic behavior of hair during the permanent waving as explained by swelling measurements, *JSCC*, **14**: 427 (1963).

155. Keil, F.: Die Quellung des Haares in kaltwellmitteln Untersuchungen in polarisiertem Lichte, *JSCC*, **11**: 543 (1960).

156. Hermann, K. W.: Hair keratin reaction, penetration, and swelling in mercaptan solutions, *Trans. Faraday Soc.*, **59**: (Part 7): 1663 (1963).

157. Haefele, J. W., and Broge, R. W.: Properties and reactions of hair after treatment with mercaptans of differing sulfhydryl acidities, *Proc. Sci. Sec. TGA*, **36**: 31 (1961).

158. Strasheim, A., and Buijs, K.: An infra-red study of the oxidation of the disulphide bond in wool, *Biochim. Biophys. Acta*, **47**: 538 (1961).

159. Rutherford, H. A., and Harris, M.: Reaction of wool with hydrogen peroxide, *J. Res. Nat. Bur. Stand.*, **20**: 559 (1938).

160. Tarbell, D. S.: The mechanism of oxidation of thiols to disulfides, in Karasch, N. (Ed.): *Organic Sulfur Compounds*, Vol. I, Pergamon Press, New York, 1961, p. 97.

161. Wallace, T. J., Schriesheim, A., and Bartok, W.: The base-catalyzed oxidation of mercaptans. III. Role of the solvent and effect of mercaptan structure on the rate determining step, *J. Org. Chem.*, **28**: 1311 (1963).

162. Stoves, J. L.: The reactivity of the cystine linkage in keratin fibres. Part III. The action of oxidizing agents, *Trans. Faraday Soc.*, **38**: 501 (1942).

163. Thunberg, T.: Auto-oxidizable substances and auto-oxidizable systems of physiological interest. III. Auto-oxidizable thio-compounds, *Lunds Univ. Arsskr. Avd. 2*, **9**: 1 (1913).

164. Kharasch, M., Legault, R. R., Wilder, A. B., and Gerard, R. W.: Metal catalysts in biologic oxidations. I. The simple system: thioglycolic acid, buffer, metal, dithiol, *J. Biol. Chem.*, **113**: 537 (1936).

165. Den Beste, M., and Reed, R. E.: U. S. Pat. 2,540,980 (1951).

166. Fava, A., Reichenbach, G., and Peron, U.: Kinetics of the thiol disulfide exchange. II. Oxygen-promoted free-radical exchange between aromatic thiols and disulfides, *J. Am. Chem. Soc.*, **89**: 6696 (1967).

167. Neville, R. G.: The oxidation of cysteine by iron and hydrogen peroxide, *J. Am. Chem. Soc.*, **79**: 2456 (1957).

168. Speakman, J. B.: Br. Pat. 453,701 (1936).

169. Speakman, J. B.: U. S. Pat. 2,261,094 (1941).

170. Patterson, W. I., Geiger, W. B., Mizell, L. R., and Harris, M.: Role of cystine in the structure of the fibrous protein, wool, *J. Res. Nat. Bur. Stand.*, **27**: 89 (1941).

171. Alexander, P., and Hudson, R. F.: *Wool—its chemistry and physics*, Franklin, Publ. Co., Inc., New Jersey, 1963, pp. 338–339.

172. Amica: Fr. Pat. 1,011,152 (1952).

173. Stoves, J. L.: The reactivity of the cystine linkage in keratin fibres. Part IV. The action of formaldehyde, *Trans. Faraday Soc.*, **39**: 294 (1943).

174. Kirby, D. H.: A method for determining the waving efficiency of cold permanent wave lotion, *Proc. Sci. Sec. TGA*, **26**: 12 (1956).

175. Stavrakas, E. J., Platt, M. M., and Hamburger, W. J.: Determination of curl strength of tresses treated with water, hair spray, and waving lotion, *Proc. Sci. Sec. TGA*, **31**: 36 (1959).

175a. Freytag, H.: Zur Erfassung von Einflüssen auf die Haardauerverformung und Bewertung haarverformungsfixierender Mittel, *JSCC*, **20**: 707 (1969).

176. Brauchoff, H.: Physikalische Untersuchungen an menschlichen und tierischen Haaren, *Parfüm. Kosmetik*, **38**: 399 (1957); continued: *ibid.*, **38**: 573 (1957).

177. Brauckhoff, H.: Etude sur les propriétés physiques et mécaniques des cheveux humains et leurs modifications sous l'influence d'actions nuisibles, *Parf. Cosm. Sav.*, **1**: 434 (1958).

178. Klemm, E. J., Haefele, J. W., and Thomas, A. R.: The swelling behavior of hair fibers in lithium bromide, *Proc. Sci. Sec. TGA*, **43**: 7 (1965).

179. Scheibner, K.: Rhodamin B in der Haardiagnostik, *Dermatol. Wochenschr.*, **140**: 1005 (1959).

180. Scheibner, K.: Hitze-Rhodaminfarbung zur Dartellung von Resten der Wurzelscheide und Schuppen an Haaren, *Dermatol. Wochenschr.*, **142**: 853 (1960).

181. Berth, P., and Reese, G.: Veränderung des Haarkeratins durch kosmetische Behandlungen und natürliche Umwelteinflüsse (Beschreibung und Anwendungsmöglichkeiten einer modifizierten Cu-Wertmethode), *JSCC*, **15**: 659 (1964).

181a. Robbins, C. R., and Kelly, C.: Amino acid analysis of cosmetically altered hair, *JSCC*, **20**: 555 (1969).

182. McDonough, E. G.: *Truth about cosmetics*, Drug and Cosmetic Industry, New York, 1937, p. 211.

183. de Navarre, M. G.: *The chemistry and manufacture of cosmetics*, Van Nostrand, New York, 1941, pp. 471–481.

184. Jannaway, S. P.: *Hair dressing preparations*, United Trade Press, London, 1946.

185. McDonough, E. G.: The development of machineless permanent waving, *JSCC*, **1**: 183 (1948).

186. Schitzler, A.: *Theorie und Hilfsmittel des Dauerwellens*, Verlag für Chemische Industrie, H. Ziolkowsky, Augsburg (Germany), 1936.

187. Grosert, T.: U. S. Pat. 1,064,901 (1913).

188. Ingrassia, L.: U. S. Pat. 1,581,577 (1926).

189. Allen, C. H.: Permanent waving preparations: a summary of standard information and developments, *SPC*, **16**: 392 (1943).

190. Kietz, A., Greenwald, W. F., and Weisberg, L.: U. S. Pat. 1,720,220 (1929).

191. Soussa, E.: Fr. Pat. 785,878 (1935).

192. Kalish, J.: Permanent wave solutions, *DCI*, **49**: 156 (1941).

193. Steinbach, P. R.: U. S. Pat. 2,002,989 (1935).

194. Rüttgers, L.: Beitrag zum Problem der Dauerwellen-wasser, *Seifen-Öle-Fette-Wachse*, **74**: 109 (1948).

195. Readers' questions (permanent wave preparations), *DCI*, **66**: 331 (1950).

196. Furman, B. N.: Br. Pat. 685,036 (1952).

197. Auch, R. A.: Permanent wave fluids, *SPC*, **10**: 43 (1937).

198. Sarensen, J. P.: Permanent wave solutions, *DCI*, **43**: 160 (1938).

199. Zotos: Br. Pat. 428,932 (1935).

200. DeMytt, L., and Reed, R. E.: U. S. Pat. 2,506,492 (1950).

201. Geier, F.: Ger. Pat. 734,889 (1943).

202. Melaro, R. E.: U. S. Pat. 2,624,347 (1953).

203. Mas et Cie.: Fr. Pat. 1,038,179 (1953).

204. Friedman, G., and Goldfarb, A. R.: U. S. Pat., 2,310,687 (1943).

205. Benk, E.: Haarschutzmittel beim Dauerwellen, *Seifen-Öle-Fette-Wachse*, **75**: 481 (1949).

206. Steinbach, P. R.: U. S. Pat. 2,095,374 (1937).

207. Coriolan G.m.b.H.: Ger. Pat. 659,120 (1938); Br. Pat. 468,845 (1937).

208. Ströher, F. A. G.: Br. Pat. 473,641 (1937).

209. Harris, M.: U. S. Pats. 2,508,713–4 (1950).

210. Speakman, J. B.: Br. Pat. 453,700 (1936).

211. Speakman, J. B.: U. S. Pat. 2,410,248 (1946).

212. Speakman, J. B.: Br. Pat. 456,336 (1936).

213. Speakman, J. B.: U. S. Pat. 2,351,718 (1944).

214. Maeder, F.: U. S. Pat. 2,068,809 (1937).

215. Reed, R. E., Tenenbaum, D., and Den Beste, M.: U. S. Pat. 2,405,166 (1946).

216. Grant, S.: U. S. Pat. 2,446,227 (1948).

217. Brown, J.: U. S. Pat. 1,681,170 (1928).

218. Maeder, F., and Sims, B. M. W.: U. S. Pat. 1,933,021 (1933).

219. Calva, J. B.: U. S. Pat. 2,390,073 (1945).

220. Nowak, G. A.: Entwicklung und Patentanlage der chemischen Welle (Kaltwelle), *Parfüm. Kosmetik*, **34**: 209,256 (1953).

221. Luckenbach, W. F.: U. S. Pat. 2,305,356 (1942).

222. Gant, V. A.: U. S. Pat. 2,643,375 (1953)

223. Gant, V. A.: U. S. Pat. 2,750,947 (1956).

224. Wilmsmann, H., and Ludwig, W.: Ger. Pat. 1,154,901 (1963).

225. Steinbach, H. H., Damm, K., and Simmler, W.: Ger. Pat. 1,174,942 (1964).

225a. Wolfram, L. J.: Modification of hair by internal deposition of polymers, *JSCC*, **20**: 539 (1969).

226. Malone, J. Y.: U. S. Pat. 2,056,358 (1936).

227. Malone, J. Y.: U. S. Pat. 2,061,709 (1936).

228. Malone, J. Y.: U. S. Pat. 2,087,953 (1937).

229. Brown, J. C.: U. S. Pat. 2,155,178 (1939).

230. Willat, A. F.: U. S. Pat. 2,180,380 (1939).

231. Willat, A. F.: U. S. Pat. 2,266,111 (1941).

232. Pye, D. J.: U. S. Pat. 2,183,894 (1939).

233. Speakman, J. B.: U. S. Pat. 2,201,929 (1940).

234. Schwarzkopf, O.: Austrian Pat. 166,240 (1950).

235. Chambers, N.: Br. Pat. 674,586 (1952).

236. Cifelli, T., Jr.: Patent thoughts and trends: "cold" wave patent found invalid by special master, *DCI*, **74**: 263 (1954).

237. Cifelli, T., Jr.: Patent thoughts and trends, *DCI*, **80**: 340 (1957) and **82**: 765 (1958).

238. Coustolle, P. P.: Fr. Pat. 784,404 (1936); Certificat d'addition 46,213.

239. Martin, H.: U. S. Pat. 2,350,178 (1944).

240. Baker, G. S.: U. S. Pat. 2,389,755 (1945).

241. McDonough, E. G.: Br. Pat. 589,956 (1947).

242. Sales Affiliates, Inc.: Australian Pat. 117,071 (1943).

243. Sales Affiliates, Inc.: Fr. Pat. 938,334 (1948).

244. McDonough, E. G.: U. S. Pat. 2,577,710 (1951).

245. McDonough, E. G.: U. S. Pat, 2,577,711 (1951).

246. McDonough, E. G.: U. S. Pat. 2,736,323 (1956).

247. Evans, R.: Cosmetics research, *JSCC*, **2**: 48 (1950).

248. McDonough, E. G.: U. S. Pat. 2,889,833 (1959).

249. Cifelli, T., Jr.: McDonough hair-waving patent invalidated, *DCI*, **74**: 781 (1954).

250. Keithler, W. R.: A review of permanent waving, *DCI*, **73**: 322 (1953).

251. Société d'Etudes et de Recherches: Fr. Pat. 921,066 (1947).

252. Freytag, H.: Über das Verhalten von Keratinfasern in thioglykolathaltigen Systemen, *Z. Naturforsch.*, **7b**: 645 (1952).

253. Heilingötter, R.: *Parfüm. Kosmetik*, **31**: 190 (1950).

254. Schnell, E. O.: U. S. Pat. 2,631,965 (1953).

255. Shansky, A.: Comparative differences in formulations between professional products and consumer products—IV. Cold wave lotions, *Am. Perf.*, **80**: 31 (1965).

256. Sluis, H. van: Cream base for mercaptan depilatories, *PEOR*, **40**: 321 (1949).

257. Brant, J. H.: Ger. Pat. 1,035,856 (1958).

258. Martin, H.: U. S. Pat. 2,876,781 (1959).

259. Superma Limited: Br. Pat. 889,572 (1962).

260. Whitman, R. C.: U. S. Pat. 3,039,934 (1962).

261. Saphir, J.: Beitrag zur Frage der Verwendung von Äthanolaminen bei der Kaltwelle und der Patentrechte auf diesem Gebiet, *Seifen-Öle-Fette-Wachse*, **90**: 116 (1964).

262. Whitman, R., and Brookins, M. G.: Toxicity studies on monoethanolamine thioglycolate cold wave lotions, *Proc. Sci. Sec. TGA*, **25**: 42 (1957); cf. Draize, J. H., *et al.*: Ethanolamines in topical preparations, *ibid.*, **27**: 12 (1957).

263. Ashe Laboratories Limited: Br. Pat. 798,674 (1958).

264. Wajaroff, T.: Ger. Pat. 1,124,640 (1962).

265. Wella, A. G.: Br. Pat. 921,543 (1963); Ger. Pat. 1,089,124 (1960).

266. Jensen, C. C., and Mittleman, F. A.: U. S. Pat. 3,230,144 (1966).

267. Schwarzkopf, H.: Ger. Pat. 1,229,980 (1966).

268. Raecke, B.: Ger. Pat. 830,095 (1951).

269. Sales Affiliates, Inc.: Fr. Pat. 1,198,718 (1959); Ger. Pat. 1,067,565 (1959).

270. Strain, R. J., and Tusa, P.: U. S. Pat. 3,025,218 (1962).

271. Wajaroff, T.: Ger. Pat. 1,139,608 (1962).

271a. Saphir, J.: Ger. Pat. 1,242,794 (1967).

271b. Wall, R. A., and Fainer, P.: U. S. Pat. 3,395,216 (1968); Fr. Pat. 1,489,011 (1967); Ger. Pat. 1,467,853 (1968).

272. Gillette Safety Razor Co.: Br. Pat. 689,641 (1953).

273. DeMytt, L. E., and Hannigan, A. M.: U. S. Pat. 2,708,940 (1955).

274. Zviak, C., and Rouet, J.: U. S. Pat. 3,157,578 (1964); Br. Pat. 890,180 (1962); Ger. Pat. 1,255,863 (1967).

275. Brown, A. E.: U. S. Pat. 2,717,228 (1955); Br. Pat. 723,917 (1955)

276. Wehr, R.: U. S. Pat. 3,071,515 (1963); Br. Pat. 902,888 (1962).

277. Orthner, L., and Reuter, M.: Ger. Pat. 974,422 (1960).

278. Grein, P. F. A.: Br. Pat. 799,432 (1958).

279. Atkinson, J. C., *et al.*: Action of mixed solvents on wool, *Nature*, **184**: 444 (1959).

280. Haefele, J. W.: U. S. Pat. 2,615,828 (1952); Br. Pat. 672,838 (1952).

281. Haefele, J. W.: U. S. Pats. 2,719,813–14 (1955).

282. Schwarz, M. H.: U. S. Pat. 2,540,494 (1951).

283. Morelle, J. V.: Fr. Pat. 1,434,991 (1966).

284. Bouthilet, R. J., and Karler, A.: Cosmetic effects of substantive proteins, *Proc. Sci. Sec. TGA.*, **44**: 27 (1965).

285. Maier, A.: Austrian Pat. 171,725 (1952).

286. Walker, G. T.: Br. Pat. 979,167 (1958).

287. Newman, A. W. T.: Br. Pat. 850,860 (1960).

288. Garternicht, E.: East Ger. Pat. 22,581 (1962).

289. Wohlfarth, C.: East Ger. Pat. 31,898 (1965).

289a. Musolf, M. C.: Fr. Pat. 1,533,503 (1968).

290. Leberl, O.: Neuzeitliche Verbesserungen an Dauerwellmitteln, *Kosm. Parf. Drogen*, **6**: (I/II): 4 (1959).

291. Morelle, J.: The chemistry of permanent waving, *SPC*, **25**: 828 (1952).

292. Gasser, E.: Coiffeurekzem, verursacht durch Thioglycerin enthaltende Kaltdauerwellenwässer, *Schweiz. med. Wochschr.*, **83**: 448 (1953).

293. Draize, J. H., Alvarez, E., and Woodard, G.: Comparative percutaneous toxicity of 3-mercapto-1,2-propanediol (thioglycerol) and ammonium thioglycolate, *Fed. Proc.*, **8**: 287 (1949).

294. Kensler, C. J., and Elsner, R. W.: The systemic actions of thioglycerol with special reference to goitrogenic activity, *J. Pharm. Exptl. Therap.*, **97**: 349 (1949).

295. Lehman, A. J.: Health aspects of common chemicals used in hair-waving preparations, *J. Am. Med. Assoc.*, **141**: 842 (1949).

296. Borelli, S.: Toxische und allergische Reaktionen auf organische Schwefelverbindungen in Dauerwellpräparaten—I, *Hautarzt*, **8**: 159; II, *ibid.*, **8**: 211 (1957).

297. Heilingötter, R. O.: Ger. Pat. 1,145,303 (1963).

298. Raecke, B.: Br. Pat. 690,866 (1953).

299. Morelle, J.: Contribution à l'étude des corps "frisants," *Parfum. mod.*, **42**: (21): 51 (1950).

300. Morelle, J.: La chimie et l'ondulation permanente, *Industrie parfum.*, **5**: 173 (1950).

301. Morelle, J.: L'acide thiolactique, *Parfum. mod.*, **42**(21): 66 (1950).

302. Morelle, J.: L'acide thiolactique et ses emplois, *Industrie parfum.*, **7**: 201 (1952).

303. Morelle, J.: Br. Pat. 743,730 (1956); Fr. Pat. 1,031,538 (1953).

304. Haefele, J. W., and Broge, R. W.: The synthesis and properties of mercaptans having different degrees of acidity of the sulfhydryl group, *Proc. Sci. Sec. TGA*, **32**: 52 (1959).

305. Voss, J. G.: Skin sensitization by mercaptans of low molecular weight, *J. Invest. Dermatol.*, **31**: 273 (1958).

306. DeMytt, L. E.: U. S. Pat. 2,976,216 (1961).

307. Gillette Company: Br. Pat. 839,923 (1960); DeMytt, L. E., and Hsiung, D. Y.: U. S. Pat. 3,066,077 (1962).

308. Finkelstein, P., *et al.*: Preparation and hair waving properties of 2,5-dimercaptoadipic acid, *JSCC*, **13**: 253 (1962).

308a. Cook, M. K.: Dry permanent waves, *DCI*, **103**:(2): 48 (1968).

308b. Mead Johnson Labs.: Br. Pat. 1,002,889 (1965).

308c. Zemlin, J. C., and Harrington, K. A.: U. S. Pat. 3,459,198 (1969).

309. Heilingötter, R.: "Versteckte" Thioglykolsäure, *Kosm. Parf. Drogen,* **May 1955,** III/IV: 25.

310. Bouvet, R. J., Welwart, Z., and Hittner, S.: Fr. Pat. 999,436 (1952).

311. Del Zoppo, M.: U. S. Pat. 2,600,624 (1952).

312. Henkel and Cie.: Br. Pat. 672,730 (1952).

313. Van Ameringen-Haebler, Inc.: Br. Pat. 771,627 (1957).

314. Van Ameringen-Haebler, Inc.: Br. Pat. 804,077 (1958); Ger. Pat. 1,096,551 (1961).

315. Mahal, A.: Swiss Pat. 322,479 (1957); Ger. Pat. 971,899 (1959).

316. Charle, R., Ritter, R., and Kalopissis, G.: U. S. Pat. 2,944,942 (1960); Fr. Pat. 1,157,158 (1958); Fr. Pat. 559,672 (1956).

317. Kalopissis, G., and Viout, A.: Fr. Pat. 1,449,595 (1966); Br. Pat. 1,067,065 (1967).

318. Kalopissis, G., and Viout, A.: Br. Pat. 1,002,455 (1963); Br. Pat. 1,095,838 (1967).

319. Evans Chemetics Inc.: Waving solutions with acid pH, Bulletin 244/4, April 1, 1958.

320. Saphir, J., Holtschak, H., and Vick, C.: Ger. Pat. 1,067,566 (1959).

321. Richter, W.: Fr. Pat. 1,174,561 (1959); Austrian Pat. 211,483 (1960); Can. Pat. 616,586 (1961); Swiss Pat. 355,896 (1961); Br. Pat. 824,426 (1959); Ger. Pat. 1,009,765 (1957); Belg. Pat. 556,292 (1957).

322. Martin, H.: U. S. Pat. 3,148,126 (1964).

323. Leberl, O.: Weiterentwicklung von Dauerwellpräparaten, *Kosm. Parf. Drogen,* **8:** 161 (1962).

324. Schultheis, W.: Ger. Pat. 1,108,703 (1961).

325. Baron, H., Hepding, L., and Hohmann, W.: Ger. Pat. 1,063,763 (1959).

326. Kadus-Werk: Austrian Pat. 210,071 (1960).

327. Wajaroff, T.: Ger. Pat. 1,104,521 (1961).

328. Kalopossis, G.: U. S. Pat. 3,063,908 (1962); Br. Pat. 859,347 (1961); Fr. Pat. 1,197,194 (1959); Ger. Pat. 1,144,440 (1963).

329. Coriolan G.m.b.H.: Br. Pat. 468,845 (1937).

330. Michaels, E. B., and Lustig, B.: U. S. Pat. 2,437,965 (1948).

331. Whitman, R., and Den Beste, M.: U. S. Pat. 2,840,086 (1958).

332. Ewald, G.: East Ger. Pat. 37,405 (1965).

333. Mora, P.: Italian Pat. 458,943 (1950); Fr. Pat. 994,703 (1951); Swiss Pat. 271,523 (1951).

334. Mora, A.: U. S. Pat. 2,783,762 (1957).

335. Warner-Hudnut, Inc.: Br. Pat. 711,060 (1954).

336. Watson, P. C.: U. S. Pat. 2,836,543 (1958).

337. Lubs, H. A.: U. S. Pat. 2,403,937 (1946).

338. Richter, W.: Ger. Pat. 1,009,765 (1957).

339. Tao, A.: Br. Pat. 976,574 (1964); Ger. Pat. 1,198,491 (1965); Fr. Pat. 1,287,699 (1962).

340. Bogaty, H., and Brown, A. E.: U. S. Pat. 2,766,760 (1956).

341. Société Monsavon-L'Oréal: Br. Pat. 835,247 (1960); Austrian Pat. 216,675 (1961); Ger. Pat. 1,199,927 (1965).

342. Société Monsavon-L'Oréal: Ger. Pat. 1,129,657 (1962).

343. Jenkins, A. D., and Wolfram, L. J.: U. S. Pat. 3,256,154 (1966); Br. Pat. 976,821 (1964); Ger. Pat. 1,196,324 (1965).

344. Edman, P., and Diehl, K.: Reduction of insulin, *Congr. Int. Biochim.*, *2e*, *Paris*, 1952: Res. com., p. 51.

345. Gillespie, J. M.: Reaction of sodium borohydride with wool, *Nature*, **183**: 322 (1959).

346. Sweetman, B. J.: The specificity of certain phosphine derivatives as reducing agents for the disulfide bond in wool keratin, *Text. Res. J.*, **36**: 1096 (1966).

347. Peterson, D. H.: U. S. Pat. 2,464,281 (1949).

348. Gershon, S. D., Goldberg, M. A., and Netzbandt, W. B.: U. S. Pat. 2,689,815 (1954).

349. Leberl, O.: Austrian Pat. 169,458 (1951).

350. Freytag, H.: Ger. Pat. 972,084 (1959).

351. Deutsch, P. A., and Temblett, A. R. S.: Br. Pat. 753,241 (1956).

352. Arnold, W. S.: U. S. Pat. 2,738,304 (1956); Br. Pat. 723,349 (1955).

353. Mace, H. W.: U. S. Pat. 2,479,382 (1949).

354. Ronk, S. O., and Hunter, L. R.: Br. Pat. 679,841 (1952).

355. Ramsey, H. R.: U. S. Pat. 2,418,664 (1947).

356. Deadman, L. L. F.: Br. Pat. 778,308 (1957).

357. Société Monsavon-L'Oréal: Br. Pat. 802,444 (1958); Fr. Pat. 1,123,721 (1956).

358. Laboratoires Scientifiques de Neuilly: Fr. Pat. 1,005,119 (1952).

359. L'Oréal: Fr. Pat. 966,988 (1950).

360. Endura Limited: Br. Pat. 632,342 (1949).

361. Sales Affiliates Limited: Br. Pat. 960,155 (1964).

362. Linke, W.: Br. Pat. 941,300 (1963); Ger. Pat. 1,155,566 (1963).

363. Rieger, M. M.: Recent developments in permanent waving, *Am. Perf.*, **75**: (VIII): 33 (1960).

364. Banker, R. D., Grounds, P. W., and Cody, R. A.: U. S. Pat. 3,099,603 (1963); Br. Pat. 959,772 (1964); Can. Pat. 637,817 (1962).

365. Shepard, W. W., Baude, M. E. R., and Moore, R. S.: U. S. Pat. 3,103,468 (1963).

366. Procter & Gamble Limited: Br. Pat. 935,530 (1963).

366a. Brechner, S., *et al.*: U. S. Pat. 3,433,868 (1969).

367. Wajaroff, T.: Ger. Pat. 1,136,057 (1963).

368. Schnell, E. O.: U. S. Pat. 2,653,121 (1953).

369. Deadman, L. L. F.: U. S. Pat. 2,751,327 (1956); Br. Pat. 748,858 (1956).

370. Martin, H.: U. S. Pat. 2,990,336 (1961); Ger. Pat. 1,186,581 (1965).

371. Schweizer, H. C.: U. S. Pat. 3,193,463 (1965).

372. Gillette Company: Br. Pat. 858,216 (1961); Ger. Pat. 1,131,362 (1962).

373. Moore, R. S.: U. S. Pat. 2,869,559 (1959).

374. Kochenrath, E.: Ger. Pat. 1,155,882 (1963).

375. Brown, A. E., and Bogaty, H.: U. S. Pat. 2,847,351 (1958); Ger. Pat. 1,050,963 (1960).

376.   Sagarin, E., and Balsam, M.: The behavior of perfume materials in thioglycolate hairwaving preparations, *JSCC*, **7**: 480 (1956).

376*a*.   Cook, M. K.: Perfuming hair waving lotions, *DCI*, **103**: 42 (1968).

377.   Cook, M. K.: U. S. Pat. 3,331,743 (1967).

378.   Freund, W. W.: Br. Pat. 674,195 (1952).

379.   Berth, P., Blaser, B., Germscheid, H. -G., and Worms, K. H.: U. S. Pat. 3,213,129 (1965).

380.   Spitz, R. D., and Prince, A. K.: U. S. Pat. 3,135,664 (1964).

381.   Hoyu Kabushiki Kaisha: Br. Pat. 1,077,758 (1967).

382.   N. V. Industriele Onderneming W. H. Braskamp: Fr. Pat. 1,257,394 (1961).

382*a*.   Isaji, T.: U. S. Pat. 3,399,682 (1968).

383.   Benk, E.: Neuartige Dauerwellmittel, *Seifen-Öle-Fette-Wachse*, **76**: 185 (1950).

384.   Kolb, G.: Ger. Pat. 1,028,293 (1958).

385.   Heilingötter, R. O.: Ger. Pat. 1,145,303 (1963).

386.   Braskamp, W. H.: Neth. Pat. 94,041 (1960).

387.   Freytag, H., and Zabel, M.: Ger. Pat. 1,034,820 (1962).

388.   Böss, J., Goldammer, R., and Vogler, W.: Swiss Pat. 320,950 (1957).

389.   Deutsch, P. A., Klap, J. M., and Temblett, A. R. S.: Can. Pat. 623,071 (1961); see also Unilever Limited: Br. Pat. 780,037 (1957).

390.   Buchi-Naef, H. E.: Swiss Pat. 327,505 (1958).

391.   Korden, M. A.: U. S. Pat. 3,227,165 (1966).

392.   Schwarzkopf Verwaltung G.m.b.H.: Br. Pat. 1,007,989 (1965); Ger. Pat. 1,157,347 (1963).

393.   Kalopissis, G., and Viout, A.: U. S. Pat. 2,976,215 (1961); Ger. Pat. 1,118,934 (1961).

393*a*.   Goble, R. W.: U. S. Pat. 3,387,379 (1968).

394.   Semco, W. B.: U. S. Pat. 2,529,886 (1950).

395.   Indle, L.: Ger. Pat. 1,126,070 (1962).

396.   McGoldrick, V., and McDonough, E. G.: U. S. Pat. 2,899,965 (1959); Br. Pat. 801,990 (1958); Austrian Pat. 198,894 (1958).

397.   Reiss, C. R., and Lichtin, J. L.: U. S. Pat 3,266,994 (1966).

398.   Freytag, H.: Ger. Pat. 968,992 (1958).

399.   R. Graf & Co.: Ger. Pat. 972,563 (1959).

400.   Meeker, R. E.: U. S. Pat. 3,208,825 (1962).

400*a*.   Grifo, R. A.: U. S. Pat. 3,394,993 (1968).

401.   Société d'Etudes et de Recherches: Fr. Pat. 931,447 (1948).

402.   Deadman, L. L. F.: Br. Pat. 827,331 (1960).

403.   Heilingötter, R.: Über die Oxydierbarkeit von Thioglykolsäure und deren Salzen in wässerigen Kaltwellösungen, *Seifen-Öle-Fette-Wachse*, **76**: 449 (1950).

404.   Schwarz, M. H., and Orgel, G.: U. S. Pat. 2,780,579 (1957).

405.   Reed, R. E., Tenenbaum, D., and Den Beste, M.: U.S. Pat. 2,564,722 (1951).

406.   Head, R. C.: U. S. Pat. 2,633,447 (1951).

407.   Saphir, J., *et al.*: Austrian Pat. 193,078 (1956); Swiss Pat. 321,325 (1957); Ger. Pat. 972,216 (1959).

408. Schwarzkopf, H.: Ger. Pat. 1,198,012 (1965).

409. Grant, S.: U. S. Pat. 3,143,476 (1964).

410. Mulinos, M. G., et al.: On the toxicity of sodium perborate, JSCC, 3: 297 (1952).

411. Gillette Company: Br. Pat. 695,797 (1953).

412. Procter and Gamble Company: Br. Pat. 699,997 (1953); McDonough, E. G.: U. S. Pat. 2,809,150 (1957).

413. Jenkins, R. G. C.: Br. Pat. 671,844 (1952).

414. Salea AG (Max Marti): Swiss Pat. 353,492 (1961).

415. Klinge, G., and Schnurch, R.: U. S. Pat. 3,253,980 (1966); Br. Pat. 896,197 (1962); Ger. Pat. 1,141,749 (1962); Can. Pat. 695,039 (1964); Austrian Pat. 221,232 (1962).

416. Henkel & Cie., G.m.b.H.: Br. Pat. 1,072,827 (1967).

417. Schwarzkopf, H.: Ger. Pat. 1,189,232 (1965).

418. Luloff, J., and Chiang, Y. -H.: U. S. Pat. 3,253,993 (1966).

419. Kalopissis, G.: U. S. Pat. 3,265,582 (1966); Ger. Pat. 1,160,984 (1964); Ger. Pat. 1,240,226 (1967); Austrian Pat. 228,938 (1963).

420. Freytag, H.: Ger. Pat. 885,128 (1953).

421. Union Française Commerciale et Industrielle: Fr. Pat. 1,004,940 (1952).

422. Demi, J. A.: Fr. Pat. 1,156,163 (1958).

423. Reed, R. E.: A new home permanent waving process, JSCC, 7: 475 (1956).

424. Bunford, L. J.: Br. Pat. 833,328 (1960).

425. L'Oréal: Fr. Pat. 928,871 (1947).

426. Shansky, A., and Tarasov, A.: U. S. Pat. 3,109,778 (1963).

427. Zviak, C., and Rouet, J.: U. S. Pat. 3,142,623 (1964); Ger. Pat. 1,166,418 (1964).

427a. Pierre, B.: Ger. Pat. 1,908,308 (1969).

428. Draize, J. H., Alvarez, E., and Whitesell, M. F.: The percutaneous toxicity of thioglycolates, Proc. Sci. Sec. TGA, 7: 29 (1947).

429. Schulte, K. E., Mleinek, I., and Hobl, H. D.: Beitrag zur Frage der percutanen Resorption der Thioglykolsäure aus Haardauerverformungsmitteln, Parfüm. Kosmetik, 45: 153 (1964).

430. Norris, J. A.: Toxicity of home permanent waving and neutralizer solutions, Food Cosmet. Toxicol., 3: 93 (1965).

431. Shansky, A.: A synthesis and evaluation of guanidine thioglycolate for cold permanent waving, Am. Perf., 78: 32 (August 1963).

432. Bogaty, H., and Giovacchini, R. P.: Toxicity and performance of guanidine salts in permanent waving, Am. Perf., 78: 45 (November 1963); cf. also Shansky, A.: ibid., 29 (December 1963).

433. Brunner, M. J.: Medical aspects of home cold waving, Arch. Dermatol., 65: 316 (1952).

434. Borelli, S., and Manok, M.: Ergebnisse von Untersuchungen bei Berufsanfängern im Friseurgewerbe, Berufsdermatosen, 9: 271 (1961).

435. Borelli, S., and Haberstroh, F.: Die Verträglichkeit organischer Schwefelverbindungen der sauren Kaltdauerwelle, Acta Allergologica, 15: 139 (1960).

436. Schulz, K. H.: Untersuchungen über die Sensibilisierung und Gruppensensibilisierung gegenüber Thioglykolsäurederivaten, *Berufsdermatosen*, **9**: 244 (1961).

437. Martin, G., Draize, J. H., and Kelley, E. A.: Local anesthesia in eye mucosa produced by surfactants in cosmetic formulations, *Proc. Sci. Sec. TGA*, **37**: 2 (1962).

438. Kramer, H.: Report on cold permanent waves, *J. Assoc. Offic. Agr. Chemists*, **35**: 285 (1952).

439. Goldemberg, R. L., and Akrongold, H.: U. S. Pat. 3,141,825 (1964).

Chapter 22

HAIR STRAIGHTENERS

George G. Kolar and Aaron Miller*

Man has for some time, been utilizing various methods and materials to modify his physical appearance. His eternal quest for beautification has manifested itself in unceasing changes in hair and clothing styles, as well as other aspects of fashion. This perpetual fashion evolution has brought about changes in the practice of cosmetic science and technology. The science, related to cosmetics, has rapidly progressed and enabled the cosmetic chemist to gain fundamental knowledge which can be applied to new formulating concepts.

Progress in the technology and art, related to the hair-straightening process, has made it possible for man to change nature to his own desire. An individual with curly hair, who wishes the opposite condition, can today, in most instances, achieve this goal with safety, ease, and effectiveness. Those who feel straight hair is aesthetically desirable no longer must accept, with stoic resignation, the physical configuration of their hair.

According to Suter, there is a change in hair fashions, from straight, to loose waves, to tight curls, which appears to be a cyclical phenomenon (1). Suter states that a complete phase for a specific hair style requires about ten years. From the time the trend starts, it takes approximately three years to reach acceptance by the public.

When this article was originally written, in 1956, women with straight hair felt it fashionable to have permanently waved hair with curls. The only individuals using hair straighteners at that time were those with relatively curly or kinky hair. At this time we may have reached the threshold of the presently popular straight hair look, and reverse trends have already been

* The authors would like to express their gratitude to Mrs. Therese M. Doob, for her invaluable assistance in the preparation of this chapter.

manifested. The Negro market, to some extent, has been influenced by proponents of the "natural" look. This has been heightened by a new awareness of race pride, but has not materially affected hair straightener usage. The natural look has been for the most part accepted by the male with short hair, whereas women with long hair, at this date, still prefer the use of a straightener.

It is interesting to observe for nonfarm families and single consumers that the nonwhite average annual consumer expenditure in 1965 for waves, shampoos, and other hair care items amounted to $30.67, whereas for whites, expenditure totaled $28.95 (1).

Morphological Considerations

In man, the cuticle, cortex, and medulla are the morphological components of the hair fiber. The cuticle, which surrounds the fiber, consists of overlapping flat scale cells, 5 to 6 cells thick and about 0.5 μ in diameter (2). The cuticle is more resistant to penetration and attack by chemical reagents, such as sodium sulfide, than the cortex (2). The resistance of human hair to chemical attack is due partly to the thickness of the cuticular layer (3).

The arrangement of the cuticle scales allows their independent motion without reducing their protective function. The free ends of these overlapping, sloping flat scales are called imbrications and point upward and

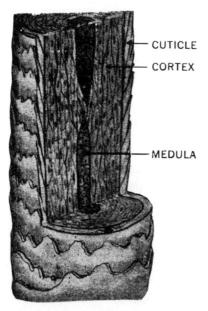

CUTICLE

CORTEX

MEDULA

Fig. 1.   Morphological components of a hair fiber. [From Reed *et al.* (19).]

outward in the direction of hair growth (Figure 1). Gaps found between these scales can act to contain sebum. The sebum helps maintain the condition of the hair as well as imparting sheen. Cuticle scales contain no natural pigments and have a frosted, translucent appearance. If investigation reveals the cuticle is colored, it generally means the hair has been artificially dyed.

Cuticle scales dislodged from the end or tip of the hair shaft can lead to unsightly frayed and split ends. This condition can follow chemical treatment such as bleaching, tinting, and cold waving. Damage to the cuticle can impart a rough, drab, and dry appearance to the hair (4).

The cortex consists of spindle-shaped cells which are oriented along the axis of the fiber. The cortex, in wool fibers, consists of an ortho, more reactive component, and a para cortex, less reactive component. In man, in specific instances, this differentiation is not evident, and the cortex appears homogeneous and similar to the para component of wool fibers in composition and properties (5). The protein, which occupies the cortical cells, is organized as microfibrils embedded within a matrix. In the para component of the cortex, the microfibrils, which are the crystalline portion of the structure, exhibit a hexagonal arrangement.

The medulla, which usually runs down the middle of the fiber and consists of hollow cells, is not invariably present in wool or hair, although it is more

1   2   3   4   5   6

Fig. 2.   Cross-section of straight, curled, and naturally curly hair. Nos. 1 and 2 are straight hair; Nos. 3 and 4 are curled hair; Nos. 5 and 6 are curly hair. [From *Today's Health* (49).]

common in hair. It is believed to make little or no contribution to the chemical and mechanical properties of the fiber (3).

At the macroscopic level, a feature which is quite obvious is the form or shape of the hair. A variation in hair forms has been observed, from spiral fuzzy in the Negroid, straight in the Mongoloid, and intermediate from straight to wavy in the Caucasoids. Usually the cross-section of the fiber will show some ethnic variation such as the elliptical Negroid and circular Mongoloid configurations. However, many cross-sectional shapes can be found even on the same scalp (Figure 2). There does not appear to be any definite relationship between straight and curly forms and their respective cross-sections. This seems to be particularly true of Caucasoid hair, in which an inverse relationship is often observed with straight hair having oval cross sections as the normal type (4).

Early investigators hypothesized the shape of the hair shaft or follicle as the cause of natural waves. However, the lack of correlation of cross-section with hair forms tends to dispute cause-effect theories of this type. It is felt that the stability of curls would be increased by an elliptical cortical configuration, assuming that the curl and major axis of the fiber were in the same plane.

A modification of the hair follicle shape has been related to the form of wool fibers by one investigator (6). Variations in the orientation of wool follicles have been theorized as being caused by cyclic action of the *erector pili* muscles over a period of 7 to 12 days. A simpler theory, related to fiber form, proposes that waves are due to rhythmic and differential biosynthesis of keratin within the follicle (7).

General Chemical Composition

A keratin fiber represents a complex chemical entity. As indicated above, individual morphological components can be further structurally differentiated. Thus the whole fiber contains a mixture of proteins, and any total quantitative studies must represent an average value (Table I).

The cuticle, upon chemical examination, exhibits more cystine than the cortex (8). Other variations in amino acids have also been found (9). The para cortex of wool has been found to contain more sulfur than the ortho cortex (5). Hair which has a greater cuticle/cortex ratio and a wholly para cortex differs significantly from wool in its amino acid composition.

A differential study of human hair with those of several breeds of wool shows hair to be richer in cystine and proline and poorer in alanine, leucine, tyrosine, phenylalanine, glutamic and aspartic acids, lysine, and arginine. Smaller variations are seen among the various wool samples (10).

Keratin

The cells of the cortex contain fibrous proteins called keratins which consist of amino acids joined into long intertwining molecules. The fibers consist of polypeptide chains each arranged in a double helical formation. In normal cortical cells, these helices are compactly folded into geometrical configurations called α-keratin. The number of amino acids per turn is approximately 3.6 and the pitch of the helix is 5.4 Å. The α-helix has a diameter of about 10.5 Å. Hair elasticity is due partly to the ability of the α-keratin chains to unfold as stress is applied. When this occurs, the helix diameter is reduced and β-keratin is formed.

Adjacent polypeptide chains as well as the polypeptide itself contain several types of linkages and bonds. The cystine or disulfide bond is one type which has been identified; hair contains about 17% of cystine. The cystine bond can be found between adjacent polypeptide chains, and it is also formed

TABLE I. Comparative Amino Acid Content of Various Fibers (μM/g)

	Side chains and amino acid present	Lincoln wool	Caucasian hair	Negro hair
I.	Aliphatic	2830	2350	2470
	Glycine	590	539	541
	Alanine	601	471	509
	Valine	570	538	568
	Leucine	740	554	570
	Isoleucine	333	250	277
II.	Aliphatic hydroxyl	1020	1520	1290
	Serine	541	870	672
	Threonine	483	653	615
III.	Aromatic	540	260	380
	Tyrosine	266	132	202
	Phenylalanine	273	130	179
IV.	Acidic	1400	1330	1350
	Aspartic acid	575	455	436
	Glutamic acid	828	871	915
V.	Basic	1040	790	800
	Lysine	310	213	231
	Arginine	662	512	482
	Histidine	71	63	84
VI.	Sulfur-containing	750	1440	1380
	Half-cystine	745	1380	1370
	Cysteic acid	6	55	10
	Methionine	0	0	0
VII.	Heterocyclic	—	—	—
	Proline	490	672	662
VIII.	Ammonia	1030	780	985

Source: Menkhart *et al.* (3).

when cystine itself is built into adjacent polypeptide chains during keratinization (11). Therefore each α-keratin chain is joined to others in many directions. It has also been postulated that hydrogen bonds, which are relatively weak, are present between adjacent chains (12). Salt linkages have also been identified between adjacent polypeptide chains whenever positively charged amino acid residues project opposite negatively charged residues. Therefore lysine and arginine groups form these linkages with aspartic and glutamic acid. Finally, peptide linkages, which are present in the chain, and van der Waals forces have also been theorized as contributing to the entire picture.

Hair Treatment Reactions

If the above concept of hair structure is accepted, it follows that the linkages and bonds discussed may be modified by chemical means, with concomitant physical change. Sulfur bonds are important in specific hair treatment operations because they may be modified by various chemical compositions. Cold-wave or permananent-wave solutions, containing a reducing agent, such as ammonium thioglycolate, are sometimes used for this purpose because they can reduce —S—S—, disulfide bonds, to —SH, sulfhydryl groups. The treatment is usually conducted on hair which has been wrapped on rods prior to the addition of the waving solution. The new configuration can be stabilized by materials which convert the —SH, sulphydryl groups, back to —S—S—, disulfide bonds. This process can also be modified to achieve a straightening effect.

A radical displacement reaction, resulting in a disulfide interchange equilibrium, has been proposed as a mechanism for the behavior of mercaptans, derivatives of which can be found in cold-wave solutions, in the presence of simple disulfides. The reaction has been said to proceed in the following manner (13):

(a) $$R'SH \rightleftharpoons R'S\cdot + H^+ + 1e$$

(b) $$R'S\cdot + R—S\cdot\cdot S—R \rightleftharpoons R'S—SR + RS\cdot$$

(c) $$R'S\cdot + R'S\cdot \rightarrow R'S—SR'$$

Various alkalis, such as sodium hydroxide, are sometimes incorporated into compositions designed to straighten curly or wavy hair. The reactions occurring during the steam treatment of wool in the presence of alkali have been studied for some time, and several theories concerning the chemical nature of the process have evolved.

On the basis of a great deal of experimental evidence, Speakman and his co-workers have stressed the importance of the covalent disulfide bonds in permanent setting. The reactions involved are illustrated by the following equations (14).

$$R—CH_2—S—S—CH_2—R \xrightarrow[\text{OH}^-]{\text{steam}} R—CH_2—S—OH + R—CH_2—SH$$

$$R—CH_2—S—OH \longrightarrow R—CH{=}O + H_2S$$

Rebuilding:

$$R—CH_2—S—OH + H_2N—R \longrightarrow R—CH_2—S—NH—R$$

$$R—CH{=}O + NH_2—R \longrightarrow R—CH{=}N—R$$

Schöberl and his co-workers suggested a scheme in which a sulfenic acid residue played an important role (15,16). This was later challenged by Mizell and Harris (17). Spectrographic evidence has been offered to support a theory of Rosenthal and Oster involving the elimination of a proton from the carbon atom adjacent to the sulfur. Their theory postulates the following reaction (13):

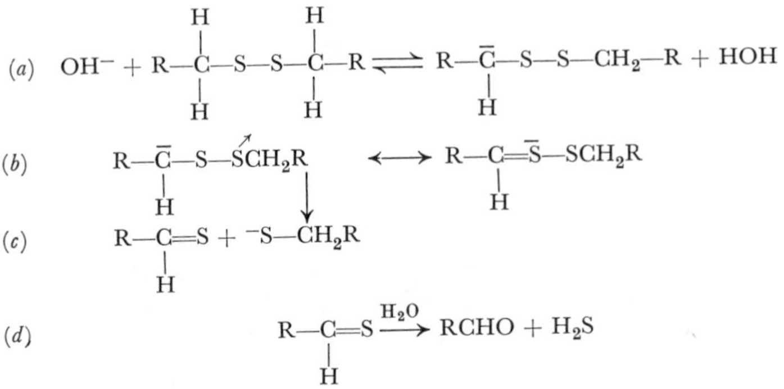

$$(a) \quad OH^- + R-\underset{\underset{H}{|}}{\overset{\overset{H}{|}}{C}}-S-S-\underset{\underset{H}{|}}{\overset{\overset{H}{|}}{C}}-R \rightleftharpoons R-\underset{\underset{H}{|}}{\overset{=}{C}}-S-S-CH_2-R + HOH$$

$$(b) \qquad R-\underset{\underset{H}{|}}{\overset{=}{C}}-S-\overset{\nearrow}{S}CH_2R \quad \longleftrightarrow \quad R-\underset{\underset{H}{|}}{C}=\overset{=}{S}-SCH_2R$$

$$(c) \qquad R-\underset{\underset{H}{|}}{C}=S + {}^-S-CH_2R$$

$$(d) \qquad\qquad R-\underset{\underset{H}{|}}{C}=S \overset{H_2O}{\longrightarrow} RCHO + H_2S$$

This mechanism involves an indirect attack on the —S—S— bond which results from a direct nucleophilic displacement on hydrogen by base. It may also be viewed as an ionization of acidic hydrogen in the presence of base to form a resonance-stabilized anion. The anion then undergoes a β-elimination reaction to form a mercaptan and thioaldehyde.

The radical displacement reaction has also been theorized to occur between sulfites and disulfides. Derivatives of sulfites are sometimes used in hair-waving and straightening formulations. The reaction, in the presence of oxygen, proceeds as follows (13):

Initiation:

$$SO_3^= + SO_3^= + \tfrac{1}{2}O_2 \rightarrow {}^-SO_3^{-}\cdot + SO_4^=$$

Propagation:

$$SO_3^{-}\cdot + RSSR \rightarrow RSSO_3^- + RS\cdot$$

$$RS\cdot + SO_3^= \rightarrow RS^- + SO_3^{-}\cdot$$

Termination:

$$RS\cdot + RS\cdot \rightarrow RSSR$$

Oxidation

Sulphydryl groups, created by reducing agents utilized in the hair-waving or straightening operations, may be converted to the corresponding disulfide by air or chemical oxidation. Chemical agents commonly used for this purpose are hydrogen peroxide, perborate, and specific bromates.

The one-electron-transfer theory, which could account for the oxidation of a mercaptan in an alkaline medium, in the presence of a metal catalyst such as Ag^+ or Cu^{2+} or Fe^{3+}, has been shown thus (13):

$$R'SH + Cu^{2+} \rightarrow R'S + H^+ + Cu^+$$

$$Cu^+ + O_2 + H^+ \rightarrow Cu^{2+} + H_2O_2$$

$$R'SH + H_2O_2 \rightarrow R'SSR' + H_2O$$

Diffusion of Reagents

In the commercial use of hair-straightening compositions, the length of time required to achieve desired results is obviously important. The rate-determining step in the reduction of keratin by mercaptans and sulfites appears to be diffusion. Therefore the rate of keratin reduction may be utilized to indirectly determine diffusion of these reagents.

Experimental results, where ammonium thioglycolate at pH 9.3 and ammonium sulfite at pH 6.0 were used as the reducing agents, are presented in Figures 3 and 4. For the two types of hair used, similar curves were obtained except for the somewhat lower reduction level of Caucasian hair at the later time intervals. As the data indicate, wool is reduced much more quickly (3).

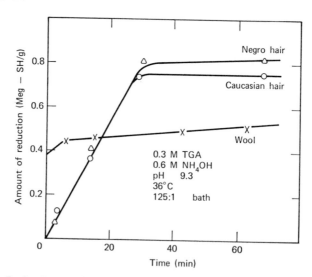

Fig. 3.   Reduction as a function of time in alkaline thioglycolate. [From Rudall (2).]

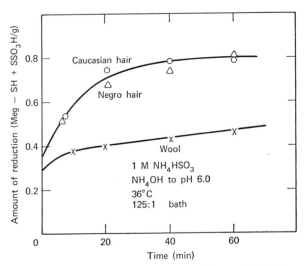

Fig. 4.   Reduction as a function of time in sulfite at pH 6.0. [From Rudall(2).]

D-Cystine Fraction

It has been established by Phillips and co-workers that about 25 % of the combined cystine in wool keratin is resistant to the attack of reagents such as sulfite, alkali, permanganate, etc. (18). This portion has been designated as the D-fraction. Other investigators have stated that only 75 % of the cystine in the hair will be reduced by commercially available alkaline thioglycolate solutions (19). This apparent resistance to various reagents of the D-fraction may be due to steric factors. It has been suggested that the success of the hair-straightening or waving process may be due to the D-fraction.

These resistant disulfide bonds may maintain the skeletal structure of the hair during treatment with materials which cleave the other disulfide bonds. If these resistant bonds were not present, all disulfide cross-links might be cleaved, and the hair would lose its fiber properties. Badly damaged hair may be characterized by a destruction of a good number of these resistant groups because the reducing solution was applied for too long a period, or because too strong a solution was used, or because of a combination of both (13).

Measurement of Physical Changes Related to Fiber Treatment

The results of chemical treatment, designed to produce a straightening effect, can be determined visually. However, this method of evaluation is generally qualitative in nature, and other procedures somewhat more quantitative might also be employed.

When fibers are deformed by twisting, stretching, or compressing, they retain their deformed state, when the deforming force is relieved, to a greater or lesser degree depending on the magnitude of the deforming force and the condition of the fiber during the deforming process. The retention of the deformation by the fiber is known as "set." Most of the data accumulated have concerned fibers deformed by stretching, and therefore set was considered the change in length after releasing the fiber. Set has been further

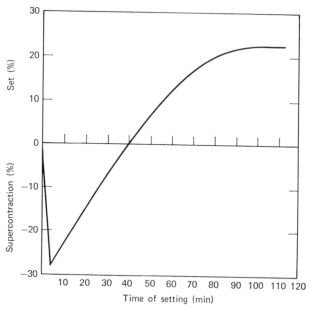

Fig. 5.   Supercontraction or set as a function of boiling time. [From Whewell (11).]

described as cohesive, disappearing when the set is released in cold water; temporary, remaining after release in cold water but disappearing on release in boiling water; and permanent, which remains even after release in boiling water. The theories which have been established in connection with the setting of stretched fibers seem to apply to fiber deformations of other kinds.

The explanation of permanent set followed from the observation that if a stretched fiber is boiled in water and then released and boiled freely in water for 1 hr, it may not take a set but contract to a length less than that of the original unstretched fiber. This phenomenon is known as supercontraction. If fiber boiling time, in a stretched position, is plotted as a function of set or supercontraction after the fiber has been boiled in water, in a relaxed state for 1 hr, a curve is obtained (Figure 5).

Values for the amount of permanent set have been obtained for hair treated, under specific conditions, with various reagents. The results indicated, under the experimental conditions utilized, that nontreated hair, when boiled in a stretched configuration for 2 hr and then released in boiling water for 1 hr, shows approximately 14.5% set. Similar experiments conducted on hair reduced with sodium bisulfite gave 16.5% set. Treatment with 0.1 N HCl gave results similar to nontreated hair, whereas 10 vol of hydrogen peroxide treatment produced 1.6% set (11).

The Function of the Setting Medium

Since the setting of hair can be regarded as a series of chemical reactions, the nature of the setting medium can be important as different reagents promote bond cleavage or bond rebuilding (11). Studies concerning the effectiveness of various reagents as setting media are of considerable importance because of the possibility of extrapolating these results to practical situations involving hair waving or straightening. Some of the general findings are summarized below.

1. Untreated hair cannot be set in solutions of strong mineral acids (20).

2. Boiling alkaline solutions are better setting agents than boiling water, the maximum setting taking place at pH 9.2 (20).

3. Hair can be set in solutions of reducing agents even at comparatively low temperatures (20). Some useful reagents for this purpose are ammonium thioglycolate, sodium and monoethanolamine bisulfites, and specific mercaptans (21).

4. Oxidizing agents, in general, are not good setting agents, but under particular circumstances certain oxidizing agents such as periodic acid, peracetic acid, and permonosulphuric acid do facilitate the process (11).

5. In general, it is difficult to set fibers in solutions of hydrogen bond-breaking agents. However, mixtures of reducing agents and hydrogen bond breakers, e.g., urea, are better setting agents than reducing agents alone (22).

Form-Related Compositions

The formulations presented are not offered as completed compositions and are intended to serve as reference points which the formulator may modify to satisfy specific needs. Basically, two functionally related types of compositions may be discussed: those designed to produce temporary and those which bring about "permanent" configuration effects.

Temporary Straightening

Some of the earliest formulations were based on the use of fats, waxes, and oils. These materials were used to lubricate and aid in fixing the hair in a desired style. Also, natural film-depositing gums were employed for this purpose. Present-day formulations of these types are represented by Formulas 1 to 3. These formulations generally rely on physical factors for functional qualities which are usually lost when the composition is removed.

Formula 1

Quince seed	3%
Ethyl alcohol	25
Water	72
Perfume, color, and preservatives	q.s.

Procedure: Soak the quince seed overnight in 50 parts of water. Strain the resulting mucilage through a cloth and add the balance of water, alcohol, color, perfume, and preservatives.

Formula 2

Gum tragacanth	1.25%
Ethyl alcohol	5.00
Water	93.75
Preservative, color, and perfume	q.s.

Procedure: Disperse the tragacanth in the water and allow hydration to proceed for 24 hr. Add alcohol, preservative, color, and perfume.

Formula 3

Petrolatum	90%
Paraffin or microcrystalline wax	10
Perfume and color	q.s.

Procedure: Melt, mix, and add perfume and color as desired.

Waxes other than paraffin, such as beeswax, ozokerite, or microcrystalline, may be used in Formula 3. However, the paraffin is more generally used because the others are more expensive. Where firmer products are required, higher-melting-point waxes can be incorporated to achieve the desired hardness.

Specific resins can also be incorporated in a product of this type. The degree of tackiness can be controlled by varying the quantity and type of wax. More than 20% is seldom used, because the product may become too hard for easy application.

Synthetic resins, such as PVP or its copolymers, can be incorporated in Formulas 1 and 2 to further modify the functional characteristics of these systems.

The use of Formulas 1 to 3 may result in shortcomings if specific qualities are desired. They do not affect the hair chemically, and their functional properties are derived as a result of their physical effect on the hair fibers. Usually, when curly or kinky hair is not acted upon by chemical means, the hair rapidly returns to its natural state. Moisture accelerates this phenomenon. Relatively unsuccessful attempts have been made to prevent this reversion by incorporating water repellents, such as aluminum stearate and specific silicone derivatives, into the pomades. A preparation incorporating a water barrier agent is described in Formula 4.

Formula 4

Petrolatum	95%
Aluminum stearate or silicone	
derivate	5
Perfume and color	q.s.

Procedure: Melt the petrolatum and add the aluminum stearate or silicone, perfume, and color.

The physical and mechanical, rather than the chemical, method of hair straightening is most graphically portrayed by hair that has been straightened by Mme. C. J. Walker's method.

Mme. C. J. Walker conceived the idea of straightening hair by the use of a petrolatum jelly product and a hot metal comb. This procedure was termed "hair pressing."

In this method, the hair is washed and dried thoroughly. Then the petrolatum product is applied to the hair. The comb, which is heated, is placed on the hair, and tension is created through its manipulation.

The petrolatum, or pressing oil as it is called, serves two purposes: It acts as a heat-modifying conductor between the hair and the comb, and it lubricates the hair so that the comb can slide through without sticking or pulling. This method of hair straightening has been utilized by both men and women.

A second pressing procedure is sometimes utilized to style the hair into waves and curls. This second press is carried out with the use of croquignole irons. Since the male usually prefers straight hair, although converse trends have been initiated, the second application is used primarily by women. There is some laboratory evidence that a cooling period is required between the first and second press to ensure a more enduring wave.

The hot comb and croquignole iron must be used with proper caution to prevent the destruction of the hair fibers and scalp burns. If treated hair is burned and broken, regrowth almost always occurs. An established practice, excercised by the experienced operator, is to test the iron on a piece of tissue paper. If the paper is visibly scorched, the iron is too hot. This method is a procedure for straightening women's hair both in beauty shops and at home.

Variations of hot combs are available; one type is shown in Figure 6. The metal combs made for this purpose are electrically as well as manually heated.

There are severe limitations to the hot-press method as regards permanency of set. Numerous cartoons have appeared, showing the embarrassment suffered by the patron when she gets into the rain. Even perspiration causes the hair to "kink." Here again, attempts have been made to incorporate water repellents into the pressing oil, but with no great success.

The hot press and use of petrolatum products are essentially an attempt to overcome the great resistance offered by very curly hair to hair styling. Whereas the person who has straight hair can achieve a certain satisfaction in

Fig. 6.   Patented steel comb utilized in Mme. C. J. Walker's procedure.

hair styling by moistening the hair with water and curling on rods with subsequent drying, very curly hair resists the treatment with water.

A simple preparation that represents a product for use with hot combs is shown in Formula 5, and a more complex preparation, along the same lines, is shown in Formula 6.

Formula 5

Petrolatum	100%
Perfume and color	q.s.

Procedure: Melt the petrolatum, and add the perfume and color.

Formula 6

Beeswax	7.00%
Ceresin	3.00
Petrolatum	60.00
Mineral oil	30.00
Perfume and color	q.s.

Procedure: Melt the oils and waxes and add the perfume and color.

Permanent Straightening

Hair treatment compositions containing "available" quantities of alkaline material, such as sodium or potassium hydroxide, are sometimes used to achieve hair configuration modifications. The range of titratable alkalinity of typical commercial products of this type can be expressed as 2.25 to 3.25% sodium hydroxide based on the weight of the formula. The lesser quantity is generally employed in formulas marketed for hair which is bleached or tinted, or which responds readily to this type of treatment, whereas the greater quantity is recommended for resistant cases. The chemical implications of this type of treatment were elucidated in reactions presented above. This type of treatment, if executed properly, can produce chemical changes in hair fibers observable to the user as a desirable form-related effect. However, the active ingredients in this type of composition can rupture various linkages and bonds found in protein molecules to an extent where problems can occur. This can cause hair damage, observable as embrittlement and tissue destruction. For this reason, all products of this type require the use of a "base" or protective creme, which is applied to the scalp and surrounding epidermis.

These protective cremes, for the most part, consist of petrolatum. With all of their obvious negative aspects, hair-straightening formulations employing alkali and treatment enjoy a certain degree of continuing popularity.

The type of hair treatment composition described above is often compounded in the form of a creme. This creme, in many instances, is an O/W emulsion system. Other sodium hydroxide containing products are not emulsions but depend on a flour paste to achieve their desired apparent viscosity. Usually these products do not have a long shelf life. Because of the excess alkali which is usually present in the compositions, the surface active agents employed to produce the emulsion, as well as other ingredients which might be incorporated, must be stable under these conditions. A patent issued to H. M. Childrey, Jr., and E. Doty, presents an approach to the formulation of an alkaline composition, in which the inventors discuss various problems associated with these compositions (23).

A formulation cited by the Childrey and Doty patent is Formula 7.

Formula 7

Cetyl alcohol	16.5 lb
Protopet (petroleum jelly)	4.0
Polyoxyalkylene lanolin	1.0
Mineral oil (70 viscosity)	8.0
Sodium lauryl sulfate (30% active)	4.0
Caustic soda	4.0
Water	62.5

Examination of the composition cited by the Childrey and Doty patent will reveal the inclusion of sodium hydroxide. The quantity of sodium hydroxide and its availability, from a chemical standpoint, can, if the formulation is mishandled, result in tissue destruction. Therefore sufficient warnings should appear on labels utilized for this type of product to inform the user of possible difficulties. A description of the product, method of use, warning to the user, and other information are often included on the label; one such label is quoted below.

> Place a little petrolatum along the hair line and on the ears before starting to use. Apply the product to the hair above the forehead. Be careful not to allow the straightener to drop on the skin. Comb the straightener through the hair in an upward movement away from the scalp. Repeat the combing until the hair becomes as straight as you desire it. When straight, wash the hair in running water to remove the hair straightener. Rinse the hair until the soapy feeling is gone. Then wash with shampoo thoroughly. Rinse with lukewarm water. Do not retain this rinse water; use fresh water for every rinsing. Be sure not to use hot water.
>
> If the hair is not thoroughly washed, the hair may break off at the scalp. When the hair is completely and thoroughly washed, apply hair pomade. It is suggested that you take the hair straightener to a barber to apply if you don't know how. The manufacturer does not assume any responsibility for the results if improperly used. The majority of the people can use the straightener without any bad results. However, some, through careless handling, will burn and discolor the hair. This is a result of the action of the individual, not the straightener.

Another approach, which is growing in popularity, toward reducing the undesirable irritation aspects of the alkaline-containing compositions, has been the so-called no-base formulation. These formulations, it is stated, can be used without the prior application of materials, such as petrolatum or mineral oil, usually utilized to protect the scalp and surrounding epidermis during treatment. The mechanism responsible for the reduced negative skin and hair effects has not been, as of this time, completely elucidated. However, the surface chemistry characteristics of this type of formulation may play a role in the stated reduction of negative attributes. These products generally contain approximately 20 % (based on formula weight) of water, the remainder consisting of emulsifying agents and other "oil" components. The emulsions, thus far studied, exhibit parameter values, conductivity, phase dilution, and dye solubility associated with O/W emulsions. However, because the oil-phase content approaches that required for inversion, it is postulated that this could produce an oil-related wetting phenomenon of the hair and scalp which, while not obviating the straightening effect, might produce a "protective" oil film. Even though the no-base formulation might effectively reduce some of the problems inherent in compositions containing an excess quantity of alkaline material, improper treatment with these can still present problems and caution must be exercised with their use.

Thioglycolate Compositions

Before launching into a discussion of the thioglycolate products, it would be well to emphasize that patents have been issued on hair-straightener formulations of this type. Since the first writing, several of these patents have been adjudicated. A list of patents that seemed, to the authors, most pertinent appears with the bibliography. In fact, thioglycolate cold-wave patents generally do not differentiate between waving and straightening, but pertain primarily to the effect of the preparation on keratin. Since the function of these formulations is influenced by quantitative values of specific parameters, it is necessary to set limits of concentration, and to provide for a careful control of pH, free alkali, and total alkali. The problems inherent in the formulation and manufacture of thioglycolate products are many.

Shortly after the thioglycolates became accepted for use in permanent waving of the hair, hair straighteners, in the form of paste-like cremes, incorporating this material were introduced to the market. Because of the greater ingredient cost, they sold at relatively high prices. The problems involved in the introduction of this type of product were great. The preparations were so different from the previous ones and the instructions for use necessarily so detailed, that a great educational job confronted the manufacturer.

With the introduction of thioglycolate-containing compositions, a great deal of discussion arose. Some authors made critical comments concerning the use of thioglycolates and the damage to the hair that might occur. The initial criticism was leveled in the medical press by Cotter (24). However, McCord countered with statements concerning his experience accumulated over a period of four years of thorough work with these compounds (25). Later, McCord published further work that touched specifically on hair-straightener formulations (26).

The Food and Drug Administration has thoroughly investigated the thioglycolate products, and at no time has any confiscation or withdrawal of thioglycolate-containing formulations been ordered. The position of the Food and Drug Administration was well stated as a result of work conducted by Draize (27). The physiological effect was also discussed by Behrman (28).

When developing a formula for a hair straightener, it is necessary to pay particular attention to the vehicle, a consideration which is secondary in a cold permanent wave composition, although both products possess the same active ingredients and involve essentially similar manufacturing and packaging problems. Whereas the hair is contained on curlers during the waving process, in straightening procedures it is necessary to develop a formulation base that, because of its apparent viscosity and adhesive qualities, will keep the hair reasonably straight. To date, there has not been developed a practical mechanism for fixing the hair in a straight position during the procedure. If and when such a mechanism is discovered, a major difficulty in hair straightening will have been overcome. This will then eliminate the periodic combing and stretching now required, which can cause a number of problems, particularly if the process is not properly executed.

The larger number of thioglycolate straighteners on the market are O/W emulsions, although gelled systems have also been introduced. A representative product is given by Formula 8.

Formula 8

Glyceryl monostearate	15.0%
Stearic acid	3.0
Ceresin	1.5
Paraffin	1.0
Sodium lauryl sulfate	1.0
Water, distilled	51.9
Thioglycolic acid	6.6
Ammonium hydroxide (26° Bé)	20.0
Perfume	q.s.

Procedure: Mix the glyceryl monostearate, stearic acid, ceresin, paraffin, and sodium lauryl sulfate in a kettle with 40 parts of the water, and heat, with constant agitation, to 95°C. Then cool to 50°C, still under agitation. While this is being done, add the thioglycolic acid to the remaining water and to this solution slowly add the ammonium hydroxide. Since considerable heat is generated, it is necessary to provide some method of cooling to keep the temperature below 50°C. For this reason it is preferable to use ammonium thioglycolate, as obtainable commercially, and reduce the amount of ammonium hydroxide by approximately 50%. Slowly add the aqueous thioglycolate solution to the stearate mixture, both being at 50°C. It is important that the thioglycolate solution not be allowed to go above 50°C. Otherwise there will be a decomposition of the ammonium thioglycolate. Care should be taken that the thioglycolate solution is mixed in thoroughly as it is added. Cool the resulting mixture to 40°C to avoid hydrolysis of the glyceryl monostearate, add perfume, and package.

The control of titratable ammonia and of thioglycolate content is most important. The final product is assayed for the thioglycolate ion, and the assay expressed as percentage of thioglycolic acid. Any adjustment due to loss of water during manufacture should be made to bring the thioglycolic acid within the range of 6.4 to 6.8%. Likewise, the titratable ammonia should be determined and adjusted to lie between 0.8 and 0.9%. The percentages of the thioglycolic acid and ammonia content can be varied to meet the specific requirements of the manufacturer. Some commercially available products contain approximately 4% thioglycolic acid.

Many variations can be instituted in the development of hair-straightening as well as waving formulations. Derivatives of thioglycerol and mercapto-propionic acid have also been used for this purpose. In place of the ammonium cation, sodium or amine salts, such as the monoethanolamine derivative, have been employed as indicated by Hollenberg (29). There are many different theories concerning the usage of these chemicals and their swelling properties.

Patents, relating specifically to thioglycolate preparations, have been granted to Ramsey, Haefele, and Schwarz (30–33).

Neutralizers

In conjunction with the thioglycolate straighteners, as with permanent wave products, many of the manufacturers recommend the use of a neutralizer. Popular neutralizers are hydrogen peroxide, sodium perborate, and sodium bromate compositions. The sodium perborate monohydrate or sodium bromate powders may be used only if proper packaging precautions are undertaken to ensure stability. Some compositions offer greater wetting, because they incorporate specific surfactants, whereas others exhibit pH variations. The safety of perborate neutralizers was pointed out by Mulinos, Higgins, and Christakis (34).

Interesting patents, such as that of Reed, DenBeste, and Tenenbaum, have also been granted covering bromate preparations (35). A liquid bromate preparation is described in Formula 9. Variations in such preparations are found in the manufacture of liquids utilizing sodium and potassium bromates.

Formula 9

Sodium bromate	14%
Propylene glycol	2
Lanolin (anhydrous)	2
Sorbitan monopalmitate	1
Sorbitan trioleate	2
Sodium cetyl sulfate	3
Water	76
Perfume	q.s.

Procedure: Dissolve the sodium bromate in the water and propylene glycol, and heat to 60°C. Add the mixture of the remaining ingredients, which have been heated to 60°C. Cool to 35°C, with agitation, add perfume and package. Because of the high concentration of salt (sodium bromate), this product may have a tendency to separate over a period of time, especially in hot weather. It is therefore advisable to state "shake well" on the label.

One of the controversial issues that have been debated within the cosmetic industry, a debate that can be applied to hair straighteners as well as waving preparations, involves the value of a neutralizer. Whitman presented the relationship of time and oxidation of thioglycolate with and without a catalyst (36). In this respect, the patent of denBeste and Reed, utilizing manganese as a catalyst for the oxidation of the thioglycolate, can be applied to hair straighteners (37). It is recognized that permanent wave products, without separate neutralizers, have been merchandised. Nevertheless there are certain benefits that can be derived by the incorporation of a neutralizer in the hair-straightening procedure.

Manufacturing and Material Specifications

As is true of all thioglycolate products, a much greater degree of control during manufacture must be exercised than with some other cosmetic products in order to ensure optimum activity. It is essential that all raw materials be assayed for metallic content, and that alkali content, where pertinent, be evaluated. Introduction of iron, manganese, or cobalt is particularly to be avoided because they may cause decomposition of the mercaptan. Accordingly, when ordering materials, it is well to require that they meet the specifications of the Toilet Goods Association which, in the case of thioglycolates, are shown in Table II.

TABLE II.   Toilet Goods Association Specification for Thioglycolates

Ash	0.05% maximum
Thioglycolate	45 to 55% as thioglycolic acid
Dithioglycolate	2% maximum as thioglycolic acid
Iron	1 ppm maximum as Fe
Copper	1 ppm maximum as Cu
Lead	1 ppm maximum as Pb
Arsenic	1 ppm maximum as As_2O_3

All ingredients in the thioglycolate preparations must be tested for metals, including the perfume. Because of this fact, it is essential that these products be manufactured in glass, earthenware, plastic, or stainless steel containers. However, it is also imperative that thorough studies be conducted on all containers prior to compounding to ascertain whether metallic contamination might result from the use of such containers. It is likewise imperative that attention be paid to valves and all contact parts. Temperature can also be an important factor in the manufacturing process, since an exothermic reaction occurs when neutralizing, to create a specific derivative, thioglycolic acid.

Care must be exercised to prevent temperatures from exceeding 50°C. Higher temperatures might induce a deterioration of the thioglycolate and concomitant reduction of functional activity. During the manufacturing and packaging procedure, caution must also be taken to protect the product against excessive oxidation. The most careful control of the free alkali and other specific parameters in the thioglycolate concentrate and compounded formula is essential during the manufacturing and packaging procedure. A thorough quantitative and qualitative analysis, after the manufacturing and filling procedures, should be instituted to determine if all operations have been adequate.

Packaging Considerations

In selecting packaging materials, care must be taken to ensure a complete seal. Therefore the closure must be such as to allow a certain relaxation of the liner and still prevent the access of air. Cork, rubber composition, and vinylite liners are materials that can be used to advantage under these circumstances.

Many products on the market today provide tamper-proof seals. Once, for any reason, such a seal is broken and the bottle opened, there is a possibility that the product will undergo loss of functional activity unless the cap is immediately replaced and tightened. Therefore a broken seal signifies that the product can no longer be guaranteed for maximum activity. Furthermore, all containers must be tested for reaction with the product. Here again, there is an opportunity for the inadvertent introduction of metallic ions. For this reason the most satisfactory containers for such products are those made of glass.

Collapsible tubes have been used for thioglycolate products in the past, but not without some difficulty. It is imperative, if a collapsible tube is to be used, that the lining of the tube be continuous and without imperfection. One pinhole in this lining may suffice to cause a breakdown of the product. Plastics, such as polyethylene, sometimes present problems. Within 24 hr, in some instances, deterioration of both ammonia and thioglycolate can result because of the plastic's permeability. Polypropylene bottles have been used in recent years by a limited number of manufacturers.

As with thioglycolate compounds, precautions must be taken with the neutralizers, of which hydrogen peroxide was one of the most difficult to handle. However, with the inclusion of specific stabilizing agents, such as phenacetin, many of the difficulties are now avoided.

The powders, such as bromate and perborate, must be packaged under dry atmospheric conditions in containers that are air- and moisture-proof. Most products of such a nature are packaged in packets of aluminum foil laminated with cellulose acetate. For a 4-fl-oz bottle of approximately 7% thioglycolate, expressed as thioglycolic acid, either 10 g of sodium perborate monohydrate or 26 g of potassium bromate is considered sufficient to ensure neutralization under practical conditions of use. Tests for pH and oxidizing potential should be made prior to and after packaging.

Method of Application

Before recommending the use of a thioglycolate product for hair straightening, the authors would recommend that similar cautionary statements appear, in the instructions as those utilized in permanent waving. One such sample is quoted here.

Do not use hair straightener if you have abrasions or cuts on your scalp. Wait until they have healed before straightening your hair.

Do not straighten your hair if you are sick or under the care of a physician without consulting him first, since your hair is affected by your physical condition.

Do not substitute hot water where directions call for warm water.

Do not use the treatment unless you use all of the neutralizer in the kit. You must use the neutralizer as directions state.

For those whose hair is in a damaged condition, the following has been added to the instructions:

Do not use hair straightener if your hair has been abused. If you have used hot combs, marcel irons, harsh lye straighteners, dyes, or bleaches, you may have abused your hair. For this condition, we recommend the use of hair conditioner, for about two weeks or longer before straightening, to help recondition the hair.

It cannot be too firmly stated that the product should not be used by any person who has experienced an allergic reaction from any thioglycolate preparation.

Because of the softening effect of the thioglycolate on the scalp, the comb must be used with great dexterity or the scalp may be abraded, causing opportunity for infection or temporary irritation to ensue. If a shampoo-type hair straightener is formulated, the effect of the composition on the eyes should be ascertained. This parameter, of course, could be evaluated in accordance with the Draize method (38).

Before proceeding with straightening, it is essential that the hair be thoroughly cleansed in order to achieve uniform processing. People with kinky or curly hair, who may use petrolatum products to excess, must remove any greasy, oily, or resinous films deposited by these.

Because the cosmetologist is often less familiar with the method of application of this product than of other cosmetics, instructions for use, with the permission of the Perma-Strate Company, are reproduced below. Accompanying these instructions were simple illustrations showing the hair being washed, the straightener applied, the hair combed, and so forth.

1. First wash your hair thoroughly with shampoo or regular soap. Soap and rinse your hair three times to be sure you remove all dirt, oil, and grease. The hair straightener won't work on oily or dirty hair. Dry slightly with a towel; leave hair a little damp.

2. Now open the jar of hair straightener and apply with your fingertips into your damp hair. Start with the short hair at the sides and on your neck. Then work up to the top of your head. Use all the creme and rub it into all of your hair thoroughly. If your hair is long and thick, separate strands with comb so that each hair is covered. Use the whole jar. Put the creme on thick!

Remember: Use the whole jar of creme. Rub it into every part of your hair.

3. Now comb your hair back with an ordinary comb. Use a comb that goes through your hair smoothly. Then smooth your hair back with your hands. Fifteen minutes later, comb the

hair again. This time comb forward, towards you. Rub back any creme that sticks to comb or hands! Repeat this process every 15 minutes, combing in an opposite direction each time, until the hair is straight. Comb with slight pull so comb pulls hair straight.

4. Look at your hair in the mirror when you start to apply the creme—it will look curly Then look at it each time you comb (every 15 minutes). Soon you will notice the curl begin to flatten. Then when you notice that hair lies down flat and straight, you are ready to rinse the creme out immediately. Comb hair for one full minute just before you rinse creme out. The creme usually should not be left on for more than 1-½ hours. Coarse resistant hair may require two hours, however.

Fine hair, dyed or bleached hair, or hair that has been weakened by hot combs and caustic straighteners, will straighten faster—usually in an hour or less.

5. As soon as you see that your hair is straight (as shown in step 4), rinse the creme completely out of your hair with lots of plain warm water (not hot). Be sure that all of the creme is washed out. The action of the creme, and the combing, and the rinsing may cause a slight smarting sensation on your scalp but skin will not burn.

6. Very important: Now mix the neutralizer in a pint (equals 2 measuring cups) of warm water (not hot) and pour it over your hair for 5 minutes, catching it each time in a basin and reusing it. Do this for 5 minutes. Time yourself. Be sure you catch all your hair. Don't miss the hair at temples, back of neck or above ears. This 5 minutes of thorough, complete rinsing is what keeps your hair straight, so do a good job. Be sure you use all of the neutralizer.

7. Rinse your hair thoroughly with plain warm water for at least 2 minutes to remove the neutralizer. Then dry it slightly with a towel.

It is recommended that, after the initial treatment, hot combs or lye products should no longer be used. Furthermore, as stated by Reed, there is a limit to the frequency of use of the thioglycolate products whether for waving or straightening (19). It is recommended that, for the latter purpose, the limit be three or four times annually.

It is considered advisable that the reaction of the hair to the straightener—rather than the texture of the hair—be the determining factor as to the processing time.

Other Materials

Within the past few years other materials have been utilized in hair straightening. A product invented by Calva has been placed on the market and differs in its chemical composition from all others known (39).

Described in Formula 10, the straightening mechanism has been postulated as due to the destruction of the hydrophilic properties of the hair fiber by the formation of condensation products with carbonyl compounds of the amino and imino groups present in the keratin molecules. These reactions occur in an environment having a pH below 7.

Formula 10

Cresol sulfonic acid	7.5%
Isopropyl naphthalene sulfonic acid	7.5
Formaldehyde	10.0
Water	75.0

Silicones

The utilization of silicone has given still another approach to straightening compounds. Many patent applications have been sought, and at least one patent of interest has been granted (40). These products, which for the most part contain organic solvents, are applied to the hair, which is then pressed into the desired style. The silicone film on the hair shaft, by virtue of its water-resistant characteristics, is reported to increase the resistance of styled hair to reversion. However, improvement in these compositions may be necessary in order to provide greater resistance to the negative effects of humidity.

In addition to the patents referred to above, the reader might wish to consult those granted to Charle *et al.*, DeMytt, Strain and Tusa, Grant, Lubs, DeMytt and Reed, and McDonough (41–46).

Sulfites

Specific sulfite or bisulfite derivatives, such as the ammonium or sodium salts, are sometimes utilized in the permanent waving or straightening of hair. This type of reducing agent, at temperatures as low as 37°C, can be successfully employed to cause disruption of the disulfide bonds of keratin. Successful waving or straightening of hair can be conducted with solutions which have a pH of about 6. It has been stated that at pH 6 the salt linkages found between the peptide chains present in animal fibers are most stable, and relaxation in fibers which have been strained is delayed. Also, since many reducing agents attack strained disulfide bonds more readily than unstrained bonds, the disruption of the bonds may be especially efficient at pH 6 (47).

The sulfite and bisulfite derivatives, usually employed in hair-straightening or waving compositions, are reducing agents, and they are subject, as are the thioglycolates, to oxidation. Therefore, in some instances, alcohol is included in formulations containing these to minimize air oxidation (47). Compositions of this type are generally prepared with a pH of approximately 8 since, during the treatment process, the hair can reduce the pH of the solution to about 6.

A representative product, deriving its functional properties from specific sulfite and bisulfite derivatives, is found in Formula 11.

In some cases, specific polymers may be included in compositions of this type to produce an increase in the apparent viscosity of the system. When these are included, the formula may be somewhat easier to apply and they also may help prevent the formulation from running off the hair.

Formula 11

Urea	10.00% w/w
EDTA (disodium salt)	0.05
Isopropyl alcohol (99%)	10.00
Ammonium bisulfite	4.13
Ammonium sulfite	3.74
Perfume	q.s.
Water	q.s. 100

Procedure: Add all the ingredients to the water with the exception of the isopropyl alcohol, which should be included after all other materials have gone into solution.

Manufacture

In manufacturing a product of this type, precautions similar to those stated for thioglycolate-containing compositions should be observed. To prevent any interaction, which might cause deterioration of the active ingredient, all equipment must be carefully considered. Furthermore, all raw materials utilized in the composition should be analyzed to ascertain their freedom from impurities, such as specific metallic ions which might cause a reduction in the activity of the reducing agent.

A successful formulation, from the standpoint of waving or straightening, generally will be determined by the nature and quantity of the reducing agent and the value of other functionally related parameters. It is advisable, during the manufacturing and filling procedures, to observe closely the value of such parameters as reducing agent concentration, hydrogen ion concentration, and titratable alkali.

Procedure

In hair-waving operations utilizing sulfite or bisulfite derivatives, Speakman makes some interesting observations which might be applied to hair-straightening procedures (47). Initially, the hair is shampooed and left either dry or wet. It is of advantage to maintain the hair saturated with the solution, at a temperature of 35 to 50°C or higher, for about 15 min. This may be accomplished by applying the necessary heat to the curlers or by heating the hair by a stream of hot air from a drying apparatus. To prevent evaporation of the solution, the entire head might be covered with a water-impermeable cover. Optimum results might be achieved by following this technique.

Neutralization

As with thioglycolate-containing compositions, it is desirable to limit the effects of the sulfites or bisulfites. To achieve this, Speakman suggests the use of a polyvalent metal compound or organic compound (47). The metal salts, which might be employed, may be the salts of metals other than the alkali metals. Calcium, barium, zinc, copper, or nickel salts may be utilized. The salts can be oxidizing agents such as specific nitrates, or they may be mixed with oxidizing agents. The concentration of salt can be determined by the formulator; however, 5% is a convenient concentration (47).

After treatment of the hair fibers with the reducing agent, the solution may be removed and the fibers washed. The reducing agent can be completely removed by the use of an agent adapted to combine, inactivate, or eliminate it. The fibers can then be treated with the metal salt or salt oxidizing agent mixture and then be washed to complete the treatment process.

General Considerations

The so-called bisulfite straightener is one of the most recent developments. Several brands have made an appearance in both professional and retail markets during the past three years.

As do the thioglycolates, reversible keratin bond-related changes are brought about by these materials. Straightening effectiveness of the bisulfite relaxers is similar to that of hot combing but is permanent (48).

However, it is felt that with present technology, there is no satisfactory way to go safely from extremely curly or kinky to completely straight hair with any of the chemical straighteners. Some degree of straightening can be obtained, but people must be realistic in their expectations.

Bisulfite-containing straighteners are reported to be mild on the scalp and skin, and the chances of hair damage might be less than that exhibited by other chemical straightening agents (48). However, as with other chemical treatments, manufacturers' directions and precautions should be carefully observed. If the hair is dry, brittle, damaged, or easily broken, bisulfites should not be applied. If bleaching, color treating, or previous straightening by other methods has been conducted, caution should accompany the use of bisulfite-containing straighteners. A test should be carried out on a small section before the product is applied to all of the hair. Manufacturers include directions for applying this test, called a "strand test," and for interpreting the results. If the scalp or skin is sore, scratched, sensitive, and scaly, this type of straightener should not be used.

Directions and precautions should be reviewed and carefully followed every time a product is used, not just the first time. Poor and harmful results frequently ensue from a failure to follow instructions (49).

REFERENCES

1. Suter, M.: Private communication to Kolar Laboratories, Inc.
2. Rudall, K. M.: The structure of the hair cuticle, *Proc. Leeds Phil. Soc.*, **4**: (Part I): 13 (1941).

3. Menkhart, J., Wolfram, L. J., and Mao, I.: Caucasian hair, Negro hair, and wool: Similarities and differences, *JSCC*, **17**: 769 (1966).

4. Powitt, A. H.: Some properties of human hair, *Am. Perf.*, **83**: 53 (January 1968).

5. Menkhart, J., and Coe, A. B.: Microscopic studies on the structure and composition of keratin fibers, *Text. Res. J.*, **28**: 218 (1958).

6. Chapman, R. E.: New ideas on the formation of crimp in wool, *Wool Technology and Sheep Breeding*, December 19, 1964.

7. Savill, A., and Warren, C.: *The hair and scalp*, E. Arnold, London, 1962.

8. Geiger, W. B.: The distribution of ortho and para cortical cells in wool and mohair, *Text. Res. J.*, **26**: 618 (1956).

9. Bradbury, J. H., and Chapman, G. V.: The chemical composition of wool, *Australian J. Biol. Sci.*, **4**: 960 (1964).

10. Crewther, W. G., Fraser, R. D. B., Lennox, F. G., and Lindley, H.: The chemistry of keratins, *Adv. Protein Chem.*, **20**: 191 (1965).

11. Whewell, C. S.: The chemistry of hair, *JSCC*, **15**: 423 (1964).

12. Whewell, C. S.: The chemistry of hair, *JSCC*, **12**: 207 (1961).

13. Rosenthal, N. A., and Oster, G.: Recent progress in the chemistry of disulfides, *JSCC*, **5**: 286 (1954).

14. Speakman, J. B., *Proc. Int. Wool Text. Conf.*, **C**: 302 (1955).

15. Schöberl, A., and Rambacher, P.: Ueber die Reaktionsfähigkeit des Keratins der Schafwolle, *Biochem. Z.*, **306**: 209 (1940).

16. Schöberl, A., Rambacher, P., and Wagner, A.: Über die Einwirkung von Natronlauge auf menschliche Haare, *Biochem. Z.*, **317**: 171 (1944).

17. Mizell, L. R., and Harris, M.: Nature of the reaction of wool with alkali, *J. Res. Nat. Bur. Stand.*, **30**: 47 (1943).

18. Middlebrook, W. R., and Phillips, H.: The action of sulphites on the cystine disulphide linkages of wool, *Biochem. J.*, **36**: 428 (1942).

19. Reed, R. E., DenBeste, M., and Humoller, F. L.: Permanent waving of human hair: the cold process, *JSCC*, **1**: 109 (1948).

20. Speakman, J. B., *et al.*: The reactivity of the sulfur linkage in animal fibres—Part I. The chemical mechanism of permanent set, *J. Soc. Dyers Colourists*, **52**: 335 (1936).

21. Davidson, A. N., and Howitt, F. O.: Experimental development of methods for the production of durable creases and crease-resistance in wool fabrics, *J. Text. Inst.*: **53**: 62 (1962).

22. Farnworth, A.: A hydrogen bonding mechanism for the permanent setting of wool fibers, *Text. Res. J.*, **27**: 632 (1957).

23. Childrey, H. M., Jr., and Doty E.: U. S. Pat. 3,017,328 (1962).

24. Cotter, L. H.: Thioglycolic acid poisoning in connection with the "cold wave" process, *J. Am. Med. Assoc.*, **131**: 592 (1946).

25. McCord, C. P.: Toxicity of thioglycolic acid in cold wave, *J. Am. Med. Assoc.*, **131**: 776 (1946).

26. McCord, C. P.: The physiologic properties of thioglycolic acid and thioglycolates, *Ind. Med.*, **15**: 12 (December 1946).

27. Draize, J. H., Woodard, G., and Calvery, H. O.: Methods for the study of

irritation and toxicity of substances applied topically to the skin and mucous membranes, *J. Pharm. Exptl. Therap.*, **82**: 377 (1944).

28. Behrman, H. T.: Cold wave lotions, their cutaneous and systemic effects, *JSCC*, **2**: 228 (1951).

29. Hollenberg, I. R.: Formation of cold wave preparations, *Proc. Sci. Sec. TGA*, **13**: 9 (1950).

30. Ramsey, H. R.: U. S. Pat. 2,418,664 (1947).

31. Haefele, J. W.: U. S. Pats. 2,615,782–83 (1952).

32. Haefele, J. W.: U. S. Pat. 2,615,828 (1952).

33. Schwarz, M. H.: U. S. Pat. 2,540,494 (1951).

34. Mulinos, M. G., Higgins, G. K., and Christakis, G. F.: On the toxicity of sodium perborate, *JSCC*, **3**: 297 (1952).

35. Reed, R. E., Den Beste, M., and Tenenbaum, D.: U. S. Pat. 2,564,722 (1951).

36. Whitman, R.: The role of the neutralizer in cold waving, *Proc. Sci. Sec. TGA*, **18**: 27 (1952).

37. denBeste, M., and Reed, R. E.: U. S. Pat. 2,540,980 (1951).

38. Draize, J. H., and Kelley, E. A.: Toxicity to eye mucosa of certain cosmetic preparations containing surface-active agents, *Proc. Sci. Sec. TGA*, **17**: 1 (1952).

39. Calva, J. B.: U. S. Pat. 2,390,073 (1945).

40. Gant, V. A.: U. S. Pat. 2,643,375 (1953).

41. Charle, R., Montmorency, S., Ritter, R., and Kalopissis, G.: U. S. Pat. 2,944,942 (1960).

42. DeMytt, L. E.: U. S. Pat. 2,976,216 (1961).

43. Strain, R. J., and Tusa, P.: U. S. Pat. 3,025,218 (1962).

44. Grant, S.: U. S. Pat. 3,143,476 (1964).

44a Lubs, H. A.: U. S. Pat. 2, 403,937.

45. DeMytt, L., and Reed, R. E.: U. S. Pat. 2,506,492 (1950).

46. McDonough, E. G.: U. S. Pats. 2,577,710–11 (1951).

47. Speakman, J. B.: U. S. Pat. 2,261,094 (1934).

48. deNavarre, M. G.: *The chemistry and manufacture of cosmetics*, Van Nostrand, New York, 1941, p. 452.

49. Guidelines for straighter hair: the look you like column, *Today's Health* (November 1967).

Chapter 23

BLEACHES, HAIR COLORINGS, AND DYE REMOVERS

Florence E. Wall

Dissatisfaction with the color of the hair has led, since earliest times, to the application of various types of natural and man-made substances in attempts to change it. From the earliest Egyptians, through many successors in their cultural heritage—Greeks, Hebrews, Assyrians, Persians—and through the contemporary cultural successors of the ancient Chinese and Hindus, the use of hair colorings is mentioned in historical records of social life and customs.

The development of hair colorings followed the traditional use of simple and complex substances from plants, metallic compounds, and mixtures of these two types, and there was little change in coloring products and processes until the last third of the nineteenth century. The discovery that some of the new synthetic organic compounds were suitable dyes for animal fibers gave impetus to the development of new-type hair colorings, but it was not until the second quarter of the twentieth century that they were successfully established in public acceptance.

As long as the use of hair dyes, either at home or in beauty shops, was kept a secret if possible, and usually limited to those that wished to conceal gray hairs, the art constituted a relatively small branch of the business and industry. Promoting hair colorings of all kinds as a feature in the ensemble of fashion—something to be used as freely and as openly as facial makeup—has quickly advanced all these products and treatments to a position of major importance. The available statistics are unsatisfactory and incomplete, but the trend is well established in published estimates. For the year 1924, the percentage of the potential market using dyes was 6%; for 1950, 43%; for 1960, 78 to 80%; and for 1968, it was estimated at 80 to 90%. The notable increase in the latest figures has been ascribed to the development and promotion of the modern-type dyes for use at home.

With the exception of a few antiquated recipes, freely published in formularies, magazines, and other periodicals, all the best-known hair colorings in current use are proprietary products of which the exact composition is not published. All recipes given in this chapter have been obtained from nonconfidential sources, and are intended merely to indicate typical compositions for the respective product.

Demands in Hair Coloring

Any modern, well-staffed beauty salon usually offers the following services in hair coloring (1):

1. Artistic bleaching of the hair.
2. Correct use of colored rinses.
3. Application of organic liquid dyes.
4. Correction of poorly dyed hair.

To meet these demands, the industry offers an ever-increasing variety of products.

Classes of Coloring Agents

Although the effect of a bleaching agent is the subtraction rather than the addition of color, bleaches for the hair are usually classed with colorings because they do, in fact, cause a change in shade. The hair colorings in general use fall into two groups, depending on whether they are intended to color the hair temporarily or permanently. The various types of products to be considered here, therefore, will be classified as follows:

1. Bleaching agents.
2. Synthetic organic dyes: temporary; semi-permanent; or permanent.
3. Miscellaneous coloring agents: plant derivatives, metallic dyes.
4. Dye removers.

Bleaching Agents

Early Bleaches

Like dyes, bleaching agents are among the oldest products on which there is a practically continuous record. Primitive recipes for bleaching the hair were common in Rome during the two centuries that bridged the beginning of the Christian era. The Roman ladies greatly admired the golden hair of many of the captives brought from northern countries and tried to imitate it.

Native minerals such as rock alum, quicklime, crude soda, and wood ash, occasionally combined with old wine (or dregs of wine) and water, served as favorite "blond washes." These preparations were left on the hair overnight or for several days, and the resulting reddish-gold shades were usually considered satisfactory. If the hair happened to be completely destroyed, the ladies frequently demanded that the hair be shorn from the captives and made into wigs for them (2–6).

Among the oldest bleaching mixtures were mullein flowers or oil of mastic in vinegar. The most popular was a soap imported from Gaul or Germany, made from goat's fat and ashes of beech wood, in a form called "Mattiac balls," from the Roman name for Marburg. So it seems that soap was invented for bleaching the hair, not for cleansing.

All the famous old books of medical (and other) "secrets" published during the Renaissance contain recipes for bleaching the hair. The alum, ashes, borax, niter, crude soap, and soda of the earliest mixtures were still used, but always with decoctions of various plants. Commonly found are birch bark, broom, celandine, lupine, mullein, myrrh, saffron, stavesacre, and turmeric; old wine and dregs of wine are also mentioned (7,8).

To produce the beautiful golden-red ("Venetian blond") shade of hair immortalized by the great artist Titian, the Venetian ladies sponged and combed a solution of soda (or rock alum, black sulfur, and honey) through their hair, spread it over the broad brim of a crownless hat, and let it dry in the sunlight. This treatment was introduced into France late in the sixteenth century by Marguerite de Valois, and with slight modifications it remained for over two hundred years the acceptable method for producing blond and reddish shades of hair for the relatively few that desired them (5,9).

Chemical Bleaches

Although the fading of colors on exposure to air and sunlight had long been recognized, the slow action and uncertain results of molecular atmospheric oxygen made continued utilization of it impractical by the nineteenth century. When interest was revived in bleaching the hair—after two centuries of perukes and powder and a period of disfavor for all red shades of hair— many oxygen-releasing compounds were available for testing.

For use on living human hair, a chemical bleaching agent must be non-toxic, mild in action, and free of harmful residue. These requirements rule out many effective oxidizing agents; in fact, they narrow the possibilities to sodium hypochlorite and hydrogen peroxide (and sources). As the former is not suitable for the bleaching of animal fibers, use of it in cosmetology is limited to the removal of dye stains. The reducing action of sulfur dioxide has also been utilized to a limited extent.

Hydrogen Peroxide

The most satisfactory bleaching agent for human hair has always been hydrogen peroxide, because after its available oxygen has been released nothing but water remains. Although it had been discovered in 1818 by Thénard, there seems to have been little practical application of it until the Paris Exposition of 1867, when the bleaching effect on hair was demonstrated (10).

The promoters were E. H. Thiellay, a "chemist-perfumer" of London, and Leon Hugo(t), a hairdresser of Paris, using a 3 % solution of hydrogen peroxide under the trade name of "eau de fontaine de jouvence golden." The treatment quickly became popular throughout Europe and in the United States, and by the turn of the century hydrogen peroxide had superseded everything else used previously for bleaching human hair.

Pure hydrogen peroxide (H_2O_2) is a colorless, syrupy liquid about one and one-half times as heavy as water. It is usually sold as solutions, identified by "volume content," that is, the number of volumes of available oxygen liberated by a given volume of the solution. It is readily decomposed by contact with metals and alkalies. In high concentrations it is corrosive to the skin.

For bleaching the hair, the preferred peroxide for many years was the one produced from barium peroxide; peroxide made electrolytically followed in favor. The product of another process, based on alkyl derivatives of anthraquinone, is now commonly utilized (11,12).

"Drug store peroxide," a 3 to 4 % solution, generating 10 to 12 vol of oxygen, is commonly used for bleaching the hair at home, but for more rapid action the 5 to 6 %, 17 to 20 vol solutions are generally used in beauty shops. These are sufficiently strong for all practical purposes; the solutions of 30 vol or more, recommended for certain treatments, offer no advantage and may actually harm the hair.

To maintain the desired quality, solutions of hydrogen peroxide must be stabilized. Among stabilizers used are acetanilide, dilute acids, colloidal silica, *p*-hydroxybenzoates, oxyquinoline sulfate, phenacetin, and tin compounds (sodium stannate, stannic hydroxide, stannous octoate). To prevent deterioration through carelessness, dealers in supplies for beauty shops recommend that peroxide be bought in one-quart bottles, rather than larger containers.

Action of Peroxide on Hair

The action and effects of hydrogen peroxide on hair have been intensively studied (13–17). The reaction is both chemical and physical, because the liquid seems to open the imbrications of the cuticle, penetrate and attack the keratin structure, and gradually lighten the shade of the hair by oxidizing the melanin that gives it color. Depending on the concentration of the solution and the time of contact, hair can be bleached to an extremely light yellow, and so softened that a strand of it can be stretched and retracted like an elastic band while wet (1).

In practice, the peroxide solution is activated by the addition of ammonium hydroxide (28%, 0.880 sp. gr.), usually 15 to 20 drops to the ounce of peroxide. The ammonia should always be used sparingly; too much may cause reddish tints to develop. It should be added at the moment of use.

Bleaching Agents and Treatments

Peroxide solutions.   Hydrogen peroxide is available as 6% solutions (some under proprietary names), which are occasionally still used in beauty salons when a "straight bleach" is required.

For this simplest method of bleaching, the hair should be well shampooed and dried. Twenty drops of ammonium hydroxide should be added to 1 oz of hydrogen peroxide, and the mixture should be applied to the full length of the hair, strand by strand. The bleaching action, which oxidizes the color to lighter shades, can be continued as long as desired by keeping the hair saturated with the peroxide; and it can usually be stopped at any stage by copious rinsing with hot water or a mild acid rinse. The great advantage of this method is that the development of color can be carefully watched and controlled. No second shampoo is required (1).

By experimenting with different proportions of peroxide and ammonia and times of contact, a wide variety of beautiful shades of hair is obtainable. Other conditions being constant, any hair can be kept the same shade by periodically applying the same bleaching mixture for the same length of time to only the new growth of hair as it appears. Repeated application to the same portion of hair leads to serious damage to the texture of the hair and the familiar strawlike, overbleached appearance.

Other ways of using peroxide solutions for quick bleaching are the following.

1. Brightening rinse. Used to bring out highlights in dull hair. The preparation of such a rinse is shown in Formula 1. This can be poured over the hair until the desired shade is seen.

Formula 1

Hydrogen peroxide (20-vol)	6 to 8 fl oz
Water	10 to 8 fl oz
Ammonium hydroxide	20 drops/oz

2. Brightening shampoo. In many beauty shops, where soap is still used for shampooing, a peroxide shampoo offers a simple method of cleansing and

bleaching the hair in one operation. The mixture given in Formula 2 will serve. This should be applied to the hair, following the customary technique for an efficient shampoo, worked into a good lather, and allowed to remain

Formula 2

Shampoo	4 fl oz
Hydrogen poroxide	2 fl oz
Ammonium hydroxide	40 drops/oz

for 3 to 5 min. It is then well rinsed with warm water. The application of the bleaching mixture and the rinsing can be repeated until the desired shade is seen. The final rinse should be with very hot water (1).

The disadvantage of these quick methods of bleaching is that they can serve only as a first treatment. New growth must be bleached by the standard method or retouching.

Simple and effective as peroxide solutions are, they have been practically abandoned in beauty shops in favor of proprietary preparations. A satisfactory brightening rinse can be prepared from a 3% (10-vol) solution, in which the bleaching action is retarded by the addition of a mild organic acid (pH 4 to 4.5), and penetration of the hair is ensured by the presence of a wetting agent (18). A typical composition for a rinse of this type is shown in Formula 3.

Formula 3

Hydrogen peroxide, 3%	97.1%
Quaternary ammonium salt, or quaternary morpholinium alkyl sulfate	1.5
Adipic (or tartaric) acid	0.8
Sodium stannate (or pyrophosphate)	0.6

Proprietary bleaching shampoos have not always been satisfactory. To incorporate hydrogen peroxide with a soap would require a stabilizer, and this would retard or inhibit the bleaching action (16). Some products contain sodium perborate with surfactants—wetting agents, detergents, foamers. In selecting any such compounds, the pH must be taken into consideration.

A satisfactory shampoo mixture can be prepared from ammonium oleate (20 to 25%) and a diethylamide (5%); hydrogen peroxide should be added at the moment of use. The pH should be adjusted around 9.

In all formulations and combinations the proportions of ingredients should be so adjusted that the amount of ammonia in the final product will be about 1%; higher percentages can irritate the skin.

The principal complaint against the use of clear peroxide solutions is that it is difficult to keep the liquid where it is applied; it may run over previously bleached hair. This is a matter of technical skill, but it has led to the development of other physical forms of bleaching agents.

Solid hydrogen peroxide. Tablets made of peroxide and urea; or mixtures of sodium perborate and, for example, tartaric acid. These are to be dissolved in water at the moment of use. Urea swells the hair and makes it absorptive.

Any package that contains urea peroxide, to be made into a solution before being added to a dye solution, should carry instructions that only distilled or deionized water should be used for this purpose, and that contact with metals should be avoided in preparing it. As a protection for both distributor and user, a date of expiration should be stamped on the package.

Newer products of this type may contain melamine peroxide and solubilizers or gums to facilitate the activity (19).

Cream bleaches. The emulsion, or creamy, form of bleaching agent is considered preferable because it can be kept in place (as on new growth). To counter this advantage, the consistency of the emulsion may obscure the development of lightening color.

The emulsion can be made of easily emulsifiable waxes, alcohols, sulfated alcohols, or fatty amides. A creamy product for quick bleaching may contain persulfates of sodium, potassium, or ammonium (sometimes all three). These compounds must always be in a separate envelope. The addition of quaternary compounds makes the hair easier to handle.

Heavier cream bleaches are really ammoniacal creams, which may contain lanolin derivatives in addition to the waxes, fatty alcohols, short-chain polyoxyethylene derivatives, and 10 to 15% of ammonia water (28%). The cream is to be mixed with 20-vol peroxide at the moment of use. This type of product is usually satisfactory because its consistency retards the loss of ammonia and the liberation of oxygen (16).

Optional additions to these bleaches are a persulfate in the peroxide solution; a blue or violet color in the base, to mask somewhat the brassy shade often seen (20); and a humectant, to ensure that the mixture remains moist as long as it is supposed to be active.

The time of contact must be watched carefully, and the cream should be rinsed away with copious applications of hot water.

Powders. Despite the prevalence of dependable proprietary products, many hairdressers still like to use so-called "white henna." This is an obsolete trade term for magnesium carbonate, to be mixed with hydrogen peroxide and ammonia and applied in the same manner as the clear liquid or cream. The claimed advantage is that the paste "stays put," but the white obscures the development of the shade, and the treatment requires a second shampoo (1,23).

In modern modifications, alkaline perborates, persulfates, or percarbonates may be added as "boosters"; also barium or magnesium oxides, silicates, and carboxymethyl cellulose, to make the paste easy to handle and prevent it from drying (21).

Proprietary products may specify mixing with either water or hydrogen peroxide. The former should be satisfactory for all practical purposes; depending on the previous treatment of the hair, use of hydrogen peroxide in such a mixture may be disastrous. Manufacturer's directions must be clear and complete and stress the importance of following them.

Oil bleach. Originally based on sulfonated castor or olive oil, the oil bleaches are now compounded with ammonium soaps, e.g., ammonium oleate, nonionic detergents, oxyethylated alcohols, and alkylphenols. Free ammonia is also present. They are usually to be mixed with twice the quantity of hydrogen peroxide.

These newer oil bleaches offer several advantages: They are easily applied; the creamy, gelatinous consistency holds the bleach in place; and they remain moist on the hair. The composition may be strengthened with persulfate, always in a separate packet.

Platinum Bleaching

Under ordinary conditions, even the lightest shade obtainable from bleaching with hydrogen peroxide is always a decidedly yellowish tinge. To produce the nearly white "platinum blond" shade which is intermittently popular, the hair must first be bleached as light as possible, and then carefully rinsed with a dilute solution of some dye that can neutralize the residual yellow. Methyl violet, methylene blue, and nigrosine have proved to be satisfactory for the purpose (22,23).

As mentioned before, some bleaching agents contain a color, intended to neutralize the yellowish tinge of bleached hair (20), but both in beauty salons and in the home, the use of "color toners" (see section on dyes) has proved satisfactory and successful.

Blanching of Hair

Besides the golden-red and yellow shades of hair, obtainable with hydrogen peroxide, there has been an intermittent and sporadic demand for some means of turning mixed gray hair an even snowy white. For such a treatment, the reducing action of sulfur dioxide has been recommended.

The blanching effect of the fumes of sulfur dioxide on animal fibers has been utilized since at least the time of Pliny, for he recorded that the Romans exposed their woolen fabrics to the smoke from burning sulfur in order to whiten them (2). Later writers mention a mixture of sulfur and lye, used similarly.

The only other suggestion to be offered since that time—to treat the hair successively with solutions of potassium permanganate and sodium thiosulfate (22–25)—seems never to give a satisfactory result (26). The desired effect can be simulated by the use of "silvery" toners, to be described in another section.

Aftertreatment of Bleached Hair

The copious rinsing with hot water, usually prescribed after a bleaching treatment, is frequently not sufficient to stop oxidation completely, especially when persalts have been used.

For clearing the hair more effectively, mild acid rinses have been recommended (27). Other compounds suggested are thiourea, pyruvic acid, and their derivatives; well-known spot tests have been adapted to testing hair for residues of peroxide and persulfates (28).

Synthetic Organic Dyes

To those in the trade and industry, the term "synthetic organic dyes" now means the group of amines, aminophenols, and related compounds called "oxidation dyes" (also called amine dyes, para dyes, and peroxide dyes). These are the so-called "permanent dyes," but synthetic colors are also used in other functional types of dyes, temporary and "semipermanent," and they are of many chemical types.

The use of colorings for the hair is of relatively ancient origin. For centuries the records of various peoples indicate that they were mainly interested in preparations and treatments that could effect a permanent change in color. Dissatisfaction with the composition and effects of all available products at the time led Dr. Herman Beigel of London to publish (1869) the following "requirements of a good hair dye" (29):

1. It must not be injurious to the general health.
2. It must dye the hair, but not the skin.
3. It must have no ill effect on the structure of the hair.
4. It must not require a long time for the production of its effect.

Modern specialists now recognize three more requirements (1):

5. It must have no irritating effect on the skin.
6. It must produce shades that are natural in appearance and reasonably lasting.
7. It must be compatible with other treatments, such as permanent waving.

Range of Shades Required

In all mammals, the natural pigment of the hair (and skin), called collectively *melanin* (Gr. *Melas*, black) (1,16,33), produces the same range of variation in shades. No matter how unusual a shade of hair on man or animal may seem to be, it can generally be grouped somewhere under tones of blond (assorted pale shades) yellow, reddish, brown, and black. Under normal conditions, hair is never any shade of true red, green, blue, or purple. True grays are seen in the hairy coat of many animals, but in man the effect of gray results from the mixture of white and colored hairs.

Anyone that wishes to specialize in either the compounding or the practical application of hair colorings must constantly study the color of hair, not only in laboratory and beauty shop but also outside, wherever it is to be seen (30,31).

To the untrained eye, for example, many heads of hair that appear black are actually a very dark brown when viewed in good light. Even true blacks vary from a dull jet black to a brilliant "blue black" or the "warm black" just mentioned, and a surprising number of lighter pigment granules may be seen when any of these is examined under a microscope.

The widest variation in pigmentation occurs in the white race. Ten distinct types can be readily identified (1). Every usual color of hair and eyes can be seen, with a gradual darkening of the skin until eyes and hair show the maximum of pigment. Once the racial border is crossed, hair and eyes (in true types) remain very dark, and further deepening of pigmentation occurs in the skin itself until, as seen in certain African tribes, skin, hair, and eyes seem to be coal-black.

The pigment in hair is found in various places: all in the medulla; partially in the medulla and in the cortex; but not normally in the cuticle, a fact which often serves to distinguish natural from dyed hair (32,33). The melanin may be in discrete granules, in shapeless masses, or as a diffused stain; and the quantity may vary even within the length of a single hair. The intensity of the shade depends on the size and number of granules of pigment. Diffused pigment produces reddish tones; if granules are present, the shade tends toward brownish, but the fewer the granules, the brighter the shade. As many as seven to 10 distinct variations in shade have been observed among the hairs on a single head, and in really artistic hair coloring this natural variation should be maintained (8).

In addition to grouping shades of hair from light to dark, it is necessary to differentiate them according to basic tone. Any shade of human hair from "tow" blond to darkest brown falls into two general classifications: *drab* or ashy tones, with neutral or silvery highlights; *warm* tones, with reddish (coppery) or golden highlights. Any line of dyes must be planned to meet

TABLE I.   Importance of Hair as a Feature of Complexion

	Skin	Blood tone[a]	Hair	Eyes	Brows and lashes	Burns	Freckles	Tans
I	Light pink; clear	Bluish	Tow, or silver blond	Blue, gray	Tow	Badly	Rarely	Some
II	Light yellow;	Orange	Silver-blond	Blue, gray	Dark brown or black	Yes	Yes	Some
III	Sallow	Orange	Golden-blond	Brown, hazel, tawny	Black	Yes	No	Yes
IV	Light yellow; clear	Orange	Golden-red, golden-brown	Blue, gray, green	Tow, golden	Badly	Badly	Rarely
V	Pinkish; clear	Bluish	Bright red	Blue, hazel brown	Red, brown	Yes	Some	No
VI	Medium Yellow	Orange	Red-brown, mahogany	Blue, hazel, gray, brown	Dark brown, black	Some	Rarely	Yes
VII	Medium pink; clear	Bluish	Ash, and medium brown	Blue, hazel gray	Black	Yes	Some	Yes
VIII	Medium yellow	Orange	Medium to dark brown	Hazel, brown	Black	Yes	Rarely	Yes
IX	Pinkish; clear	Bluish	Dark brown, black	Blue, gray, hazel	Black	Yes	Yes	Yes
X	Dark yellow; "olive"	Purplish	Dark brown, black	Green, hazel, dark brown	Black	Yes	No	Yes

[a] *Blood tone* means the tinge of natural color brought up when the skin is pinched. It should serve as a guide to correct choice of powder and rouge.

the demands of these basic variations; hence the compounder must know how to produce natural-looking golden, silver, and ash blonds; golden, ash, and reddish light and medium browns; golden, coppery, and auburn reds; warm and dull dark browns; even a bluish and a dull black.

Table II shows the average basic requirement in a typical line of shades. The designations are merely descriptive. To the trained eye, each of these is a distinctly separate and recognizable shade. Variations may be added without limit according to the whim of the manufacturer or distributor,

TABLE II.   Typical Shades of Hair Dyes

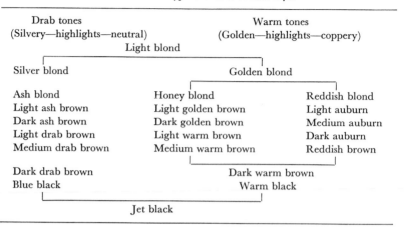

Drab tones (Silvery—highlights—neutral)		Warm tones (Golden—highlights—coppery)
	Light blond	
Silver blond		Golden blond
Ash blond	Honey blond	Reddish blond
Light ash brown	Light golden brown	Light auburn
Dark ash brown	Dark golden brown	Medium auburn
Light drab brown	Light warm brown	Dark auburn
Medium drab brown	Medium warm brown	Reddish brown
Dark drab brown		Dark warm brown
Blue black		Warm black
	Jet black	

but they are not really necessary, because any given shade can produce slightly different results, depending on the original shade of the hair to which it is applied (1,32).

As arranged in the table, shades on the same horizontal line are of about the same depth or value, although they differ appreciably in reflection, or highlights. The quality of color can best be learned by careful study of standard samples, obtainable from dealers in postiche, rather than from shade cards. Many of the latter have, in the past, been assembled from hair dyed with vegetable colorings, or with anything that would produce what were supposed to be "all the standard shades." Lithographed shade cards, also, may be a cause of surprise and disappointment to both hairdressers and home users of hair dyes, because the representation may be different from the reality.

Special Shades

In addition to the range of shades that resemble those of human hair "high style" colorings may be in demand from time to time. Other innovations which will probably always be popular to some extent are some true gray shades in a range of values from off-white through silver- and iron-gray to almost black. These are used both to modify a natural color and to enhance the appearance of all-gray or mixed gray hair by neutralizing any yellowish tinge that may develop.

Temporary Colorings

As temporary colorings, we here consider the miscellaneous assortment of products used at any time to effect a physical change in the shade of the hair.

These include colored rinses, powders, crayons, etc. All should be easily removable and have no appreciable effect on the texture or other characteristics of normal hair.

Colored Rinses

The use of temporary colorings for the hair is of comparatively recent origin. The original colored rinses (also called *tint rinses*) were introduced by the Nestle-Lemur Company in 1922, patterned on a similar product commonly used in the household to restore or change the shade of lingerie, curtains, or other articles.

These early tinted rinses are similar in composition to the proprietary acid rinses commonly used as accessories to a shampoo. The base is usually tartaric acid (adipic, citric, acetic, and other acids have also been used) with which water-soluble azo dyes are blended.

The modern temporary rinses based on acid dyestuffs are applied at pH 7. They must be fast to light rinsing with water, and be removable with the next shampoo.

Theoretically, several types of dyes are suitable for these products: azoic, azinic, indigoid, triphenylmethane dyes, and derivatives of anthraquinone, thiazine, and xanthine (16). In the United States, however, most manufacturers restrict themselves to the official list of colors suitable for cosmetics, as published by the Food and Drug Administration (34). Several other colors, not certified, are still used by manufacturers who have long found them to be both satisfactory and quite harmless. It must be understood, however, that any use of noncertified colors is always at the risk of the manufacturer. Any product or item that contains a noncertified color should bear a label stating that the contents may be harmful.

Many lines of colored rinses are offered in a wide range of shades, from golden and silver blond, through various reddish and brownish shades, to black. The average range of shades must be produced from the certified colors in Table III.

Very few of these colors are suitable for use alone. To produce the desired tints, dilute solutions (1 g/liter) should be tested, singly and in mixtures, on white wool or hair and also on naturally colored hair of shades on which the rinses might appropriately be used. For satisfactory compositions, prospective manufacturers should keep closely in touch with the sources of supply.*

In manufacturing these rinses, solutions of the colors—especially mixtures—must be thoroughly worked into the crystals of the acid. The finished

---

* It seems futile to give any suggestions for formulations. Of the eight colors shown in the first edition of this book as ingredients of satisfactory and acceptable shades for rinses, only two were still certified at the time that this edition went to press. The missing ones must be made by mixing others that are allowed.

TABLE III.   Certified Colors for Hair Dyes

FDA designation	Common name	FDA designation	Common name
FD&C blue No. 1	Brilliant blue FCF	D&C red No. 22	Eosin YS
FD&C blue No. 2	Indigotine IA	D&C red No. 30	Helindone pink CN
D&C blue No. 4	Alphazurine FG	D&C red No. 33	Acid fuchsin D
	erioglaucine		Fast acid fuchsine B
D&C blue No. 6	Indigo		Napthalene red B
FD&C green No. 3	Fast green FCF	FD&C violet No. 1	Acid violet 6B
D&C green No. 5	Alizarin cyanine	D&C violet No. 2	Alizurol purple SS
	green		
D&C green No. 6	Quinizarin green	FD&C yellow No. 5	Tartrazine
	SS		
Ext D&C green	Naphthol green B	FD&C yellow No. 6	Sunset yellow FCF
No. 1			
D&C orange No. 4	Orange II	D&C yellow No. 10	Quinoline yellow WS
FD&C red No. 2	Amaranth	D&C yellow No. 11	Quinoline yellow SS
FD&C red no. 3	Erythrosine	Ext D&C yellow	Metanil yellow
		No. 1	
FD&C red No. 4	Ponceau SX	Ext D&C yellow	Naphthol yellow S
		No. 7	

product may be solid in envelopes of cellophane or translucent, moisture-proof paper, containing about 1 oz, in capsules, compressed tablets, or—currently most popular—solutions. When dissolved in 1 quart of hot water, the specified quantity of the product usually makes a solution in good concentration. This should be poured over the hair repeatedly until practically all the tint has disappeared from the water (1).

As originally presented, these colored rinses were intended to clear away any scum that might be left on the hair after shampooing it with soap (especially prevalent where hard water must be used), and impart tinted highlights. They became extremely popular, both for this purpose and for restoring faded, sun-bleached, or permanently waved hair to its natural shade. When the texture of the hair is normal, the tinted film deposited by these rinses usually disappears in the next shampoo. If the texture of the hair has been altered in any way, however, as in bleaching or permanent waving, it is usually more absorptive and the effect of the colored rinse is considerably more lasting. The use of wetting agents and other surface-active agents in synthetic ("soapless") shampoos also makes the hair more receptive and thus tends to make it retain these colors.

The addition of surface-active agents to the rinses themselves presents complications which should not be ignored by manufacturers and distributors of these products. Sodium lauryl sulfate, commonly used, tends to degrease the hair too thoroughly. The alkylated aryl sulfonates, besides making the color more lasting, also tend to coat the hair with a deposit which impairs texture and may interfere with other treatments, such as permanent waving and the use of penetrating (oxidation) dyes.

The amount of surface-active agents to be added to products of this type should be controlled most critically. Any excess of quaternary ammonium or other cationic compounds, for example, in compositions based on acid dyes, may cause the coloring to diffuse within the mixture itself and thus not color the hair properly. Sulfonated oils may actually strip color from the hair (see section on *dye removers*).

Attempts to incorporate these temporary colors in setting fluids and spray lacquers have for many and various reasons seemed so far to be impracticable. Anyone that considers these necessary to a line is undoubtedly continuing research on them.

Rinses for Gray Hair

To neutralize the yellowish tinge that commonly appears in natural gray and white hair, and to produce "platinum blond" in bleached hair, rinses based on methylene blue, acid violet 6B, and water-soluble nigrosine are sold both as dry crystals and in solution. The accompanying instructions should ensure a concentration of no more than 0.001 % in the solution, as prepared for application to the hair.

For all practical purposes, it is well to prepare these dyes as 1 % solutions. The methylene blue and nigrosine should be dissolved in 1 part alcohol and 3 parts water; the acid violet 6B, in alcohol alone. A rinse of the proper concentration can then be prepared by mixing 1 part of any of these solutions with 1000 parts of water. When they are made and used correctly, these rinses can produce some beautiful effects, from bright white through all values of gray from silvery to charcoal.

To make them more effective, these bluing or silvering rinses are usually compounded with adipic, citric, or tartaric acid, as described for the lines of colored rinses. They may also be based on quaternary compounds, fatty acid alkanolamides, ethylene glycols, or nonionic waxes.

Importance of Instructions

To protect consumers against unexpected results and possible disappointment from the use of supposedly temporary colored rinses, instructions for use should be most explicit and comprehensive. To prevent shock or disappointment, the depth of color to be obtained and the time of immersion should be tested on a strand of hair before the whole head is rinsed.

The warning on preliminary tests for allergy should also be included as a protection for both distributors and consumers.

Colored rinses that utilize basic dyes with phosphate buffers are really too lasting to be considered true rinses. Moreover, they stain the scalp and eventually cause unpleasant discoloration which cannot be removed.

Color Shampoos

In addition to the rinses in various forms, already described, azo dyes have been combined with shampoos of different types. The earliest products of this kind used a base of neutral soap in small cakes, and were applied like any soap for shampooing. As they proved to be unsatisfactory, they were soon abandoned.

Many "color shampoos" (all proprietary products) have been based on synthetic surface-active agents with wetting and detergent properties; for example, the sulfonated oils. The average line offers four to ten shades, in 0.5 to 2.0% concentrations (of approved color), suitable for blond, reddish, light- and dark-brown hair. Because the actual amount of color imparted is relatively slight, permitting the natural shade of the hair to show through it, the few colors in these lines seem to be satisfactory for hair of any shade.

Newer color shampoos use nitro dyes and anionic detergents, with the addition of a fatty acid alkanolamide to increase solubility. Urea compounds also aid solubility.

Also used are polyhydroxy naphthquinone, e.g., 2-hydroxy-1,4-naphthquinone, plus triethanolamine lauryl sulfate and a fatty acid alkanolamide. Lauryl sulfate degreases the hair and makes it receptive to anything that may follow (35–37).

Steadily popular is the "shampoo-in" type of colorings, because they can be used conveniently in either shop or home, and they require no retouching. The formulation can allow for either stronger bleaching action or an increased amount of surface-active agent to ensure better penetration. The shade gradually lightens so that the whole head can be redyed after a few weeks without causing variegated effects.

Powders

The use of powder for changing the color of the hair is now largely confined to theatrical makeup, masquerades, and pageants in which it might be required for historical costumes; otherwise it is limited to the occasional freshening of the wigs worn by judges and certain other participants in courts of law.

During the seventeenth and eighteenth centuries, when powders were commonly used as a mark of fine dress, the basis of a good product was wheat starch, ground very fine. Potato starch became popular later, and the basic starches were frequently adulterated with plaster of paris, burnt alabaster, flour, and chalk (4).

The first colored powders consisted of starch with various percentages of burnt sienna, umber, or other pigments for brownish shades, and Chinese (India) ink for black. Pastel shades, popular during the nineteenth century, were made by adding other pigments, or by moistening the white base with solutions of dyes, and grinding the tinted mass after it dried.

The use of powdered gold, which dates from Biblical times and imperial Rome, was revived as a sensational extravagance by the ladies of the imperial court of Louis Napoleon in France, about 1853. The gold was adopted by the brunettes, and powdered silver was introduced by the Empress Eugénie for blondes. Political economists of the day seemed to regard the fad favorably, as a means of utilizing some of the surplus gold that was being produced in California and Australia (38). It soon passed, however, and metallic powders of similar colors but much lower value were substituted. Metallic powders in a wide range of shades are currently available from dealers in supplies for hairdressers and the theater.

In marketing metallic powders that utilize copper or brass for golden effects, instructions should warn users about the necessity for completely removing such powders from the hair and scalp. Even a slight residue would turn black in contact with a thioglycolate solution in permanent waving, and might even cause deterioration of the solution itself.

In any basic mixture for a hair powder, extremely light powders, e.g., magnesium carbonate, should not be included. A mixture of half potato meal and one-fourth each of talc and starch should make a satisfactory basis. Powders of different colors can be made by adding various mineral pigments and lampblack to this basic mixture (25).

Crayons

Sticks or blocks of coloring, similar to the mascara used as makeup for the eyes, have been made in several handy shapes, offered primarily for retouching new growth of hair between applications of permanent dyes.

These colored crayons for the hair are made of mixtures of natural or (usually now) synthetic waxes, with a soap, such as triethanolamine stearate, into which dyes or pigments are thoroughly incorporated. They are generally offered in a good range of shades, simulating the average colors of natural hair, so that they will blend with other dyes that may be on the hair. The consistency of the finished product must be such that the color can be easily applied as the wet stick is rubbed directly over the hair, or transferred from the stick to the hair by a moist brush.

Typical compositions for two types of crayons for the hair are shown in Formula 4 (a soap-base product), and Formula 5 (a wax-base product).

Formula 4

Water	26.5%
Gum arabic	27.5
Sodium stearate	15.0
Glycerol	15.0
Color	16.0

Procedure: Mix water and glycerol; add gum arabic to half of this solution and allow it to stand; add sodium stearate to remainder and warm mixture until it is dissolved. Mix all ingredients, add color, mill thoroughly, run paste into molds, and dry in heat.

Formula 5

Triethanolamine	7.5%
Glyceryl monostearate	4.0
Stearic acid	15.0
Beeswax	46.0
Microcrystalline wax	10.0
Coconut fatty acid diethylamide	7.5
Paraffin	10.0
Color	q.s.

Procedure: Heat the first two ingredients to 70°C; add stearic acid and raise temperature to 75°C. Melt waxes at 75 to 80°C; add them to other mixture and stir until well blended. Add color and mix thoroughly; pour into moulds at 68 to 70°C (22).

Semipermanent Dyes

The semipermanent dyes are so called because they remain on the hair through several shampoos. They are actually direct dyes without bleaching action. On the principle that "what puts it on takes it off," these products, which are usually applied as a shampoo, are gradually washed out by successive ordinary shampoos. They can color a good percentage of white hairs, but their prime effect is to impart highlights to naturally colored hair. They also are good for neutralizing a yellowish tinge in gray or white hair.

The degree of penetration of the hair depends on solubility; concentrated solutions are often necessary. Various direct dyes, azo and nitro, and derivatives of naphthalene and anthraquinone, can serve as bases. Many certified dyes are not suitable for light shades. All colors must be carefully blended so that the various components will not separate into different tints on hair of uneven texture. The final product must be thoroughly tested for fastness to light, friction, shampooing, and other treatments for the hair, especially permanent waving (16,39–44).

Several types of compounds used for semipermanent colorings are discussed below.

Nitro Dyes

Aromatic nitro and amino compounds are usually included in the oxidation dyes (see next section), but many derivatives of amines and aminophenols are also usable without a developer or oxidizer. The simple compounds give red and yellow shades, but the range of colors can be broadened to the whole spectrum by substituting alkyl, aminoalkyl, hydroxy, and other radicals in the amino groups of ortho- and paraphenylenediamine, nitro- and aminophenols.

Most of these compounds are not readily soluble in water, but a 1 to 2% solution will usually suffice. Many of the substitution products are darker in shade and show less affinity for the keratin of the hair; of these the solutions are used in higher concentrations.

Formula 6 is a typical semipermanent dye of this type.

Formula 6*

Monoethanolamine lauryl sulfate	20.0%
Ethylene glycol monostearate	5.0
2-Nitro-*p*-phenylenediamine	1.5
Diethanolamine coconut fatty acid	3.0
Perfume/preservative	q.s.
Water	q.s. 100.0

* Ref. 35.

Basic dyes can also be used in an amphoteric system. Amphoteric compounds increase viscosity, and they are good wetting agents in hard water; optimum *p*H, 7 to 9, in as shown Formula 7.

Formula 7*

Amphoteric surfactant	8.0%
Lauric isopropanolamide	1.0
Nonionic surfactant	6.0
Dye combination	2.0
Oleyl alcohol	0.5
Nonionic thickener (gum)	2.0
Perfume/preservative	q.s.
Water, to make	100.0

* Ref. 42.

Self-Oxidizing Dyes

Besides the preceding, several other compounds usually listed with oxidation dyes are worthy of special consideration because they can be successfully oxidized in the air at room temperature.

Some of these compounds, e.g., 1,2,4-triaminobenzene, develop a satisfactory color quickly, but continuing action causes this to change into an entirely different shade. The best colors have been obtained from derivatives of 1,2,4-trihydroxybenzene, aminohydroquinone, and 2,4-diaminophenol. By mixing the products, many beautiful colors, natural looking and of good coverage, have been obtained (16,40).

Solvent-Assisted Dyes

One of the ideas for dyeing hair, taken over from the textile industry, is that of utilizing some of the nonoxidation dyes for wool and other fibers. A hindrance has always been the boiling temperature required. Textile dyes are ineffective on living human hair at the preferred room temperature unless they can be solubilized by surfactants or other chemicals (45-48).

Among organic solvents that have been studied for the purpose of solubilizing dyes for both wool and hair are amyl, benzyl, butyl, furfuryl, and phenylethyl alcohols, cyclohexanol, dioxane, ethyl acetate and formate, and the monobutyl ethers of ethylene and diethylene glycol. The quantities used varied from 1 to 10% (16).

In these experiments the chosen dye was dissolved in the solvent. All were effective, especially benzyl alcohol, which caused a 2 to 4% increase in the rate of dyeing. The color penetrated the hair rapidly but it also stained the skin.

The use of these organic solvents makes available many dyes that formerly would not have been considered possible. Two types of dyes that are helped by solvents are the following:

Disperse dyes. The disperse dyes are a group of azo, anthraquinone, and nitro dyes, insoluble in water, used for dyeing cellulose acetate, nylon, and other synthetic fibers. They are sold as mixtures of insoluble dyes and especially selected surfactants, and they are applied as aqueous suspensions of finely divided powders which dissolve in the fibers.

When first proposed as a coloring for human hair (1953), the dye solution was heated on the head by a cap, but this was not practical or popular. Sulfated wetting agents and other solubilizing surfactants, in either solution or cream form, are helpful adjuvants for insoluble dyes. Another combination has, e.g., polyoxyethylene lauryl ether, urea, and a fatty alcohol. The addition of 3% of benzyl alcohol greatly increased substantivity.

Metallic complexes. These "premetallized" dyes are metallic complexes of azo dyes, used for dyeing wool and polyamide fibers. As they contain chelating groups, which satisfy all the valences of the metallic ions, they show no characteristics of metals but still furnish the advantages of mordants. The dyes take well on the hair, and the chelate structure ensures good stability to light and oxidation. As azo dyes they can offer the whole range of colors, but the presence of the metal dulls them somewhat (16,39,40).

The so-called *chrome dyes* are azos with acid and phenol groups. The metal most commonly used is chromium (sodium dichromate, hence "chrome dyes"), but aluminum, bismuth, iron, cobalt, and nickel have also served (49).

It has been suggested that the application of these metallic chelates should be in two steps: first to immerse the hair in the metallic solution and then in the chelate. In days when everything strives for more speed and efficiency, such a treatment would not be popular either in beauty shops or in the home (44).

The pioneer research of Peters and Stevens (45) showed that many premetallized dyes formerly used only for wool can also be applied successfully to human hair when benzyl alcohol is added to the dyebath.

The same types of dyes can be effectively utilized on hair in practically any condition by the addition of tetraethyl urea to the dyebath (50).

Anion-Cation Complexes

The value of a complex of anionic with cationic compounds was discovered accidentally through adding a so-called conditioner intended to improve a product. These complexes can be made in several ways: anionic (acid)dyes, reacted with cationic surfactants; cationic (basic) dyes, reacted with anionic agents; also reaction products of both together. They can be used for textiles and for many types of cosmetic products besides hair colorings.

The combination of ingredients must be most carefully balanced to ensure maximum substantivity. If the complex is not completely redissolved in the nonionic compound included for this purpose, it may form a heavy coating on the hair, which will dull it and rub off.

These complexes are especially good for rinses for gray hair. They can be so formulated that they can be completely removed in one shampoo (16,39, 51).

Reactive Dyes

Under investigation in Europe since 1949, the reactive dyes have only recently become of interest elsewhere. They represent another idea taken over from the textile industry, in this case the dyeing of cellulose fibers.

A reactive dyestuff is one that colors fibers chemically through covalent bonding between fibers and dye. The group of dyes comprises monoazo dyes, azo metal complexes, polysulfites, derivatives of anthraquinone and phthalocyanine. They must be readily soluble in water. The reactive groups with which they combine may be acrylamide, chlorinated pyrimidine or triazinyl, or vinyl sulfones.

These reactive dyes have proved useful on animal fibers—wool, silk, leather, and hair. In substances with —S·SO$_3$H, the —S·S— bond can be split by a reducing agent and the dye will be taken up by the keratin. The possibilities here suggest a combination of dyeing and permanent waving (41,52,53).

Aminoanthraquinone Dyes

A worthy group of basic colors has been prepared from aminoanthraquinone as tertiary and especially quaternary derivatives by substituting an alkyl group for a hydrogen in the amino group. They take very well to keratin fibers, are stable to light, and offer a variety of shades (48,54).

Also widely used in semipermanent dyes for the hair are 1,4-diaminoanthraquinone and its derivatives, which are insoluble in water but soluble in organic solvents.

It has been found that amine oxides, e.g., dimethyl dodecylamine oxide, make good ingredients of the direct cationic semipermanent dyes. The resulting colors are more intense than those produced with sulfates, and the mixture both foams and cleanses well (55).

Permanent Dyes
(Oxidation Dyes)

As mentioned before, to those who dye hair professionally, the best-known synthetic organic dyes are the *oxidation* dyes, popularly called peroxide dyes. Because they are easily applied, produce natural-looking shades, and have no adverse effect on normal hair, they have entirely superseded the hennas in beauty shops and seem to be increasingly popular for use at home. Essentially the same compounds and compositions are available under many trade names; all are applied cold, and most depend for development of the shade on hydrogen peroxide or some other noncontaminating compound that readily liberates oxygen.

Research on all products of this type has been intensive and prolific in recent years. There is little resemblance to the products and treatments of earlier years in the bewildering profusion of products now offered both to beauty shops and to consumers. The following historical review, though possibly of only academic interest, has been retained because it is accurate as to names and dates, and it shows how far this branch of the industry has advanced.

Pyrogallol

The first synthetic organic dye to be used on human hair was pyrogallol. The effectiveness of many of the natural substances used to produce brown shades is due to the pyrogallol, $C_6H_3(OH)_3$, 1,2,3,-trihydroxybenzene (mistakenly called "pyrogallic acid"), which is either present or developed in one or another of the ingredients.

The existence of pyrogallol was first observed by Scheele in 1786, and it was isolated and identified by Bracconot in 1832. Use of pyrogallol for dyeing the hair—that is, the synthetic compound, not as nutgalls or walnut shells—was first suggested in 1845 by one Wimmer, and it was long a common ingredient of vegetable mixtures and also a developer for metallic dyes.

Used alone, pyrogallol acts as a progressive dye. A freshly prepared aqueous solution gradually darkens when exposed to the air; the addition of dilute ammonia hastens the development of the color. Until comparatively recent times pyragollol was often mixed with henna, to take advantage of the strong dyeing property of the latter but produce brown shades. Without still other ingredients to fix the color, these so-called "vegetable packs" are usually unsatisfactory, because the brownish tinge soon wears off, leaving the characteristic red shade of the henna. (See section on plant derivatives.)

The earliest reports of irritation of the skin and toxic effects when it was taken internally seem to have been published in 1880, and these eventually led to the restricting of pyrogallol as a hair dye in several European countries. Various authorities seem to disagree on this point. The 0.5% solution suggested as safe is often ineffective as a developer, whereas the more concentrated solutions commonly used in dyeing processes in the trade seem to cause no untoward results. Pyrogallol is frequently listed among substances that are harmful when and if they are absorbed by the skin; yet contents of 4 to 10% are found in recipes for many accepted antiseptic and medicinal products (25). It has been strongly hinted that all the early laws should be reviewed in the light of modern knowledge and experience, in order that many of them might be rescinded or amended.

The toxic effects of pyrogallol are said to be lessened by sulfonating it; and the fleeting brown shade may be made more lasting by the addition of an alkaline sulfide, either in the dye solution itself or as a preliminary treatment of the hair. Many of the standard works of reference offer recipes containing pyrogallol—alone, with plant derivatives, and as a developer for metallic dyes (4,5,23,56–60).

In modern cosmetic chemistry, pyrogallol serves principally as a modifier in certain formulations for permanent hair dyes.

Introduction of Amino Dyes

The first of the amino dyes to be used on human hair was phenylenediamine [$C_6H_4(NH_2)_2$]. The first compound of the kind—now known as the *meta* compound—had been discovered in 1854 by A. W. Hofmann, then working in London, during a long program of research on polyammonias. He obtained it by the reduction of nitroaniline, and predicted the possibility of an isomeric compound. This prediction was fulfilled several years later (1861), and the first complete description of *para-* (then called β-) phenylenediamine was published by Hofmann and J. S. Muspratt in 1863 (61). When in 1872 Peter Griess discovered a third isomer, now called the *ortho* compound, he made a complete study of the three phenylenediamines so that they could be readily distinguished and identified (62).

The possibilities of *p*-phenylenediamine as a dye for hair, fur, and feathers were first made known in 1883, when a patent was granted to P. Monnet et Cie., of Paris (63). Continuing research and prolific discovery made a great number of related compounds available, and in 1888 the first of several German patents was granted, to E. (and/or H.) Erdmann, for the use of diamines, aminophenols, and related compounds in the dyeing of fur, hair, and feathers (64).

For the dyeing of living hair, the earliest method of applying *p*-phenylenediamine was in two steps: (*a*) Wet the hair with a 1 to 3 % solution of the dye, made alkaline with caustic soda, sodium carbonate, or ammonia water; (*b*) apply oxidizing solution to bring out color. Many familiar oxidizing agents were used: ferric chloride, potassium dichromate and permanganate, and hydrogen peroxide. It was soon found, however, that the best results were obtained with ammonia as the alkalizer and hydrogen peroxide as the developer, and by mixing the two solutions immediately before use. *p*-Phenylenediamine became popular as a hair dye because it could produce a good, natural-looking, lustrous black, in contrast to the dull, harsh, dusty, greenish or iridescent blacks produced by the prevalent metallic dyes. *o*-Aminophenol and *p*-aminophenol (or their hydrochlorides) were soon adopted for, respectively, orange-red and medium brown shades; and it was found that by varying concentrations and mixing these basic colors several other shades could be produced.

Early Commercial Development

By the mid-1890's, many proprietary products were available, both for the retail trade in France and for export. Through the activities of these manufacturers and the migration of hairdressers, the amino dyes were introduced into England, other European countries, and North and South America

(65).* For many years, however, most hairdressers prepared their own solutions, mixing colors to meet the needs of individual patrons.

In 1910 the first standardized product (called *Inecto*), accompanied by workable instructions for use, was offered in a range of 11 controlled, reproducible shades. This was the work of Gaston Boudou, a well-known hairdresser with salons in London and Paris, with the cooperation of Emile Rousseau, a chemist and pharmacist, and Raimond Sabouraud, an experienced dermatologist. The immediate success of this product led in 1917 to an increase in the number of available shades to 18, which fairly well simulated the average shades of human hair (see Table II). The time for a treatment was cut by the addition of a little soap and ammonia to the solution of dye (66).

To designate his series of 18 shades, Boudou numbered them from 1 to 9, in order from black to lightest blond, using odd fractions for modifications of the original numbers. Originally in French, they were so poorly translated that many of the descriptive names did not accurately describe the respective shade produced by a given number (67). For many years, nevertheless, as competitive proprietary lines of amino dyes were introduced, these names were perpetuated—with or without the same numbering system—as if there were no other way to designate such a line.

The 18 standardized shades which served satisfactorily for about 25 years have been augmented to 30, 40, and more in some lines. In others, the number has been reduced to about a dozen average shades which can be modified to blend with intermediate tones. The descriptive names, however, were long ago abandoned for fancy words that roughly designate a shade and certainly tax the ingenuity of a promotion department.

Boudou's line of oxidation dyes (then called *Inecto Rapid*), merchandised through an English company, was introduced into the United States in 1919. The type of dye was not new here (65), but the new line was immediately successful, both because of the superiority of the products and because hairdressers and public were reacting against the compound hennas then in vogue.

Quest for Substitutes; Protective Measures

The voiding of his German patents (1892–1893) turned Erdmann against the promotion of *p*-phenylenediamine as a hair dye and prompted him to undertake intensive research for substitutes (5). He worked out a good code of safety regulations for workers in the fur dye industry, but supported a

---

* The first hairdresser to use oxidation dyes in the United States was A. F. Godefroy, who brought them back from Paris in 1892 to the salon he had established in 1882 in St. Louis. P. V. Sénégas introduced them in New York in 1894.

concerted effort on the part of physicians and scientists to have p-phenylenediamine banned as a hair dye (26). Because of accumulating reports on dermatitis from hair dyes and serious toxicity as shown in pharmacological tests on animals, the official pronouncement banning p-phenylenediamine from unrestricted retail sale in Germany and including it in the official list of poisons was made in 1906 (68). This does not mean, as is so often stated in the literature, that the use of p-phenylenediamine was forbidden in Germany. The wording of the section is as follows:

Paraphenylenediamine, its salts, solutions, and derivatives must be withdrawn from unrestricted sale, and placed in Division 3 of the official list of poisons.

Henceforth all receptacles containing paraphenylenediamine and destined for the retail trade must be marked with the label "POISON" in red letters on a white ground, and give a statement of the contents.

Division 3, as listed in supplements, comprises the following:

Compounds of antimony, barium, cadmium, copper, iodine, potassium, hydrofluoric acid, gold, cresol, phenol, ipecachuana, caffein, colocynth, creosote, lobelia, sea onions, ergotin, mercury, sodium, phenacetin, paraphenylenediamine, picric acid, silver, tin, and zinc.

This ordinance did discourage German manufacturers from using p-phenylenediamine in their hair dyes, but it did not prevent the importing of products from France and Austria, many of which were labeled and advertised as "pure" or "only" vegetable.

By the time this regulation was functioning, Erdmann had successfully introduced two new types of dyes for both hairdressers and the retail trade (26). A few years later (1911) another type of dye, a mixture of p-tolylenediamine, $CH_3C_6H_3(NH_2)_2$, and sulfites, was introduced by Colman and Löwy and widely publicized (69).

Meanwhile, use of the amino dyes had continued unabated in France, and the French physicians who reported increasing numbers of "accidents" among those that had used p-phenylenediamine as hair dye were the first to recognize any connection between dermatitis or other unpleasant effects and a predisposition, personal susceptibility, idiosyncrasy, or (now) allergy in the individual (70–72).

In 1911, following years of agitation by physicians, p-phenylenediamine was officially restricted in France. In the same year, however, Sabouraud, whose increasing clinical experience had convinced him of the importance of personal susceptibility of allergy in cases of dermatitis from the amino hair dyes, succeeded in publicizing the necessity for a preliminary test by those that wished to use these dyes. This "skin test" had been known and used for many years (4,65), but what became the standard test for susceptibility was introduced in 1895 by Josef Jadassohn, a dermatologist of Breslau, Germany. Considering the large numbers of application of hair dyes, Sabouraud believed the percentage of susceptible persons to be relatively small, but in order to protect them his directions for making the preliminary tests were disseminated widely both in scientific journals and in the public press (73,74).

Research for Improved Products and Processes

Within a short time after the new line of oxidation dyes was introduced, cases of dermatitis were reported from all over the United States. The American company immediately took steps to protect users of their products and to initiate intensive research toward improving both compositions and methods of application. The pioneer in all this work was Ralph L. Evans of New York, then an instructor on the staff of Marston T. Bogert, in the Department of Chemistry at Columbia University. He first modified the compositions by addition of acetone sodium bisulfite (75) to inhibit the irritating effects of the p-phenylenediamine dyes; and then, in 1924, he introduced a new line of colorings based on p-tolylenediamine in place of this ingredient. He standardized methods for applying the oxidation dyes, including requirements for a preliminary shampoo and pretreatment with hydrogen peroxide and ammonia to make the hair more receptive to the dyes. Among other improvements were the establishing of strict specifications for raw materials and for hydrogen peroxide (5 to 6%, 17 to 20 vol) and the development of products and treatments for removing old dyes so that prospective users of amino dyes could change from the metallic varieties with the minimum of damage to their hair (76).

In 1926 Dr. Evans' company established the first center for formal instruction on hair, hair dyes, and dyeing methods, and a staff of well-trained technicians who carried this instruction to all parts of the country, through public and private demonstrations to the trade and in schools of general beauty culture.

Advantages and Disadvantages of Oxidation Dyes

Even in the earliest days, with all the impure materials, caustic oxidizing agents, relatively crude shades, and uncertain methods of application, it was obvious that the effects of dyeing hair with the new organic products were better than those obtainable with earlier dyes. The colorings penetrated the hair shaft instead of coating it, and when the excess was washed away, the hair retained its natural luster and was not adversely affected by curling irons. As compositions were improved and the range of shades was extended beyond those of black poodle, red fox, and brown bear, it became obvious that any natural shade of hair, from palest blond to deepest black, can be duplicated with the oxidation dyes. For anyone with the requisite knowledge and technical skill they are easily applied.

It would seem, therefore, that the amino dyes meet the requirements of the "ideal dye" except for the one disadvantage of their irritating effects on the skin of allergic individuals. Although the importance of personal susceptibility and the benign nature of the condition in most cases were recognized in the earliest reports (70,71), it was many years before it was recognized that "hair dye poisoning" is simply contact dermatitis (*dermatitis venenata*). The amino dyes should thus be classified with poison ivy, and several other plants, chemicals, and other dermatitis-producing allergens (77–80).

Reports of death and of serious internal disturbances from *p*-phenylenediamine hair dye can usually be traced to medical writers who were not themselves investigators and who mistook early reports of troubles in the fur industry, or pharmacological studies on animals, as general truth. The bulk of adverse medical literature on *p*-phenylenediamine alone is impressive, but little of it is of value at this stage in the development of the whole broad field of oxidation dyes in cosmetics and cosmetology. Searches in the literature are now best limited to recent publications by physicians and scientists who have both clinical experience and first-hand knowledge of contemporary procedures in the manufacture and application of this type of product.

European products introduced to the American market are invariably based on *p*-tolylenediamine, because of the restrictions against *p*-phenylenediamine in Europe. The former base has been found to offer no advantages over the latter, and as the regulations apply equally to all, many of the imported products have been changed accordingly.

Composition of Modern Oxidation Dyes

During the past decade many articles and patents have been published on the composition and reactions of the modern oxidation dyes. All the good dyes in current use are proprietary products. The recipes in newer books and articles show the trends in composition and in research in science, medicine, and technology. Patents divulge the acceptable minimum; otherwise all technical knowledge is carefully guarded (81–85).

Far from being the simple dye-alkali-oxidizer of the early days, the modern dye of this type also contains color modifiers, antioxidants, stabilizers, and other adjuvants. It must be so compounded that it will not only color the hair but also cleanse and recondition it, and "take" properly with controlled development of the desired shade. Anyone that wishes to do research on hair dyes therefore must have a thorough knowledge of hair (including its characteristics and its typical reactions), a good eye for color, and familiarity with many other types of contributory substances and compounds.

Essential points to be considered include the following (83):

1. Shade to be produced.
2. Properties of dyebath: color, pH, viscosity, stability.
3. Action of dye on hair: color, rate of development, bleaching, fastness, staining, levelness.

Dyes (intermediates). The compounds used as bases for the synthetic organic hair colorings are more properly called *intermediates*, because their actual dyeing properties are developed only on oxidation.

Only relatively few of the long list of currently available amino and related compounds are suitable for use on human hair under practicable conditions. These are, in general, the *o-* and *p-*diamines and amino hydroxy compounds of benzene, certain homologues, and related coal tar derivatives, which pass through a quinoid stage during oxidation. A few other nitro, hydroxy, and alkyl derivatives may also be used, but many nitro and alkyl compounds with desirable dyeing properties are automatically barred from hair dyes because they are known to irritate the skin. (Most nitro compounds are not oxidized.)

The typical reaction for *p-*phenylenediamine has been presented in great detail in many works of reference (86,87). The assumed conversion from the simple amine through a transitory imino compound to Bandrowski's base, and the actual nature and function of the latter, are still being studied (88–91).

Table IV, compiled from many sources, offers suggestions, from which any variety of combinations and modifications can be assembled.

It has been suggested that no one really needs all the intermediates shown in Table IV, but just a few choice ones, sorted out for the primary colors, as shown in Table V (85).

To become familiar with the basic shades obtainable with the acceptable intermediates, the investigator should make up 1 to 3 % solutions (1:1 water/alcohol) of various compounds and dye small samples of (preferably) light gray natural hair. To simulate further the actual conditions of application, sufficient ammonium hydroxide should be added to each solution to adjust the pH between 9 and 10; and for each test the solution of intermediate should be mixed with an equal quantity of 6 %, 20-vol hydrogen peroxide at moment of use. The development of the shade should be noted at intervals until 1 hr has passed. To stop it at any time, the sample should be well rinsed with hot water.

Once the possibilities of the single dyes have been ascertained, various shades can be prepared by altering the concentration of the basic colors and mixing them. The number of intermediates used in any one composition should be kept at the minimum for the result desired.

TABLE IV. Suggested Compositions for Oxidation Dyes

Compound	Black	Drab brown	Warm brown	Reds	Drab blond	Gold blond	Blues, grays
p-Aminodiphenylamine	×	×	×	—	×	×	×
p-Aminodiphenylamine hydrochloride	—	—	—	—	—	—	×
p-Aminodiphenylamine–sulfonic acid	×	×	—	—	×	—	×
2-Amino-5-hydroxytoluene	—	×	×	×	×	×	—
5-Amino-2-hydroxytoluene	—	×	×	×	×	×	—
2-Amino-4-nitrophenol	—	—	×	×	—	×	—
4-Amino-2-nitrophenol	—	—	—	×	—	×	—
o-Aminophenol	—	×	×	×	×	×	—
m-Aminophenol	—	—	—	—	—	×	—
p-Aminophenol	—	×	×	×	×	×	—
p-Aminophenol hydrochloride	—	—	×	—	×	—	—
p-Aminophenol sulfate	—	—	—	—	—	×	—
2-Aminophenol-4-sulfonic acid	—	—	—	—	—	×	—
4-Aminophenol-2-sulfonic acid	—	—	—	—	—	×	—
N-(p-Aminophenyl)glycine	—	—	×	—	—	—	×
o-Anisidine	—	×	—	—	—	—	—
p,p'-Diaminodiphenyl amine	×	×	—	—	—	—	×
p,p'-Diaminodiphenyl amine sulfonate	×	×	—	—	×	—	×
p,p'-Diaminodiphenyl methane	—	—	—	—	—	—	×
1,8-Diaminonaphthalene	×	—	—	—	—	—	×
2,4-Diaminophenol	—	×	×	×	—	—	—
2,4-Diaminophenol hydrochloride	×	×	—	—	—	—	—
2,5-Diaminophenol-4–sulfonic acid	×	—	×	—	—	×	—
3,4-Diaminotoluene	—	×	×	×	×	×	—
N,N-Di-sec-butyl-o-phenylenediamine	—	×	×	×	×	×	—
N,N-Dimethyl-p-phenylenediamine	—	×	—	—	—	—	—
4,6-Dinitro-2-aminophenol	—	—	—	×	—	×	—
5-Nitro-o-aminophenol	—	—	—	×	—	×	—
2-Nitro-p-aminophenol	—	—	—	×	—	×	—
N-(p-Hydroxyphenyl)glycine	—	×	—	—	×	—	—
N-(2-Hydroxy-5-nitrophenyl)glycine	—	—	—	—	—	×	—
p-Methylaminophenol sulfate	—	×	×	×	—	×	×
1,5-Naphthalenediol	—	—	—	—	—	—	×
4-Nitro-o-phenylenediamine	—	×	×	×	×	×	—
2-Nitro-p-phenylenediamine	—	×	×	×	—	—	—
N-(p-Nitrophenyl)glycine	—	—	—	—	—	×	—
o-Phenylenediamine	—	—	—	—	×	—	—
m-Phenylenediamine	—	×	×	—	—	×	×
m-Phenylenediamine hydrochloride	—	—	—	—	—	×	—
p-Phenylenediamine	×	×	×	×	×	×	×
p-Phenylenediamine hydrochloride	—	—	×	—	—	—	—
p-Phenylenediamine sulfate	—	—	×	—	×	—	—
m-Tolylenediamine	—	×	—	—	×	—	×
p-Tolylenediamine	×	×	×	—	×	—	×
p-Tolylenediamine sulfate	×	×	×	—	×	—	×
2,4,6-Trinitroaniline	—	—	—	×	—	—	—

TABLE V.   Essentials for Oxidation Dyes (85)

Red	Yellow	Blue
Nitro-p-phenylenediamine	o-Aminophenol	(No single compound)
p-Aminophenol +	5-Nitro-2-aminophenol	p-Phenylenediamine OR
$\quad$ m-Phenylenediamine	o-Phenylenediamine	2,5-Tolylenediamine
m-Tolylenediamine	4-Nitro-2-aminophenol	$\quad$ sulfate +
2,4-Diaminoanisole	Nitro-m-phenylenediamine	$\quad$ m-Phenyl-enediamine
1,5-Naphthalenediol		$\quad$ m-Tolylenediamine
α-Naphthol		4-Methoxy-6-methyl-
		$\quad$ m-phenylenediamine
		2,4-Diaminoanisole sul-
		$\quad$ fate
		α-Naphthol
		p-Aminodiphenylamine +
		$\quad$ m-Aminophenol
		$\quad$ m-Phenylenediamine
		$\quad$ m-Tolylenediamine

For reproducibility and uniformity of results, the intermediates used in any one shade should always be the same, that is, from the same supplier and of the same quality. Comparative studies, made with various grades of the same compound from technical to chemically pure, may show considerable variation in the results (81). Whatever the quality, the products currently obtainable are infinitely better than those that were available even a quarter of a century ago.

For correct control, all solutions of single intermediates and mixtures should be tested on samples of clean hair, both untreated and after a preliminary bleaching and softening with 20-vol hydrogen peroxide (1 fl oz) and ammonium hydroxide (28%, 0.880; 15 to 20 drops). The effective time for this preliminary treatment may be from 15 to 30 min, depending on the texture of the hair. In the compounding of blond shades, this pretreatment is a most important consideration.

For highlights in certain shades, direct dyes may occasionally be included in the basic mixture.

Modifiers. In addition to the direct color-forming intermediates, other related compounds are commonly included in dye compositions because they serve to modify or stabilize certain shades. Most useful among these are metadiamines and several polyphenols, listed in Table VI. Many other compounds have been suggested in various patents; for example, α- and β-naphthols, diaminoanisole and -phenetole. The naphthols impart a purple or mahogany cast to brown shades; the others bring out a bluish tinge.

TABLE VI.   Modifiers for Oxidation Dyes

Compound	Drab Black	Warm brown	brown	Red	Drab blond	Gold blond	Blue gray
2,4-Diaminoanisole	—	×	—	×	×	×	×
2,4-Diaminoanisole sulfate	—	—	—	—	—	—	×
2,4-Diaminophenetole	—	—	—	—	—	—	×
Hydroquinone	—	—	×	×	—	×	—
Hydroxyhydroquinol	—	—	—	—	×	×	—
m-Phenylenediamine	—	—	—	—	—	—	×
Phloroglycinol	—	—	—	—	×	×	—
Pyrocatechol	—	—	—	×	—	—	×
Pyrogallol	—	×	×	×	—	×	×
Resorcinol	—	×	×	×	×	×	—
m-Tolylenediamine	—	—	—	—	—	—	×
Xylenol	—	—	—	—	—	—	×

The compounder must be careful in the use of modifiers, because the different amounts of time required for complete oxidation of various intermediates may cause the development of transitory greenish, violet, or other "off-shades." All such factors must be considered in preparing instructions for use of a product, so that sufficient time for complete development of color will always be allowed.

Modifiers are essential in all compositions intended to produce the currently popular grayish tones (called variously silver, smoke, steel, etc.). When used as drabbing agents in normally brownish or golden mixtures, the dull effect produced is directly proportional to the amount used. One disadvantage in the addition of the bluing modifiers is that some of them are not so fast to light as the basic intermediates are, and the drabbed tinge may fade out of the dyed hair (81).

The hydrochlorides, sulfates, and sulfonic acid derivatives of many of the diamines and aminophenols also serve as modifiers in certain compositions. Originally introduced in the belief that they diminish toxicity (69,75), the sulfonic compounds were found to inhibit dyeing properties, so they are now used more commonly in the lighter shades.

Antioxidants. Because of the readiness with which the amino dyes oxidize and darken even in the air, all manufacturing processes must be carefully guarded against access of air. This may be avoided by maintaining an atmosphere of nitrogen during compounding and filling, but the simplest precaution is the addition of a chemical antioxidant.

The most generally used preservative is now sodium sulfite, the same compound that was proposed as a protection against toxicity in some of the early compositions. The amount to be included is calculated on the total amount of oxidizable intermediates; hence more of the sodium sulfite should be added to composition for dark shades than for light ones.

Ascorbic acid, 0.2 to 0.5 %, in the dye solution, has been suggested as an effective antioxidant. It is said to permit capping and reuse of material in a bottle (92). Thioglycolic acid has also been suggested (93).

Alkalizers. Very early in the application of the amino dyes it was found that they work most efficiently in an alkaline medium. Erdmann's solution of *p*-phenylenediamine (2 %) and sodium hydroxide (1.4 %), although it was satisfactory for fur, feathers, and dead hair, was admitted to be risky—if not dangerous—for living human hair and skin. In one of the dye compositions based on substitutes for *p*-phenylenediamine, the same investigator used 1.5 % sodium carbonate.

Ammonium hydroxide has for many years been considered the best alkali for dye compositions because it helps efficiently in softening the hair and leaves no foreign residue as the fixed alkalies do. The amount to be added to the dye solution must be carefully controlled so that it will not exceed 1 to 2 % in the final mixture that is to be applied to the hair. That is, it may be 3 to 4 % in compositions that are to be mixed with an equal volume of liquid developer; but if urea peroxide tablets are to be added directly to the dye solution, only 1 to 2 % of ammonium hydroxide should be present.

The *p*H of most compositions of amino dyes is 9 to 10, although it may be a little lower in certain shades from which reddish glints are undesirable. It is impossible to generalize on this point because the *p*H must be adjusted for each ingredient of every mixture.

Alkanolamines have been used in dye compositions, with or without ammonia, and also glycerol or some other polyhydroxy compound (94). The use of all such compounds must be carefully checked and controlled, lest they occlude some of the active ingredients or inhibit complete penetration of the hair.

Other adjuvants. For various reasons, it has become common manufacturing practice to "improve" a good dye solution by adding one or more substances with special properties. Small amounts of soaps, ammonia, and certain wetting agents will help dyes to penetrate the hair and "take" more quickly.

Everything to be added for any other purpose, such as a sequestering, foaming, or thickening agent, should be most carefully checked, both as to quality and as to quantity, to ensure that it is compatible and effective for the purpose intended. Sulfonated oils, for example, strip color from both natural and dyed hair; therefore they must be used with caution. The alkylated aryl sulfonates are adsorbed on the hair and consequently may prevent adequate penetration of the hair for complete coverage or full development of the shade. Certain other surface-active chemicals make the hair feel harsh (almost metallic) and difficult to handle. The amount of adjuvant to be added should be kept at the minimum, because even a slight excess of some substances may cause the dye to diffuse through themselves and not affect the hair at all.

The use of organic catalysts, such as enzymes, to control the hydrogen peroxide developer, or for any other alleged benefit, is of questionable value.

In the prevailing preference for dyes that will not run on the hair, the effects of all types of surfactants, chelating agents, fatty acids, alcohols, amines, and related compounds should be checked and not be indiscriminately added to a basic mixture (95).

Details of developments in this phase of research and activity on hair dyes should be sought in the appropriate patents and technical journals, and from the suppliers of raw materials.

Formulas 8 to 15, compiled from various sources, offer some typical compositions for different shades.

Developer. Although all the amino or oxidation dyes darken on exposure to light and air, a developer is always used with permanent dyes to bring the time required for the oxidizing process within reasonable limits. Of all the oxidizing agents that have been used since the early days, only hydrogen peroxide has been consistently used to the present time.

Under normal conditions, hydrogen peroxide in a solution of 5 to 6%, generating 20 vol of oxygen, suffices for the usual procedures in hair dyeing. Even on normal hair, the use of peroxide of 25 or 30 vol, or higher, may prove deleterious. It is even more damaging to hair that is not in good condition.

Most products offer the developer as a clear, cloudy, or creamy fluid in a relatively large bottle into which the dye is to be poured for mixing. Certain others require that the peroxide, in whatever form, be bought separately.

The tablets of urea peroxide, more common in European than in American products, are available in various sizes and shapes. They may be individually wrapped in plastic or metal foil or contained in a small vial. The minimum number to be included in the package should be determined by tests for proper development of the dye (see section on *solid hydrogen peroxide* for special precautions to be observed).

Melamine peroxide or other "nonperoxide" powdered developers may be packaged in envelopes (19).

For most efficient action of a dye, the intermediates and the developer should be mixed at the moment of use and applied to the hair as quickly as possible. It is an advantage to have as much as possible of the oxidation process develop right on the hair. Normally it starts rapidly and continues slowly, which fortunately equalizes the development of color in spite of the time required for applying the dye. This is another argument against the

Formulas for Oxidation Dyes

	8	9	10	11	12	13	14	15
	Black	Dark brown	Medium brown	Medium red	Ash blond	Golden blond	Blue violet	Steel gray
Intermediates								
p-Aminodiphenylamine	0.05%	0.05%	0.20%	—	0.005%	0.10%	—	—
4-Amino-2-nitrophenol	—	0.10	—	1.50%	—	—	—	—
o-Aminophenol	—	—	0.20	—	0.100	0.13	—	—
p-Aminophenol	—	0.21	0.20	1.00	0.025	—	0.90%	—
1,5-Dihydroxynaphthalene	0.06	—	—	—	—	—	0.80	0.40%
p-Methylaminophenol sulfate	—	—	0.15	—	—	0.40	—	0.25
4-Nitro-o-phenylenediamine	—	—	0.15	0.50	—	—	—	—
2-Nitro-p-phenylenediamine	—	0.10	0.15	2.50	0.065	—	—	—
p-Phenylenediamine	3.75	2.10	2.00	1.00	0.150	0.20	0.55	0.60
Fusion mixture No. 1*	0.80	0.20	—	—	0.020	—	—	—
Fusion mixture No. 2†	0.05	—	—	0.15	0.110	—	—	—
Modifiers								
2,4-Diaminoanisole sulfate	—	—	—	—	—	—	0.25	0.55
Hydroquinone	—	—	—	—	0.007	—	—	—
Pyrocatechol	—	—	—	—	—	0.20	0.80	—
Pyrogallol	—	0.20	—	2.00	0.500	0.50	—	2.50
Resorcinol	—	1.40	1.00	1.30	0.130	0.20	—	—
Antioxidant								
Sodium sulfite	0.20	0.20	0.40	0.80	0.050	0.20	0.30	0.60

* 2,4-Diaminoanisole,1; p-phenylenediamine,2.
† 4-Nitro-o-phenylenediamine,1; 2-nitro-p-phenylenediamine,1.

use of extra-strong peroxide, which generates heat in the dye mixture and may cause too rapid oxidation in the dish or on the head. Studies of the chemical reactions in the oxidation of the amino intermediates have proved that mixed solutions lose their efficacy after standing for any length of time (81).

Forms of Oxidation Dyes

The amino or oxidation dyes have been predominantly products intended for application by experienced and skillful hairdressers, but more and more of them are also being sold directly to consumers. Regardless of the physical form in which these dyes are offered to the trade and to the public, the underlying principle in all is the same. The materials for any single application are packaged in two units, one containing the dye, and the other containing the developer. The latter may be either a bottle of clear or creamy stabilized hydrogen peroxide, or a well-protected, measured quantity of urea peroxide in tables or powder.

The clear liquids. The oldest form of the modern successful dyes, which remain thin clear liquids when mixed with the developer, from the market. They permit good control of development of the shade, but this is outweighed by the requirements for preliminary bleaching of the hair and for considerable skill to prevent overlapping of color during a retouch. They have been superseded in popularity by the following.

Shampoo tints. Theoretically a combination of coloring and shampoo, these dyes originated in Germany and were introduced into the United States from France and England in the early 1930's. The first products were thickened with a simple fatty acid or (usually) a synthetic soap, and the (usually) dry oxidizer was added either directly or as a freshly prepared solution. This type of product offered few, if any, advantages over the clear ones. The techniques of application were exactly the same; that is, the hair had to be shampooed and pretreated with hydrogen peroxide before the dye mixture was applied. The viscous liquid tended to dry quickly, especially around the hair line, and the dark crusting obscured and prevented the development of the proper shade.

In none of the early products of this type were the detergent ingredients sufficient for a satisfactory final cleansing of the hair after it was colored. Even now, most of the assorted surface-active agents, oils, and other materials, which have been added to various compositions in more recent developments, have not completely obviated the necessity for a good shampoo following the treatment with many so-called shampoo tints.

The basis of shampoo tints may be a soap or a synthetic (anionic or nonionic) detergent. The ammonium or ethanolamine compounds of various fatty acids (coconut oil, lauric, myristic, oleic, palmitic fatty acids) make satisfactory bases. Small quantities of surface-active agents may be added if it is desired to enhance the wetting, foaming, conditioning, or other special properties of the product, but everything of the kind should be carefully checked to ensure that it is completely compatible with the intermediates that are to be present in each individual shade of a line of dyes. The diphenylamino compounds, for example, are not readily soluble in aqueous solutions, so compositions that contain any of them should also contain a water-miscible organic solvent, such as an alkanolamine, or some other suitable solubilizer (81).

Formula 16 could be used as a base for Formulas 8 to 15.

Adjustments in the base may be necessary for different combinations of intermediates. As the developer, an equal quantity of 20-vol hydrogen peroxide should be specified.

Formula 16

Solvent for intermediates

Ammonium hydroxide (26°Bé)	10.00%
Isopropanol	2.50
Perfume (e.g., lavender) to	0.50

Base

Oleic acid	35.00
Polyoxyethylene sorbitan monooleate	10.00
Nonionic surfactant	3.50
Water-soluble lanolin	1.75
Lecithin	1.25
Chelating agent(s)	0.25
Water to make	100.00

Procedure: In compounding any product, all the ingredients shown should be added to the solvent mixture. This serves also as an emulsifier for the base. The ingredients of the latter should be combined, and added to the dissolved intermediates, cold. Any heavy-duty apparatus may be used; agitation should be steady but slow. The *p*H of the final mixture should be adjusted at 9.0 to 9.2.

In compounding a shampoo tint, the viscosity of both the dye base and the mixture of base and developer must be taken into consideration. The base must flow freely from its bottle so that there will be the minimum of residual waste, and the mixture must still be sufficiently viscous to adhere to the hair and give the impression of soapiness or oiliness.

Cream dyes. This form of amino dyes was introduced from Germany in 1950. The dye was first incorporated in a paste (or, later, a softer creamy base) with surface-active agents, essentially like those used in cream shampoos. The product is sold in tightly sealed tubes from which a measured amount is squeezed out and mixed with hydrogen peroxide. Because the technique for mixing and applying these products demands a good eye for color and a high degree of technical skill, they are best planned for use only in beauty shops.

The cream dyes have been much more popular in Europe than in the United States. They offer the advantage that whatever is applied stays wherever it is put, but also the disadvantage that the thick deposit obscures the development of the color.

The composition of the creamy base must be perfectly water-miscible, to ensure complete removal.

Color shampoos. The most popular form of coloring, both in the beauty salon and in the home, is the "shampoo-in" type. A measured amount of developer—clear, cloudy, or creamy—is supplied in a relatively large bottle, into which a measured amount of dye solution is poured. The top of the bottle is a nozzle which serves both to part the hair and to apply the coloring as rapidly as possible. As it must be worked into a lather all over the head, gloves must be worn.

Formulas for Oxidation Dyes*

Intermediates, modifiers	17 Black	18 Dark brown	19 Light brown	20 Auburn	21 Blond	22 Gray
p-Phenylenediamine	2.7%	0.8%	0.56%	0.08%	0.15%	0.15%
p-Aminophenol	—	0.2	0.20	—	0.20	0.15
o-Aminophenol	0.2	1.0	0.28	0.04	0.20	0.15
Nitro-p-phenylenediamine	—	—	0.04	0.40	0.04	—
2-Nitro-4-aminophenol	—	—	—	0.40	—	—
2,4-Diaminoanisole sulfate	0.4	—	—	—	0.01	0.07
Resorcinol	0.5	1.6	0.80	0.10	1.00	0.80

* Ref. 85.

The directions should specify the time of contact and should state that the application can be repeated until the desired shade is seen (see Formulas 17 to 22).

As a base for the preceding compositions, Formula 23 is recommended.

Formula 23

Oleic acid	20.0%
Oleyl alcohol	15.0
Solubilized lanolin	3.0
Propylene glycol	12.0
Isopropanol	10.0
EDTA	0.5
Sodium sulfite	0.5
Ammonium hydroxide, 28%	10.0
Deionized water	19.0

Procedure: Dissolve dye in propylene glycol and heat. Mix oleic acid and lanolin and stir dye solution into this. Dissolve oleyl alcohol in isopropanol and add to preceding. Add remaining ingredients.

Toners. The admiration and desire for blond hair are fairly widespread, but in recent years there has been a decided reaction against the brassy tones usually obtained with plain hydrogen peroxide. This has led to the development of a series of delicate pastel "toners," which successfully change the unpleasant yellow to a whole gamut of more natural-looking blond shades.

In another type of product, the vehicle is a thin, translucent cream, through which the development of shade can be observed. Most important, the time for development has been cut to about 10 min.

Most compositions for toners are dilutions of the appropriate darker shades. Examples of a few of the popular toners are shown in Formulas 24 to 30.

Formulas for Bleach-Toners*

Intermediates, modifiers	24 Platinum	25 Silver	26 Smoke	27 Platinum blond	28 Ash blond	29 Golden blond	30 Red blond
p-Phenylenediamine	0.08%	—	0.06%	—	0.10%	0.050%	0.06%
p-Tolylenediamine sulfate	—	0.100%	0.04	0.08%	—	—	—
p-Aminophenol	0.08	—	—	0.04	—	—	—
2-Nitro-4-phenylene-diamine	—	—	—	—	—	0.008	0.02
2-Nitro-4-amino-phenol	—	0.002	—	0.02	0.04	0.200	—
6-Chloro-4-nitro-phenol	—	—	—	—	—	0.040	—
4-Nitro-2-phenylene-diamine	—	—	—	—	—	—	0.03
2,4-Diaminoanisole sulfate	0.04	0.050	0.04	0.02	0.01	—	—
Pyrogallol	—	—	0.04	—	—	—	0.40
Resorcinol	0.50	0.020	0.08	0.08	0.70	0.400	—

* Ref. 85.

As a base for the preceding compositions, Formula 31 is suggested.

Coloring gels. Another popular type of product has the dye in a light oily liquid which, on addition of hydrogen peroxide, becomes the consistency of a clear thin gelatinous mass. It is sufficiently viscous to remain where it is applied, and its transparency permits watching the development of shade (16).

The base of this type of product may be a soap (e.g., ammonium oleate) or ethoxylated nonionic surfactants.

Bleach-dye combinations. In this type of product, introduced in 1950 exclusively for use at home, the intermediates were combined with a relatively high percentage of surface-active agents. The original product was developed by a strong hydrogen peroxide, said to be controlled by an enzyme catalyst. It was thus possible to avoid all preliminary treatment of the hair and shorten considerably the time of application for untrained users.

Formula 31*

Oleic acid	20.0%
Bis-2-hydroxyethyl soybean amine	9.0
Hydroxyethyl stearyl amide	6.0
Propylene glycol	12.0
Isopropanol	10.0
EDTA	0.5
Ammonium hydroxide	10.0
Sodium sulfite	0.5
Deionized water	22.0

Procedure: Blend hydroxy compounds. Dissolve dye in propylene glycol, heat, and add to preceding. Dissolve oleic acid in isopropanol and add to preceding. Add remaining ingredients.

* Ref. 95.

The paradoxical double action of bleaching and coloring was at first accomplished by an excessive amount of ammonium hydroxide in the dye and proportionately more (1 to 1.5 or 2) of the hydrogen peroxide developer. Contrary to all inherited beliefs, it became possible to dye hair a lighter shade; nevertheless, claims for the degree of this lightening effect should be restrained. With 20-vol hydrogen peroxide, varying the content of ammonium hydroxide can lighten the hair two or three shades. A 30-vol peroxide is also used, but it is not recommended, for fear of irritating the scalp. Continued research has indicated that the addition of enzymes serves no purpose.

The actual degree of lightening the shade of any head of hair depends largely on the original color of the hair and the nature of the pigment. Hair in which the granular form of pigment predominates seems to react to bleaching more readily than hair that contains much diffused pigment (see section on range of shades required). Hair dyed with certain shades of the amino dyes sometimes is extremely resistant to bleaching. All these factors must be considered and covered in instructions for the application of the bleach–dye products, especially those that are intended primarily for use by amateurs at home.

Among the compounds that have been used to impart the requisite properties—foaming, wetting, detergent, emulsifying—to the base for a bleach–dye product are alkanolamine esters and salts of fatty acids, saponin, polyoxyalkylene esters, and alkanolamides of fatty acids. Very little information has been released on the composition of these products; Formula 32 has been suggested as a satisfactory base (81).

The secret of success for this type of product is the exact proportions of ingredients. They must release just enough oxygen to ensure bleaching the

Formula 32

Triethanolamine lauryl alcohol sulfate	20 to 40%
Fatty acid alkanolamide (high)	5 to 10
Water-soluble lanolin derivative	1 to 3
Water	q.s. 100

hair while the dye is entering it, yet not too much to bleach color from the dye itself.

Miscellaneous. In addition to the more usual types described, the amino dyes have, from time to time, been offered in novel forms. In one of the oldest in the United States, the intermediates are in the form of powder in capsules, the contents of each to be added directly to a specified quantity of hydrogen peroxide. In another, the solution of intermediates is highly concentrated, and the shade produced varies according to the number of drops to be added to the developer. There also have been cakes of soap, from which the color was supposed to develop directly on hair previously treated with hydrogen peroxide (96). Dyes have also been offered as sticks (97).

There are many changes in packaging. For bleaching hair in streaks, the purchaser finds a whole set of utensils in the box—bowl, brush, gloves, perforated cap, even a crochet hook for pulling out the strands of hair. Other devices include a double dispenser for simultaneous release of dye and developer, and aerosol containers for coloring agents in spray form.

A new package for cream dyes offers three items: (*a*) tablets of dye, alkali-free filler, and binder; (*b*) tablets of urea peroxide; (*c*) vehicle, containing ammonium hydroxide, alcohol, fatty acid, and emulsifier. All are to be blended into a smooth cream. This is said to keep well in storage (41).

The future will undoubtedly bring many other innovations.

Complete dyes in powder form are not new. The intermediates are combined with melamine peroxide, with methyl cellulose as thickener and anhydrous sodium pyrophosphate as the stabilizer. Other stabilizers that can be used with melamine peroxide are monosodium phosphate, magnesium silicate, and phenacetin (19).

Regulations Governing Amino Dyes

Following the agitation in Europe, and always singling *p*-phenylenediamine from the large group of amino dyes, the medical authorities in the United States sought from time to time to have it banned from use as hair dye, or subjected to special control by the Food and Drug Administration. This was first proposed in 1924, but no action was taken. In 1926 the Health Department of the City of New York amended its Sanitary Code to prohibit the sale and distribution of cosmetic preparations containing p-phenylene-diamine, lead, and mercury. This was never effectively enforced.

In the face of multiplying competitive lines of products and consequent rapid increase in the use of amino dyes, as well as a better understanding of the trouble caused by these dyes—in which none proved to be completely blameless—some of the authorities took a more lenient attitude toward them. Ethical manufacturers had for years included in the directions for use of their products the instructions for making the preliminary test for allergy; but the injunction was directed to those that knew they were allergic to foods, drugs, or other chemical substances. For better protection of all users, the Sanitary Code of the City of New York has been amended (1959) to make the preliminary test mandatory, 24 hr before any general application of any amino dye, and to require all beauty shops where such dyes were applied to post a sign acknowledging the use of amino dyes and stating the requirement for the test.

The Federal Food, Drug, and Cosmetic Act, as enacted in 1938, specifically requires every user to make the preliminary skin test before every application of any hair dye that contains a coal-tar compound. It also specifically prohibits all use of dyes of this type on the eyebrows and eyelashes. Most of the state laws governing foods, drugs, and cosmetics include similar regulations for use of amino hair dyes; a few have extended them also to hair colorings based on the certified colors.

Responsibility in claims for damages in the now occasional cases of dermatitis from an application of hair coloring is today determined by whether or not the preliminary test was made as prescribed by law.

Research for New Dyes

After many decades of vicissitudes with the amino dyes, investigators undertook to study some of the nitrogen analogues of the benzene series, e.g., pyridine (C_6H_5N) and quinoline (O_9H_7N), and their amino and hydroxy analogues.

It was found that 2,5-diaminopyridine (analogue of p-phenylenedia-mine) can be oxidized to a beautiful red dye. Other compounds that can serve as intermediates or modifiers are 2,6-diaminopyridine, and both the 2,5- and 2,6-dihydroxypyridines (98).

Out of intensive research on the nature of the pigment in human and animal hair, the cause and process of graying hair, and related phenomena, there has developed a project for coloring the hair by actual repigmentation. Among compounds that have been studies are 1,2,4-trihydroxybenzene, and 2,4,5-trihydroxytoluene. Many patents have been issued on these ideas (99).

Plant Derivatives

The earliest known hair dyes were natural organic substances obtained from plants, used as infusions, decoctions, or packs. Many of the decoctions of woods are still utilized for the dyeing and blending of postiche (hair pieces), but of all the vegetable substances used by the ancients, only henna persists, and this has practically been abandoned by schools of cosmetology and beauty salons. The following information is given for its possible historical interest.

Henna

The first recorded user of henna as a hair dye was the Egyptian Queen Ses, mother of King Teta (Third Dynasty). The plant has always popularly been called Egyptian henna, although it is found abundantly also in Tunis, Arabia, Persia, India, and other tropical countries. It was commonly used by the inhabitants of all these countries to dye not only human hair but also the nails, the palms and soles of dancers, and the manes and tails of horses. Because all red shades of hair were out of favor for several centuries, and because the plant was obtained from foreign countries only with great difficulty, the use of henna died out in Europe. It was revived by the Spanish-born Italian singer Adelina Patti (1843–1919), after the middle of the nineteenth century, and it was introduced into the United States when the great singer made her debut in New York in 1859. The dark purplish red or so-called mahogany shade she favored remained popular for decades.

The henna plant is a shrub, *Lawsonia alba* Lam. or *Lawsonia inermir* L., similar to the familiar privet. The active coloring principle in henna is lawsone, and its chemical constitution is 2-hydroxy-1,4-naphthoquinone, $C_{10}H_6O_3$ (100).

The principal advantage offered by henna as a modern hair dye is that it is harmless to the system and causes no irritation of the skin. The disadvantages are that the shades it produces are not altogether natural. For keratin, henna acts as a substantive dye. A few applications impart a slight amount of color to the entire hair shaft, and depending on the original color of the hair, a whole range of reddish, "pure vegetable" shades—tomato, beet, eggplant—can be produced. In continued use, however, it seems to accumulate on the outside of the hair. If it is applied often enough, therefore, hair of any and all original shades can be brought to the same characteristic orange-red (1).

To modify the often unpleasing reddish shades that are the only ones possible with henna alone, this plant was long ago mixed with indigo, or one or more of the wood extracts.

Lawsone has been used for highlights in color shampoos based on sulfated detergents in acid solution (35).

Indigo

Another natural substance known and utilized as a dye since the dawn of history is *indigo*, which occurs in the leaves of woad (*Isatis tinetoria*) and several plants related to the pea family. As early as the mid-seventeenth century, indigo was cultivated as an important crop in the American southern colonies.

The best source of indigo for hair is *Indigofera argentea*, long cultivated in Persia for this purpose, but it cannot be used alone. The dried and powdered leaves (known as reng) are either mixed with ground henna or applied alternately with henna, to produce a good black. Shades from light brown to black can be produced by varying the proportions of henna and reng. Many recipes are available, but indigo has no place in modern cosmetology.

Camomile

Another of the few plants that have been used for coloring the hair is camomile (also chamomile). Several varieties of related plants are found in Great Britain, western Europe, and the United States, of which only two—*Anthemis nobilis*, called Roman camomile, and *Matricaria chamomilla*, called German or Hungarian camomile—are used to any extent. The coloring principle has been identified as *apigenin* (4',5,7-trihydroxyflavone), $C_{15}H_{10}O_5$, both free and as a glucoside, found in an extract of the florets. Repeated applications of camomile on any shade of hair eventually bring out a brilliant chrome yellow, similar to the coating commonly seen on lead pencils (1).

Wood Extracts

Decoctions of woody fibers, bark, or nuts from several trees are also used as dyes. Of all that have been tried since earliest times only the following survive, and these are now used for dyeing postiche (hair pieces) rather than hair on the head.

Brazilwood (redwood, pernambuco wood). Wood from *Caesalpinia brasiliensis* or *C. echinata*. The active dyeing principle is *brazilin*, $C_{16}H_{14}O_5$, a yellowish color which in oxygen or with alkalies becomes the red dye brazilein, $C_{16}H_{12}O_5$. It is used in various combinations to produce warm brown shades (6,24,25).

Catechu (gambier, cutch). Name given to colored product of two distinct species—*Ouroparia (Uncaria) gambir*, which is yellow, and *Acacia catechu*, which is brown to black. The active principles in both are catechin ($C_{15}H_{14}O_6 \cdot 4H_2O$) and pyrocatechol (1,2-dihydroxybenzene, $C_6H_5O_2$) (catechuic acid). One product or the other is used in combinations to produce all shades from blond to black.

Fustic. The wood of *Chlorophora* (*Morus, Maclura*) *tinctoria,* a tree similar to the mulberry, common in the western hemisphere. The active principle is *morin* (2',3,4',5,7-pentahydroxyflavone), $C_{15}H_{10}O_7$; it also contains *maclurin* (morintannic acid), $C_{13}H_{10}O_6 \cdot H_2O$. Under different conditions and in various combinations, it produces yellow-to-brown shades. This dye is often called "old fustic" (from Arabic, *fustuq*) to distinguish it from "young fustic" (from Portuguese, *fustet*), obtained from an entirely different European tree, *Cotinus coggygria* (*Rhus cotinus*), which yields another dyestuff (23,25,101).

Logwood (*bluewood, campeche wood*). The heartwood of *Hæmatoxylon campechianum* L., a tree common in the West Indies and Central America. Its active principle is *hematoxylin* (hydroxybrasilin), $C_{16}H_{14}O_6 \cdot 3H_2O$, which is readily oxidized to hematein, $C_{16}H_{12}O_6$. It is specifically good for producing black and for drabbing or neutralizing reddish tones in the hair.

Nutgalls (*gallnuts*). Pathological excrescences on leaves, twigs, or other tissues of the white oak tree (*Quercus infectoria* or *Q. lusitanica*), caused by the invasion of bacteria, certain worms, or insects. These little "nuts," which are especially rich in tannin and gallic acid, serve as a source of pyrogallol, and are priced according to the content of tannin. The European variety yields about 30%; the Turkish (Aleppo), 60%; and the Chinese (obtained from *Rhus semialata*, related to cashew nut), 70% or more. Nutgalls are of interest because they form the basis of "rastiks," a very ancient type of hair dye, still used in the Orient (23,25).

Quercitron: The inner bark of another species of oak tree (*Quercus tinctoria*), common in North America. Its active principles are quercetin (3,3',4',5,7-pentahydroxyflavone), $C_{15}H_{10}O_5$, and its 3-L-rhamnoside, quercitrin, $C_{21}H_{20}O_{11}$. This bark is usually combined with fustic and logwood to produce dark-brown shades. With alkalies it gives an intense yellow which was prized by the European textile industry as an import from the American colonies.

Walnut. The leaves of walnut trees (*Juglans cinerea, J. nigra, J. regia*), but more commonly the green nutshells, have been used from most ancient times as a source of brown hair coloring. The active principle is *juglone* (5-hydroxy-1,4-naphthoquinone or *nucin*), $C_{10}H_6O_3$; pyrogallol is also believed to contribute to the dyeing property. Because the color produced on the hair is not lasting, walnut extract is now rarely used alone; the ground shells are usually mixed with other color-producing substances.

Because the walnut tree was not native to England, and the relatively small growth of it there was cultivated for timber and for the edible nuts, quantities of the shells were among the earliest exports from the American Colonies, for the fabric-dyeing trade in England (102).

Mixed Wood Dyes

In practice, none of the wood extracts is ever used alone. Except for the dyeing of postiche, and of samples for some shades on shade cards that show a line of dyes in natural hair, the use of these wood extracts is of merely historical or academic interest. Anyone interested in reviving them (possibly as a line of "hypoallergenic" hair colorings) can find numbers of recipes in many of the older books (6,24,25,56–60) as a starting point for research.

Henna is occasionally mixed with some of the wood extracts, e.g., nutgalls or walnut shells, to modify the natural orange-red shade from henna alone. These mixtures may satisfy the request for a "pure vegetable pack," but they are usually not satisfactory because the brownish tinge is not lasting.

Hematoxylin has been mixed with certain oxidation dyes (22).

Miscellaneous Plant Products

Among other plants commonly listed as sources of hair colorings are *rhubarb* and *sage*. The former has occasionally been combined with henna, black tea, and camomile. Its active principle is *chrysophanol* (chrysophanic acid) (1,8-dihydroxy-3-methylanthraquinone), $C_{15}H_{10}O_4$, which gives good blond shades, especially with alkaline mordants. Whatever its possibilities, it seems not to be utilized commercially (25).

Sage, prepared as an infusion ("sage tea"), is now too antiquated to be considered, though it may still be used in homemade preparations. Its principal effect is to dull white hairs, making them less conspicuous, but long-continued application makes the hair look grayish and dingy.

Use of Mordants

To hasten the development of the desired shade or to modify it, the natural organic colorings have commonly been mixed with solutions of metallic salts. These adjuvants, so-called mordants, have been used in the dyeing of fabrics since the days of the early Egyptians. One of the earliest successful dyes for the hair was the *rastik* (from Turkish and Persian words for hair dye), made originally by roasting nutgalls, copper and iron filings, and a small quantity of oil in a copper vessel. The roasted mass was then ground, made into a paste with a little water, and this was applied to the hair for several hours. It produced a beautiful permanent black.

Variations have been introduced into this process, but the underlying principle is still the same; the occasionally weak organic dye is intensified and, by a choice of metallic salts, modified into a fair range of shades. Copper and iron sulfates and alum were the earliest recorded mordants. All may be found in products still in use, especially ferrous sulfate, which is commonly used with the wood extracts in dyeing postiche (23,57).

The newer use of mordants (as described in the section on semipermanent dyes) differs from the older method in that the metals are chelated, not free to interact with other ingredients (16,49).

Advantages and Disadvantages of Natural Dyes

The principal advantage in the use of vegetable dyes is that they are harmless to the system. The disadvantages are in the relatively harsh and unnatural shades produced, especially after repeated applications, and in changes in texture whereby the hair becomes stiffer, more wiry, occasionally brittle, and adversely affected by permanent waving (1).

Metallic Dyes

As hair colorings, metallic salts are almost as old as the plant derivatives. Both men and women in ancient Rome darkened their hair by passing through it a lead comb that had been dipped in vinegar. Many other metals have been used from time to time, and a wider variety of salts is still to be found in the European market than in the United States. Because the metallic dyes are incompatible with permanent waving and oxidation dyes, the great increase in popularity of these two treatments has practically stopped use of the metallic dyes by women. Large quantities of some types, however, are still used by men.

Most of the metallic dyes have been sold with the appeal that they are "not dyes—just color restorers for the hair." At one time many of these preparations were compounded by pharmacists and hairdressers, to be sold under their own name. The metallic dyes are almost never applied in modern beauty shops, but as "color restorers" they are sold in retail stores for application at home.

The old theory underlying the use of metallic dyes is that solutions of various salts act on the sulfur in the keratin of hair, thus forming a deposit of the corresponding sulfides. This seems not to be substantiated by more recent investigations and practical research. The fact that several applications may be required before any change in color is noted seems to support the idea that these products restore the natural coloring of the hair. The new substance, however, is mainly deposited on the outside of the hair like a plating which, through the action of light and air or a developer, consists of an insoluble oxide, or sulfide, or both. The range of shades obtainable is necessarily restricted to the colors of the oxides and sulfides of the metals used.

The metallic dyes in the American market are based principally on (in order of prevalence) lead, silver, and copper. Nickel alone is rare, but it is found with silver; iron is found in most copper preparations. Bismuth (comparatively rare in the U.S.), manganese, and cobalt are fairly common in foreign products.

Men should be warned against use of these "restorers" on the moustache because of the danger of taking the metallic compound internally, with food or drink.

Like the preceding section on natural organic dyes, the following information is included "for the record" and for possible historical interest.

Lead Dyes

Most of the "color restorers" based on lead consist of dilute solutions of lead acetate or nitrate. Many compositions call for the addition of finely powdered sulfur, glycerol, and either rose water or bay rum as perfume.

All lead dyes are classified as "progressive." Ordinarily, the color develops from dull yellow through greenish, violet, and mahogany shades to a dull black. As these products are to be applied "day after day until the desired shade is seen," the accumulated coating is likely to rub off on hatbands and bedding. Repeated applications to new growth of hair frequently cause unsightly variation in the shade produced, both near the scalp and throughout the length of the hair, because of dribbling.

Since the sixteenth century, when the favorite recipe was a mixture of equal parts of litharge (or lead carbonate), quicklime, and chalk, which was to be made into a thick paste and left on the hair overnight, the lead dyes appeared as one-bottle and two-bottle liquids, one or two powders, and, most recently, pomades or creams.

All formularies, handbooks, and works of reference on cosmetics offer numbers of recipes for these types of products. Despite all that has been published about the defects of lead dyes and how to recognize them, new proprietary products are occasionally introduced. Many are made and sold locally by pharmacists or dealers, or by mail order.

In some new variations, lead acetate (1 to 2%) and precipitated sulfur have been added to compatible recipes for scalp pomade and also to bases for cream shampoos, in which surface-active agents are also incorporated to ensure quicker development of color and some penetration of the hair. There seems to be no reason for adding inert, elemental sulfur to recipes for these lead "restorers," except the fact that since about 1860 it has been traditionally included.

Another innovation has been the use of ammonium thioglycolate and lead solutions in two-bottle preparations, the former to serve as an accelerator if it is desired to produce the color instantaneously, rather than gradually.

Silver Dyes

Although a recipe for silver nitrate in rose water appears in several publications of the sixteenth century, the regular use of it as a hair dye seems to date from the beginning of the nineteenth century (103). Various physicians and pharmacists investigated and promoted it, usually under such exotic names as "water of" Egypt, Greece, Persia, or China.

The principle underlying the so-called restorers based on silver is twofold: (a) All silver salts darken on exposure to the light, and (b) silver combines readily with protein, forming a brown stain.

In the oldest preparations, silver nitrate was used alone, in a dilute solution which was combed through the hair and allowed to remain undisturbed during exposure to strong sunlight. If the shade obtained was too light, the application of solution and exposure were repeated on successive days. Used in this way, silver dyes may be classified as "progressive."

To hasten the action, silver dyes were offered in two bottles, one containing silver nitrate, the other containing a developer, such as sodium (potassium, ammonium) hydrosulfide or pyrogallol. With the former, silver sulfide is deposited directly on the hair; with the latter, there is believed to be double action, the silver salt acting as a mordant for the coloring property of the pyrogallol, and the latter serving to reduce the silver.

As this new type of hair coloring was quickly adopted by hairdressers and barbers, in whose establishments the offensive odor of the sulfide discouraged trade, the use of pyrogallol, after 1845, became almost universal. Most of the hairdressers prepared their own solutions, and possibly because the pyrogallol looked like the still-popular nut extracts, the silver dyes were soon being called, and openly sold as, "vegetable colorings."

As later improvements in these compositions, varying amounts of ammonia were added to the silver solution, thereby inhibiting certain proportions of silver and producing a fair gradation in shades. Sodium thiosulfate (photographers' "hypo") was adopted as a developer, presented either as a second solution or as a powder in packets or capsules.

Restorers of this type are classified as "instantaneous." For best effect, the hair must be absolutely clean, so it should be freshly shampooed and dried. The developer or mordant is combed through the hair first and distributed evenly; this may be allowed to dry, or is followed immediately by the silver solution which brings out the color.

Any desired range of shades can be produced by varying the concentration of the silver nitrate. The usual suggestions for developers are 3% sodium thiosulfate; also 1% pyrogallol in a 24% alcoholic solution.

Products containing sodium thiosulfate should be well protected from light and air.

In general, silver dyes are more satisfactory than those based on lead. The user may have trouble, however, after the first application, in the retouching of new growth of hair. Repeated applications to the same portion of hair cause the hair to become progressively darker and to develop an unpleasant metallic or iridescent sheen.

Recipes for silver dyes of all types are available in most of the old standard works of reference. Current products, when found, show nothing new.

Dyes for Eyebrows and Eyelashes

Although silver dyes for the hair are not so popular among women as they were a generation ago, they are still widely used, both at home and in beauty shops, for dyeing the eyebrows and eyelashes. Silver nitrate, in concentrations to 5 %, is the only coloring agent commonly used for this purpose in the United States.

The use of pyrogallol as a developer is not to be recommended, for two reasons: (a) The application of pyrogallol—a synthetic organic compound— in the orbital area is questionable, and (b) the consumer may avoid the product under the impression that two small bottles, one containing a brown liquid, the other a colorless liquid, may indicate an amino dye (strictly forbidden).

Sodium thiosulfate (3 to 5 %) is quite satisfactory as the developer. Instructions should specify that the solutions are to be applied separately, the developer first and, after 2 or 3 min, the silver nitrate. The caps of the twin bottles should be fitted with soft pencil brushes. If desired, a third bottle may be included in the package, containing a light mineral oil which may be tinted and perfumed. This is to be carefully applied to the skin above and below the eyebrow, to prevent staining of the nonhairy surface and any chance of the liquid running into the eyes (1,23).

Copper Dyes

The modern hair dyes based on copper have been evolved from the ancient rastiks, in that the same essential ingredients are combined in liquid form instead of powders. The earliest were in two bottles, one containing copper sulfate (with or without ammonia), the other potassium ferrocyanide. Later, a hydrosulfide or pyrogallol was the developer. Like the older silver preparations, these were to be applied successively. Many of these were made and applied in hairdressing salons, but they have practically disappeared.

Use of copper sulfate alone with pyrogallol produces a rather good brown; if ferric chloride is added, the resulting shade is a much darker brown or black. By varying the proportions of the metallic salts, some variation can be effected in the shade produced, but copper is usually not to be considered for lighter shades of hair. One or two applications may be satisfactory, but repeated retouching, especially if any of the dye is allowed to run over previously treated hair, makes the hair a dusty harsh black, and eventually harsh, stiff, and brittle.

An improvement in copper dyes was offered by Schueller of Paris, in 1907, whereby the metallic salts (copper and iron) were combined in one bottle with both pyrogallol and, for example, sodium sulfite. This type of product starts to darken on the hair as soon as it is exposed to the air. Anything that remains unused must be tightly closed at once because it soon separates into a clear green supernatant fluid with a brown sediment (5,23,60,72).

Undoubtedly because the bright blue ("unnatural") color of copper solutions might raise questions about composition, most of the proprietary dyes based on copper are sold as one-bottle products, to which the pyrogallol gives the color of black coffee.

Compound Hennas

The dyes called compound hennas were once considered sufficiently important to constitute a separate classification of modern hair dyes. They are now used so relatively little, however, that they need be considered merely another form of metallic dyes.

Theoretically, any mixture of henna with indigo, logwood, or any other natural coloring derived from plants is a compound henna, and many such mixtures have been prepared and applied as "vegetable packs" by hairdressers. The term came officially into prominence commercially about 1914 when both Broux and Schueller of Paris introduced to the trade a line of products, based on henna, which promised to produce any desired shade from blond to black.

The compound hennas immediately became popular because they answered the growing demand for "something like henna but not red," to be used in place of the synthetic organic dyes. Both the trade and the public were in fear of the latter because of increasing complaints of irritation following their use both by hairdressers and in the home.

These colorings are based on the formation of lakes from phenolic compounds, for example, aminophenol, pyrogallol, and hematoxylin, with salts of cobalt, copper, iron, and nickel. The preferred reducing agent is sodium thiosulfate. In the liquid preparations (Schueller), interaction of the ingredients is prevented by excess of the reducing agent; in the powdered mixtures, the reducing salt is packed separately, to be added as the paste is made with water. Henna is the carrier for all the shades, but its carroty-red tone is successfully masked by additions of tannin, pyrogallol, nutgalls, lampblack, and other coloring agents.

These modern rastiks maintained their vogue for about twenty years. The more natural tints obtainable with the newer synthetic organic dyes drew many users away from compound hennas as well as from other metallic dyes; and the rapid increase in permanent waving, which has been steady since 1934, gradually finished the use of them in the United States.

Miscellaneous Metallic Dyes

To produce a wider range of shades, and especially lighter shades, than is possible with the metallic dyes just described, various metals have been employed from time to time, and may still be found in current proprietary products. Details of the composition and effects of all types can be found in several works of reference (5,23,25,56–60).

Working on different principles, bismuth, cadmium, cobalt, iron, manganese, nickel, and tin have been utilized (26). Most of them have been applied by hairdressers, but a few have gone into the retail trade.

As described in the section on semipermanent dyes, of the metals listed, cobalt, iron, and nickel are among those used with chelating agents in "premetallized" dyes (39,49).

Advantages and Disadvantages of Metallic Dyes

The principal advantage that can be claimed for metallic dyes is that, unless they are accidentally taken internally, by mouth or through abrasions in the skin, they are generally considered harmless to the skin and to the system.

The disadvantages of metallic dyes as now constituted are numerous: The range of shades normally produced is limited to harsh dark browns and black; lighter shades darken unpredictably or fade; all metallic dyes deposit a plating on the hair which frequently causes unpleasant off-shades; and eventually makes the hair stiff and brittle. Because of abnormal changes in texture, such hair frequently is adversely affected by permanent waving.

Toxicity of Metallic Hair Dyes

Modern science, especially in the applications of biochemistry to medicine, has reappraised, through tests in clinics and laboratories, many of the inherited notions about the toxicity of metallic compounds. It is now known that, in the low concentrations commonly used in hair dyes, there is practically no danger of absorption through the unbroken skin. The danger of cumulative effects by allowing the metallic compounds to enter the system by mouth still exists; hence the need for caution by men who might apply the so-called "color restorers" to the moustache.

Manufacturers and distributors of metallic dyes should ensure protection of users by most explicit instructions both for applying their products and— equally if not more important—for observance of all sanitary precautions, so that there will be no slightest danger of carrying the metallic compounds to the mouth through handling foods or otherwise.

Despite all the dire implications of poisoning and death from the use of lead hair dyes, continually promulgated by medical authorities and lay reformers, there seems to be no authenticated case of lead poisoning traceable to any of these products when they were applied correctly. A decision of the Supreme Court in New York (1934), based on the kind of evidence usually brought forth in such trials, deserves to be more widely known (104).

Copper dyes also seem never to have been a cause of trouble when they have been used correctly. This is in spite of the fact that the pyrogallol commonly used as a developer has been alleged to cause allergic reactions in sensitive persons.

A statement common in the medical literature, that the application of silver dyes can cause the bluish-gray diffused pigmentation of the skin known as argyria, has been the subject of two investigations (26). There seems to be far less danger of argyria from metallic hair dye than from the silver–protein compound commonly used as an antiseptic for mouth and throat.

The Federal Food, Drug, and Cosmetic Act imposes no restrictions on metallic hair dyes. The Sanitary Code of the City of New York originally required that all such proprietary products carry a statement that the product contains a metallic salt (unnamed), that it was for external use only, and must be used with care. As amended in 1959 to conform with the federal law, this caution must state that the product contains "ingredients which may cause skin irritation on certain individuals and should be used with care." This seems to be too similar to the requirement for amino dyes, and it implies that users may be allergic to metallic compounds.

Need for Research

Admitted that many, if not most, of the metallic "restorers" on the market may be antiquated, and that the results obtained with them are at best unaesthetic, there is still a definite place for such products, to meet the needs of those that are allergic to organic dyes.

Researchers in the United States should take full advantage of the freedom from antiquated and unrealistic legal restrictions—such as those that throttle the cosmetics industry in some European countries and really work on some new metallic hair dyes. The fact that a certain type of product has been used since the dawn of recorded history is no good reason for perpetuating it. Study of new combinations, for single specialties or a range of shades that can approximate a fair selection of the shades of human hair, should be most rewarding.

It must be remembered, however, that by the standards of the latter part of the twentieth century, the primary test for a successful hair coloring is the mutual effect between it and permanent waving. In any program of research on metallic compounds, therefore, whether they are to be used alone or as mordants for "hypoallergenic" organic intermediates, this point merits special consideration. All such tests should be made on strands of real hair, not on bits of fur, wool, cotton, or synthetic fibers.

Tests for Metallic Dyes

Although the analysis of a sample of hair, for the purpose of identifying dyes previously used, is not so necessary now as it was a generation ago, it might occasionally be advisable. Many persons have their hair dyed abroad without knowing or caring what type of product is used on it, and if an oxidation dye should be applied directly over a metallic preparation, the results might be disastrous. From too high a content of certain surface-active agents—or the wrong kind—in the synthetic organic colorings, the hair may feel as it does with metallic preparations. If there is any doubt, therefore, it is best to analyze it.

Although the number of proprietary hair dyes on the market is bewildering, familiarity with the physical appearance of the various types described here is the first long step toward identifying them. In a simple method, devised by Evans in 1926 and widely utilized in early educational programs, a small sample of hair is completely incinerated in a silica crucible. Plain henna and other natural organic substances are readily identified by characteristic earthy or "woody" odors, and leave no appreciable residue. Most compound hennas leave a black residue, which is dissolved in a little nitric acid and tested for copper and iron. Nickel, silver, cobalt, lead, and any other metal suspected from the appearance of the residue, can be easily identified by standard qualitative analytical tests.

The idea of the test for allergy seems to have been anticipated in a test for metallic dyes, in which possible irritation could be ascertained by painting a small spot of the product on the skin of the arm. This was published in 1868 (4).

Dye Removers

Just as, several years ago, a skeptical cynic remarked that "the only thing to do for gray hair is admire it," a more recent one, well versed in this lore, suggested that "the only thing to do for metallic-dyed hair is let it grow out" (22).

With the exception of the solutions suggested for the removal of spots from the skin after the application of silver and copper dyes, relatively little information has been available on dye removers. There was evidently little demand for products that could take coloring off hair that had been dyed.

Most of the "decolorizers" mentioned in a few publications before 1925 (105) consisted of hydrogen peroxide or a hypochlorite, either of which was ineffective or disastrous on dyed hair. In that year, however, coincident with the intensive promotion and wide acceptance of the new American amino dyes, there was a spontaneous demand, both from hairdressers and from the public, for products and treatments by which dyed hair could be cleared of the hennas and metallic "restorers" and converted to the new dyes. The truth of an axiom, then being taught to hairdressers, soon became obvious: *There is no single safe quick-removal process for all old dyes.*

General Precautions

The removal of a hair dye, especially if there is any question of changing from one type of dye to another, is a serious and important undertaking, which is best left to an experienced hairdresser or cosmetologist. Certain products can be prepared for retail sale directly to consumers, but practically everything for this purpose is best applied by the cosmetologist in a beauty shop or sold there to the patrons if supplementary treatment at home is advisable. Instructions for use should be absolutely foolproof.

With the exception of a chance patron who may unknowingly have had a metallic preparation applied somewhere abroad, most problems of dye removal are a radical change of shade or lightening of one that has gone too dark.

Even the rare user of henna does not count. If she uses it, it is because she is allergic to the oxidation dyes, so she will probably not be changing.

Sodium Sulfoxylates

Among contributions to cosmetics from the textile industry are certain reducing agents, used to remove fast dyes from fabrics.

Sodium sulfoxylate (or hydrosulfite, $Na_2S_2O_4$) and sodium formaldehyde sulfoxylate ($NaSO_2OH_2OH$) are reducing agents. For textiles the solution should be strongly alkaline, but for hair, the pH should be below 9. The product should be supplied as a powder, to be dissolved in water at time of use. An accompanying packet may contain citric acid, lanolin or protein derivatives, a thickener, a wetting agent—any combination of ingredients intended to remove dye without harming the hair (106).

Other suggested adjuvants are ethoxylated fatty alcohols, and polyvinyl-pyrrolidone (PVP), which, being cationic, forms an effective complex with anionic dyes. An acidified sulfoxylate solution with the addition of activated carbon successfully lightens very dark hair.

Formamidine Sulfinic Acid

Another type of dye remover, which can be prepared as a liquid, is a viscous 1 to 2 % solution of formamidine sulfinic acid ($HN:C:NH_2SO_3H$; thiourea dioxide; amino imino methane sulfinic acid). This product contains a thickening agent, polyvinylpyrrolidone, and ammonia and ammonium bicarbonate, in an ethylene glycol monoalkyl ether (butyl suggested in patent). This mixture is said to decolorize oxidation-dyed hair in 15 to 20 min at room temperature (106).

Hot Oil Treatments

An old and entirely safe treatment for all dyed hair is a good soaking in oil. Various mixtures have been used successfully, e.g., crude oil (petroleum) alone or, preferably, mixed with olive oil or any readily available less expensive vegetable oil. Mineral oil may also be mixed with olive oil or any good blend (1).

Many hairdressers have an assortment of suitable oils at hand and prepare their mixtures at moment of use, but suitable, well-proportioned proprietary mixtures should be good salable items, welcome to the trade.

Instructions should specify that the hair be completely covered with the oil and heated by slipping strands of it through a heated iron. A marcel iron, a pressing iron (as used for paper curls), or a metal pressing comb, such as is used for straightening the hair, serves the purpose. The hair is then to be rubbed, strand by strand, with a Turkish towel, and thoroughly shampooed.

This treatment seems actually to soften any coating that may be on the hair so that much of it can be removed mechanically without causing breakage. It should be repeated until a test with whatever product is to follow shows that the latter can be safely blended into the previously dyed hair.

Other means of heating the oil on the hair, such as an electrically wired cap, an infrared lamp, or other device may optionally be promoted for these treatments. The mechanical action of the iron in actual contact with the hair however, seems to be more rapid and efficient.

Metallic dyes. Of the three metals still occasionally found in hair dyes ("restorers"), lead is most easily removed. Silver and its combinations are more resistant but usually tractable. Copper, either alone or with other metals, is often troublesome because the mordants that are now always used with it fix the dye tenaciously in the hair. Treatments should be continued until tests with another type of dye show no discoloration or breakage.

Synthetic organic dyes. As originally presented, that is, as clear ammoniacal solutions, to be applied to hair that was shampooed and pretreated with hydrogen peroxide, the oxidation dyes penetrated the hair and colored it throughout. Hair so dyed usually does not respond to treatments with the simple oil mixtures described.

The increasing use of surface-active agents in compositions for oxidation dyes has presented problems for both cosmetologists and home users in effects on the texture of the hair. If it has become stiff and harsh, such hair can be restored to normal texture by a few treatments with these simple oil mixtures.

Color rinses (azo dyes, certified colors), presumably temporary, have been made more lasting by the addition of surface-active chemicals. If there is any appreciable deposit on the hair after the application of these rinses, it can be removed by one or two heating treatments with oils.

In summary, hot oil treatments are good for the removal of any type of dye that has coated the hair.

Sulfonated Oils

At any time, sulfonated oils must be used with caution, because they can strip even the natural color from the hair and frequently bring out unpleasant streaks.

Sulfonated castor oil has long been used to remove all types of dyes. It offers an advantage over the simple oils because, being water-miscible, other chemicals for special purposes can be added to it.

To remove henna and metallic dyes, sulfonated castor oil is more effective with the addition of 1 to 2% of salicylic acid, other (polybasic) acids, or chelating agents. For amino dyes, a portion of the oil may be mixed with half its quantity of 20-vol hydrogen peroxide.

The sulfonated oils may be applied to the hair and heated by any of the means mentioned in the preceding section.

Aqueous Solutions

All the dye removers mentioned by the early authorities were aqueous solutions, recommended for removing the dark spots caused by silver nitrate. Among the compounds recommended by certain medical writers, and even sold with some proprietary hair dyes, was potassium cyanide. As this is a deadly poison, it should never be considered for this purpose.

Other compounds tried for removing silver were citric and oxalic acids, potassium oxalate, potassium bichromate, potassium iodide—both as a saturated solution and containing dissolved iodine—sodium hydrosulfide (NaSH), and sodium "hyposulfite" (now sodium thiosulfate, $Na_2S_2O_3$). All but the last, used as a 5% solution, have been discarded as unsuitable or unsafe.

Sodium hydrosulfite ($Na_2S_2O_4$). This compound is the basis of several proprietary dye removers. Slightly acidified with sulfuric acid, it is most effective with the dyes based on certified colors, but it can also be used to lighten the shade of the oxidation dyes, and especially to remove dye stains from the skin.

Sodium hydrosulfite should be sold as a powder, with instructions for making a 2 to 5 % solution.

Hydrogen peroxide. Light shades of amino dyes, and some of the medium drab shades, can be lightened to some extent by hydrogen peroxide (15 to 20 vol). Black, dark browns, and certain reddish shades do not react so favorably. Products compounded with certain surface-active agents may be especially resistant.

The so-called quick bleaches are usually more effective than the liquid hydrogen peroxide, but the effect on the hair is so drastic that these preparations must be used with great care.

Hydrogen peroxide should not be used in products or treatments for removing metallic dyes. Relatively inoffensive shades might be darkened or discolored, and hair that is black or very dark might break off completely at the line of the last application.

Acid rinses. An early recommendation for removing plain henna was a good rinsing with dilute hydrochloric acid (105). This was succeeded by dilute solutions of organic acids, such as acetic, citric, oxalic, or salicylic. The last-named seems to be more satisfactory as an addition to sulfonated oils than as the solution of acid alone.

Need for Research

In the long succession of preparations and treatments for the hair, dye removers came relatively late. Because all those on the market are proprietary products, very little has been published about them, but this class of hair preparations offers a good field for study.

Although improved dyes and permanent waving have practically driven the vegetal and metallic dyes off the market, there still may be some that wish to change from henna or a "restorer" to an oxidation dye, or at least change the shade of the latter. So there is still a need for research on safe and effective dye removers. Any product of this kind should always be accompanied by explicit instructions for applying it over different types of dyes, and especially for simple tests by which the user can check progress toward the results desired.

Importance of Clinical Testing

If all hair that is to be dyed were gray and in good condition, the compounding of a good product would be a comparatively simple problem. The production of good hair dyes, however, only begins with the blending of shades. Ideal colors on natural gray human hair in good condition may look entirely different on heads of hair that have been permanently waved, bleached, treated with remedial preparations, even only shampooed.

The clinical testing of any new hair dye is of the utmost importance to both manufacturer and user. There must be some assurance that the proposed product really is a good hair dye; that is, that it will give good results, not only of itself, but also in combination with other customary treatments. Factors that must be considered before any line of dyes is thought to be satisfactory include the following (107):

Shampoos.   Many beautiful shades of hair coloring may be spoiled through shampooing with the strongly alkaline soaps and water softeners that may still be used in regions where hard-water prevails. The correcting of these "off-shades" is one reason for the new group of colors in grayish tones which are to be mixed with, or applied over, the shade selected for a given head of hair.

The rapid increase in the use of synthetic detergents instead of soap for shampooing, in order to circumvent the unpleasant effects of hard water, has caused a corresponding increase in other effects in the hair, e.g., extreme dryness or harshness. These, too, must be considered in the compounding and testing of many shades (see the section on other adjuvants), using different types of products and in different kinds of water.

In any testing program of this kind, the effects of the synthetic shampoo should be checked against those of some acceptable standard soap shampoo, used on opposite sides of the head. The mutual effects should be closely observed, that is, both for any stripping of color by the shampoo or any film of "conditioner" that may be left on the dyed hair, and for any effect on the texture of the hair that might adversely affect the next application of the dye.

To complete the record of all studies, there should be a set of samples showing all significant steps in the tests.

Permanent waving.   When permanent waving with heat was the rule, there were frequent complaints of discoloration and streaking of dyed hair, and occasionally breakage from the waving process.

In general there is less trouble from cold waving over certain forms of oxidation dyes, but those that contain any appreciable amount of surface-active agents should be checked on two points:

1. Coating of the hair, which might prevent proper penetration of the hair by the waving solution.

2. Relaxing of the wave, or making the hair feel harsh and wiry when coloring is applied after waving.

Biological tests.   In view of the generally broader enlightenment now on the nature of the dermatitis caused in allergic individuals by some of the oxidation dyes, the necessity for, and the value of, some of the many elaborate programs of biological and pharmacological tests on the internal effects of these dyes (especially *p*-phenylenediamine) on small animals may be questioned (107).

There should be good reason for undertaking biological tests. The mere producing of a set of good colors on cut swatches of hair, wool, fur, or cotton should not be considered sufficient justification for embarking on a long and expensive program. The researcher should have the further assurance that his product will withstand permanent waving of hair on the living head, and average shampooing with a wide variety of modern products. If the product should prove to be unsatisfactory in the slightest detail as a hair coloring—its primary function—biological tests would be quite superfluous (108).

The introduction of new and untried materials is another matter. Unless— and even when—accompanied by certified reports from the manufacturer, all new materials should be carefully tested, both for compatibility with all other ingredients of any composition in which they substitute for a familiar one, and for any change they might cause in the effects on the user of the product.

Dermatological tests.   Much has been learned in recent years about methods of testing for allergy in individuals that use hair dyes. Any manufacturer of these products should have access to a good roster of reliable subjects (usually called "models") who call regularly for control applications of their usual coloring, or can be called at any time for special tests.

Any test should follow every step of a complete application of a hair dye; that is, with products that require a preliminary shampoo and treatment with hydrogen peroxide, the area to be "patch-tested" should be swabbed with the customary shampoo (soap or synthetic detergent) and with the customary mixture of peroxide and ammonia if any is required, before the solution of intermediates and developer is applied. Such a precaution often serves to prevent what occasionally happens—a negative result of the patch test with dye alone and dermatitis after the complete hair-coloring treatment.

Should dermatitis occur, separate tests must be made to ascertain which substance was the actual cause of the irritation. In one case, for example, it was found that a woman with an extremely sensitive skin could safely have her hair dyed by having her hair shampooed on one day, allowing the resultant inflammation until the next day to subside, and then adding no ammonia (or bare minimum) to the hydrogen peroxide for the preliminary bleaching (109).

Few hairdressers or consumers object to the requirement for a patch test before a first application of hair coloring. Knowing that every successful retouch is its own test for allergy, many consider the requirement for a test before every application at least a nuisance. As long as it remains in the law, however, everyone charged with testing programs or instructions for use of products must conscientiously see that it is fulfilled.

Any test for allergy must be made with the shade or shades of hair dye that actually will be used on the subject's hair. A small quantity should be freshly prepared, and the mixture of intermediates and developer should be in the proportions specified in instructions for a full application.

Too many testing programs in the past have been limited to the use of p-phenylenediamine alone, and too little has been published on the results of similar programs with other intermediates now used in most of the successful commercial products on the market.

Carefully controlled programs of tests for the external effects of all of them on large numbers of human beings of all types and under all conditions would help to establish the actual percentage of sensitive individuals, and thus resolve much of the long-standing doubt and fear of these valuable chemical compounds.

Official tests for identifying the oxidation dyes, especially in combinations, have been published by the Cosmetics Division of the Food and Drug Administration (110), and these can be obtained by application to this office.

The steady publication of books and articles on medicine, science, and technology related to cosmetics serves not only as a record of what was and what was done, but also as an inspiration for future developments in products and treatments in hair coloring.

REFERENCES

1.   Wall, F. E.: *The principles and practice of beauty culture*, 4th ed., Keystone Publications, New York, 1962.

2.   Pliny: *Historiae naturalis, libri XXXVII*, Tauchnitz, Leipzig, 1930.

3.   Rimmel, E.: *The book of perfumes*, Chapman and Hall, London, 1865; *Le livre des parfums*, E. Dentu, Paris, 1870.

4.   Cooley, A. J.: *The toilet in ancient and modern times*, Hardwicke, London, 1868.

5.   Monsegur, P. A.: *Hair dyes and their application*, Osborne, Garrett, London, 1915.

6.   Müller, F.: *Die moderne Friseur und Haarformer in Wort und Bild*, 4th ed., Killinger, Berlin, 1926.

7.   Marinello, M. G.: *Gli ornamenti delle donne*, 1st ed., Francesco, Venice, 1562.

8.   Pagel, J.: "Geschichte der Kosmetik," in Joseph, M.: *Handbuch der Kosmetik*, Von Veit, Leipzig, 1912.

9.   Vecellio, G.: *Habiti antichi e moderni*, Didot, Paris, 1859.

10.   von Schrötter, A.: Wasserstoffsuperoxyd als Kosmetik, *Ber.*, **7**: 980 (1874).

11.   Chadwick, A. F., and Hoh, G. L. K.: Hydrogen peroxide, in *Encyclopedia of chemical technology*, Vol. 11, 2nd ed., Wiley-Interscience, New York, 1966.

12.   Powell, R.: Hydrogen peroxide manufacture, *Chem. Process Rev.*, No. 20, Noyes Development Corporation, Park Ridge, N. J., 1966.

13. Edman, W. W., and Martin, M. E.: Properties of peroxide-bleached hair, *JSCC*, **12**: 133 (1961).

14. Zahn, H.: Chemische Vorgänge beim Bleichung von Wolle und Menschenhaar mit H_2O_2 und Per-acids, *JSCC*, **17**: 687 (1966).

15. Böllert, V., and Eckert, L.: Quantitative Verfolgung des Blondiervorgangs von Humanhaar, *JSCC*, **19**: 275 (1968).

16. Sidi, E., and Zviak, C.: Décoloration des cheveux, in *Problèmes capillaires*, Gauthier-Villars, Paris, 1966.

17. Mentecki, A. D.: Hydrogen peroxide in cosmetic preparations, *Am. Perf.*, **78**(3): 17 (1963); *JSCC*, **13**: 362 (1962).

18. Peroxide bleaching of hair under alkaline and acid conditions, *Schimmel Briefs*, No. 225, 1956.

19. Melamine peroxide for oxidation of hair dyes, *Schimmel Briefs*, No. 337, 1963.

20. Lustig, B.: U. S. Pat. 2,991,228 (1961).

21. New versions of "white henna," *Schimmel Briefs*, No. 331, 1962.

22. Harry, R. G.: *Modern cosmeticology*, 5th ed. (revised by J. B. Wilkinson), Chemical Publishing Co., New York, 1962.

23. Redgrove, H. S., and Foan, G. A.: *Hair dyes and hair dyeing*, 3rd ed. (with J. Bari-Woollss), Heinemann, London, 1939.

24. Klein, F.: *Die moderne Kosmetik*, Carsten, Berlin, 1919.

25. Winter, F.: *Handbuch der gesamten Parfümerie und Kosmetik*, 2nd ed., Springer, Wien, 1932.

26. Wall, F. E.: Bleaches, hair colorings, dye removers, in *Cosmetics: science and technology*, 1st ed., 1957.

27. Flesch, P.: Protection of H_2O_2-bleached hair with acid solutions, *Proc. Sci. Sec. TGA*, 321 (1959); *Am. Perf.*, **75**: 25 (May 1960).

28. Treatment of bleached hair to destroy peroxide and persulfate residues, *Schimmel Briefs*, No. 382, 1967.

29. Beigel, H.: *The human hair*, Renshaw, London, 1869.

30. Preisinger, S.: Microscopic examinations on hair coloring, *Am. Perf.*, **75**: 27 (July 1960).

31. Den Beste, M., and Moyer, A.: An instrumental description and classification of natural hair coloring, *JSCC*, **19**: 595 (1968).

32. Evans, R. L.: Some technical aspects of coloring hair, *Proc. Sci. Sec. TGA*, **10**: 9 (1948).

33. Nicolaus, R. A.: *Melanins*, Hermann, Paris, 1968.

34. *Federal Register*, February 27, 1971.

35. Color shampoos, *Schimmel Briefs*, No. 332, 1962; Gillette: Br. Pat. 889,327 (1962).

36. Den Beste, M.: U. S. Pat. 2,763,269 (1956).

37. Peacock, W. H.: Certified colorants: some color and colorant problems met when revising formulas, *Am. Perf.*, **75**: 25 (November 1960).

38. Rowland, A.: *The human hair*, Piper, London, 1853.

39. Goldemberg, R. L.: Hair coloring—modern formulation considerations, *JSCC*, **10**: 291 (1959); Keratin substantivity, *DCI*, **85**: 618 (1959).

40.   Heald, R. C.: Methods of dyeing hair without the use of an oxidizing agent (a review of recent patents), *Am. Perf.*, **78**: 40 (April 1963).

41.   Alexander, P.: New trends in hair dyes, *Am. Perf.*, **82**: 31 (June 1967).

42.   Cook, M. C.: Semi-permanent hair colorings, *DCI*, **102**(6): 38 (1968).

43.   Goldemberg, R. L.: Hair coloring and bleaching, *DCI*, **89**:(4): 446 (1963).

44.   Corbett, J. F.: Recent developments in the synthesis of hair dyes, *JSDC*, **84**(11): 556 (1968).

45.   Peters, L., and Stevens, C. E.: Effect of organic solvents on absorption of dyes by wool..., *JSDC*, **72**: 100 (1956); **73**: 23 (1957), **76**: 543 (1960).

46.   Beal, W. *et al.*: The dyeing of wool by solvent-assisted processes, *JSDC*, **76**: 333 (1960); Br. Pat. 826,979 Peters, L., and Stevens, C. B.: (1960).

47.   Solvent-assisted dyeing of hair, *Schimmel Briefs*, No. 365, 1965.

48.   Sardo, F.: Shampoo hair coloring compositions for direct dyeing of human hair; *JSCC*, **20**: 595 (1969).

49.   Heald, R. C.: Hair coloring compositions based on chrome dyes, *Schimmel Briefs*, No. 312, 1961.

50.   Lustig, B.: Br. Pat. 979,405 (1965).

51.   Stead, C. V.: Recent developments in the chemistry of hair dyes, *Am. Perf.*, **79**: 31 (February 1964).

52.   Broadbent, A. D.: Reactive dyes as colourants for human hair, *Am. Perf.*, **78**: 21 (March 1963).

53.   Shansky, A.: Reactive dyes, *Am. Perf.* **81**(11): 23 (1966).

54.   Kalopissis, G., Bugaut, A., and Bertrand, J.: Synthesis and use in hair preparations of some new quaternary dyes derived from anthraquinone, *JSCC*, **15**: 411 (1964).

55.   Lake, D. B., and Hoh, G. L. K.: Recent advances in fatty amine oxides: I, *J. Am. Oil Chem. Soc.*, **40**: 623; II (Matson, J. P.), **40**: 640 (1963); through *Schimmel Briefs*, No. 348 1964.

56.   Poucher, W. A.: *Perfumes, cosmetics and soaps;* Vol. I, *Dictionary of raw materials*, 4th ed., Van Nostrand, New York, 1936.

57.   Foan, G. A. *et al.*: *The art and craft of hairdressing*, 3rd ed., N. E. B. Walters, (Ed.), Pitman, London, 1950.

58.   Mann, H.: *Die moderne Parfümerie*, Verlag Chem. Ind., Augsburg (Germany), 1912.

59.   Bachstez, M.: Haarbleich- und Haarfärbemittel (Chemischer teil), in Truttwin, M.: *Handbuch der kosmetischen Chemie*, 2nd., Barth, Leipzig, 1924.

60.   René le Florentin: *Cosmétiques et produits de beauté*, Desforges, Paris, 1938.

61.   Hofmann, A. W.: Research on poly-ammonias, *Proc. Roy. Soc. Lond.* (VIII) **10**: 495 (1860); (XIX) **11**: 518 (1861); (XXIV) **12**: 639 (1863).

62.   Griess, P.: Decomposition of isomeric diamino-benzoic acids, *Chem. Jahrg.*, **3**; 143 (1871). *Ber.* **5**: 201 (1872).

63.   Monnet et Cie.: Fr. Pat. 158,588 (1883).

64.   Erdmann, E., and Erdmann, H.: Ger. Pats. 47,349, 51,073, 64,908, 80,814 (supp. 47,349), 92,006 (1888–1896).

65.   Godefroy, A. -F.: *Mémoires*, Long, Paris, 1933.

342   FLORENCE E. WALL

66. Rousseau, E.: *Les teintures capillaires à la p-phénylènediamine*, LeGrand, Paris, 1914.

67. Boudou, G.: Personal communication, 1939.

68. *Bundesratsbeschluss*, Berlin, February 1, 1906.

69. Colman, J., and Löwy, A.: Über Primal—ein neues unschädliches Präparat zum Farben von Haaren, *Deut. med. Wochenschr.*, **37**: 926 (1911).

70. Cathelineau, H.: Accidents provoqués par une teinture pour cheveux à base de chlorhydrate de paraphénylènediamine, *Bull. soc. franc. derm. syph.*, **6**: 16 (1895); *Ann. derm. syph.*, **9**: 29 (1898).

71. Brocq, L.: Les éruptions eczématiformes provoquées par une teinture pour cheveux à base de chlorhydrate de paraphénylènediamine, *Bull. méd.*, **12**: 237 (1898).

72. Schueller, L.: Les teintures pour cheveux, *Rev. sci.*, **9**: 417 (1908).

73. Sabouraud, R.: Sur les préjugés concernant les cosmétiques, dépilatoires, et teintures pour cheveux, *La Clinique (Paris)*, **6**: 694 (1911); What the doctors say: about cosmetics and toilet dyes, *New York Herald* (Paris ed.), November 26, 1911.

74. Sabouraud, R.: Les teintures de cheveux, *Presse. méd.*, **32**: 819 (1924); *Tiba*, **2**: 729 (1924).

75. Evans, R. L.: U. S. Pat. 1,497,262 (1924); Br. Pat. 234,971 (1924).

76. Snell, F. D.: Ralph Liggett Evans, Medalist, *JSCC*, **2**: 43 (1950).

77. Sabouraud, R.: *Diagnostique et traitement des affections du cuir chevelu*, Masson, Paris. 1932.

78. Goodman, H.: *Cosmetic dermatology*, McGraw-Hill, New York, 1936.

79. Wall, F. E.: Fifty years of paraphenylenediamine, *J. Tech. Assoc. Fur Ind.*, **5**: 118 (1934).

80. Schwartz, L., and Barban, C. A.: Paraphenylenediamine hair dyes, *AMA Arch. Dermatol. Syphilol.*, **66**: 233 (1952).

81. Kass, G. S.: Technology of modern oxidation hair dyes. *Am. Perf.*, **68**: 25 (July 1956); **68**: 34 (August 1956); **68**: 47 (September 1956).

82. Heilingötter, R.: The constitution, coloring power, and toxicity of hair dyes, *Am. Perf.*, **63**: 345 (1954).

83. Kass, G. S. and Hoehn, L.: Color reactions of oxidation dye intermediates, *JSCC*, **12**: 146 (1961).

84. Tucker, H. H.: Hair coloring with oxidation dye intermediates, *JSCC*, **18**: 609 (1967).

85. Tucker, H. H.: The formulation of oxidation hair dyes, *Am. Perf*, **83**: 59 (October 1968).

86. Bandrowski, E.: Über die Oxidation Paraphenylendiamins, *Wien. Monatsh. Chem.*, **10**: 123 (1889); *Ber.*, **27**: 480 (1894).

87. Cox, H. E.: The chemical examination of furs in relation to dermatitis, *Analyst*, **59**: 3 (1934): **60**: 350 (1935).

88. Altman, M., and Rieger, M. M.: Function of Bandrowski's base in hair dyes, *JSCC*, **19**: 141 (1968).

89. Brody, F., and Burns, M. S.: Studies concerning the reactions of oxidation dye intermediates, *JSCC*, **19**: 361 (1968).

90. Dolinsky, M., Wilson, C. H., Wisneski, H. H., and Demers, F. X.: Oxidation products of p-phenyldiamine in hair dyes, *JSCC*, **19**: 411 (1968).

91.   Corbett, J. F.: Intermediates and products in oxidative hair dyeing, presented at Joint Conference on Cosmetic Sciences, April 1968; published by TGA, 1968.

92.   Stabilization of hair dye solutions, *Schimmel Briefs*, No. 349 (and *cf.* No. 275), 1964.

93.   Schueller, E.: U. S. Pat. 2,610,941 (1952).

94.   Lever Brothers: Br. Pat. 712,451 (1954).

95.   Goldemberg, R. L., and Tucker, H. H.: Effect of base composition on the properties of oxidation hair dyes, *JSCC*, **19**: 423 (1968).

96.   Kritchevsky, W.: U. S. Pat., 1,663,202 (1928).

97.   Heilengötter, R.: Oxidation hair-dyes in stick form, *Am. Perf.*, **75**: 19 (May 1960).

98.   Pyridine derivatives as oxidation hair dyes, *Schimmel Briefs*, No. 377, 1966.

99.   Lange, F. W.: Ger. Pats. 1,141,748 (1962), 1,142,045 (1963).

100.   Cox, H. E.: The chemistry and analysis of henna (hair dyes. I), *Analyst*, **63**: 397 (1938).

101.   *Merck Index*, 8th ed., Merck, Rahway, N. J, 1968.

102.   McDonough, E. G.: *Truth about cosmetics*, Drug & Cosmetic Industry, New York, 1936.

103.   Trommsdorf, J. B.: *Kallopistria, oder die Kunst der Toilette für die elegante Welt*, Erfurt, 1805.

104.   Wall, F. E.: Metallic restorers under fire, *Am. Perf.*, **35**: 34 (December 1937).

105.   Müller, C.: Quoted from *Dermatol. Wochschr., 1918:* No. 40, p. 674; in Bachstez (ref. 59).

106.   Preparations for removing dyes from hair, *Schimmel Briefs*, No. 355, 1955.

107.   Wall, F. E.: The clinical investigation of cosmetics, *Proc. Sci. Sec. TGA*, **5**: 28 (1946).

108.   Goldemberg, R. L.: Cosmetics and the general population—safety aspects, *Proc. Sci. Sec. TGA*, **38**: 34 (1962).

109.   Wall, F. E.: Cosmetics—outcast of medical science, *Med. Times and L. I. Med. J.*, **61**: 334 (1933).

110.   *J. Assoc. Offic. Agr. Chemists*, Washington, D.C.

Chapter 24

HAIR CONDITIONERS, LACQUERS, SETTING LOTIONS, AND RINSES

Frank J. Berger and George H. Megerle

Hair Conditioners

The words "hair conditioning" are all-encompassing. They can be applied to almost every hair product on the market. Indeed, every hair product on the market must claim to condition hair in order to help ensure consumer acceptance. Thus hair conditioning, even though an entity of its own, is in reality closely interwoven with each of the other specialty products under discussion, namely, hair lacquers, setting lotions, and rinses.

In order to understand hair conditioning *per se*, it is important first to understand the basic structure of hair and the ways in which the structure and appearance may be affected.

Basically, hair consists of three layers. These are the cuticle, the cortex, and the medulla (1). Each of these regions plays a specific role insofar as the "condition" of the hair is concerned.

The cuticle of the hair is the layer on the outside of the hair shaft which consists of separated, flattened, horny scales of hard keratinized protein. These scales overlap and completely cover the hair shaft (2). The scales can move independently over each other, resulting in a flexible covering which protects the remainder of the hair shaft. The gaps between the scales may act as reservoirs for natural sebum which helps maintain flexibility and sheen. These gaps are also responsible for some of the optical properties of hair and give the hair a greater three-dimensional appearance.

The cortex makes up the major portion of the hair fiber. Indeed, the strength and elasticity of the hair is due predominantly to the structure of the cortex. The cortex consists of amino acids such as tryptophan and tyrosine, various inorganic moieties, and sulfur-containing compounds such as cystine (3). The amino acids are condensed during the keratinization process into long, intertwined molecules whose exact chemical structure is to date unknown.

The central column of the hair fiber is called the medulla and usually consists of soft keratin. Two of these regions play a specific role insofar as the "condition" of the hair is concerned. The soft keratin may be totally missing in sections of the medulla, but this results in little or no loss of strength or flexibility of the fiber.

Although this overall view of the hair shaft has been greatly simplified for the purposes of this discussion, it should prove sufficient in helping to point out some of the problems encountered in hair conditioning. Additional information on the structure of hair may be found elsewhere (4).

The parameters involved with the mechanical and surface properties of hair which enter into the conditioning of the hair are listed by Shansky as being lubricity, management, substantivity and sheen (5). Powitt, however, defines these parameters as being related to handle, feel, texture, quality, body, and manageability (2).

The cuticle represents the portion of the hair shaft best suited for the simple changes which can affect some of the above parameters associated with hair conditioning. The sheen, feel, and lubricity of hair may be improved by the simple addition of certain classes of chemicals to this portion of the hair. Many of the methods which attempt to improve these specific attributes involve the deposition of a film on the hair. This film may be continuous or discontinuous, substantive or not, monomolecular or thick. There is no doubt, however, that the thicker films are less effective, because they tend to adversely influence the feel of the hair. The sheen and feel of the hair may also be improved by physically ordering the scales of the cuticle, which results in a smoother hair surface with improved optical properties. This ordering of the scales may be accomplished by the use of a simple acidic rinse. Most severe hair treatments such as bleaching, coloring, and curling or straightening are carried out in an alkaline or strongly alkaline medium. This results in the swelling of the hair and the raising of the scales of the cuticle. Acidic treatments tend to reverse these effects.

Adsorption of chemical compounds onto the cuticle of the hair also results in improved sheen, gloss, and lubricity. Chemical compounds which are capable of adsorbing onto the hair are said to be substantive. Although the mechanisms involved in adsorption are probably very complex, this phenomenon may be treated as a simple Lewis acid-base reaction. This is true, in particular, for the adsorption of cationic surfactants on hair. The free amino groups of the keratin of the hair are electron-rich, but chemically carry no charge. These groups can electrostatically bond with electron-poor compounds to form rather strongly bonded complexes. Cationic surfactants would, in this scheme, be considered as Lewis acids, and the hair as a Lewis base. This simple explanation does not, however, explain why a wide variety of materials which cannot be considered Lewis acids can adsorb to the hair shaft.

Wilmsmann and Marks have shown that nearly all surface-active compounds react with keratin and with various enzymes found on skin (6). From this work, it can be seen that surface dirt on hair will reduce substantivity, and that substantivity is greater on damaged hair, or porous hair, than it is on hair in good condition. From this it would appear that the films formed on hair by substantive compounds are neither monomolecular nor continuous. These films do, however, alter surface characteristics to the extent that conditioning effects are apparent.

The cortex of the hair is often exposed near the ends of the hair when it has been treated with strong chemical reagents during the bleaching, tinting, straightening, or permanent waving operations. As with the cuticle of the hair, films may be deposited on the exposed portions of the cortex. However, it is extremely difficult to repair or to strengthen this part of the hair. Ideally, one would like to replace proteins which may have been washed out of the cortex or may have been chemically removed or destroyed. It is, however, not possible to chemically react conditioning materials with the keratin fibers of this portion of the hair using presently known technology and the simple one-step conditioning treatments which are employed by the trade at this time. Some of the limitations placed on this system arise from the fact that penetration or diffusion of molecules is limited by the size of the molecules that are to be used as conditioning materials. Furthermore, any molecule which can diffuse into this portion of the hair can also diffuse out of the hair again unless fixed or altered within the hair cortex itself. The size limitations of the molecules which can be employed have been determined by Wilmsmann in a paper dealing with the coloring of hair by oxidative dyestuff intermediates (7). These limitations can be overcome in part, for example, by the use of such treatments as permanent waves containing added materials such as cystine which can be made to chemically combine with the hair during the neutralization step. Such treatments are, however, best considered as permanent waves, and will not be covered here.

The medulla of the hair is least accessible to conditioning treatments. Because the structure of the hair is such that the soft keratinized protein of the medulla may be missing, with no apparent ill effect on the hair, little may be gained by attempting to chemically alter this portion of the hair. The inaccessibility of this portion of the hair shaft also makes the treatment untenable without first altering the cuticle and cortex of the hair.

A large number of materials have been used for the purpose of conditioning hair. These materials generally fall into a limited number of classes including proteins, surfactants, and materials which can broadly be classified as oils and waxes.

The broad class of proteins includes such materials as collagen-derived polypeptides whose sorption properties on human hair have been demonstrated (8). Sorption includes, by definition, not only adsorption but also absorption into the hair shaft. Such factors as time and pH greatly affect sorption rates, sorption being optimum at a pH of about 6.0 on damaged hair. The quantities of protein incorporated into the hair also increase with increasing contact time (9). Although the polypeptides are water-soluble, they cannot be totally removed by washing the treated hair with water. This has been shown by the presence of hydroxyproline on protein-treated, washed hair—a compound not found in human hair but present in collagen-derived proteins (10). The various substantive effects of proteins have been adequately described by Karjala, Johnsen, and Chiostri (11). The conditioning effects of these materials in cosmetic products, such as permanent waves and bleaches, have also been determined (12,13).

Protein additives to cosmetics are not limited to collagen-derived proteins. A wide variety of protein sources may be used for these purposes, including such materials as processed milk casein (14) and albumin.

The treatment of hair with surface-active agents has been discussed by Wilmsmann and Marks (6). Included in this work are a number of types of surface-active agents. However, the most generally used materials for hair conditioning are the cationic surfactants and the amphoteric surfactants. These materials are keroplastic and substantive as well as surface-active (15,16). Care must, however, be taken in the formulation of conditioners because these materials absorb strongly and are irritating to the eyes. The maximum recommended levels for various cationic compounds are given by Roehl (17) and by Finnegan and Dienna (18). The in-use concentrations should generally remain at levels between 0.1 and 1.0 % in order to be both safe and effective. The use of higher levels may be not only dangerous but also unnecessary, since only specific amounts of these materials will be adsorbed by the hair (19,20). The amounts adsorbed do, however, depend on the condition of the hair as well as the chemical structure of the cationic material employed (5).

The chemistry of various cationic surfactants has been discussed by Kling (21), and information is also available from various manufacturers. These materials also need not be restricted to the simple tetraalkyl substituted amine salts and can include compounds where the nitrogen is part of a ring system. The quaternization of the nitrogen can be carried out using a number of reaction routes (22).

The use of oils and waxes to condition hair, insofar as merely lubricating the hair shaft and improving the optical properties of the hair are concerned, is probably the oldest method employed for hair conditioning. Preparations, such as brilliantines and oil-based hairdressings employing mineral oil, replaced natural oils and waxes. Formulations for this type of conditioning effect can be found in many of the older texts dealing with cosmetics (23). These types of conditioners do, however, have severe limitations because the hair is coated with a sticky film which holds dirt and dust. Many of the more modern formulations employ such compounds as cholesterol, lanolin, silicone oils, and panthenol, the precursor of pantothenic acid, a vitamin of the B-complex group.

The products based on cholesterol and lanolin probably most closely duplicate the action of sebum naturally present on hair (24). Little can be said, however, on the adsorptive or absorptive properties of these compounds and their derivatives. The derivatives of these compounds are generally more cosmetically acceptable, since they can be made with less color and odor and are available with various degrees of solubility.

Silicone oils also find application in hair conditioners. These materials are chemically stable and are available in a wide variety of viscosities and with various solubility characteristics. Certain of these compounds also exhibit an inverse solubility relationship, being more soluble in cold water than in the heated solvent (25). This makes these compounds excellent candidates for conditioning hair rinses which are applied to the hair and then rinsed out with warm or hot water. A certain amount of this oil then remains on the hair after rinsing.

Panthenol is also widely used in hair-conditioning preparations. This material is oxidized to pantothenic acid, which has been shown to be essential for the growth and normal maintenance of the skin and hair (26). Increases of the pantothenic acid content of the hair have been reported after the application of panthenol to human hair (27). The stability of the vitamin and vitamin precursor has been studied by Rubin (28), and the chemical properties of these materials are readily available (29).

A number of other materials may also be used for conditioning hair. Glycerol, for example, is believed to have a good chance of penetrating the hair shaft and remaining there after rinsing. The effect of this compound in the hair is said to result in a softening and swelling of the hair as well as providing a humectant action in the hair (30).

The bitter components of gentianaceous plants such as *Swertia Japonica Makino* diluted with an innocuous vehicle are said to improve the luster of the hair. Chemically, these compounds are glucosides and alkaloids, and are used at a level of 0.001 to 1.0 % (31).

The resin polyvinylpyrrolidone is also substantive to hair. When hair is placed in contact with a PVP solution, then rinsed with water, not all of the resin is removed. The presence of this compound on the hair can be demonstrated by treating the hair with a weak solution of iodine. The resulting mahogany color is characteristic of the PVP-iodine complexes (32). Microscopic examination of hair which has been treated with PVP shows that the hair has an improved appearance and seems to be smoother.

The strength and elasticity of hair can also be improved by treating hair with a suspension of the reaction products of urea or thiourea and formaldehyde (33). This product is said to react with the free amino groups on keratinous material to form a film which can strengthen the hair. This film can also make hair more resistant to moisture.

A wide variety of other materials have also been used to condition hair. These materials also have their limitations as well as outstanding characteristics for use in hair conditioners. It may, however, be of use to consider some basic formulations employing the various classes of compounds. These formulas are essentially starting formulations, and such factors as product stability, proper methods of product preservation, and the perfuming of the product must also be considered.

Proteins and amino acids may be formulated into hair conditioners in a number of ways. The substantivity of these compounds has been demonstrated (8,34). A typical starting product is shown in Formula 1 (35).

Formula 1

WSP-X250	400.0 g
Triton X-100	14.0
Tween 20	56.5
Methylparaben	7.5
Propylparaben	1.5
Propylene glycol	50.0
Roccal	20.0
Perfume	q.s.
Water to 1.0 gal	

The formulations which include proteins must be adequately preserved. Acceptable preservatives include potassium sorbate and formaldehyde (36). Formaldehyde may, however, react with protein and result in a darkening of the solution and loss of the preservative. Bronopol (37) has also been recommended for the preservation of proteins in cosmetic products (38).

Protein-containing hair conditioners may also be formulated with other conditioning agents such as PVP (39).

Proteins can also be incorporated into cream rinse products or instant conditioners. The concentrations of the protein must, however, be relatively high because these conditioners or rinses must achieve effects in a short period of time. A typical formula is Formula 3.

Formula 2

PVP K-30	3.0%
Neobee M-20	5.0
Drewmulse 1128	5.0
Water	73.5
Triethanolamine	1.0
Carbopol 934	1.0
WSP-X250	5.0
Amerchol L-101	3.0
Lipal 15 CSA	3.0
Preservative	q.s.
Perfume	0.5

Procedure: Dissolve the PVP in part of the water. Dissolve the Carbopol in the rest of the water. Combine and heat the remaining ingredients exclusive of the protein, perfume, and preservative. Combine the two aqueous solutions, then add the oil mixture with vigorous agitation. Add the remaining ingredients while mixing.

Formula 3

Tegamine S-13	4%
Citric acid	1
WSP-X250	10
Water	85
Preservative, perfume, color	q.s.

Conditioners broadly classified as oil-based, or conditioners containing oils or waxes, can be formulated in a wide variety of ways. A typical formulation employing lanolin derivatives is shown in Formula 4 (40).

Formula 4

Acetylated lanolin	3.5%
Glyceryl monostearate, self-emulsifying	13.5
Spermaceti	1.5
Amerchol L-101	9.0
Mineral oil, 70 viscosity	8.5
Glycerol	4.5
Water	59.5
Perfume, color, preservative	q.s.

Procedure: Melt the oils at about 85°C and dissolve the water-soluble materials in the water. Mix the two phases at about 85°C and cool slowly while mixing.

Other conditioning ingredients can, however, also be incorporated into these systems. Formula 5 is an example of a conditioner employing proteins, panthenol, and surfactant (41).

Formula 5

Skilro	2.0%
Polawax	3.0
Novol (Croda)	1.0
Petroleum jelly	4.0
Stearyl alcohol	4.0
Sorbic acid	0.2
Ammonyx-4	10.0
Lanasan CL	1.0
Glycerol	3.5
DL-Panthenol	0.5
Water	70.8
Perfume	q.s.

Procedure: Mix the first six ingredients and heat to about 85°C. Dissolve the next four compounds in the water at about 80°C and add this to the oil phase. Mix at high speed and cool to 40°C before adding the perfume.

Surfactants, and in particular, cationic surfactants may also be used as hair conditioners (20,42). A large number of chemical types of surfactants are available (43). These materials, however, are generally incorporated into cream rinses, which are covered in a later part of this chapter. Cream rinses provide a basis for the formulation of instant hair conditioners. Materials such as proteins, silicone oils, and panthenol can be added to these formulations for additional conditioning effects. These products are then employed as instant conditioners although they are basically improved cream rinses.

Hair conditioners can incorporate an almost unlimited number of conditioning ingredients. These materials and the formulations into which they are compounded must all improve the visual appearance of the hair. Hair conditioners cannot repair hair that has been chemically or mechanically damaged.   Well-formulated hair conditioners can, however, improve such hair by changing the physical properties of damaged hair to result in improved feel, combability, and visual attributes.

Hair Lacquers or Hair Sprays

Hair spray has become a staple commodity in the cosmetic market along with such items as hair conditioners, permanent waves, and hair coloring. Indeed, this form of cosmetic has been available for a number of years before the advent of the aerosol system. This particular means of dispensing a hair spray, and the particular formulation requirements that must be considered for an aerosol system, are not within the scope of this chapter and will be covered elsewhere in this text.

The dispensing of hair sprays may be accomplished by a wide variety of methods other than the aerosol system. The simplest of these methods include the rubber-bulb and plunger-type sprays, corresponding to perfume atomizers or throat sprays and to window sprays, respectively. In addition, sprays powered by compressed air have found use in many beauty shops. This system is similar to the operation of the Innovair (44) aerosol system, in which the propellant and the product to be sprayed are contained in separate concentric containers. Other methods make use of electrically operated pumps, such as diaphragm pumps, to atomize or spray the desired product. Indeed, almost any type of device can be used for hair sprays as long as the spray pattern and the particle size distribution of the sprayed fluid are satisfactory. The latter is, however, not only a function of the type of sprayer used but also of the materials being sprayed. For example, for a given sprayer, a smaller range of particle sizes can be obtained if an alcoholic solution is sprayed rather than an aqueous solution.

Hair sprays, or hair lacquers, are formulated to coat the hair with a polymeric film which helps to keep the coiffure in place. The mechanism of the formation of this film is, again, not within the scope of this chapter but has been discussed by Shansky (45) and by various other authors cited in this paper.

Films, such as those employed in hair sprays, can be cast from a wide variety of naturally occurring and synthetic resins or from an almost unlimited combination of these resins. Formulations employing only natural resins have, however, been replaced to a great extent by formulations employing synthetic or chemically modified resins. Formulations for hair sprays as well as setting lotions employing natural gums and resins have been adequately publicized, and the limitations of these formulations, such as flaking and susceptability to moisture, are well known. Of the naturally occurring resins, however, one material in particular is still widely accepted and used. This material is shellac, or the resinous secretion of the parasitic lac insect.

Shellac is generally regarded as a mixture of polyhydroxy acids, mostly aliphatic, present in the form of lactones, lactides, and interesters, varying in molecular weight from about 300 to about 3000. The mean molecular weight is of the order of 1000, and the empirical formula of this material can be represented as C_4H_6O. The resin is acidic, with an ionization constant of about 1.8×10^{-5} (46). The average molecule contains at least one free carboxyl group, five hydroxyl groups, and three ester groups (47). This raw material is a very complex mixture of chemical compounds whose true chemical composition has defied exact chemical analysis. Shellac is, however, nonirritating to skin and nonhygroscopic, and has the ability to form smooth, relatively flexible films. The material is also compatible with a wide variety of other resins. These properties, as well as vapor permeability, alcohol solubility, and ease of saponification with subsequent solubility in water, make this resin suited for hair sprays (48).

The formulation of a shellac-based hair spray is, however, not an easy matter. The very versatility of this resin permits a wide variety of formulations, each with specific attributes as well as limitations. For example, the solvent system employed can be water, alcohol, or any mixture of these two commonly used solvents. Plasticizers can range from glycerol to esters of citric acid, and the pH of the product can be adjusted with a number of bases including borax. Martin (49) has listed various compatible plasticizers and synthetic resins which can be used with shellac. Included in this work is the basic formula for a water-based hair spray reproduced in Formula 6.

Formula 6

Refined, wax-free, bleached shellac	15.00%
Borax, 10 M	3.45
Water	81.55

Procedure: Heat the water to 63°C and dissolve the borax. Dissolve the shellac at this temperature using a high-speed stirrer, cool, and filter. Adjust the pH to about 8.5 with ammonium hydroxide. Dilute as follows:

Shellac-borax concentrate	80.0%
Citroflex-type plasticizer	1.0
Perfume	0.2
Water	18.8

The borax in Formula 6 can be replaced with either ammonia or morpholine. Ammonia (26°) and morpholine are generally used at a level of 16 parts per 100 parts shellac, and borax (10 M) is used at about 23 parts per 100 parts shellac. Alcohol can also be added in the final dilution step to speed drying time and to help preserve the product (48).

Shellac formulated in an alcohol-based hair spray can be neutralized with triethanolamine, 2-amino-2-methyl-1-propanol (AMP) or with 2-amino-2-methyl-1,3-propanediol (AMPD). The amounts that are used depend on the properties desired in the finished formula. In general, the greater the degree of neutralization of the shellac, the poorer is the humidity resistance and the better the removability of the spray (48). The optimal degree of neutralization should be determined after a plasticizer and perfume have been incorporated into the product.

Shellac-based formulations should also be preserved against microbial or bacterial attack, particularly when an aqueous-based product is formulated. Preservatives such as the methyl or propyl esters of 4-hydroxybenzoic acid suffice for these products, although other preservatives can also be used(47).

Shellac is, however, not the only resin which finds use in hair sprays. From a historical point of view, PVP was one of the first of the synthetic resins to find wide application in hair sprays. The resin is nontoxic (50) and has excellent holding power for hair, attributable in part to its substantivity (32).

This resin does, however, have various shortcomings for use in hair sprays. The rather strong substantivity of polyvinylpyrrolidone to hair makes it difficult to remove quantitatively. Furthermore, the film formed by this resin is rather brittle and hygroscopic. Consequently the film tends to flake or to become tacky and dull in humid atmospheres (51). Although the brittleness and, in part, the dullness of this film can be changed by the use of a suitable plasticizer such as isopropyl myristate, the hygroscopicity of the film is not so easily changed. Basic hair-spray formulations employing PVP as the film-forming resin can be found in the first edition of this text.

The copolymers of polyvinylpyrrolidone and vinyl acetate (PVP/VA) have properties resembling those of PVP itself. The vinyl acetate moiety, being water-insoluble, reduces the hygroscopicity of this resin. However, as the vinyl acetate content of the resin increases, the hair-holding ability of the resin decreases while the resin becomes less sensitive to humidity (52). A number of resins with various PVP/VA ratios are available commercially. The choice of the specific resin depends on the specific properties that are desired for the hair spray.

Other modifications of the basic PVP resin also exist. For example, PVP combined with dimethyl hydantoin formaldehyde resin (DHFR), results in a film which imparts less stiffness to the hair. These films are also less tacky in moist atmospheres, but perfume oils tend to plasticize the films. This effect can be overcome by the addition of various substituted diphenyl compounds to strengthen the resinous film. These additions are, however, covered by a patent (53).

The various additives that can be employed in hair sprays have been adequately covered in the literature (52,54,55). These additives fall into various chemical classes and include materials which plasticize the resin film, improve the luster and gloss of the film, impart antistatic properties to the film, or harden the film. Other possible additives can color the film, absorb ultraviolet light, or provide disinfectant or conditioning properties. All these materials should, however, be checked for efficacy in the particular formulations employed.

A wide variety of other resins can also be employed in hair sprays. These include polyester-based resins which are said to be excellent for use on oily hair (56,57). A partial listing of various resins usable for such purposes follows:

PVP, PVP/VA, polyvinylacetate, acrylic resins, polyacrylic acid resin, polyvinylimidazole, cellulose ethers, vinyl acetate crotonic acid resins, acryl sulfonamide formaldehyde resin, vinyl terpolymers, methylvinyl ether/maleic anhydride copolymers, and cyclohexanone resins.

Formula 7

Gantrez EZ-425, 50%	19.62%
Ammonia, 28%	1.89
Water	78.49
Perfume and color	q.s.

The use of some of these types of resins can best be illustrated by several formulations. The spray based on Formula 7 has good holding properties over a high humidity range. The film is nontacky and flexible and does not result in a stiff or lacquered appearance (39).

The wide latitude permissible for the formulation of hair sprays can be illustrated by Formula 8. The resin content of this hair spray has been drastically reduced in comparison to that contained in the previous formula (58).

Formula 8

Gantrez ES-425	4.0%
Ethyl alcohol	21.2
Water	74.0
Emulphor EL-719	0.3
AMPD	0.2
Perfume	0.3

Another aqueous alcoholic hair spray is shown in Formula 9 (59).

Formula 9

Polypeptide LSN, 40%	15%
Maypon UD	1
Maypon UD acid	1
Water	45
Ethyl alcohol	38

The ingredients are mixed in order as shown, aged, and filtered. The product is said to provide good holding power and is said to have an excellent conditioning action.

The effects of plasticizers and perfume oils on hair spray formulations can best be illustrated by Formula 10.

This is a hard-to-hold formulation which forms a very strong resin film. The omission of the silicone oil and part of the perfume, both plasticizers,

Formula 10

Resyn 78-3305 (20% neutralized)	3.5%
Silicone oil SF-1075	0.2
Plasticizing perfume oil	0.4
Ethyl alcohol	q.s. 100.0

results in a product which has far too much holding power. The resin film formed in the latter manner is also brittle and flakes when the hair is brushed. The degree of neutralization of this resin largely determines the water solubility and thus the subsequent ease with which this material can be removed by washing the hair.

Certain resins can tolerate relatively high concentrations of plasticizers without becoming tacky (60). This permits the formulation of hair sprays ranging from minimal holding power to maximum holding power by simply changing the concentration of the plasticizer. Formula 11 is an example of a regular or normal-hold spray.

Formula 11

Dicrylan 325-50	1.40%
Acetylated lanolin alcohols	0.04
Silicone fluid	0.01
Perfume	q.s.
Color	q.s.
Ethyl alcohol	q.s. 100.00

Procedure: The components may be mixed in almost any order at room temperature.

Another avenue of approach is, however, open for the formulation of hair sprays with varying degrees of holding power. The concentration of the film-forming resin as well as that of the plasticizer can be decreased. This can also be accomplished by the simple dilution of a product with the appropriate solvent.

It would not be practical to present starting formulations for each of the types of resins available for use in hair sprays, particularly because many of these resins can be used in combination with each other. The utility of these formulations is also limited. Although these formulations can serve as starting points on which to base further experimental work, the effects of various perfumes as plasticizers could necessitate an increase in the resin content or a change in the solvent system employed. Formulations are given by most of the manufacturers of the resins employed in these systems. Even aerosol formulations can be used as starting points for nonaerosol sprays by replacing the propellants with a suitable solvent such as ethyl alcohol. Thus the approximate concentrations of the various resins employed in hair sprays are readily available. The perfume oil should then be chosen, and checked for its ability to plasticize the particular resin. Other additives can then be incorporated into the system. These can include proteins for added crispness of the film as well as for the conditioning claims that can then be made. Panthenol can be added for conditioning and lubricity effects, and plasticizers to modify the resin film can also be employed. Plasticizers such as isopropyl myristate and diisopropyl adipate also give a certain conditioning effect insofar as they generally result in added gloss, even though their primary function is to make the resin film more flexible. The resin content must then be adjusted according to the holding power desired in the hair spray. The process is generally one of trial and error until the effects of the additives on the amount of resin used are balanced properly to result in an acceptable product.

Setting Lotions

When a primitive woman pushes back her almost unkempt hair and attempts to keep it in place with some water, she is actually seeking to use a waveset or, to be more exact, a hairset. Nearly all of the effect that she desires is lost when the water evaporates, but not before a minute amount of swelling has taken place, and hence some setting has been brought about.

The rate of swelling of hair in pure water was experimentally determined by Valko and Barnett (61). In addition, Heilingötter (62) has shown the difference in elasticity and stretch between dry and wet hair. Probably one of the first advances in wavesets was the use of seeds which were pounded into a mucilaginous mass, a method which seems to date back to ancient Egypt.

It was not too long ago that natural materials such as gum karaya, acacia, tragacanth, and quince seed were being utilized to manufacture finger wavesets and hairsets. Alginates also saw a period of use, and some of these are still around today. Formulations for this type of product will be found in Bennett (63), Thomssen (64), and Keithler (3). Today, however, even the name of the product has been changed. One no longer talks of wavesets and hairsets. In line with the thought of built-in hair conditioning, the product is now a setting lotion, because "lotion" denotes richness and desirable qualities. In general, setting lotions and so-called conditioning setting lotions are based on the same type of synthetic materials used in aerosol and nonaerosol hair sprays. The differences between these products, sprays and sets, lie in the manner of compounding.

Films formed by hair lacquers are definitely not continuous in nature and are far stronger than those used in setting lotions. Once a hair lacquer has been sprayed on the hair, very little further styling or rearrangement of the hair takes place. Quick drying is essential. With a setting lotion, on the other hand, the hair is well moistened before being put up on some sort of curler or rod. Once dried, however, the hair is handled extensively. Since the film formed should not rupture or flake during this handling, the use of larger amounts of plasticizer is called for. In a hair lacquer, on the other hand, plasticizers are kept at a minimum so as to make the final film more moisture resistant. Thus the same basic materials used in hair lacquers can be used in setting lotions. The difference is purely a matter of concentration of film former and the amount of plasticizer used. Naturally this relationship will vary according to the resin used but, in general, it will hold true.

The advent of polyvinylpyrrolidone, its copolymers, and monoester resins, revolutionized the hair-setting field. Many types of polymers, sizing agents, and even starches can be used as setting lotion resins.

Some enterprising females utilize beer to set hair, and this works well up to a point. Fine hair, set with beer, will gain quite a bit of body, which is attributable to the protein content of the beer. From basic experiences such as this, it was a simple step to refine setting lotions to the point where they could be considered as conditioning setting lotions. The addition of materials such as proteins, panthenol, lanolin derivatives, and even small amounts of quaternaries to the basic film former, did, in fact, result in more body and lubricity. Indeed, these conditioning agents are enhanced when used in setting lotions because of the method of use. The product is applied liberally, not rinsed from the hair, and then completely dried on it. Thus many forces including relatively high temperatures are brought into play in the formation of the final film produced.

In general, the modern-type setting lotion is applied from a plunger dispenser, or small individual application bottles. Because of the extensive handling of the hair when dried, it is advisable to incorporate static charge reducers into setting lotion formulations. The use of polyethylene glycols for this purpose has been recommended (65). The compounder, however, must be careful not to upset the balance of his formulation in attempting to overcome static electricity. Excellent basic formulations utilizing PVP, PVP/VA, and monoester resins in hair-setting lotions are found in commercial literature (58). Formulas 13 and 14 are representative.

Formula 15 utilizes Dicrylan, which is an acrylic copolymer resin (66).

A good basic formula for an alcoholic setting lotion utilizes Lanexol, an alkoxylated lanolin emollient, as a conditioning agent (67).

Formulas for Setting Lotions

	13	14
Gantrez ES-225 or ES-425	4.0%	—
PVP/VA E-635	—	5.0%
Ethyl alcohol	21.2	20.0
Water	74.0	74.5
Emulphor EL-719	0.3	0.5
Aminomethylpropanediol (AMPD)	0.2	—
Perfume	0.3	q.s.

Formula 15

Dicrylan 325 WA-50	4.00%
Solulan 16	0.40
Silicone fluid SF-1066	0.01
Ethyl alcohol	19.90
Water	75.69
Perfume	q.s.
Color	q.s.

Hair setting or styling gels are of rather recent vintage and are quite popular. Since setting lotions have a low solids content, a thickener is required for this type of gel which will do the job at low concentrations. For this reason, just about all setting gels are based on Carbopol (68) resins. These resins are carboxy vinyl polymers of extremely high molecular weight. They

Formula 16

Lanexol	0.2 to 0.4%
PVP K-30	2.5 to 4.5
Dimethyl phthalate	0.1
Ethyl alcohol, 95%	47.0
Water	q.s. to 100.0
Perfume	q.s.

Procedure: Blend all components by warming and mixing together with mechanical agitation. Filter if necessary for clarity.

are supplied as powders in the acid form and require neutralization by a base to develop their thickening, suspending, and emulsifying properties. One of the attributes attendant to their use is the availability of clear types, which results in a very attractive product.

Formulas 17 to 24 are taken from an extensive review in which a comprehensive survey of hair styling is found (39).

Formulas for Setting Lotions

	17	18
Carbopol 940 resin	0.5%	0.30%
PVP K-30 (45% water solution)	5.7	2.89
Gafanol E-550B	3.3	—
Water	89.4	86.50
Ammonia, concentrated	0.6	—
Ethyl alcohol	—	9.61
Diisopropanolamine	—	0.30
Dimethyl phthalate	—	0.40
Perfume, water-soluble	q.s.	q.s.
Dye	q.s.	q.s.

Formulas for Setting Lotions

	19	20	21	22	23
PVP/VA E-735 (50% solids)	5.0%	—	—	—	—
PVP/VA E-635 (50% solids)	—	5.00%	—	—	—
PVP/VA E-535 (50% solids)	—	—	5.00%	—	—
PVP/VA E-335 (50% solids)	—	—	—	5.00%	—
PVP/VA S-630 (100% solids)	—	—	—	—	2.50%
Carbopol 940 resin	0.5	1.00	1.00	1.00	1.00
Water	90.2	80.11	68.52	56.94	80.11
Gafanol E-550B	3.8	3.80	3.80	3.80	3.80
Diisopropanolamine	0.5	1.00	1.00	1.00	1.00
Ethyl alcohol	—	9.09	20.68	32.26	11.59
Perfume, water-soluble	q.s.	q.s.	q.s.	q.s.	q.s.
Dye	q.s.	q.s.	q.s.	q.s.	q.s.

Procedure: Dissolve the Carbopol in water at room temperature.

Add the neutralizer (diisopropanolamine or ammonia) with agitation to pH 7. In formulations utilizing PVP K-30 and PVP/VA E-735, blend in the resin with moderate agitation. (PVP-/VA E-635, E-535, E-335, S-630 must first be mixed with the appropriate amount of ethyl alcohol.)

Melt the Gafanol E-550B and blend into the gel until a clear uniform effect is obtained. Where dimethyl phthalate is used, dissolve it in the ethyl alcohol; add this solution to the PVP K-30 solution.

Blend in the perfume and dye. Since such dilute solutions of PVP may support mold growth, 0.1% sorbic acid or 0.1% sodium benzoate should be added to the system.

Formula 24

PVP K-30	1.5%
Carbopol 940 resin	0.4
Lipal 20 OA	2.5
Ucon 50-HB-2000	10.0
Triethanolamine	1.5
Water	83.6
Perfume	0.5

Procedure: Dissolve the PVP K-30 in a portion of the water. Dissolve the Carbopol in the remaining water. Add the other ingredients together and heat until homogeneous. Add the PVP K-30 solution to the Carbopol solution and mix thoroughly. Add the oil mixture with vigorous agitation until homogeneous. Add the perfume with thorough mixing.

Note: Hair styling gels often undergo viscosity reduction when exposed to ultraviolet radiation.

Formula 25

WSP-X 250	5.00%
Carbopol 940	1.00
Triethanolamine	1.00
Solulan 98	2.00
Tegosept M	0.10
Tegosept P	0.05
Formalin	0.20
Perfume, color	q.s.
Water	q.s. 100.00

Procedure: Slurry Carbopol in about 90% of the required water. Add Tegosepts to the WSPX-250 and heat until dissolved, then mix with Carbopol slurry. Add and mix Solulan, formalin, remaining water, and finally the triethanolamine.

Formula 26

Solulan C24	1.0%
PVP K-30	2.0
Soleonic HAP	1.5
Triethanolamine	1.0
Carbopol 940	1.0
Water	93.5
Perfume and preservative	q.s.

Formula 27

Part A

Lanexol	0.50%
Ethyl alcohol	5.00
PVP K-30	2.00
Diisopropanolamine	0.35
Perfume	q.s.
UV light absorber	q.s.

Part B

Carbopol 940	0.40
Sodium ethylenediaminetetraacetate	0.01
Water, distilled	91.74
Colors and preservatives	q.s.

Procedure: Disperse the EDTA in the water, followed by the Carbopol to form a smooth 0.5% solution. Blend part A, warming slightly to speed solution. When dissolved, add A to B with slow mechanical agitation. The product will gel and become cloudy temporarily, then it will clear upon continued agitation.

Proteins can also be utilized in this type of formulation, as shown in Formula 25 by Karjala *et al.* (10).

The use of lanolin derivatives in hair styling gels is pointed out by DeRagon (69).

Lanexol is suggested as a plasticizer for a clear transparent gel, as illustrated in Formulas 27 to 29 (67,70).

Formulas for Setting Lotions

	28	29
Part A		
Uvinul D-50, UV absorber	0.1%	0.1%
Perfume	q.s.	q.s.
SD 40 alcohol	10.0	15.0
Carbopol 940	0.4	0.4
Water, distilled	29.6	29.6
Part B		
Solan	0	2.5
Volpo 20	2.5	0
Sequestrene Na$_4$	0.1	0.1
Water, distilled	42.8	37.6
Polyox resin WSRN 3000	0	0.2
Ucon 50-HB-660 fluid	10.0	5.0
PVP K-30	1.5	1.5
Glycerol	0	5.0
Diisopropanolamine, 10% aqueous solution	3.0	3.0

Procedure: Disperse the Carbopol in the water shown in part A. Dissolve the Uvinul and perfume in the alcohol of part A. When Carbopol slurry is completely uniform, add the alcoholic solution to complete A. Warm the water in B and dissolve the Solan or Volpo, then the Sequestrene, Polyox, and PVP in that order. Add the glycerol to that. Combine solutions A and B, then when uniform add the Ucon. Complete the gel by neutralizing with the amine. For ease of handling, some of the water from B may be used to make a more dilute solution.

Hair setting gels or styling gels are convenient to apply and allow the user further to style or manipulate the hair after application. The proper use of an FD&C color here is important because the appearance of the gel can be made quite attractive. Proper plasticizing in this type of formula is most important since the film formed should not rupture on brushing or combing of the coiffure. Improperly plasticized films will dust or flake, giving a virtual snowfall of resin flakes which look like dandruff.

The patent literature abounds with various materials and compounds used for the setting of hair.

Manzke (71) claims good gloss, elasticity, and nonbrittle films for setting lotions based on condensation products of proteins and maleic acid anhydride.

A British patent (72) states that derivatives of maleic anhydride and ethylene copolymers have an affinity for keratin and act as softening agents which improve the appearance and make the combing of hair easier.

Blance and Cohen (73) claim a lustrous flexible film adhering to the hair through the use of Poly (N-propyl-N-vinyl acetamide).

Rieger and Berenbom (74) claim the use of an aqueous dispersion of a tertiary aminoalkyl ether of starch for use in hairsets. These ethers are used at a 1 to 2% level.

Richardson and Hoss (75) patented the use of aqueous solutions of polyvinyl alcohol and a carboxy vinyl polymer which is said to leave a microscopically thin film on the hair. A typical illustration given in the patent is presented in Formula 30.

Formula 30

Polyvinyl alcohol (Elvanol, type 52-22)	0.100%
Carbopol 934	0.030
Butyl p-hydroxybenzoate	0.018
Triethanolamine	0.036
Water	99.816

Shansky (76) has patented the use of amides such as behenamide, arachidamide, stearamide, and palmitamide to opacify PVP solutions. Generally, opacifiers soften or plasticize the resins used as film formers. These compounds are claimed not to effect the formation of the film on the hair shaft. Formula 31 is given in the patent. This formula is claimed to have the maximum opacity desirable.

Formula 31

PVP	2.50%
Urea	1.25
Triethanolamine	1.25
Water	85.00
10% by weight of stearamide in Igepal CO-430	10.00

Wilmsmann and Ludwig (77) claim long-lasting sets in humid conditions. They utilize a two-application system. First, hair is treated with a solution of a polyepoxy compound, and then a solution of a hardener such as a polyamine or polycarboxylic acid is used.

A British patent (78) was obtained on the use of copolymers capable of forming salts which are soluble in water and which form a water-insoluble film when dried from an aqueous solution at 18 to 60°C. One of the salts cited is the formate salt of the copolymer from 40 % ethyl acrylate, 55 % of N-tert-butyl acrylamide and 5 % of N:N-diethylamine methacrylate.

Bechmann and Lukesch (79) claim low hygroscopicity, good elasticity, gloss, good film forming, as well as low toxicity and ease of removal, from a diisocyanate-modified gelatin.

To show that alginates are still used in the hairset field, we cite a British patent (80). Hair is set with an aqueous solution of a water-soluble alginate which has a pH of from 3.6 to 7.0. Thereafter the pH is lowered below 3.6. This precipitates alginic acid from solution as a film on hair. An example is Formula 32.

Formula 32

Sodium alginate	0.30%
Ethyl alcohol	15.00
Citric acid	0.10
Methyl ester of p-hydroxy-benzoic acid	0.12
Propyl ester of p-hydroxy-benzoic acid	0.06
Polyethylene glycol 4000	0.06
Water	q.s. 100.00

Hair Rinses

Hair rinses have had a long and varied history. Both vinegar and lemon juice were used for many centuries to remove limesoap after shampooing. Even before the advent of the synthetic detergents and the entrance of the cold wave upon the scene, improvements had been made to give impetus to the rinse field. Solutions of polybasic hydroxy acids were widely used. First, weak solutions of citric and tartaric acids, and later those of malic acid, were suggested as rinses. Although used as liquids, they were usually sold in powdered form, and the consumer was instructed to dissolve a small package in water in order to make the rinse, approximately ½ oz usually sold to make up 1 quart of rinse. This type of package was later to become the basis of the temporary color rinse. Actually, what these weak acids did was to remove any film left by the limesoap which had dulled the hair. The resultant gloss, however, was mostly due to the freshly washed hair, and was therefore a natural gloss.

The next step in the development of the hair rinse was the use of available types of cationic material, as exemplified by quaternary ammonium salts. Many of these materials had been developed for various industrial uses and were available under a variety of trade names. These gave a smooth and soft feel to the hair and contributed toward greater hair manageability, combability, and gloss.

Although clear solutions of some quaternary ammonium salts can easily be made, these preparations were usually marketed as emulsions or dispersions, not only for esthetic reasons, but also to heighten the impression that the hair is thus softened or "creamed." In this way the cream rinse, both the product and the term, may have arisen.

Since cationic agents form salts with anionic materials, application is usually a separate wash, or a rinse applied after other processes have been completed on the hair.

One of the earliest quaternaries used was an alkyl dimethyl benzyl ammonium chloride which had been developed as a germicide and deodorant. A simple cream rinse utilizing this material is shown in Formula 33. This is a preparation with a 3% concentration of active material. In use, the product is diluted about 15 times with water to form the actual rinse used on the hair.

Formula 33

Glycerol monostearate, acid-stabilized	3%
Alkyl dimethyl benzyl ammonium chloride, 50% solution	6
Water, distilled	91
Perfume and color	q.s.

Procedure: Place all the ingredients except the perfume into one vessel and heat to about 63°C, with agitation. Mix while allowing to cool until room temperature is reached.

One of the most widely used quaternaries in the hair-rinse field is Triton X-400. Chemically this material is 20% stearyl dimethyl benzyl ammonium chloride and 5% stearyl alcohol. This material seems to be well adsorbed on the hair shaft. A stable product can be made by following the procedure for the preparation of a Triton X-400 dispersion, as shown in Formula 34.

Formula 34

OPE-1	1.0%
Triton X-400 (Rohm and Haas)	12.5
Water, distilled	86.5
Perfume and color	q.s.

Procedure: Add the OPE-1 to the Triton X-400 with slow but thorough stirring until uniform. The total amount of water in the formulation is then divided into two equal parts. Heat the first part of the water to 63°C and add this heated water slowly to the previously formed paste, mixing thoroughly but not too rapidly. Cool to room temperature with a very slow, even mix. Chill the remaining portion of the water to 2°C and add slowly to the aqueous OPE-Triton solution. Stir until smooth and consistent, add perfume, color, and acidifying agents such as citric acid to bring the finished product to the desired pH.

A preparation made up along the lines above gives good results at an acid *p*H. The finished product will contain about 12% of the quaternary or about 3% of active product, which is diluted about 10:1 for final use. Most preparations of this type can also be formulated with acid-stabilized glyceryl monostearate to make preparations that appear creamier than those which are merely dispersed.

Another excellent formula for a concentrate product of this type has Emcol E 607 S as the active quaternary (81). This is a stearic acid homologue of N-acyl colamino formyl methyl pyridinium chloride. Formula 35 is noteworthy for the fact that it also includes a sterol-type product.

Formula 35

Oil phase	
Amerchol L-101	5%
Solulan C 24	2
Emcol E 607 S	2
Arlacel 165	4
Cetyl alcohol	1
Water phase	
Water	86
Preservative	q.s.
Perfume	q.s.

Procedure: Add the water phase at 70°C to the oil phase at 70°C while stirring. Cool with stirring to 40°C and add the perfume. Continue mixing and cool to 35°C.

Modifications: Solulan 25 or Solulan 16 may be substituted for the Solulan C 24 to give slightly higher or lower viscosities.

Hilfer (82) patented specific quaternaries which are not supposed to relax hair but to do all other things expected from a cationic hair rinse. One of the examples given and its use in a formulation is Formula 36. This

Formula 36

Quaternary compound	3.0%
Cetyl alcohol	0.8
Sodium bromate	0.5
Water, distilled	95.7

Note: The quaternary compound is formed by quaternizing a condensation product of 1 M of pyrrolidine and 30 M of propylene oxide with *o*-chlorobenzyl chloride.

product is in concentrate form. Gradually the concept of concentrate type of cream rinses faded in the direction of finished or ready-for-use products. This was due in part to packaging directed toward customer and consumer

Formulas for Hair Rinses

	37	38
Part A		
Arquad 2HT-75	2.67%	—
Arquad DM18B	—	1.87%
Aromox DM18DW-L-25	to 100.00	to 100.00
Sodium chloride	1.00 to 1.50	1.00 to 1.50
Water	—	0.25
Part B		
Ethomeen TD/25	0.50	—
Kessco PEG-600 distearate	—	1.00
Stearyl alcohol	2.50	0.60
Neo-Fat 18-55	0.72	—
Part C		
Perfume	q.s.	q.s.

Procedure: Heat parts A and B to 70°C. Add part A to part B under continuous agitation. Mix and cool to 40°C and add part C. Adjust evaporation losses with water.

convenience. Also, because of the almost complete replacement of soaps in the shampoo field by synthetic detergents, an acid-type rinse, as such was no longer needed.

Formulas 37 and 38 are ready-to-use products that have been suggested by a commercial supplier (83).

The cosmetic formulator is not limited to the use of finished quaternary salts in the preparation of hair rinses. Fatty amidoalkyldialkyl amine bases can be used to form salts with the addition of a preferred acid; many of these bases have been commercially marketed (84). A simple pearlescent cream rinse ready for use is shown in Formula 39.

Rinses are not limited to the use of this type of material. A recent British patent (85) claims superior manageability, a softer feel, and a more lustrous appearance through the use of a organosilicon compound.

Because of the germicidal activity that these types of quaternary compounds exhibit, they have been utilized as dandruff inhibitors. Although anti-dandruff preparations are beyond the scope of this chapter, we make mention of this one compound because it is used in the form of a rinse. Lauryl isoquinolinium bromide is of interest because of its activity against *Pityrosporum ovale*, the so-called bottle bacillus, which has been considered as a

Formula 39

Tegamine S-13	4.00%
Phosphoric acid, 85%	0.63
Water, distilled	95.37

possible cause of dandruff. A simple rinse can be made by mixing 1 oz of a 10% solution of lauryl isoquinolinium bromide with 1 gal of distilled water.

With the development of newer and more effective types of quaternary ammonium compounds, the cosmetic formulator has had a much bigger arsenal of material to choose from. Roehl (17), in an excellent review, lists 29 quaternary ammonium compounds from 14 manufacturers as being suitable for use in cosmetic preparations. Wilmsmann and Marks (6) attempted to measure the degree of adsorbtion of various quaternary ammonium compounds on hair and tabulated some excellent data.

Some of the newer quaternary ammonium compounds are discussed by Egan and Hoffman (20), who also ran adsorption studies, according to the method of Hughes and Koch (86).

Scott, Robbins, and Barnhurst (87) describe a depletion method for measuring the sorption of quaternary ammonium compounds on hair. Zwicker (88) mentions the downward trend of the cationic rinse as such, while pointing out possible variable uses of this type of product. There can be no doubt that the use of many of the cationic bases is indicated, together with compatible dyestuffs for the utilization of color rinses of the temporary as well as the semipermanent variety.

Increases in the efficiency of this type of product with regard to sorption, plus the use of compatible cosmetic adjuncts, lead the way toward newer product concepts. Gradually this type of product gained acceptance as a true hair conditioner (which has been discussed previously).

Thus it can be seen that hair conditioning is indeed all-encompassing. Hair conditioners as such can be formulated not only for various types of hair, but especially for the pretreatment or the posttreatment of other products, such as hair bleaches, hair colorings, and hair-waving products. Indeed, such is the state of the art today that hair conditioning items can be built into each of these products, as well as materials to reduce irritation and hair damage without affecting the efficacy of these types of products to a large degree.

Therefore, in today's multimillion-dollar hair market, all of the above types of products could conceivably be used on one and the same consumer or patron of a beauty salon. To this one can add a conditioning shampoo, and also a conditioning hair rinse followed by a setting lotion with conditioning effects, which is to be followed by the final hair spray.

REFERENCES

1. Kuczera, K.: *Seifen-Öle-Fette-Wachse*, **42**: 243 (1969).

2. Powitt, A. H.: Some properties of human hair, *Am. Perf.*, **83**: 53 (January 1968).

3. Keithler, W. R.: *The formulation of cosmetics and cosmetic specialties*, Drug and Cosmetic Industry, New York, 1956, Chapter 25.

4. *Ibid.*

5. Shansky, A.: Evaluation methods for some of the mechanical surface properties of hair: a theoretical consideration, *Am. Perf.*, **83**: 25 (March 1968).

6. Wilmsmann, H., and Marks, A.: Reactions of surfactants with keratin and enzymes, *Fette, Seifen, Anstrichmittel*, **61**: 965 (1959).

7. Wilmsmann, H.: Beziehungen zwischen der Molekülgrösse aromatischer Verbindungen und ihrem Penetrationsvermögen für das menschliche Haar, *JSCC*, **12**: 490 (1961).

8. Karjala, S. A., Williamson, J. E., and Karler, A.: Studies in the substantivity of collagen-derived polypeptides to human hair, *JSCC*, **17**: 513 (1966).

9. Karjala, S. A., Karler, A., and Williamson, J. E.: The effect of pH on the sorption of collagen-derived peptides by hair, *JSCC*, **18**: 599 (1967).

10. Karjala, S. A., Bouthilet, R. J., and Williamson, J. E.: Factors affecting the substantivity of proteins to hair, *Proc. Sci. Sec. TGA*, **45**: 6 (1966).

11. Karjala, S. A., Johnsen, V. L., and Chiostri, R. F.: *Substantive proteins in cosmetics*, Wilson Pharmaceutical and Chemical Corp., Chicago.

12. Bouthilet, R. J., and Karler, A.: Cosmetic effects of substantive proteins, *Proc. Sci. Sec. TGA*, **44**: 27 (1965).

13. Bouthilet, R. J., Karler, A., and Johnsen, V.: Cosmetic effects of substantive proteins. Part II, *Proc. Sci. Sec. TGA*, **45**: 27 (1966).

14. Salzberg, H. K.: Processed milk casein for hair and skin cosmetics, *Am. Perf.*, **82**: 41 (1967).

15. Roehl, E. L.: Quaternary ammonium compounds in cosmetics, *Dragoco Report*, **10**: 36 (1963).

16. Scott, G. V., Robbins, C. R., and Barnhurst, J. D.: Paper presented at the annual scientific meeting of the Society of Cosmetic Chemists, Washington, D.C., December 1967.

17. Roehl, E. L.: The properties of quaternary ammonium compounds and their possible cosmetic applications, *Specialties*, **1**(6): 8 (1965).

18. Finnegan, J. K., and Dienna, J. B.: Toxicological observations on certain surface active agents, *Proc. Sci. Sec. TGA*, **20**: 16 (1953).

19. Kluge, A.: The properties of quaternary ammonium salts and their uses as cosmetic products for treatment of hair, *Parfüm. Kosmetik*, **42**: 341 (1961).

20. Egan, R. R., and Hoffman, B. J.: New quaternary ammonium compound hair conditioning agent, *Am. Perf.*, **83**: 55 (October 1968).

21. Kling, W.: Surfactant chemistry. II. Nonionic, cation-active, and ampholytic surfactants, *Parfüm. Kosmetik*, **45**: 29 (1964).

22. Lang, E. W., and McCune, H. W.: U.S. Pat. 3,313,734 (1967).

23. Thomssen, E. G.: *Modern cosmetics*, Drug & Cosmetic Industry, New York, 1947, Chapter 34.

24. Schrader, K.: Hair lotions: a general survey, *Dragoco Report*, **15**(5): 89 (1968).

25. Silicones: cosmetics and toiletries handbook, General Electric Corp., Waterford, N.Y.

26. Rubin, S. H., Magid, L., and Scheiner, J.: Panthenol in cosmetics, *Proc. Sci. Sec. TGA*, **32**: 6 (1959).

27.   Stangl, E.: Pantoethenic acid content of the hair of the human head, *Intern. Z. Vitaminforsch.*, **24**: 9 (1952).

28.   Rubin, S. H.: The comparative stability of pantothenic acid and panto-thenol, *J. Am. Pharm. Assoc.*, **27**: 502 (1948).

29.   Roche panthenol in pharmaceuticals, Hoffman La Roche, Nutley, N.J.

30.   Thomsen, E.: Advertisement, *Am. Perf.*, **75**: 41 (April 1960).

31.   Hagiwara, Y.: U.S. Pat. 3,133,864 (1964).

32.   Shelanski, H. A., Shelanski, M. V., and Cantor, A.: Polyvinylpyrrolidone (PVP): a useful adjunct in cosmetics, *JSCC*, **5**: 129 (1954).

33.   Joos, B.: Br. Pat. 1,057,104 (1967).

34.   Herd, J. K., and Marriott, R. H.: The sorption of amino acids from shampoos on to hair, *Am. Perf.*, **74**: 31 (November 1959).

35.   Model formulations, Wilson Pharmaceutical and Chemical Corp., Chicago.

36.   WSP preservation, Wilson Pharmaceutical and Chemical Corp., Chicago.

37.   Bronopol technical bulletin, Boots Pure Drug Co., Nottingham, England.

38.   Chiostri, R. F.: Personal communication.

39.   GAF guide to cosmetics, General Aniline and Film Corp., New York, 1967.

40.   Amerchol lanolin derivatives, American Cholesterol Products, Edison, N.J., 1964.

41.   Typical cosmetic formulations containing DL-panthenol, Roche Chemical Division, Hoffman La Roche, Nutley, N.J.

42.   Carter, P. J.: Basic emulsion technology, *Am. Perf.*, **71**: 43 (June 1958).

43.   Sisley, J. P.: *Encyclopedia of surface-active agents*, Vol. 2., Chemical Publishing Co., New York, 1964.

44.   Roth, W.: U.S. Pat. 3,289,949 (1966).

45.   Shansky, A.: Polymeric substances in hair spray film-forming compositions, *Am. Perf.*, **83**: 31 (May 1968).

46.   Kamath, N. R., and Potnis, S. P.: Constitution of lac: Part III. Acid value of lac, *J. Soc. Ind. Res.*, **14B**: 437 (1955).

47.   Shellac, Angelo Bros., Cossipore, Calcutta, 1965.

48.   Shellac product data bulletin 55-4-15, William Zinsser and Co., New York.

49.   Martin, J. W.: Shellac in hair sprays, *DCI*, **94**: 828 (1964).

50.   Burnette, L. W.: A review of the physiological properties of polyvinylpyrrol-idone, *Proc. Sci. Sec. TGA*, **38**: 1 (1962).

51.   Schrader, K.: Aerosol hair cosmetics, *Dragoco Report*, **14**: 267 (1967).

52.   Perfumes for hair sprays bulletin, Haarman and Reimer Corp., Union, N.J.

53.   Haefele, J. W.: U.S. Pat. 3,068,151 (1962).

54.   David, L. S.: Additives for hair sprays, *DCI*, **97**: 502 (1965).

55.   Roehl, E. L.: Raw materials for hair grooming products with a setting effect, *Dragoco Report*, **9**: 247 (1962).

56.   Chemische Werke Albert: Br. Pat. 995,175 (1963).

57.   Sender, H., and Lukesch, H.: U.S. Pat. 3,234,098 (1966).

58.   GAF Chemfo #35 technical report, General Aniline and Film Corp., New York.

59.   Consumer service communiqué, Stepan Chemical Co., Maywood Division, Maywood, N.J.

60. Cosmetics bulletin for Dicrylan, Ciba Chemical and Dye Co., Fairlawn, N.J.

61. Valko, E. I., and Barnett, G.: A study of the swelling of hair in mixed aqueous solvents, *JSCC*, **3**: 108 (1952).

62. Heilingötter, R.: Die elastischen Kräfte des Haares, *Seifen-Öle-Fette-Wachse*, **79**: 603 (1953).

63. Bennett, H.: *The cosmetic formulary*, Chemical Publishing Co., New York, 1937.

64. Thomssen, E. G.: *Modern cosmetics*, 3rd ed., Drug & Cosmetic Industry, New York, 1947.

65. Fluids and emollients for cosmetic preparations, Union Carbide Corp., New York, 1967.

66. Dicrylan 325-50, Dicrylan 325WA-50 for hair care products, Ciba Chemical and Dye Co., Fairlawn, N.J.

67. 1967 Cosmetic and pharmaceutical formulary, surfactants and emollients, Croda, Inc., New York.

68. Carbopol formulary issue No. 5, B. F. Goodrich Co., Cleveland.

69. DeRagon, S. A.: Enriching cosmetics with lanolin derivatives, American Cholesterol Products, Edison, N. J., 1968.

70. Satulan, hydrogenated lanolin specification, Croda, Inc., New York, 1968.

71. Manzke, O.: Ger. Pat. 1,238,158 (1967).

72. L'Oréal: Br. Pat. 1,021,400 (1966).

73. Blance, R. B., and Cohen, S. M.: U.S. Pat. 3,285,819 (1966).

74. Rieger, M. M., and Berenbom, G.: U.S. Pat. 3,186,911 (1965).

75. Richardson, E. L., and Hoss, G. C.: U.S. Pat. 3,133,865 (1964).

76. Shansky, A.: U.S. Pat. 3,171,786 (1965).

77. Wilmsmann, H., and Ludwig, W.: U.S. Pat. 3,250,682 (1966).

78. Ciba Ltd.: Br. Pat. 1,031,540 (1966).

79. Bechmann, G., and Lukesch, H.: Ger. Pat. 1,192,370 (1965).

80. Schwarzkopf Verwaltung: Br. Pat. 1,036,497 (1966).

81. Amerchol laboratory handbook, American Cholesterol Products, Inc., Edison, N.J., p. 54.

82. Hilfer, H.: U.S. Pat. 3,155,591 (1961).

83. Bulletin No. 67-18F, Armour Industrial Chemicals Co., Chicago.

84. Data bulletin No. 524, Goldschmidt Chemical Co., New York, 1968.

85. Dow Corning Corp.: Brit. Pat. 999,222 (1965).

86. Hughes, G. K., and Koch, S. D.: Evaluation of fabric softeners, *Soap*, **41**(12): 109 (1965).

87. Scott, G. V., Robbins, C. R., and Barnhurst, J. D.: Sorption of quaternary ammonium surfactants by human hair (abstract), *Am. Perf.*, **83**: 56 (February 1968).

88. Zwicker, A. I.: Hair rinses cationic: the market today, *Am. Perf.*, **83**: 79 (October 1968).

Chapter 25

ANTIPERSPIRANTS AND DEODORANTS

SOPHIE PLECHNER

Cosmetic deodorants are preparations which mask, remove, or decrease perspiration odors, prevent their development, or do all of these. Many such products which have a satisfactory deodorant efficacy and cosmetic characteristics appear on the market each year.

According to the U.S. Census of Manufacturers, manufacturers' sales of deodorants in 1963 amounted to $94,306,000. The June 1967 issue of *Drug and Cosmetic Industry* pointed out that a leading aerosol deodorant for men jumped from 4% of the deodorant market in 1962 to 25% in 1966. By 1968 the sales of antiperspirants and deodorants was in excess of $300 million. This growth was due largely to the development of aerosol deodorant sprays and the subsequent development of aerosol antiperspirants. In 1968, when aerosol antiperspirants came into the market, aerosol products accounted for more than half of the 300 million antiperspirant-deodorant sales. By mid-1969 more than 60% of antiperspirant-deodorant sales were aerosol products. Antiperspirant aerosols constituted more than half of these and were growing at the rate of 10 to 15% a year.

The increasing use of cosmetics and toiletries by teen-agers has also created a new and rapidly growing market. A survey reported by LeVan (1) shows that in a cosmetic market survey 94% of junior high school students and 99% of senior high school students use deodorants.

The tremendous increase in the sale of these products and the variety of forms in which deodorants are now available are indications of the consciousness on the part of both men and women of the offensiveness of malodorous perspiration.

The odor of fresh perspiration varies with the individual and, in the same individual, with physical conditions, activity (2), emotional state, and diet (3). Fresh perspiration from clean skin has a mild and generally not objectionable odor. Perspiration undergoes considerable change on standing, mainly if not entirely because of bacterial decomposition. In a study of the source of perspiration odors, Killian and Panzarella (4) found that fresh perspiration, when filtered through a Berkefeld filter, showed no change in odor for 24 hr, whereas an unfiltered portion of the same perspiration developed a strong, objectionable odor during the same interval.

The intense body odor sometimes noted under stress has been thought to be the result of apocrine sweating. Shelley and Hurley (5) noted the lack of odor in pure apocrine sweat as it initially appears on the skin surface. On standing, such sweat developed a definite foul odor. In a subsequent study of axillary odor (6), the authors concluded that normal apocrine sweat is initially sterile and that bacterial action on apocrine sweat produces typical axillary odor. They found that eccrine sweat is also sterile and odorless upon secretion. Odor which develops with bacterial action on pure eccrine sweat differs from the characteristic axillary odor and is not as strong or offensive. The authors pointed out that the presence of hair greatly increases axillary odor since it acts as a collecting site for axillary secretions, debris, and bacteria.

It is evident that in order to control objectionable odors of perspiration, it is necessary either to check the flow of excess perspiration or to eliminate its odor, or both.

Antiperspirants

A variety of substances which have astringent action inhibit the flow of perspiration. The mechanism by which such antiperspirants act has not been clearly defined. The narrowing of the openings of the sweat ducts and the formation of a keratotic plug in the sweat duct orifice to obstruct the flow of sweat were suggested by dermatologists as the possible causes of anhidrosis (7). Papa (8) found that the sweat suppression by formalin is due to a high-level obstruction of the eccrine duct, but that aluminum chloride anhidrosis results from increased transductal absorption of sweat. Papa and Kligman (8) studied the histology of the eccrine duct and concluded that aluminum salts increase the permeability of the duct, resulting in complete dermal resorption of the sweat. They give as an example of this mode of action a long hose with innumerable perforations which drain off water before it reaches the nozzle.

Salts of metals such as aluminum, iron, chromium, lead, mercury, zinc, and zirconium have astringent properties which may be demonstrated by protein precipitation (9). However, some of these, because they produce discoloration, and others, because of possible toxic effects, are not suitable for cosmetic preparations. Salts of aluminum and zinc are those most commonly used. The astringency of these salts is also dependent on the anion. Sulfate, chloride, chlorhydroxide, and phenolsulfonate have been most widely used, although basic formate, lactate, sulfamate, and the alums are also found in antiperspirant products. Acetates are generally unsatisfactory because of odor. At the acid pH necessary for astringency, the acetic acid odor is very definite and is a difficult one to cover. Formates are to be avoided since they tend to produce skin sensitization. Aluminum salts are generally used in concentrations of 12 to 20%. Tannins and tannic acid have been used as antiperspirants.

The aluminum compound that has been most widely used in antiperspirant compositions is aluminum chlorhydroxide complex. Commonly referred to as aluminum chlorhydrate or aluminum chlorhydroxide, the product is a 5/6 basic aluminum chloride complex with the atomic ratio of aluminum to chlorine of 2:1. In dry form it is a glasslike rather than a crystalline substance, readily soluble in water. A 20% solution has a pH of approximately 4.2 with good buffering capacity. It is not irritating or sensitizing to normal skin and causes little or no damage to fabrics. The concentration recommended for use in antiperspirant products is 20%. Aluminum chlorhydrate complex is sold under a variety of trade names, dry in granular or powdered form, or as a 50% solution. The solution is stable and remains clear with little change in pH over long periods of standing. The solid is insoluble in 95% ethyl alcohol, but the 50% solution is miscible in all proportions with 95% ethyl alcohol.

Recently aluminum compounds which have antiperspirant properties and are soluble in anhydrous alcohol have been developed for use in aerosol products. Rehydrol A.S.C., an aluminum chlorhydroxide-propylene glycol complex (10) (Reheis Chemical Co., Division of Armour Pharmaceutical Co.), A.C.H.-A-23, a two-thirds basic aluminum chloride in anhydrous alcohol, and P.G.A.-C-112, 2 chloro-4-methyl-1,3,2 dioxalumolane (Chattem Chemical Co.) (11) are compounds which have been tested for such use.

Most of the salts which show good astringent properties have a low pH (2.5 to 4.2) and because of this may be irritating to the skin and corrosive to fabrics. Irritation owing to low pH may occur where minor abrasions are present or on normal, unbroken skin. The presence of a wetting agent in a preparation with a low pH will produce even greater incidence of skin irritation, presumably because of the better contact of the active ingredient with the skin surface. The pH of such a product may be satisfactorily modified to reduce skin irritation by addition of small amounts of zinc oxide, magnesium oxide, aluminum hydroxide, or triethanolamine.

The probability of the development of skin irritation from use of a cream or ingredient may be predicted by the patch-test method. A new product should first be tested for primary irritation on the abraded and intact skin of the albino rabbit (12). If the product shows little or no irritation on rabbit skin, it should be tested on human subjects before marketing. The method described by Schwartz and Peck (13), in which a small amount of the test substance is applied to the skin of 200 subjects under a closed adhesive patch for 48 hr, is simple and convenient for determining whether or not the substance in question is a primary irritant. A subsequent patch applied on the same skin area after a 10-day interval will give an indication of the sensitizing properties of the test material.

Since a new combination of ingredients may be sensitizing, any new product should be tested first on guinea pigs and subsequently on humans for sensitization. The repeated patch technique of Draize and Shelanski (14,15) gives an indication of the incidence of irritation and sensitization which may be expected when the skin is subjected to frequently repeated applications of the product to the same area. This type of test may be valuable in judging the dermatological properties of a product, such as an antiperspirant, which will probably be applied daily or even more frequently to the same skin area, often without washing between applications.

In some subjects, sensitization to adhesive tape may develop in the course of patch testing and must be distinguished from irritation caused by the product under test. With any patch-test method, a large number of subjects must be used to obtain results which reliably predict the incidence of irritation that will occur if the product is marketed. A "use test," in which the product is used in the normal way by large groups of subjects who may be checked by a trained observer, is the most satisfactory method of final product evaluation.

The problem of fabric protection is also a complex one. If a readily hydrolyzed salt, such as aluminum chloride, is used in the product, direct rotting of the fabric will occur after contact with the solution or cream. When the salts used are not so strongly acid, immediate fabric damage may not occur. However, damage will occur on longer contact, and severe damage will be produced at the higher temperatures used in ironing. Such damage may occur on ironing even after the fabric has been laundered, if the astringent is not modified by an adequate inhibitor. Dry-cleaning solvents will remove the waxy portion of a cream and leave the astringent salts deposited in the fabric. It is obvious that ironing after dry cleaning will cause severe damage with such a preparation. Maximum damage will occur if a garment which has picked up cream is pressed without cleaning or laundering. Linen, cotton, and viscose rayon are particularly susceptible to damage by the acid creams or liquids. Silk, wool, polyesters, nylon, and acetate rayon are more resistant.

The action of antiperspirant creams on fabrics is discussed in a paper by Bien (16), and both a laboratory and a practical use procedure for predicting fabric destruction are outlined. The percentage of damage to a fabric may be determined from the difference between the tensile strength of a strip of fabric treated with an antiperspirant cream and that of an untreated control strip. Bien found that although pH is an important factor in fabric damage, the presence of humectants and buffers and the type of emulsion also influence the effect of the cream on the fabric. Another similar method of testing the damaging effect of creams on fabric has been devised by the American Institute of Laundering. Since cotton is the fabric most susceptible to acid damage, fabric tests are generally carried out with cotton strips. Bien found that antiperspirant creams which produced a reduction in tensile strength of less than 20% by the laboratory method caused negligible damage in use. The American Institute of Laundering considers that a loss of tensile strength up to 10% by their method is not excessive. The ideal product, of course, would cause no damage to the fabric, and no loss in tensile strength would be observed.

Discoloration or staining of clothing can occur when antiperspirants are applied to or wipe off on the fabric. One source of such discoloration is the use in clothing of acid-sensitive dyes which will change color (blue to red, brown to orange) in contact with an antiperspirant or an antiperspirant mixed with perspiration. This change of color is often reversible and may be corrected by sponging the area with a diluted solution of ammonia.

Another type of staining occurs in the washing of clothing which has a deposition of aluminum salt from an antiperspirant in the fabric. If the salt is first removed by washing in cool, soft water or hard water with a water softener, no stain occurs. If soap is applied directly to the area, insoluble aluminum soaps which form cause yellow discoloration, are sticky, and are difficult to remove. Bluing or colored detergents may also combine with the aluminum salt to give a blue or gray stain. Optical whiteners used in some laundry detergents may increase the discoloration.

Some incidence of staining of clothing by antiperspirant products is unavoidable. Most stains on washable clothing can be removed by washing first with cool, soft water with no soap or detergent, and then with a laundry soap or toilet soap which contains no added deodorant. New home laundry products containing proteolytic enzymes are said to remove perspiration stains which cannot be removed by ordinary soap or detergents.

The importance of the effect of antiperspirant on fabric is indicated by the large number of patents that have been granted which are concerned with various means of protecting fabric.

Among the early patents on antiperspirant preparations is one issued to Tate (17) in 1921 for a compositon of matter for stopping perspiration "without irritation or prejudice to the health." The formula in this patent is as follows:

    1 part aluminum chloride (crystals)
    3 parts distilled water
    ½ part borax
    ½ part powdered alum

According to the patent, "the office of the aluminum chloride is to stop the perspiration, of the alum and borax to prevent the irritation to delicate skins that would be caused by the aluminum chloride alone, and of the water to dissolve the ingredients."

In 1934 a patent granted to Taub (18) described a solid waxlike preparation which acts to absorb and deodorize perspiration and which is nonirritating to the skin and noninjurious to fabrics. The product is essentially a waxlike stick comprising an absorption base, a mildly astringent perspiration deodorant, zinc sulfocarbolate with or without aluminum palmitate, and a cutaneous sedative, zinc oleate, and wax.

A patent for a perspiration-inhibiting material issued to Teller (19) in 1940 described the corrosive effect of commonly used astringents on fabrics and claimed to prevent such rotting by use of an inhibitor, basic aluminum formate, which is said to overcome the potential acid effect of the aluminum salts. Formula 1 is an example of the compositions described in the patent.

Formula 1

Basic, aluminum formate solution	35.0 parts
Aluminum sulfate, crystallized	8.0
Ammonium alum, crystallized	5.0
Boric acid, powdered	3.0
Tegacid	20.0
Stearic acid	2.5
Petrolatum	2.5
Water	25.0
Total	*101.0 parts*

Several patents issued to Montenier (20) in 1941 are concerned with compositions to overcome the defects of excessive acidity in astringent preparations, including an astringent material and a soluble chemical compound represented by the formula

$$R{-}\underset{\underset{O}{\|}}{C}{-}N\underset{R''}{\overset{R'}{<}}$$

wherein R is an alkyl, cycloalkyl, aralkyl, or aryl group, and R' and R" are hydrogen, alkyl, cycloalkyl, aralkyl, or aryl groups. Formulas shown in the patent use aluminum chloride with urea, acetamide, or urethane.

The second and third Montenier patents use certain imide and nitrile compounds, such as succinimide and pyrazole, to overcome the acidity of aluminum chloride. These materials are to be used in astringent preparations as cream, liquid cream, stick, or impregnated cotton cloth pads.

A patent granted to Wallace and Hand (21) in 1941, and subsequently invalidated, described an improved perspiration-retarding or -inhibiting compound which "while retaining in full the desirable inhibitory action of previously known preparations, avoids their corrosive action and is harmless to the skin and fabrics." The perspiration-inhibiting compound described includes an astringent, one or more water-soluble heavy-metal salts of strong acids, usually mineral acids, and a normally neutral water-soluble amino compound possessing one or more intact reactive amino groups. The metallic salt preferably comprises a water-soluble strong-acid salt of one or more metals such as aluminum, zinc, cerium, zirconium, titanium, iron, or bismuth, although salts of tin, lead, and cadmium are also effective. The normally neutral protective ingredient is an aliphatic amide or aliphatic amino acid, soluble in water or in the strong-acid salt. Examples which are named as suitable are formamide, acetamide, carbamide (urea) and derivatives thereof, and aminoacetic and aminopropionic acid.

A patent obtained in 1944 by Klarmann and Gates (22) related to the incorporation in an antiperspirant composition of a water-insoluble basic compound of a metal, such as an oxide, hydroxide, or carbonate of zinc, magnesium, or aluminum, to inhibit fabric corrosion. The hydrolysis of the astringent salt of the antiperspirant develops sufficient acidity to bring about solution of the insoluble basic compound. The insoluble metallic base is used in proportions ranging from 4 to 15 % of the amount of soluble astringent salt in the cream. An example selected from those in the patent is given in Formula 2.

Formula 2

Aluminum chloride	15.0%
Tegacid	15.0
Spermaceti	3.0
Beeswax	2.0
Magnesium oxide	2.5
Water	62.5

Another invention related to the composition of a perspiration-inhibiting preparation which does not have a corrosive effect on fabrics is described in a patent granted to Richardson and Russell (23) in 1946. This invention relates

to the use of salts which have the effect of inhibiting or preventing the corrosive action of the antiperspirant on fabrics and which also serve as additional
antiperspirant agents in the preparation. Aluminum orthophosphate and
aluminum pyrophosphate are suitable for such a purpose. In a cream the
amount of aluminum phosphate introduced may exceed that which will
dissolve. The method used for determining the value of these phosphates
in protecting fabrics from either an antiperspirant cream or solution is
described in this patent. A cream preparation selected from the examples in
the patent is given in Formula 3.

Formula 3

Glyceryl monostearate, acid-stabilized	16%
Spermaceti	5
Aluminum sulfate, crystallized	19
Aluminum phosphate (AlPO$_4$)	7
Water	53

This cream, it is claimed, caused a 32 % reduction of the tensile strength
of a cotton fabric when subjected to the test described in the patent. Using
no inhibitor—that is, substituting water for the aluminum phosphate—the
same test gave an 84 % loss in tensile strength. The use of 10 parts of aluminum phosphate with a subsequent reduction of water gives an antiperspirant
which shows no corrosive effect on cotton fabric when tested by the method
described in the patent.

In 1949 Anderson (24) was granted a patent (since declared invalid) on a
composition having an astringent action containing an aluminum chlorohydrate. It is stated that aluminum chlorohydrate in the cream composition
described is substantially without action on various textile materials, as
evidenced by negligible loss in tensile strength when the materials were
tested with the cream by the regular procedure as outlined by the American

Formula 4

Stearic acid	200 parts
Mineral oil	20
Beeswax	20
Glyceryl monostearate	120
Propylene glycol	100
Sodium lauryl sulfate	24
Water	800
Aluminum chlorohydrate	300
Water	300
Total	*1884 parts*

Society for Testing Materials. Several examples of cream formulas are given in the patent. One of these is described in Formula 4.

A patent granted to Van Mater (25) in 1950 relates to compounds of zirconium which are said to be of particular value as deodorants and anti-perspirants. These substances, chemically neutral or nonacid compounds of zirconium, are described as being nonirritating to the human skin and posses-sing properties which make them suitable as deodorants and antiperspirants. The compounds described are formed by the neutralization with an alkali of the product formed from an inorganic zirconium salt and a hydroxyali-phatic acid. The zirconium salts of the hydroxyorganic acid, which are insoluble, form water-soluble salts with alkali, alkaline-earth, or ammonium hydroxides as well as with amino or amino hydroxy compounds. Examples cited are neutralized zirconium lactate, citrate, and tartrate.

Three patents granted to Berger and Plechner (26) describe the use of zirconium salts of hydroxyorganic acids alone and in combination with alu-minum salts of inorganic acids. Antiperspirant solutions, creams, and lotions are described using combinations of these salts in an aqueous carrier. Typical examples are shown in Formulas 5 and 6.

Formula 5

Petrolatum	1%
Spermaceti wax	3
Glycerol monostearate	13
Glycerol	10
Triethanolamine salt of alkyl aryl sulfonate	10
Sorbitan monostearate polyoxyethylene derivative	3
Water	44
Titanium dioxide	1
Sodium zirconium lactate	10
Aluminum chloride	5

Formula 6

Sodium zirconium lactate	12.0%
Zirconium lactate	3.0
Polyethylene glycol ether of higher fatty alcohols	2.5
Propylene glycol	10.0
Cetyl alcohol	0.5
Water	72.0
Perfume	q.s.

Weiss (27), in an earlier German patent, described a powder which is said to be particularly nonirritating to the skin even in the presence of perspiration. It has excellent spreading properties, so that small amounts are effective, and it is pure white in color. It consists of zirconium compounds, such as the oxide, hydroxide, basic sulfate, or basic carbonate, alone or in mixture with other materials commonly used in powders.

A patent granted in 1950 to Grote (28) disclosed an aluminum compound which is said to be a deodorant but does not irritate the skin or rot clothing. It is described as a dichloro aluminum aminoacetate hydrate containing less than 10 % by weight of free aluminum chloride. This may be used with water or organic solvents.

Another type of astringent salt which is claimed to have very little corrosive effect on fabrics was described by Govett and Almquist (29) in a patent granted in 1951. A double complex, calcium aluminum basic chloride, is said to have good astringent properties without the deteriorating effect on fabrics observed with other antiperspirants. The patent described this compound, its method of synthesis, and cosmetic preparations containing it.

Three patents granted to Apperson and Richardson (30) described an aluminum sulfamate antiperspirant preparation. In the first of these, the antiperspirant effect of aluminum sulfamate was noted and a method of measuring its efficacy was described. This patent related to antiperspirant compositions as emulsions, solutions, salves, creams, lotions, and the like containing aluminum sulfamate together with suitable compounds to prevent discoloration of cellulose fabrics which occurs with the use of aluminum sulfamate. An example of a cream preparation is given in Formula 7.

Formula 7

Glyceryl monostearate, acid-stablized	10.00%
Glyceryl monostearate	4.00
Spermaceti	4.00
Mineral oil	2.00
Igepon T	2.50
Titanium dioxide	0.50
Thioglycolic acid	1.59
Aluminum sulfamate	18.00
Water	57.41

The second patent is concerned with those compounds used for preventing discoloration. Neutral amides, amino carboxylic acids, and compounds which are salts of nonhydroxy nonmercapto saturated carboxylic acids and ammonia bases, and soluble in aqueous aluminum sulfamate, are used. Many examples of creams and solutions are given.

An antiperspirant cream using aluminum sulfamate as the active ingredient stabilized by use of a polypropylene glycol is described in a patent granted to Henkin and Messina in 1957 (31).

A patent granted to Thurmon (32) in 1953 described antiperspirant and deodorant creams containing weakly acidic cation exchange resins together with aluminum phenolsulfonate. The deodorant properties of the creams are attributed to the capacity of these resins for absorbing amino acids present in perspiration and thereby preventing the bacterial decomposition of these which causes perspiration odor. The resins also are capable of absorbing ammonia and organic amines formed by the decomposition of amino acids. The creams, which contain 15 to 20% of cation exchange resins and 10 to 20% of aluminum phenolsulfonate, are said to be nonirritating to the skin and to have no weakening effect on fabrics.

It is apparent that many of the more obvious materials suitable for preventing damage to fabric or for buffering these creams have been used. Since the use of urea as a buffer is no longer covered by a patent it is now available for general use. It is soluble in water in all proportions and is probably the most effective modifier for reducing fabric damage. It has two amino groups which are available to react with acid formed by hydrolysis of the astringent salt. It does not appreciably increase the pH of the cream or react as an alkali to precipitate aluminum hydroxide, which would reduce the efficacy of the antiperspirant. At ironing temperatures it gives excellent fabric protection, probably because it decomposes to form ammonia, which neutralizes the acidity of the astringent salt. Since it has a low molecular weight, only small amounts are required for the necessary buffering effect. However, it has some undesirable properties. Since at elevated temperatures it decomposes to form ammonia and carbon dioxide, in creams containing aluminum sulfate such urea breakdown will result in the formation of ammonium aluminum alum crystals. Carbon dioxide formed in the cream will cause swelling and sponginess.

Buffered antiperspirant compositions using aqueous solutions of zirconium salts and hafnium salts of strong monobasic mineral acids in combination with certain basic aluminum compounds and urea are described in a patent granted to Daley (33) in 1957. Formulas for cream, lotion, and liquid antiperspirants are given. Test procedures for evaluating the antiperspirant efficacy and the effect on tensile strength are described. The type of composition described is shown in Formula 8.

Glycine may be used as a buffer for fabric protection and is more stable than urea under normal storage conditions. Buffered antiperspirant compositions using glycine or an amino acid in which the number of amino groups is equal to the number of carboxyl groups in the molecule with a mixture of astringent salts of zirconium and aluminum are described in patents by

Daley (34) and by Grad (35). The Grad patent describes two methods for testing the reduction in sweating produced by an antiperspirant as well as a test for measuring the effects of such products on the tensile strength of textiles. An example of the type of product described in these patents is shown in Formula 9.

Formula 8

Zirconyl chloride octahydrate	7.5%
Aluminum chlorhydroxide complex	7.5
Urea	25.0
Veegum	2.5
Ethyl alcohol	3.5
Water	46.8
Mineral oil, light	4.2
Stearyl alcohol	1.0
Atlas G-2152	1.0
Atlas G-2160	1.0
Perfume	q.s.

Formula 9

Zirconyl hydroxy chloride	5.0%
Aluminum chlorhydroxide	7.5
Glycine	2.0
Water	59.1
Perfume	0.2
Glyceryl monostearate	10.0
Spermaceti wax	4.0
Mineral oil (about 70 SSU at 100°F)	4.0
Polyoxyethylene 50 stearate	
(Myrj 53)	5.0
Glycerol	3.0
Titanium dioxide	0.2

Formula 10

Solution of calcium monolacto	
titanylate adj. to pH 3.1	30.0%
Nonionic emulsifier	3.0
Polyethylene glycol 300	10.0
Isopropyl palmitate	3.0
Glyceryl monostearate	12.0
Spermaceti	2.5
Cetyl alcohol	2.5
Titanium dioxide	0.5
Water	36.5
Perfume	q.s.

Titanic acid complexes of aliphatic carboxylic acids have also been found to have antiperspirant properties. A patent granted to Berger and Plechner (36) describes antiperspirant compositions containing such complexes as active ingredients. An example of one of these is given in Formula 10.

A patent granted to Siegal (37) describes an antiperspirant aluminum titanium lactate. This product was found to be nonirritating and non-sensitizing to the skin of human subjects and caused little or no loss of tensile strength of fabric. The patent gives examples of antiperspirant compositions in the forms of spray, cream, lotion, and stick containing aluminum titanium lactate.

Sodium aluminum lactate (38) and sodium aluminum chlorhydroxy lactate (39) are compounds which cause no primary irritation, are non-destructive to fabrics, and are moderately effective antiperspirants. These compounds are alkaline in solution and are therefore useful in formulating antiperspirant sticks.

An aluminum sulfate-aluminum chlorhydrate combination is used as the astringent in creams described in a patent granted to Messina (40). Glycine is used in some of the formulas to prevent fabric damage. One of the compositions of the patent is given in Formula 11.

Formula 11

Ethylene glycol monostearate	15.0%
Polyethylene glycol ether of mixed fatty alcohols (about 25 M of ethylene oxide)	5.0
Mineral oil	3.0
Propylene glycol	5.0
Sodium lauryl sulfate	2.5
Titanium dioxide	1.0
Aluminum sulfate (hydrate)	10.0
Aluminum chlorhydroxide, 50% solution	10.0
Glycine	2.5
Perfume	0.2
Water	q.s. 100.0

A number of investigators (41–44) have studied the effects of oral administration of parasympatholytic compounds on sweating. Such drugs as Banthine, Prantal, and scopolamine hydrobromide were found to be effective to varying degrees in different subjects, and some unwanted side effects were always observed. Shelley and Horvath (44) introduced eleven anticholinergic compounds into the skin by iontophoresis. Of those tested, scopolamine hydrobromide was the most effective. A topical application of an aqueous solution of the scopolamine hydrobromide also produced anhidrosis. Such a solution would be much too toxic for general antiperspirant use. Stoughton (45) reported effective suppression of eccrine sweat delivery by topical application of an anticholinergic AHR 483 (Glycopyrrolate; 1-methyl-3-pyrrolidyl-D-phenyl cyclohexaneglycolate methobromide, A. H. Robins Co.). This agent was reported to be of a low order of toxicity.

MacMillan and co-workers (46) evaluated 25 commercially available anticholinergic agents and a large number of research compounds from several pharmaceutical companies and university laboratories for their ability to inhibit sweating by topical application. Esters of scopolamine hydrobromide were found to be much more effective than any other type of structure studied. Evaluation of safety of the benzoyl ester of scopolamine showed that 0.025 % solution can be used on a limited area of the body without incidence of systemic effects. A British patent (47) covers the use of some esters of scopolamine hydrobromide as antiperspirants in water-dispersible dermatologically acceptable vehicles. The esters were tested in several concentrations in various vehicles; no evidence of skin irritation, fabric damage, or side effects was found in the compositions tested.

No antiperspirant using an anticholinergic drug as its active ingredient has been marketed at this time. However, because of the low concentration needed, a parasympatholytic agent with low toxicity and a low level of systemic effects would lend itself to many types of antiperspirant formulation. Extensive pharmacological testing would be necessary to ensure the safety of such a product for marketing.

Antiperspirant Creams

Antiperspirant creams which have had the widest consumer acceptance are of the vanishing-cream type. Since such a product must contain 15 to 20 % of an acid astringent salt, the usual type of vanishing cream formulated with a soap emulsifier generally does not give a satisfactory product. An acid-stable emulsifier which is compatible with astringent salt must be used. Satisfactory creams can be made with acid-stabilized glyceryl monostearate with or without additional emulsifier. Several anionic emulsifiers, such as sodium lauryl sulfate, sodium cetyl sulfate, triethanolamine lauryl sulfate, or an alkyl aryl sulfonate, can be used. Nonionics such as the hexitol esters of the common fatty acids (Spans) and their polyoxyethylene derivatives (Tweens), the polyoxyethylene ethers and some of the cationic emulsifiers form acid-stable emulsions. The type of emulsifier and the wax with which it is used will determine the amount of emulsifier required. The concentration of emulsifier helps to determine the final consistency of the cream. A high concentration of the emulsifier tends to give a soft cream. Creams which are made with cationic or nonionic emulsifiers generally have a softer consistency than those with an anionic emulsifier. A nonionic emulsifier can be combined with either a cationic or an anionic to form a cream of the desired consistency and texture.

Glyceryl and glycol esters of stearic or other fatty acids are used alone or blended to give desirable consistency and texture to the cream. Stearic acid used with the esters has a stiffening effect. Smaller amounts of spermaceti or cetyl alcohol are often used with the glyceryl or glycol esters. White petrolatum or mineral oil, 1 to 3%, may be used. The total fatty content of the cream is generally between 15 and 20%.

The concentration of fabric-damage inhibitor needed will depend on the amount and type of astringent used. Urea in concentrations of 5 to 10% will give good fabric protection with an aluminum salt. Some of the complex amides give good results in lower percentages. Glycine in concentrations of 3 to 10% will give satisfactory fabric protection and is somewhat more stable than urea.

A number of satisfactory humectants are available for cosmetic creams. Griffin (49), in a paper on hygroscopic agents for cosmetics, lists 11 desirable properties of a humectant:

1. High hygroscopicity.
2. Narrow humectant range, i.e., minimum change in water content and with change in relative humidity.
3. Desired viscosity.
4. Low viscosity index, i.e., little change in viscosity with temperature with water content.
5. Good compatibility.
6. Low volatility.
7. Low cost.
8. Low toxicity.
9. Good color and odor.
10. Lack of corrosive action.
11. Low freezing point.

Glycerol, propylene glycol, sorbitol, and polyethylene glycol 400 are the humectants most commonly used in antiperspirant creams. Of these glycerol is the most hygroscopic, and sorbitol syrup (85% solution) is the least hygroscopic under equilibrium conditions. Sorbitol, however, gains or loses moisture more slowly with changes of relative humidity. Of the group listed, polyethylene glycol 400 has the lowest viscosity and sorbitol syrup the highest.

Propylene glycol and polyethylene glycol 400 are more volatile than glycerol, and sorbitol is essentially nonvolatile. All of these have low toxicity.

The polyethylene glycols 200 to 400 have been found to be innocuous for external use (50,51). Any of these can be obtained as colorless liquids with little or no objectionable odor.

Humectants are used in concentrations of 3 to 10 % of the formula weight. Higher concentration will cause the creams to have an excessively wet feel on the skin. The concentration needed to prevent drying out of the cream will vary with the humectant used and the other ingredients present. In some formulations, sorbitol will give good moisture-retaining properties to the cream in a concentration as low as 2 %, whereas glycerol or propylene glycol is more effective in the 5 to 10 % range (49).

In addition to their hygroscopicity, humectants have an effect on the consistency, texture, and crusting properties of the cream. The consistency of a vanishing cream generally increases to a maximum with increasing content of humectant and then decreases with further increase of polyol content. Vanishing creams in which propylene glycol is the humectant generally show more crusting on long exposure to air than those in which sorbitol syrup or glycerol is used.

Creams prepared with nonionic emulsifiers are generally softer in consistency and show less tendency to crust formation than anionic creams even after loss of moisture. The choice of humectant in a nonionic cream is not so critical as in an anionic cream. Although nonionic creams lose water on exposure to air, they generally remain soft to the touch and show less lateral shrinkage and splitting.

Titanium dioxide is added to creams as a whitener and opacifier. Between 0.5 and 1.0 % of a fine particle size is adequate, if well dispersed. The color of a cream may be made lighter or brighter by an increased amount of titanium dioxide. However, excessive amounts of whitener in cream which rubs off on clothing will cause staining, and such stains are difficult to remove. Finely divided titanium dioxide adheres to the fibers of the fabric when the cream is removed.

The perfume to be used in an antiperspirant must be acid-stable and must be compatible with the emulsion. Occasionally a perfume will cause bleeding or separation in an otherwise stable cream. It is therefore important to check the perfume in the cream, even though it is compatible with the individual ingredients of that cream. Karas (52), in a study of the effect of some aromatic chemicals and essential oils on the stability of cosmetic emulsions, found that they occasionally lengthen but more often shorten the life of the emulsion. The effect varies with the emulsifier.

Formulas 12 to 14 may be used in developing satisfactory antiperspirant creams. The manufacture of the creams requires careful control, in order to incorporate the high concentration of astringent salt.

It is generally desirable to mill a cream of this type, in order to ensure a smooth and homogeneous product. The temperature at which the cream is

Formula 12

Part A

Glyceryl monostearate, acid-stabilized	16.0%
Spermaceti	5.0

Part B

Sodium lauryl sulfate	1.5
Propylene glycol	5.0
Water	46.0
Titanium dioxide	0.5
Water	3.0
Urea	5.0
Aluminum sulfate	18.0
Perfume	q.s.

Procedure: Heat together the waxes (part A) to a temperature of approximately 75 to 80°C. Heat the water phase (part B) in another container to the same temperature, and add B to A, with stirring. Maintain the temperature for 10 to 20 min, and continue agitation until the emulsion is formed. Cool the emulsion with continuous stirring; form a slurry of the titanium dioxide, and incorporate. Continue to agitate, and cool to 35 to 40°C. Add the aluminum sulfate slowly, with agitation and cooling. The temperature of the cream will rise as the salt dissolves; this rise should be kept at a minimum, because too rapid incorporation of the salt causes the emulsion to break or become too soft. Agitate until the cream is uniform and the salt dissolved. Add urea at 40°C or lower, after incorporation of the aluminum sulfate. The urea should be powdered and added slowly, with continuous agitation, until it is completely dissolved. Finally, add the perfume slowly, and continue to agitate until it is uniformly distributed.

Formula 13

Part A

Glyceryl monostearate, acid-stabilized	18%
Spermaceti	5

Part B

Glycerol	5
Water	53
Titanium dioxide	1
Aluminum chlorhydroxide	18

Procedure: Prepare same way as Formula 14, except that glycerol is the humectant, and urea has been omitted.

Formula 14

Water	29%
Aluminum chlorhydroxide, 50%	
solution	40
Glycerol	8
Mineral oil, light	2
Polyethylene glycol 1000 monostearate	3
Cetyl alcohol	14
Spermaceti wax	3
Titanium dioxide	1
Perfume	q.s.

Procedure: Add the ingredients in the order listed to the kettle, holding water to slurry the titanium dioxide before it is added. Heat with agitation to 75°C. Hold the temperature for 20 min or until the mixture is uniform. Cool with continuous stirring to 56°C. Add the perfume and continue to cool with stirring to 30°C.

milled will be determined by the formula and the consistency which is desired. The temperature at which filling takes place will also help to determine the consistency and stability of the final product. Creams filled at too high a temperature will generally become hard in the jar after cooling. Those filled too cold may become too soft and may bleed. The temperatures most suitable for milling and filling should be determined for a given formula by laboratory test batches.

Since the efficacy of antiperspirant creams is greater when the water phase separates on the skin surface, very strong emulsions show less efficacy (53). It is therefore desirable in this type of product to have an emulsion which is reasonably stable but will release the active ingredient on application. The finished product should be stable when stored at a slightly elevated temperature (45 to 50°C) for 30 to 40 days or at refrigerator temperature (5°C) for several days. It should remain stable and cosmetically attractive in appearance under these conditions if it is to withstand the shipping and storage which are involved in marketing a cosmetic product.

Only antiperspirant creams of the oil-in-water emulsion type have been considered here. Better efficacy is generally obtained with this type of cream since the astringent salts are water-soluble and will remain in the continuous phase of the emulsion.

The formulas given are not intended as finished cosmetic preparations, but will serve as a guide in formulating.

Antiperspirant Lotions

Antiperspirant lotions still maintain a reasonable share of the antiperspirant market. For an emulsion type of lotion, the manufacturing procedure is basically the same as that used for a cream. The incorporation of astringent salts in a lotion must be carefully controlled to avoid breaking the emulsion. The finished lotion requires careful testing for emulsion stability with age and change in temperature. The lotion should also be tested for change in viscosity with age. Formula 15 is a basic formula for an antiperspirant lotion.

Formula 15

Glyceryl monostearate	5%
Polyethylene glycol 1000 monostearate	3
Polyethylene glycol 400	5
Water	67
E-607 Special	5
Aluminum chlorhydroxide	15
Perfume	q.s.

Procedure: Dissolve the aluminum chlorhydroxide and the E-607 Special (a cationic emulsifier) in the water. Add the polyethylene glycol and heat the mixture to 70 to 75°C, with slow agitation. Add the waxes and maintain the temperature at 70°C until completely melted and incorporated into the solution. Cool the lotion with agitation to 45°C. Add the perfume and continue to agitate until the temperature reaches 35°C. The rate of agitation should be controlled in order to avoid the incorporation of air in the lotion.

Formulas for emulsion-type lotions with various active antiperspirant ingredients are given in the patents cited (33–37).

Clear roll-on lotions may be prepared using an aqueous alcohol solution of such thickeners as methyl cellulose or hydroxyethyl cellulose in a concentration of 0.4 to 0.7% with aluminum chlorhydroxide 18 to 20%. S.D. 40 alcohol may be used at 15 to 20%. Addition of small amounts of humectant and emollient prevents formation of aluminum salt crystals on the ball and improves the feel of the product on the skin. A nonionic emulsifier will be needed to disperse the perfume.

The final viscosity of a roll-on-type product should be established before a package is selected. The amount of clearance required between the ball and fitment will depend on the viscosity of the product. If the lotion is too thin, it will drip and run as it is applied. If it is too thick, it will be scraped off on the fitment as the ball turns and will feel wet after application. The exact viscosity desired and the amount to be applied will vary with the kind of thickener used in the lotion and with the astringent selected.

Antiperspirant Sticks

Antiperspirants have also been developed in stick form. Probably the earliest antiperspirant stick was the waxy type described in the Taub (18) patent in 1934. This preparation, using zinc phenolsulfonate and oleate in a waxy base, is not strongly astringent. Another type of antiperspirant stick is made from a solidified alcohol gel composition, as described in a patent by Moore (54). A combination of an astringent chloride with a hard wax dissolved by heating in alcohol will form a solid gel on cooling. Addition of a higher fatty acid or an ester improves the texture and rigidity of the product. The following is a sample composition described in the patent:

22.5 g of aluminum chloride, hydrated ($AlCl_3 6H_2O$), 12 g of candelilla wax and 16.0 g of stearic acid were boiled under a reflux condenser with an alcoholic menstruum consisting of 108 cc of isopropyl alcohol (90% by volume). When solution of the solid ingredients had been effected, 0.7 cc of perfume base was added and the mixture was run off into small containers. When cold it formed a solid mass of smooth salvelike consistency, small portions of which could be removed in discrete quantities from the container, by means of the fingers, and applied to the human skin, as under the arms, to control or deodorize perspiration.

Sodium aluminum chlorhydroxy lactate complex and sodium aluminum lactate have been found to be useful as antiperspirant ingredients in sticks. Kalish (39) describes Chloracel (sodium aluminum chlorhydroxy lactate) as an effective antiperspirant which causes no irritation and is nondestructive to fabrics. This compound is compatible with the components of the alcohol gel type of stick. This type of stick, when properly formulated and packaged, is attractive in appearance and gives a pleasant cooling application. Since the gel of this stick consists principally of sodium stearate and alcohol, it must be packaged to prevent evaporation of alcohol and consequent drying and hardening of the product.

The stick may be made by dissolving sodium stearate and other ingredients in the alcohol under a reflux or by forming the soap from sodium hydroxide and stearic acid. Kalish gives the two basic formulas, 16 and 17.

In this type of product the proportions of alcohol, water, and aluminum complex must be maintained between narrow limits if the stick is to be clear

Formula 16

Sodium aluminum chlorhydroxy lactate complex, 10% solution	50 g
Ethyl alcohol	12 cc
Propylene glycol	3 g
Sodium stearate	6 g
Perfume	q.s.

Procedure: The aqueous aluminum salt solution is heated to 60 to 65°C and the alcohol and propylene glycol slowly added with agitation, heating to maintain the temperature at 60 to 65°C. The sodium stearate is added with stirring until the soap dissolves. The perfume is finally added, mixed thoroughly, and the solution poured into molds.

Formula 17

Sodium aluminum chlorhydroxy lactate, 10% solution	50.00 g
Ethyl alcohol	12.00 cc
Propylene glycol	3.00 g
Sodium hydroxide	0.75 g
Stearic acid	5.25 g
Water	2.00 g
Perfume	q.s.

Procedure: Dissolve the stearic acid in the warmed alcohol. In another container dissolve the sodium hydroxide in the water and combine this with the aluminum complex solution. This combined aqueous solution is heated to 65 to 70°C. The alcohol solution is heated in a separate container to the same temperature and added to the aqueous solution with stirring. The soap forms rapidly. When the reaction is complete, the mixture is cooled slightly and the perfume mixed in. The product is then poured into molds.

and firm. Glycerol may be used in place of propylene glycol, and plasticizers and emollients may be added.

A patent granted to Bell (55) refers to a two-phase deodorant-antiperspirant stick using sodium aluminum chlorhydroxy lactate as the active antiperspirant ingredient. It describes a stick comprising a solid antiperspirant core surrounded by a solid deodorant stick body comprising an alcohol-soap gel containing an antiseptic deodorant and perfume. In the illustration given, the antiperspirant ingredient is 40% w/w sodium aluminum chlorhydroxy lactate in water and the antiseptic deodorant is hexachlorophene.

Sodium aluminum lactate, which is supplied as a 40% solution, can also be used as the antiperspirant in alcohol-soap sticks (38).

Patents assigned to Kolmar (38) describe the preparation of such compounds as sodium aluminum gluconate complex, sodium aluminum glycolate complex, sodium aluminum lactate complex, and others and their use in sodium soap-alcohol gel antiperspirant sticks.

Two patents granted to Teller (56) relate to antiperspirant sticks. The earlier patent claims an antiperspirant stick comprising a sodium stearate-aqueous alcohol gel base with sodium zirconium lactate as the antiperspirant agent. Although sodium zirconium lactate was found to be a satisfactory antiperspirant, producing no primary irritation to skin and no damage to fabric, its use in the soap-stick type of product resulted in a large number of reports in the medical and dermatology journals of a new type of "granulomatous reaction" (57–60).

The skin eruption reported followed the use of this type of deodorant for periods of a few days to several months. Shelley (58) described the eruption as an allergic granuloma, the result of a hypersensitivity to zirconium. As a result of the reactions, deodorant sticks containing zirconium were withdrawn from the market.

The patent granted to Teller (56) in 1960 claims an alcohol-sodium stearate gel antiperspirant stick containing aluminum hydroxide as the active antiperspirant agent. This patent claims the use of isopropyl palmitate or isopropyl myristate as a physical stabilizer. An example from the patent is shown in Formula 18.

Formula 18

Ethyl alcohol	62.00 to 69.00%
Water	12.25 to 17.00
Sodium stearate	6.50 to 9.20
Aluminum hydroxide gel	
(10% Al_2O_3)	5.00 to 13.16
Isopropyl myristate or pal-	
mitate*	1.50 to 6.00

* The ratio of sodium stearate to isopropyl myristate and/or isopropyl palmitate should exceed 1.75:1.00.

Another patent granted to Teller (56) in 1966 describes stable aluminum hydroxide gel or aluminum hydroxide-sodium zirconium lactate antiperspirant sticks. It is claimed that the proper proportions of aluminum hydroxide 10 % gel with sodium stearate, alcohol, and water depresses the ionization of the aluminum hydroxide and thus prevents the formation of aluminum stearate and forms a stable antiperspirant stick.

A patent granted to Barton (61) relates to the method of preparing stable rigid nontacky antiperspirant products in stick form. An aluminum chlorhydrate-alkaline salt containing irreversible gel is prepared and modified by inclusion of chemically inert particles to produce a stick. The particles can be oily or waxy substances, wood pulp cellulose fibers, polystyrene and the like. The term "chemically inert" is used to denote particles which will not react with the aluminum chlorhydrate or alkaline components of the gel.

A wax-water emulsion is prepared and cooled to room temperature. This emulsion is then added to the astringent gel which is comprised principally of aluminum chlorhydrate solution and an alkaline salt, an acetate, propionate, or lactate of an alkali metal or an alkaline-earth metal. The pH of the sticks described is between 4 and 6. The active antiperspirants used in the gel are aluminum chlorhydrate, anticholinergics such as n-butyl scopolamine, and zirconium-aluminum complex.

Eleven examples are shown, none of which is considered to be excessively sticky or brittle. They are all said to have reduced perspiration significantly.

The soap-gel sticks in general are not as effective as a well-formulated antiperspirant cream or lotion. However, they have reasonably good astringency and constitute a convenient form of application.

Antiperspirant Powders

Antiperspirant powders are the least effective of the perspiration inhibitors. This is probably because an insufficient quantity of powder adheres to the skin to check the flow of perspiration appreciably. A patent issued to Jones (62) described an antiperspirant powder which contained 25 parts of an alkali-metal phosphate, such as Graham's salt, suitably mixed with lycopodium (5 parts) and talc (70 parts).

The use of finely divided aluminum chlorhydroxide in a powder will give a product of reasonably good antiperspirant effect for use on the body or the feet. An antiseptic deodorant is generally included in a powder-type product to increase the deodorant effect. Ten to twenty percent of impalpable aluminum chlorhydroxide may be used in a talc base.

Antiperspirant powder may also be pressed into a powder cake. For such a product, binders must be used to hold the powder and to allow it to rub off easily. Aluminum chlorhydrate with or without an added antibacterial agent may be used as the active ingredient in a talc base.

A compressed antiperspirant powder is described in a British patent (63). Aluminum chlorhydroxide powder is coated with high-molecular-weight (1000 to 6000) polyethylene glycol to reduce dusting and moisture absorption. This coated powder is then made into a cake with a talc base. A small amount of oil binder is used if it is needed. Other waxlike materials suitable for coating the particles of astringent are listed.

A cake using kaolin up to 40% in the talc base is described in a British patent granted to Fiedler (64). The proportions of ingredients recommended in the patent may vary between the limits shown in Formula 19.

Formula 19

*Compressed Antiperspirant Powder**

Talc	0.40 to 70.00%
Aluminum chlorhydroxide, impalpable	0.12 to 25.00
Kaolin	0.15 to 40.00
Zinc stearate	0.50 to 5.00
Magnesium carbonate	0.10 to 0.50
Pigment	0.25 to 1.00
Mineral oil	1.00 to 8.00
Perfume	0.10 to 1.00

* Ref. 105.

The kaolin absorbs moisture and acts as a binder along with the mineral oil. The ingredients are mixed dry and compacted into a cake under a pressure of 110 psi.

Antiperspirant powders have been adapted to an aerosol powder spray. A patent granted to Goldberg (65) describes an aerosol spray antiperspirant powder using aluminum chlorhydroxide, aluminum sulfate, or aluminum chloride as the antiperspirant ingredient with a suitable antiseptic agent.

Aerosol powders will be discussed in further detail in the section on aerosol products.

Liquid Antiperspirants

Liquid antiperspirants, generally applied as a spray, are probably the most easily formulated. For the most part they consist of an aqueous or a hydroalcoholic solution of an astringent salt, a small amount of humectant, a perfume, a dispersing agent for the perfume, a deodorant, and a buffer if it is needed. The astringent salt may be any one of those previously discussed. If aluminum sulfate or chloride is used, a buffer is needed. Aluminum chloride is quite irritating and should be used only in low concentrations. Aluminum chlorhydroxide complexes can be used in a spray without a buffer. A small amount of nonionic emulsifier will serve to disperse the perfume in the solution. The deodorant used must be soluble in the astringent solution. Alcohol tends to prevent hydrolysis of the antiperspirant in solution and to increase the rate of evaporation. A small amount of humectant helps prevent drying of the solution in the spray nozzle and subsequent stoppage of the opening if the product is to be sold in a spray container. A workable liquid or spray antiperspirant is shown in Formula 20.

Formula 20

Part A

Alcohol	50.0%
Propylene glycol	5.0
Hexachlorophene	0.1
Perfume	q.s.

Part B

Aluminum chlorhydroxide	15.0
Water	29.9

Procedure: Dissolve hexachlorophene and perfume in alcohol to make solution A. Dissolve the aluminum chlorhydroxide in the water, and add this solution slowly to A. Keep in a closed container for at least 48 hr, then filter before filling. If a lower concentration of alcohol is used, a small amount of nonionic emulsifier may be needed to disperse the perfume. Less alcohol produces a warmer spray.

Because it is retained on the skin, hexachlorophene has been widely used in deodorant products to give longer deodorant protection.

Quaternary ammonium salts, such as cetyl pyridinium chloride, cetyl trimethyl ammonium bromide, or the Hyamines [diisobutyl cresoxy (or phenoxy) ethoxy ethyl dimethyl benzyl ammonium chloride] (Rohm and Haas Co.), are also suitable for use as deodorants in antiperspirant solutions. Before a quaternary is selected, its compatibility with the astringent salt and with any emulsifier in the preparation should be checked. Quaternary ammonium salts are generally inactivated or precipitated by anionics, and their efficacy may be affected by some nonionics.

A simple antiperspirant preparation using a quaternary for deodorant effect is given in Formula 21.

Formula 21

Aluminum chlorhydroxide	18.0%
Hyamine 1622	0.5
Propylene glycol	5.0
Ethyl alcohol	35.0
Water	41.5
Perfume	q.s.

Procedure: Add the water, propylene glycol, and alcohol to a container with an agitator. Slowly add the Hyamine and aluminum chlorhydroxide with agitation which is continued until a clear solution is obtained. Then add the perfume.

If chlorophyll is used as a deodorant in a preparation of this type, then one of the chlorophyllins soluble in 50% alcohol must be used. The solution should be carefully checked for fabric-staining properties.

Antiperspirants are acid in reaction, and any perfume selected should therefore be acid-stable with regard to odor and color. The perfume should be tested in the product on the shelf at room temperature and at a slightly elevated temperature (40 to 45°C) for a reasonable period of time (3 to 6 months) to observe any change in color or odor which may take place.

Since perfumes may be irritating to the skin, it is important that the finished product, including the perfume, be tested on the skin for possible irritation or sensitization.

Because of its flexibility, inertness, and general attractiveness, polyethylene has been widely used in spray-type containers for liquid antiperspirants. In selecting such a package, the spray nozzles and fittings should be carefully tested for spray characteristics with the solutions to be packaged. Suppliers of polyethylene containers can furnish such parts with an orifice suitable to the viscosity of the spray.

Although polyethylene containers are inert to spray ingredients, they are somewhat permeable to many perfume ingredients. It is therefore important, in selecting a perfume for a product to be packaged in such a container, to choose one which shows the least diffusion through the container. Wight (48) listed common perfume materials in the order of their transmission rates through polyethylene. He found that the nature of the cosmetic vehicle plays a strong role in determining the rate of permeation. Alcoholic preparations lost less perfume than preparations in mineral oil or water vehicles. This may be due to a "blocking effect" by the vehicle or may be a partition of the perfume between the polyethylene and the solvent vehicle. It is obvious that shelf-life tests are important in selecting a perfume for a product to be packaged in polyethylene.

With the development of alcohol-soluble astringent salts, more effective antiperspirants can be formulated for sale in aerosol containers. The formulation and evaluation of aerosol antiperspirants requires a study of valves, cans, and linings, as well as product stability. These problems are discussed in the section of this book concerned with cosmetic aerosols.

Antiperspirant Test Methods

Several methods of testing the efficacy of antiperspirant products have appeared in the literature. Probably the simplest of these is a method described by Govett and deNavarre (9) in which the precipitation of egg albumin is used to measure astringency. Christian and Jenkins (66) used a casein solution in a similar precipitation test. This method is satisfactory for screening the relative efficacy of astringents in solutions or lotions. An *in vivo* procedure more applicable to product testing is one first described by Fredell and Read (67). Antiperspirant efficacy is determined by a use test in which axillary perspiration is absorbed and weighed on gauze pads.

A method of measuring the quantity of sweat excreted on small areas of skin by use of a specially prepared indicator paper was described in Herrmann, Prose, and Sulzberger (68). This method, which is a modification of one described by Manuila (69), is suitable for use only on skin areas where sweat excretion is slight (5 to 30 mg/2 cm^2) since with increased sweating coalescence of the droplets prevents a satisfactory print on the test paper from being taken.

Richardson and Meigs (70) described a method in which areas of the forearm are used in a comparative evaluation of antiperspirants. A method of evaluating antiperspirant effectiveness using frog membrane and radioactive tracer techniques has been described by Urakami and Christian (71).

The method described by Fredell, Read, and Longfellow (72) has been the one most widely used with various modifications. By this method subjects with shaved axillae use no antiperspirant or deodorant for a 1-week conditioning period. At 9:00 A.M. on Monday morning of the next week tared absorbent pads are placed in the dried axillae and held in place by keeping the arms close to the body. The pads are retained for 15 to 30 min or until a minimum of 100 mg of perspiration is obtained from the least perspiring axilla. This procedure is repeated 3 to 6 times to collect control data, the number of collections being determined by the reproducibility of the ratios of the weight of perspiration from the right axilla to that from the left. This value is designated pR (perspiration ratio).

On the afternoon of the same day the product to be tested is applied to the axilla which perspired the most. At the end of 1 hr, pads are applied to both axilla and 3 to 6 collections of perspiration on preweighed pads are taken as before and the pR calculated.

This method of treatment and subsequent collection of perspiration is repeated on the next three days, applying product at 9:00 A.M., always to the same axilla, and starting collection of perspiration on weighed pads at 10:00 A.M. The pR is calculated for each collection of perspiration.

The data are tabulated for the control perspiration, and averaged for the perspiration after 1, 2, 3, and 4 applications of antiperspirant. The effectiveness of the product may be calculated by comparing the data obtained on each day of the test with the control data.

To increase perspiration flow without altering the pR, subjects may be placed in a temperature- and humidity-controlled room during the collection period.

Buffered aluminum chloride solution is suggested as a benchmark preparation for use in checking the test procedure. Under controlled conditions of test and a carefully standardized routine, this method will give a good evaluation of the antiperspirant efficacy of a product.

Daley (73) developed a back-test method which is more rapid and more convenient than the Fredell and Read procedure. This method proved unreliable with certain cream-type products which form a surface coating on the skin, resulting in large reductions of perspiration on the back which could not be duplicated in the axilla. Another back-test procedure has been described and statistically evaluated by Wooding, Jass, and Ugelow (74). Small aluminum desiccators containing silica gel were preweighed and applied to treated areas of the back. The position of the treatment areas was predetermined by the use of a "Latin square." The desiccators were removed after 3 hr, placed in a container, and weighed. Each treatment was repeated on five successive periods a day apart. For a time prior to and during the application of the desiccators, the subjects were exposed to controlled temperature and humidity conditions.

The paper describes the method used for statistical analysis of the data obtained. The statistical evaluation showed that the method described yielded reliable and precise results in the study of antiperspirant effects.

O'Malley and Christian (75) describe a continuous-recording *in vivo* method of measuring sensible perspiration over a limited area. They used a specially constructed electrical circuit and electrolytic cell containing a methanol acetone-oxalic acid mixture which was sensitive to small amounts of water. Dry nitrogen was used to pick up moisture from the skin and deliver it to the cell containing the moisture-sensitive solution. If two such systems were used simultaneously, two skin areas could be studied. Bullard (76) describes a similar continuous technique for recording sweat rates. Air of controlled temperature and humidity was used to pick up sweat from an area of skin under a special skin capsule. The amount of water evaporated from the skin was calculated from the relative humidity change in the air as it passed over the skin, the air flow rate, and the temperature.

Jenkins (77) described a method for quantitative measurement of perspiration in the axillary area. The method proposed requires a specially constructed thermal room with controlled elevated temperature and humidity for the subject area. The axillary moisture is picked up from a specially fitted cup by a controlled flow of dry air. The air from the axilla is passed through a coil immersed in a dry ice-alcohol bath which permits complete condensation of the moisture in the tube. The method appears to give meaningful results but would be costly to set up.

A paper by James (78) describes still another method for continuously recording perspiration electronically. Perspiration is collected from the axillae of a subject maintained in a prone position in a chamber with controlled air supply, temperature, and humidity. The effectiveness of the antiperspirant is determined by recording the ratio of the values obtained from the two axillae for various rates of perspiration when no product has been applied and the ratios obtained at various rates of perspiration when a product has been applied to one axilla. This test requires bulky and highly specialized equipment.

The tests using continuous recording methods allow evaluation of products using a small number of subjects. Methods using absorption of perspiration on pads will give a satisfactory measure of effectiveness with less elaborate equipment if a reasonably large number of subjects is used.

Deodorants

Since many people who do not perspire excessively are concerned with body odor, a wide variety of products which are deodorant and not antiperspirant have been successfully marketed. Since most body odor is the result of bacterial action on perspiration, a satisfactory deodorant product must contain an effective antibacterial agent.

Many of the aluminum salts which have astringent properties have been found to have antiseptic properties. Shelley (6) noted that aluminum chloride inhibited bacterial growth in the axilla. Alum, aluminum potassium sulfate, and aluminum acetate are reportedly antiseptic (79). Aguilar (80) found that basic aluminum formate and aluminum formoacetate had antibacterial properties when tested in several types of ointment base. Aluminum chlorhydroxide and aluminum aminoacetate showed little or no zones of inhibition in the ointment bases studied. Killian (4) reported that the addition of aluminum sulfate to stale specimens of collected perspiration reduces the bacterial count as well as its odor. Klarman (81) reported that a single application of antiperspirant cream containing aluminum sulfate brought about a reduction of the total bacterial count in the axilla of over 95%. Blank (82) found that solutions of aluminum chloride, aluminum sulfate, aluminum phenolsulfonate, and aluminum chlorhydroxide, respectively, are effective in reducing the cutaneous bacterial population on the hands and in the axillae. It is therefore apparent that products which contain astringent aluminum salts will act as deodorants as well as antiperspirants.

Compounds of metals other than aluminum have been used in deodorant products. Zinc oxide, zinc peroxide, and zinc stearate are mildly astringent and antiseptic and can be used in deodorant powders. Zinc phenosulfonate, also antiseptic and mildly astringent, is a useful ingredient in liquid deodorant products.

The quaternary ammonium compounds are effective in controlling perspiration odor. Their alcohol solubility and lack of color and odor make them useful in various forms of deodorants.

Quaternary ammonium compounds are not satisfactory deodorants for use in a dusting powder since talc, which is the normal base for such a product, has a tendency to absorb quaternaries, thus reducing the antibacterial effect.

The bisphenols, particularly hexachlorophene, have been most widely used in deodorant products. The retention of such products on the skin (83) makes them peculiarly effective with repeated use.

Gump (84) described a series of tests using hexachlorophene in a powder and a stick. The efficacy of these products was determined by application to the axillae of a group of subjects and measuring the decrease in both odor and cutaneous bacteria. The powder containing 0.5% and the stick containing 0.25% hexachlorophene were found to effect over a 99% reduction in the bacterial population of the axillary fossae. The odor of the perspiration collected from the axillae was reduced to levels which were not objectionable. Other phenolic antibacterials which have been found effective in body deodorants are PCMX (parachlorometaxylenol) and DCMX (dichlorometaxylenol). The phenolic odor of these products is objectionable in many products.

TCC (3,4,4'-trichlorocarbanilide) has also been found to be effective against skin organisms (85). TCC is very sparingly soluble in water and in most commonly used organic solvents.

Many of the antibacterial agents which have been found to be effective against skin organisms have proved to be irritating or sensitizing. Formaldehyde and other compounds that may release formaldehyde can be both irritating and sensitizing. The chlorinated salicylanilides have been reported to produce photodermatitis in many people (86). Bithionol has been found to produce the same type of reaction (87). Cross-reactions with these among each other and with brominated salicylanilide and with hexachlorophene have also been reported (88).

Because of the high incidence of photosensitization resulting from the use of bithionol, the Food and Drug Administration announced in the *Federal Register* of February 14, 1968, that any cosmetic in interstate commerce containing bithionol will be subject to regulatory action as an adulterated product.

Some cases of photoallergic dermatitis from use of the bromosalicylanilides (bromsalans) have been reported. Epstein (89) points out that this form of sensitivity from use of the tribromsalans is rare. Vinson and Flatt (90) found that subjects photosensitized to soap containing trichlorosalicylanilide were not cross-sensitized to soap containing tribromosalicylanilide. Vinson (91) and Peck (92), in photosensitization tests on guinea pigs and on human skin, have found that dibromo- and tribromosalicylanilide are not photosensitizing. Since the bromosalicylanilides are more effective against a wider spectrum of organisms than the bisphenols, they are well suited to deodorant use.

Topically applied antibiotics will also inhibit axillary odor. Ferguson (93), using an ointment which contained 3% of aureomycin, found good deodorant effect and total suppression of bacterial growth in the axillae of 8 out of 10 subjects. Similar deodorant properties have been attributed to tyrothricin and neomycin. Shelley and Cahn (94) found that a cream and a lotion containing 3.5 mg of neomycin/g were effective in preventing the appearance of any axillary odor. Shehadeh and Kligman (95) studied the prolonged use of deodorant preparations containing antibiotics. They found that a lotion containing 0.175% of neomycin was an effective deodorant over an extended period of use. No sensitization was observed.

Of the antibiotics not too toxic for human use, many are used in topical medicinal products and by oral administration. Since when used topically there is greater danger of sensitization than when used orally, some of the antibiotics such as penicillin are not suitable for cosmetic use. Consideration of the toxicity, stability, and compatibility would indicate that neomycin and possibly tyrothricin are the most suitable of the antibiotics for deodorant use.

The deodorant action of ion-exchange resins has been investigated by Ikai (96). Both anion- and cation-exchange resins were found to absorb *in vitro* the odorous substances found in axillary sweat. When an anion- or a cation-type resin in the form of a fine powder was used in the axilla of a human subject, an "acid" or an "alkaline" type of odor was observed. A mixture of both types of resin was more effective but was readily washed from the area by the perspiration. When a mixture of a cation- and an anion-exchange resin was incorporated in an ointment or emulsion, good deodorant efficacy was observed. In these preparations, 20% of the mixed resins was used, and deodorization was complete for 1 to 3 days after application, if the axilla was not washed. No product using ion-exchange resins as deodorants has been marketed.

Deodorant Powders

In making a deodorant powder, it is essential that the active ingredient be uniformly dispersed. This can be accomplished by mixing and grinding. The dry ingredients should be well blended in a ribbon-type mixer. The active ingredient can be dissolved in a suitable solvent and distributed through the powder mix. The perfume can be well blended with a part of the talc and then incorporated into the batch. Since the active ingredients in deodorants may affect the fragrance, any perfume that is to be used should be tested in the finished formula over a period of 2 or 3 months. Samples should be checked at intervals for change or loss of odor. A much more uniform and better-textured product will be obtained if after preliminary mixing the powder is passed through a micropulverizer. Basic formulas for deodorant powders are given in Formulas 22 and 23.

Two patents granted to Thurmon (32,97) refer to the use of ion-exchange resins in deodorant compositions. One of these describes the use of aluminum phenolsulfonate and a cation-exchange resin in a hydrophylic ointment base. The second covers the use of a cation-exchange resin of the carboxylic type in dusting powders. The use of cation-exchange resins in dusting powders is discussed in *Schimmel Briefs*, No. *283*, Oct. 1958.

Formula 22

Talc	70%
Chalk, light precipitated	10
Boric acid	10
Zinc oxide	9
Zinc phenolsulfonate	1
Perfume	q.s.

Procedure: Follow instruction in text in section on deodorant powders.

Formula 23

Talc	84.0%
Boric acid	3.0
Chalk	12.0
Cetyl alcohol	0.5
Hexachlorophene	0.5
Perfume	q.s.

Procedure: Dissolve the hexachlorophene and the cetyl alcohol in a minimum quantity of alcohol, and add to the dry powders. Follow instructions in text, in section on deodorant powders.

A patent granted to Melton (98) in 1939 described a composition of deodorant powder of zinc peroxide and kaolin used together, with or without talc and calcium carbonate.

A deodorant foot powder may contain some alum for its astringent effect and/or a fungicidal agent to aid in preventing dermatophytosis. Undecylenic acid and its salts and propionic acid and propionates have been found to be effective against the organisms which cause ringworm of the feet (99). Sulzberger (100) found that combinations of undecylenic acid and zinc undecylenate are especially effective in the prevention of fungus infections of the feet. Sulzberger and Kanof conducted a series of experiments to ascertain as accurately as possible the efficacy of certain fatty acids and their salts in the prophylaxis and treatment of dermatophytosis. A powder containing 20% zinc undecylenate, 2% undecylenic acid, 76% talc, and a red color, dibenzo thio indigo 2%, was found to reduce the incidence of dermatophytosis by 85%. The color was used to facilitate the test.

Sodium propionate with propionic acid, as well as calcium propionate has been found suitable for use in powder for prevention of superficial fungus infections of the skin (99). The fatty acids and their salts have been found to be quite effective and sufficiently nonirritating for continued use.

Liquid Deodorants

Several of the quaternary ammonium compounds have been tested and found relatively nontoxic and sufficiently nonirritating for use in cosmetic preparations (101,102). Some of these, such as the Hyamines [diisobutyl phenoxy (or cresoxy) ethyl dimethyl benzyl ammonium chloride] (Rohm & Haas Co.), have been used in concentrations as high as 2% in deodorant liquids and creams (103,104). Trimethyl long-chain alkyl ammonium bromide (such as Cetab and others) and alkyl dimethyl benzyl ammonium chloride (Benzalkonium chloride USP) may also be used. Quaternaries are particularly suitable in aqueous deodorant preparations since they tend to affix themselves to the skin and are not readily washed off by perspiration.

The antibacterial and deodorant effect of these substances is therefore prolonged. A simple deodorant liquid or spray without antiperspirant properties may be prepared by dissolving 0.5 to 2% of a quaternary in water or in 50% denatured alcohol of cosmetic grade. Perfume and humectant may be added, if desired.

If a nonalcoholic spray is used, a small amount of nonionic wetting agent must be used to disperse the perfume. The perfume selected must be tested for compatibility with the quaternary and the dispersing agent used. Witch hazel may be substituted for water or alcohol in this type of product.

The water-soluble derivatives of chlorophyll, potassium-copper chlorophyllin, sodium-potassium-copper chlorophyllin, sodium-magnesium chlorophyllin, and others, have been found to have reasonable deodorant efficacy. Killian (4) found that water-soluble chlorophyllins, in a concentration of 0.05%, completely deodorized samples of malodorous stale perspiration. A simple *in vitro* test (105) for the deodorant activity of chlorophyllins has been devised. A numerical score for efficacy is obtained by measuring the capacity of several dilutions of the sample to decrease the intensity of odor of a fresh onion extract as determined by observations of a panel of five judges. The relevance of this test to the control of perspiration odors, however, is open to question. With chlorophyll, as with most deodorants, the efficacy is determined by the concentration used. Chlorophyll used in an effective concentration can give an objectionable color to the product and may cause staining of fabric. Chlorophyll derivatives may be used in combination with quaternaries in a deodorant solution or cream.

If one of the chlorinated phenol derivatives, such as hexachlorophene, is used in a liquid product, it must be carefully dissolved in propylene glycol and alcohol before the water is added. It has been observed that iron causes discoloration with this as with other phenol derivatives. It is therefore essential to eliminate iron contamination from other ingredients or from equipment when phenolic antibacterials are used. These deodorants give satisfactory antibacterial activity and deodorant effect in a concentration of 0.25 or less.

Polyethylene-spray-type containers are a suitable and inexpensive package for a deodorant spray. The finished product should be tested in the polyethylene for loss of perfume through the wall of the container.

Many deodorant liquids are currently packaged as aerosol sprays. These sprays generally consist of an anhydrous alcohol solution containing zinc phenolsulfonate, 2 to 5%; an antibacterial such as hexachlorophene, 0.1 to 0.3%; a small amount of a humectant such as propylene glycol; and perfume pressurized with propellant 12 or 12/114. The problems associated with the packaging of this type of product are discussed in the section on aerosol products.

Deodorant Creams

The deodorant cream has almost disappeared from the market as the result of the development of many more convenient forms for deodorant application. However, a well-formulated cream has excellent deodorant properties. A satisfactory deodorant cream may be prepared by incorporating an antibacterial such as hexachlorophene or a quaternary in a vanishing cream base. It has been observed (106) that many nonionic emulsifiers inhibit the activity of these antibacterials. It is therefore important to select an emulsifier that is compatible with the antiseptic to be used.

A potassium stearate-stearic acid vanishing cream serves as a satisfactory base. For a cream of this type, Formula 24 may be used.

Formula 24

Part A

Hexachlorophene	0.5%
Glyceryl monostearate	10.0
Stearic acid	4.0
Cetyl alcohol	2.0
Isopropyl myristate	4.0

Part B

Potassium hydroxide	1.0
Water	66.5
Propylene glycol	12.0
Perfume	q.s.

Procedure: Heat together ingredients A, with agitation, to 75°C. Continue to stir at this temperature until the hexachlorophene is completely dissolved. Heat ingredients B in another container to 75°C, and then add to A, with agitation. Continue to stir until cream reaches room temperature. Add the perfume at 45°C. Allow to stand overnight and stir again before filling.

A deodorant cream of a heavier body serves to absorb odorous materials from the skin surface and prevents subsequent development of odor by inclusion of antibacterial ingredients. A cream of this type may be prepared by use of zinc oxide and stearate as absorbents and a phenolsulfonate as an antibacterial agent, as shown in Formula 25.

Water-in-oil-type emulsions are satisfactory deodorant creams since some of the more commonly used deodorants are oil-soluble. A biphasic cream which rubs in smoothly may be preferred since it does not leave a greasy feel on the skin. A satisfactory cream of this type may be prepared from Formula 26.

Formula 25

Part A

Sorbitan sesquioleate	4.00%
Ceresin	6.00
Petrolatum, white	8.50
Mineral oil	20.00
Lanolin	4.50

Part B

Magnesium sulfate	0.15
Water	21.85
Zinc oxide	15.00
Zinc stearate	10.00
Aluminum phenolsulfonate	10.00
Perfume	q.s.

Procedure: Heat the ingredients in portion A to about 80°C, heat those in B in a separate container to about the same temperature, and then add B to A, with stirring. Continue to stir until the emulsion reaches 50°C. Add the zinc oxide and the zinc stearate slowly, with stirring. When the temperature is reduced to 40°C, add the aluminum phenolsulfonate slowly, with stirring. As soon as the phenolsulfonate has been completely incorporated in the cream, add the perfume. This cream will have a smoother texture if it is passed through a colloid mill.

Formula 26

Part A

Hexachlorophene	0.5%
Amerchol L-101	10.0
Glyceryl monostearate	13.5
Spermaceti	1.5
Tween 61	8.5
Isopropyl palmitate	3.0
Water	63.0
Perfume	q.s.

Procedure: Heat ingredients in part A to 70°C and stir slowly until the hexachlorophene is dissolved. Heat water in a separate container to 70°C, and then add water slowly, with rapid agitation, to the wax phase, part A. Add the perfume when the cream reaches 40°C and continue to agitate until it reaches 35°C. Allow to stand overnight and mix again before filling.

Stick Deodorants

Deodorants which contain no antiperspirant have also appeared in stick form. A waxy type of deodorant stick is described in a patent granted to O'Neil (107) in 1928. This preparation, according to one of the patent claims,

is composed of substantially 25 % zinc oxide, 25 % boric acid, 33 % spermaceti, 16.75 % petrolatum, and 0.25 % pertolatum oil, molded into stick form
and adapted to rub off in a smooth, even film when applied.

An increasingly popular type of deodorant for summer use is the alcohol gel
deodorant stick, or deodorant cologne stick. This may be prepared with the
hexachlorophene type of deodorant, or with a quaternary as the active
ingredient. The basic formula for such a product consists of 5 to 10 % of
sodium stearate, or sodium stearate with other hard soaps; 2 to 5 % of humectant; and the necessary alcohol and perfume. The gel may be formed by
dissolving soap in warm alcohol, or by saponification of an alcoholic solution
of stearic acid with sodium hydroxide. Agitation should be rapid enough to
dissolve the soap as it is formed. The solidifying temperature will vary with
the soap content. The alcoholic soap solution is poured hot into molds.
According to a patent by Slater (108), addition of small amounts of triisopropanolamine myristate to a sodium stearate-alcohol gel stick improves the
application properties by producing a stick with a lower yield point. Tackiness is also said to be reduced. Basic formulas for preparing deodorant sticks
are given in Formulas 27 to 29.

Formula 27

Sodium stearate	8.00%
Sorbitol syrup	5.00
Tribromosalicylanilide	0.25
Water	8.75
Ethyl alcohol	75.00
Perfume	3.00

Procedure: Add the water, sorbitol, and Hyamine to the alcohol in a container
with a source of heat and agitation to give a clear solution. Add the soap and heat
the mixture with stirring until the soap is dissolved. Loss of alcohol by evaporation
should be avoided. When the soap is in solution, add the perfume and pour the
mixture into molds.

Formula 28

Soap, hard	8.00%
Isopropyl myristate	10.00
Glycerol	3.00
Hexachlorophene	0.25
Ethyl alcohol	75.75
Perfume	3.00

Procedure: Proceed as for Formula 27, adding the hard soap after the other
materials (except the perfume) are in solution. The isopropyl myristate improves the
feel of the product on the skin and reduces the rate of evaporation.

Formula 29

Part A

Stearic acid	6.00%
Ethyl alcohol	64.75

Part B

Sodium hydroxide	1.00
Water	5.00

Part C

Hexachlorophene	0.25
Propylene glycol	2.00
Perfume	3.00
Ethyl alcohol	18.00

Procedure: Mix ingredients A in a container with a source of heat and agitation. Heat the mixture with stirring until the temperature reaches 70°C. Weigh portions B and C into separate containers, stir to give clear solutions, and heat each to about 60°C. When solution A is at 70°C, add B to A with rapid agitation. A clear soap solution is obtained. As soon as the addition of B is complete, add portion C and pour the mixture into molds.

The sticks may be wrapped in foil and packed in glass containers with screw caps. It is particularly important that a container which allows a minimum of evaporation be available for this type of product. A swivel-type case similar to those long identified with lipsticks has been successfully used by some manufacturers.

Deodorant Test Methods

Various procedures have been described for evaluation of deodorants. Barail (109) describes a procedure for measuring perspiration odor by use of an osmometer. A procedure in which the axilla is swabbed with a cotton ball to absorb odor is described by Fredell and Read (110). The swabs were dropped into a clear glass jar with an unlined cap and then evaluated by three observers. Odor was recorded at four levels from slight to very objectionable. In a later report (72), Fredell and Longfellow describe a modification of this method using an untreated axilla as a control and the other axilla treated with the product to be evaluated. A pretest conditioning period using no deodorant of any kind is recommended. On the first day of test the axillary odor is recorded for both untreated axillae using a scale of 0 to 3. Odor levels are obtained by direct sniffing. The test product is applied to one axilla, nothing to the control area. At the end of 6 hr, both axillae are again sniffed and the odor recorded. The test may be repeated on succeeding days using the same axilla as a control. The control axilla must have a definite characteristic axillary odor in order to judge the effectiveness of the product under test. The authors recommend that a benchmark preparation be used on the test subjects as an aid in evaluating the product to be tested. A marketed product may be used or a standard solution of an effective antibacterial. This procedure is probably the simplest and most satisfactory method for evaluating deodorant products. A detailed description of a modification of this method was presented in a paper by Whitehouse and Carter (111) at the December 1967 meeting of the Toilet Goods Association.

Labeling of Antiperspirants and Deodorants

The Federal Food, Drug and Cosmetic Act includes certain provisions which apply to the labeling of preparations for which claims are made as to their antiperspirant action (112). A product falls under the definition of a "drug" as stated by this act if it is claimed to be capable of reducing the flow of perspiration, because in so doing it affects a bodily function. The active ingredient of such a preparation must be disclosed on the label, along with other information required by the act, such as adequate directions for use and in some cases warnings.

Deodorant preparations for which no claim is made other than that of eliminating or reducing odor are classified as cosmetics and must comply with the requirements for such products. If a cosmetic preparation is claimed to be antiseptic, it falls under the portion of the act which states the antibacterial performance required of a preparation so designated (112).

The classification of a product as a drug or cosmetic is determined by its intended use. However, many products commonly regarded as cosmetics are also drugs as defined in the act. Statements made in the labeling of the product or the effect of any ingredient, even when no claims are made, may cause the product to be classified as a drug.

"New drugs" are (a) substances not previously used as drugs, (b) well-known substances not previously used under the conditions recommended, or (c) a mixture of known drugs not previously used in combination (113). The sale of a "new drug" requires a "new drug application" (114). The opinion of the Food and Drug Administration may be obtained in determining whether or not a product will be considered a "new drug."

According to the Federal Food, Drug and Cosmetic Act, a drug or a cosmetic shall be deemed to be misbranded if its labeling is false or misleading in any particular (112,113). An antiperspirant preparation may be claimed to "check" or "reduce" perspiration rather than "stop" it. Deodorant preparations may be said to "decrease" or "stop" temporarily the odor of perspiration. The term "labeling" means all labels and other written, printed, or graphic matter upon any article, upon any of its containers or wrapper, or accompanying such article.

A drug or a cosmetic shall be deemed misbranded if in a package form unless it bears a label containing (a) the name and place of business of the manufacturer, packer, or distributor, and (b) an accurate statement of the quantity of the contents in terms of weight, measure, or numerical count. Under the second clause reasonable variations are permitted.

The ultimate test of a deodorant, as is that of any product, is its acceptance by the consumer. The ideal product is probably one which has optimum antiperspirant and deodorant efficacy, an attractive perfume, and a well-designed package. It should be easy to apply and leave no uncomfortable residue on the skin. It must cause no discomfort after continued use. It should not stain or corrode fabrics. It must not change appreciably during storage or in the partly filled containers. Very few products, if any, possess all of these characteristics. However, with the new types of antiperspirants, the varied emulsifiers, and the interesting innovations in packaging supplies now available, it is inevitable that many new and interesting deodorant preparations will make their appearance.

REFERENCES

1. Le Van, P.: *The evaluation of therapeutic agents and cosmetics,* Sternberg, T. H. and Newcomer, V. D. (eds.), McGraw-Hill, New York, 1964 p. 143.

2. Way, S. C., and Memmesheimer, A.: The sudoriparous glands, *Arch. Dermatol. Syphilol.,* **41**: 1086 (1940); **34**: 797 (1936); **38**: 373 (1937).

3. Berry, E.: Über die Abhängigkeit des Stickstoff- und Chlorgehaltes des Schweisses von der Diät, *Biochem. Z.,* **72**: 285 (1916).

4. Killian, J. A., and Panzarella, F. P.: Comparative studies of samples of perspiration collected from clean and unclean skin of human subjects, *Proc. Sci. Sec. TGA,* **7**: 3 (1947); Killian, J. A.: Evaluation of *in-vitro* and *in-vivo* methods of testing deodorants with particular reference to chlorophyll and its derivatives, *JSCC,* **3**: 30 (1952).

5. Shelley, W. B., and Hurley, H. J.: Methods of exploring human apocrine sweat gland physiology, *Arch. Dermatol. Syphilol.,* **66**: 156 (1952); The physiology of the human axillary apocrine sweat gland, *J. Invest. Dermat.,* **20**: 285 (1953).

6. Shelley, W. B., Hurley, H. J., and Nichol, A. C.: Axillary odor, *Arch. Dermatol. Syphilol.,* **68**: 430 (1953).

7. Sulzberger, M. B., Zak, F. G., and Herrmann, F.: Studies on sweating, II, *Arch. Dermatol. and Syphilol.,* **60**: 404 (1949); Shelley, W. B.: The effect of poral closure on the secretory function of the eccrine sweat gland, *J. Invest. Dermatol.,* **22**: 267 (1957).

8. Papa, C. M.: The action of antiperspirants, *JSCC,* **17**: 789 (1966); Papa, C. M., and Kligman, A. M.: Mechanism of eccrine anidrosis, *J. Invest. Dermatol.* **47** (1): 1 (1966); Mechanisms of eccrine anidrosis. II. The antiperspirant effect of aluminum salts, *ibid.,* **49**(2): 139 (1967).

9. Govett, T., and deNavarre, M. G.: Aluminum chlorhydrate, new antiperspirant ingredient, *Am. Perf.*, **49**: 365 (1947).

10. Martin, J. J., Jones, J. L., and Lawrence, J. A.: A new aerosol antiperspirant, *DCI*, **99**(5): 54 (1966).

11. Beekman, S. M., Holbert, J. M., and Schmank, H. W.: Studies of new aluminum compounds for antiperspirant use. I., *JSCC*, **18**(2): 105 (1967).

12. Draize, J. H.: *Appraisal of the safety of chemicals in foods, drugs and cosmetics*, The association of Food and Drug Officials of the United States, 1959, p. 151.

13. Schwartz, L., and Peck, S. M.: *Cosmetics and dermatitis*, Hoeber, New York, 1949, p. 151; Schwartz, L.: The skin testing of new cosmetics, *JSCC*, **2**: 321 (1951).

14. Draize, J. H.: *op. cit.* (ref. 12), pp. 51, 52; Maibach, H. I., and Epstein, W. L., Predictive patch testing for sensitization and irritation, *Am. Perf.*, **80**: 55 (October 1965).

15. Shelanski, H. A., and Shelanski, M. V.: A new technique of human patch tests, *Proc. Sci. Sec. TGA*, **19**: 46 (1953).

16. Bien, R. R.: The action of antiperspirant creams on fabrics, *Proc. Sci. Sec. TGA*, **4**: 8 (1945).

17. Tate, L. H.: U. S. Pat. 1,371,822 (1921).

18. Taub, H.: U. S. Pat. 1,984,669 (1934).

19. Teller, W. K.: U. S. Pats. 2,210,013–14 (1940).

20. Montenier, J.: U. S. Pats. 2,230,082–84 (1941).

21. Wallace, J. H., and Hand, W. C.: U. S. Pat. 2,236,387 (1941).

22. Klarmann, E. G., and Gates, L. W.: U. S. Pat. 2,350,047 (1944).

23. Richardson, E. L., and Russell, K. L.: U. S. Pat. 2,412,535 (1946).

24. Anderson, C. N.: U. S. Pat. 2,492,085 (1949).

25. Van Mater, H. L.: U. S. Pat. 2,498,514 (1950).

26. Berger, F. M., and Plechner, S. L.: U. S. Pat. 2,734,847 (1956), 2,790,747 (1957), 2,889,253 (1959).

27. Weiss, L.: Ger. Pat. 237,624 (1910).

28. Grote, I. W., and Holbert, J. M.: U. S. Pat. 2,508,787 (1950).

29. Govett, T., and Almquist, M. L.: U. S. Pat. 2,571,030 (1951).

30. Apperson, L. D., and Richardson, E. L.: U. S. Pats. 2,586,287–88 (1952).

31. Henkin, H., and Messina, R. P.: U. S. Pat. 2,783,181 (1957).

32. Thurmon, F. M.: U. S. Pat. 2,653,902 (1953).

33. Daley, E. W.: U. S. Pat. 2,814,584 (1957).

34. Daley, E. W.: U. S. Pat. 2,814,585 (1957).

35. Grad, M.: U. S. Pat. 2,854, 382 (1958).

36. Berger, F. M., and Plechner, S. L.: U. S. Pat. 3,090,728 (1963).

37. Siegel, B.: U. S. Pat. 3,018,223 (1962).

38. Kolmar Cosmetics (Europa), A. G.: Ger. Pat. 1,083,502 (1960); Joseph Lucas Industries: Br. Pat. 843,865 (1960).

39. Kalish, J.: Aluminum chlorhydroxy lactate complex, *DCI*, **79**(3): 319 (1956).

40. Messina, R. P.: U. S. Pat. 3,235,458 (1966).

41. Grimson, K. S., *et al.*: Successful treatment of hyperhidrosis using Banthine, *J. Am. Med. Assoc.*, **143**(15): 1331 (1950).

42. Nelson, L. M.: Prantal in the treatment of hyperhidrosis, *J. Invest. Dermatol.*, **17**: 207 (1951).

43. Zupko, A. G.: An evaluation of Banthine in hyperhidrosis, *J. Am. Pharm. Assoc., Sci. Ed.*, **41**: 212 (1952); Zupko, A. G., and Prokop, L. D.: A comparative study of Prantal and Banthine in hyperhidrosis, *ibid.*, **41**: 651 (1952); The newer anticholinergic agents. I. Effectiveness as anhydrotics, *ibid.*, **43**: 35 (1954).

44. Shelley, W. B., and Horvath, P. N.: Comparative study on the effect of anticholinergic compounds on sweating, *J. Invest. Dermatol.*, **16**: 267 (1951).

45. Kernen, R., and Brun, R.: Experiences sur la transpiration. V. Examens pharmacodynamiques au niveau de la glande sudoripare de l'homme, *Dermatologica*, **106**: 1 (1953); Stoughton, R. B., Chiu, F., Fritsch, W., and Nurse, D.: Topical suppression of eccrine sweat delivery with a new anticholinergic agent, *J. Invest. Dermatol.*, **42**: 151 (1964).

46. MacMillan, K., Reller, H. H., and Snyder, F.: The antiperspirant action of topically applied anticholinergics, *J. Invest. Dermatol.*, **43**: 363 (1964).

47. Procter and Gamble, Ltd.: Br. Pat. 940,279 (1963).

48. Wight, C. F., Tomlinson, J. A., and Kirmeier, S.: Difficulties encountered in the use of polyethylene as a packaging material, *Proc. Sci. Sec. TGA*, **19**: 30 (1953).

49. Griffin, W. C., Behrens, R. W., and Cross, S. T.: Hygroscopic agents and their use in cosmetics, *JSCC*, **3**: 5 (1952).

50. Smyth, H. F., Jr., Carpenter, C. P., and Shaffer, C. B.: The subacute toxicity and irritation of polyethylene glycols of approximate molecular weights of 200, 300, and 400, *J. Am. Pharm. Assoc., Sci. Ed.*, **34**: 172 (1945); Smyth, H. F., Jr., Carpenter, C. P., and Weil, C. S.: The toxicology of the polyethylene glycols, *ibid.*, **39**: 349 (1950).

51. Shaffer, C. B., Critchfield, F. H., and Nair, J. H., III: The absorption and excretion of a liquid polyethylene glycol, *J. Am. Pharm. Assoc., Sci. Ed.*, **39**: 340 (1950).

52. Karas, S. A.: The effect of some aromatic chemicals and essential oils upon the stability of cosmetic emulsions, *JSCC*, **1**: 374 (1949).

53. Hilfer, H.: Deodorants and antiperspirants, in *Drug cosmetic review 1950–51*, Drug & Cosmetic Industry, New York, pp. 85–8.

54. Moore, W. C.: U. S. Pats. 2,087,161–62 (1937).

55. Bell, S. A.: U. S. Pat. 2,970,083 (1961).

56. Teller, W. K.: U. S. Pat. 2,732,327 (1956): U. S. Pat. 2,933,433 (1960); U. S. Pat. 3,259,545 (1966).

57. Pinkus, M., and Botwinick, I.: Deodorant stick eruption (zirconium granuloma) of axillae, *Arch. Dermatol.*, **75**: 756 (1957).

58. Shelley, W. B., and Hurley, H. J.: Experimental evidence for an allergic basis for granuloma formation in man, *Nature*, **180**: 1060 (1957); Shelley, W. B., and Hurley, H. J.: The allergic origin of zirconium deodorant granulomas, *Brit. J. Dermatol.*, **70**: 75 (1958).

59. Prior, J. T., Rustad, H., and Cronk, G. A.: Pathological changes associated with deodorant preparations containing sodium zirconium lactate: an experimental study, *J. Invest. Dermatol.*, **29**: 449 (1957).

60. Epstein, W. L.: Contribution to the pathogenesis of zirconium granulomas in man, *J. Invest. Dermatol.*, **34**(3): 183 (1960).

61. Barton, S.: U. S. Pat. 3,255,082 (1966).

62. Jones, K. K.: U. S. Pat. 2,144,599 (1938).

63. Kolmar Labs.: Br. Pat. 1,001,690 (1965); French Pat. 1,370,247 (1964).

64. Fiedler, J. G.: Br. Pat. 996,560 (1965).

65. Goldberg, M. A., and Netzbandt, W. R.: U. S. Pat. 3,288,681 (1966).

66. Christian, J. E., and Jenkins, G. L.: A comparison of a new astringent agent with such agents now commonly used, *J. Am. Pharm. Assoc., Sci. Ed.*, **39**: 663 (1950).

67. Fredell, W. G., and Read, R. R.: Antiperspirant—axillary method of determining effectiveness, *Proc. Sci. Sec. TGA*, **15**: 23 (1951).

68. Herrmann, F., Prose, M. D., and Sulzberger, M. B.: Studies on sweating. IV. A new quantitative method of assaying sweat delivery to circumscribed areas of the skin surface, *J. Invest. Dermatol.*, **17**: 241 (1951); V. Studies of quantity and distribution of thermogenic sweat delivery to the skin, *ibid.*, **18**: 71 (1952).

69. Manuila, L.: Traitement medicamenteux local de la transpiration, *Dermatologica*, **100**: 304 (1950); Nouveau test de transpiration, *ibid.*, **102**: 302 (1951).

70. Richardson, E. L., and Meigs, B. V.: A method for comparative evaluation of antiperspirants, *JSCC*, **2**: 308 (1951).

71. Urakami, C., and Christian, J. E.: An evaluation of the effectiveness of antiperspirant preparations using frog membrane and radioactive tracer technique, *J. Am. Pharm. Assoc., Sci. Ed.*, **42**: 179 (1953).

72. Fredell, W. G., and Read, R. R.: Axillary perspiration, *DCI*, **79**(4): 468 (1956); Fredell, W. G., and Longfellow, J. M.: Evaluating antiperspirant and deodorant products, *JSCC*, **9**(2): 108 (1958).

73. Daley, E. W.: Antiperspirant testing: a comparison of two methods, *Proc. Sci. Sec. TGA*, **30**: 1 (1958).

74. Wooding, W. M., Jass, H. E., and Ugelow, M. S.: Statistical evaluation of quantitative antiperspirant data I, *JSCC*, **15**: 579 (1964).

75. O'Malley, W. J., and Christian, J. E.: An evaluation of the ability of compounds to reduce perspiration flow, *J. Am. Pharm. Assoc., Sci. Ed.*, **49**: 398 (1960).

76. Bullard, R. N.: Continuous recording of sweating rate by resistance hygrometry, *J. Appl. Physiol.*, **17**: 735 (1962).

77. Jenkins, J. W., Ouellette, P. A., Healy, D. J., and Della Lana, C.: A technique for perspiration measurement, *Proc. Sci. Sec. TGA*, **42**: 12 (1964).

78. James, R. J.: A new and realistic electronic approach to the evaluation of antiperspirant activity, *JSCC*, **17**: 749 (1966).

79. Cromwell, H. W., and Leffler, R.: Evaluation of skin degerming agents by a modification of the Price method, *J. Bacteriol.*, **43**: 51 (1942); Miguel, I., and Wood, H. C. *The dispensatory of the U.S.A.*, 22 ed., J. B. Lippincott, Philadelphia, 1937, p. 101.

80. Aguilar, T. N., Blaug, S. M., and Zopf, L. C.: A study of the antibacterial activity of some complex aluminum salts, *J. Amer. Pharm. Assoc., Sci., Ed.*, **45**(7): 498 (1956).

81. Klarman, E. G.: Chemical and bacteriological aspects of antiperspirants and deodorants, *JSCC*, **7**(2): 94 (1956).

82. Blank, I. H., Moreland, M., and Dawes, R. K.: The antibacterial activity of aluminum salts, *Proc. Sci. Sec. TGA*, **27**: 24 (1957).

83. Fahlberg, W. J., Swan, T. C., and Seastone, C. V.: Studies on the retention of hexachlorophene (G-11) in human skin, *J. Bacteriol.*, **56**: 323 (1948); Manowitz, M., and Johnston, V.: Deposition of hexachlorophene on the skin, *JSCC*, **18**: 527 (1967).

84. Gump, W. S.: Hexachlorophene in cosmetics, *DC*, **72**: 622 (1953).

85. Roman, D. P., Barnett, E. H., and Balske, R. J.: Cutaneous antiseptic activity of 3,4,4-trichlorocarbanilide, *Proc. Sci. Sec. TGA*, **28**: 12 (1957).

86. Wilkinson, D. S.: Photodermatitis due to tetrachlorosalicylanilide, *Brit. J. Dermatol.*, **73**: 213 (1961).

87. Jellson, O. F., and Baughman, R. D.: Contact photodermatitis from bithionol, *Arch. Dermat.*, **88**: 409 (1963); Baughman, R. D.: Contact photodermatitis from bithionol. II. Cross-sensitivities to hexachlorophene and salicylanilides, *Arch. Dermat.*, **90**: 153 (1964).

88. Harber, L. C., Harris, H., and Baer, R. L.: Structural features of photoallergy to salicylanilides and related compounds, *J. Invest. Dermatol.*, **46**(3): 303 (1966).

89. Epstein, S., and Enta, T.: Photoallergic contact dermatitis, *JAMA*, **194**(9): 168 (1965); Harber, L. C.: Photoallergic contact dermatitis due to halogenated salicylanilides and related compounds, *Arch. Dermatol.*, **94**: 255 (1966).

90. Vinson, L. J., and Flatt, R. S.: Photosensitization by tetrachlorosalicylanilide, *J. Invest. Dermat.*, **38**: 327 (1962).

91. Vinson, L. J., and Borselli, V. F.: A guinea pig assay of the photosensitizing potential of topical germicides, *JSCC*, **17**: 123 (1966).

92. Peck, S. M., and Vinson, L. J.: Use of two prophetic patch tests for the practical determination of photosensitization potential, *JSCC*, **18**: 361 (1967).

93. Ferguson, E. H.: A note on axillary odor, *J. Invest. Dermatol.*, **24**: 567 (1955).

94. Shelley, W. B., and Cahn, M. N.: Effect of topically applied antibiotic agents on axillary odor, *JAMA*, **159**: 1736 (1955).

95. Shehadeh, N. H., and Kligman, A. M.: The effect of topical antibacterial agents on the bacterial flora of the axilla, *J. Invest. Dermatol.*, **40**: 61 (1963).

96. Ikai, K.: Deodorizing experiments with ion-exchange resins, *J. Invest. Dermatol.*, **23**: 411 (1954).

97. Thurmon, F. M.: U. S. Pat. 2,838,440 (1958).

98. Melton, H. E.: U. S. Pat. 2,144,632 (1939).

99. Lesser, M. A.: Newer fungicidal products, *DCI*, **69**(4): 468 (1951).

100. Sulzberger, M. B., and Kanof, A.: Undecylenic and propionic acids in the prevention and treatment of dermatophytosis, *Arch. Dermatol. Syphilol.*, **55**: 391 (1947).

101. Alfredson, B. V., Stiefel, J. R., Thorp, F., Jr., Baten, W. C., and Gray, M. L.: Toxicity studies on alkyl dimethyl benzyl ammonium chloride in rats and in dogs, *J. Am. Pharm. Assoc., Sci. Ed.*, **40**: 263 (1951).

102. Draize, J. H., and Kelley, E. A.: Toxicity to eye mucosa of certain cosmetic preparations containing surface-active agents, *Proc. Sci. Sec. TGA*, **17**: 1 (1952).

103. Lesser, M. A.: Quaternary ammonium compounds, *Drug cosmetic review*, *1950–51*, pp. 130–13.

104. Botwright, W. E.: Quaternary ammonium compounds in cosmetic and related products, *JSCC*, **3**: 118 (1953).

105. deNavarre, M. G.: Desiderata (*in-vitro* onion test), *Am. Perf.*, **61**: 353 (1953).

106. Lawrence, C. A., and Erlandson, A. L., Jr.: A new inactivating medium for hexachlorophene (G-11), *J. Am. Pharm. Assoc., Sci. Ed.*, **42**: 352 (1953); Wedderburn, D. L.: Preservation of toilet preparations containing nonionics, *JSCC*, **9**(4): 210 (1958).

107. O'Neil, E. H.: U. S. Pat. 1,669,016 (1928).

108. Slater, J. N.: U. S. Pat. 2,900,306 (1959).

109. Barail, L. C.: Measurement of odors in the cosmetic industry, *JSCC*, **2**(1): 4 (1950).

110. Fredell, W. G., and Read, R. R.: Axillary perspiration—odor and deodorization, *Proc. Sci. Sec. TGA*, **25**: 32 (1956).

111. Whitehouse, H. A., and Carter, O.: Deodorant efficacy of toilet bars, *Soap*, **44**(2): 64 (1968); Evaluation of deodorant toilet bars., *Proc. Sci. Sec. TGA*, **48**: 31 (1968).

112. deNavarre, M. G.: *The chemistry and manufacture of cosmetics*, Vol. I., Van Nostrand, 1962, pp. 145–71.

113. The Federal Food, Drug and Cosmetic Act, Section 201.

114. Theile, F. C.: Proper procedure for the approval of a new drug or drug cosmetic, *Am. Perf.*, **80**: 45 (October 1965).

Chapter 26

AEROSOL COSMETICS*

Morris J. Root

The toiletries and cosmetic business has been identified as one of the growth pace setters in American industry by the United States Department of Commerce. Predictions are that it is likely to continue at its high rate of growth. Although this forecast is due partially to the increasing affluence of the American consumer and the growing acceptance of personal products by men, it is also due to the new aerosol technology which started in the early 1950's.

Aerosol personal products are now an important segment of the toiletries and cosmetic industry. The rate of growth of the aerosol segment of the industry compared to the industry as a whole is shown in Figure 1. In addition to antiperspirants and deodorants, men's products showed noticeable activity during 1968. Aerosol fragrances and hair grooms for men frequently were on new products lists.

Aerosol feminine hygiene products, which were introduced only recently, have moved ahead to about 17 million units per year. Hot dispensed products went into national distribution in 1968. Industry researchers are still busy with new product possibilities employing the codispensing concept used in the hot products. Using the codispensing principle, we already have a hot shaving lather, a hair conditioning foam, a facial cleanser, a moisturizing cream, and a hair dye.

Currently, the cosmetic aerosol products are divided up as follows:

Hair sprays	41%
Antiperspirants and deodorants	28
Shaving lathers	13
Colognes and perfumes	9
Medicinals and pharmaceuticals	5
Others	4

* Some of the material in this chapter is based on the chapter on the same subject, by H. R. Shepherd, that appeared in the first edition of this book.

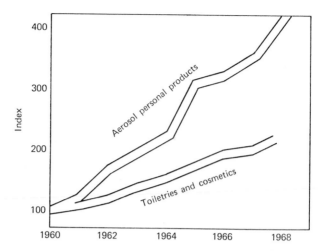

Fig. 1.   Index comparison of aerosol personal products with toiletries and cosmetics.

Breath fresheners have contributed significantly to the strong growth of the "Others" category.

The antiperspirant and deodorant market in aerosols should reach 400 million units in 1970; this may surpass hair sprays, which have been number one in the field since the beginning of the aerosol development.

Definitions

The word "aerosol" is a generic term that was first employed (as it still is) in the field of colloid chemistry. It was used by Gibbs, and later defined by Whytlaw-Gray and Patterson (1) as a colloidal system consisting of very finely subdivided liquid or solid particles dispersed in and surrounded by a gas. This definition was made more specific by Sinclair (2), who stated that such particles should be smaller than 50 μ in size and usually less than 10 μ. The definition of Sinclair is accepted by Avy (3) in his recent study of this subject.

The "aerosol" industry got its name from the first space insecticides that were packaged in the early 1940's and which were truly "aerosols". With the advent of products that are dispensed as coarse sprays, foams, pastes, and gels, the term "aerosol" is no longer confined to the scientific definition. Perhaps a more accurate description of these packages is "pressurized products." The Chemical Specialties Manufacturers Association in 1966 adopted a definition for an aerosol or pressurized product—"may be a liquid, solid, gas or mixture thereof discharged by a propellent force of liquefied and/or non-liquefied compressed gases usually from a disposable type dispenser through a valve."

Pressurized packages can be classified by at least three different categories:

1. The physical characteristics of the products as it is emitted from the package:
   a. Space sprays $<50\ \mu$.
   b. Surface sprays $>50\ \mu$.
   c. Streams.
   d. Droplets.
   e. Foams.

In all cases, the product can be homogeneous, an emulsion, or a dispersion.

2. Another manner of characterizing a pressure package is by the propellant, which can be a liquid, gas, or mechanical system:
   a. Liquid:
      (1) Fluorocarbons.
      (2) Hydrocarbons.
   b. Gas:
      (1) Nitrogen.
      (2) Nitrous oxide.
      (3) Carbon dioxide.
   c. Mechanical System:
      (1) Spring.
      (2) Rubber or elastomer bladder.

3. A third manner of describing a pressure package is by the type of container:
   a. Metal:
      (1) Tin plate.
      (2) Aluminum.
      (3) Stainless steel.
   b. Glass.
   c. Sepro.
   d. Preval (Precision).
   e. Innovair (Geigy).

The official Chemical Specialties Manufacturers Association (4) definitions of all terms used in the aerosol industry are given in the appendix to this chapter.

Historical Background

A comprehensive treatise on the development of the "aerosol" or pressure package, and the growth of the tremendous industry, has not as yet been written. Perhaps the first record for a self-dispensing package is in a patent granted to Lynde (5) for a valve, complete with dip tube, for dispensing an aerated liquid from a bottle. The first reference to the use of liquefied gases appears to be contained in Helbing and Pertsch's patent of 1889 (6). These inventors used methyl and ethyl chlorides, making a solution with either or both of these alkyl chlorides, with or without the addition of alcohol or ether. "This solution", the patent states, "is placed, with a considerable excess of methyl or ethyl chloride, if required, in a receiver of glass or metal, having a suitable orifice which can be hermetically sealed at will with a close-fitting cap." When it is required to make use of the solution, the cap is removed and the vessel inclined as may be necessary. The heat of the hand holding the vessel, or even of the surrounding atmosphere, if of the ordinary temperature of a room, immediately causes the ethyl or methyl chloride inside the receiver to begin to evaporate. This evaporation causes internal

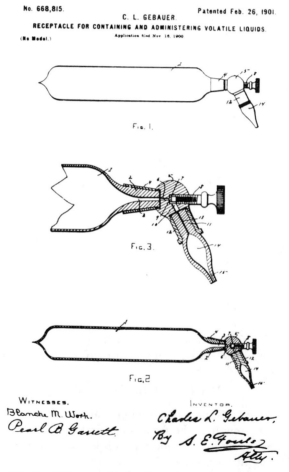

Fig. 2.   Valve mechanism in earliest Gebauer patent.

pressure, and the solution is thereby ejected through the orifice in a fine jet or spray. The reader familiar with modern aerosol technology will recognize how close Helbing and Pertsch were to the products destined to appear on the market almost half a century later.

Soon after the contribution of Helbing, there appeared the first fruits of the creative talents of Gebauer (7), who, in his earliest patent of 1901, proposed "to eject the liquid in the form of a spray" and described a receptacle for containing and ejecting such a spray. Gebauer followed this up in 1902 with a second patent (8), in which he described certain improvements in the original receptacle, to make it cheaper in construction and permit ejection in either a spray or a stream. As a matter of historical interest, the diagrams accompanying the Gebauer patents are reproduced in Figures 2 and 3.

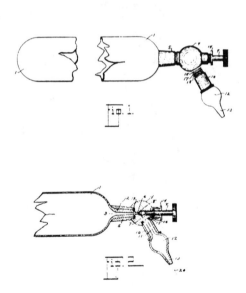

No. 711,045.   Patented Oct. 14, 1902

C. L. GEBAUER.

RECEPTACLE FOR CONTAINING AND ADMINISTERING VOLATILE LIQUIDS.

Application filed Aug. 21, 1901.

(No Model.)

Fig. 3.   Valve mechanism in second Gebauer patent.

On December 15, 1903, R. W. Moore (9) was granted a patent, entitled "Perfumery Atomizer" in which he described in some detail an aerosol perfume dispenser somewhat similar to that in present-day usage. Moore employed carbon dioxide as the propellant. The valve used by Moore is shown in Figure 4.

Although Gebauer and others continued their studies and were marketing their products, there is little record in the literature, patent or other, of further advance until 1921, when Mobley was granted a patent (10) in which he described a method of applying liquid antiseptics in an aerosol form,

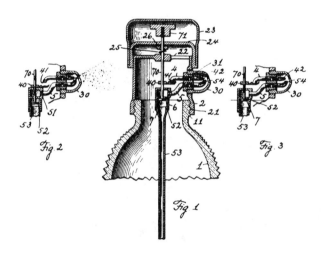

No. 746,866   PATENTED DEC. 15, 1903.

R. W. MOORE.
PERFUMERY ATOMIZER.
APPLICATION FILED APR. 9, 1903.

NO MODEL.

Fig. 4.   Cross-secion of valve in Moore patent.

using carbon dioxide as the propellant. For many years, however, the work of the early scientists in this field went largely unnoticed, until Rotheim, in Norway, brought up to date and reintroduced the art of Gebauer and Helbing, and for the first time (except for the unnoticed and forgotten efforts of Moore) saw implications for this work outside of the pharmaceutical and specifically the local anesthetic products, and particularly in cosmetics.

In two patents granted to Rotheim, in 1931 and 1933, respectively, (11,12), the possibilities for perfume and cosmetics were recognized. "When the desired method is to be used for example for eau de Cologne," the earlier of the two patents states, "this material obtains the novel property of giving a spray which is considerably cooled in relation to the atmospheric temperature as a consequence of the expansion of the added condensed gas." A little later in the same document, Rotheim claimed that "cosmetic products such as for example liquid or solid brilliantines, pomatums, vaselines, creams, toilette liquids and the like are in accordance with the present method handled in a more practical and hygienic manner than at present." In the following patent, Rotheim offered certain refinements and mentioned several condensable gases that could be used in his products. In summary, Rotheim suggested new gases, envisaged new uses, and designed valves of a complex nature, all of which contributed considerably to the growth of the art. Among his contributions, Rotheim put liquid soap into his package and, one may assume, obtained a foam.

The propellants used in the early aerosol work included ethyl chloride, dimethyl ether, isobutane, and other hydrocarbons and halogenated hydrocarbons (particularly chlorinated compounds), but there was to that time no use of fluorinated hydrocarbons. In the early 1930's a research project was instituted at General Motors aimed at the development of new and improved refrigerants to replace sulfur dioxide and ammonia, which were the commonly accepted refrigerating gases, but which had the serious shortcoming of being extremely irritating and toxic in high concentrations. The basic problem was to develop a nontoxic, nonirritating compressible gas which would be harmless if it leaked into a kitchen or apartment house. Toward this end, a group of fluorinated hydrocarbons were synthesized, and in 1933, Midgley, Henne, and McNary (13) obtained a patent for the use of these materials as fire-extinguishing agents, stating that they "may be used in pressure devices in which a low-boiling compound is employed, which creates sufficient pressure to expel itself from the apparatus." The patent, assigned to the Frigidaire Corporation, specifically mentions dichlorodifluoromethane, later to become well known under the trade names of Freon-12 and Genetron-12.

The first chemical compounds corresponding to the modern aerosol propellants were, therefore, developed as refrigerating gases, in which they offered the unquestioned superiority of freedom from toxicity. Inasmuch as they were nonflammable, they were subsequently used in fire extinguishers, resulting in the first aerosol product closely corresponding to the fluorinated hydrocarbon aerosols now on the market.

Work on self-propelling fire-extinguishing compounds was continued, not only by Midgley and his associates, but particularly by Bichowsky, who obtained a patent in 1935 (14), assigned to General Motors, which made significant advances. Whereas Midgley had used the fluorinated hydro-carbons to propel themselves and act as fire extinguishers, Bichowsky used these propellants to expel other fire-extinguishing substances, which had no capacity for self-propulsion. These other substances were mainly powders.

In the latter 1930's and after the outbreak of World War II, the United States Department of Agriculture showed extreme interest in the insecticidal possibilities of a solution "sprayed on hot surfaces to form a vapor which immediately condenses into a fog when cooled by contact with air." These materials were known as thermal aerosols. In 1942 Goodhue and Sullivan announced their work on the utilization of liquefied refrigerant gases, which had hitherto been of interest only in the fire extinguisher aerosols, for the dispensing of insecticidal materials, in order to produce a fog similar to that produced by means of thermal vaporization and condensation. This classical work, which has been recognized generally as the most significant contri-bution to modern aerosol technology, was reported by Goodhue in the chemi-cal literature (15) and became the basis of a patent (16) made freely available by its assignee, the United States government.

The work of Goodue and Sullivan at the United States Department of Agriculture made it possible to produce during wartime millions of aerosol dispensers which were accepted by the armed services. This became a great impetus for the nascent aerosol industry, for not only did these "bombs" (as they came to be known) serve a public health function, so far as the war was concerned, but they also introduced this method of dispensing to a cross-section of the American public in the armed forces, laying the ground-work for the peacetime acceptance of aerosols during their later civilian life.

As a result of the work of Goodhue and Sullivan, who used what has come to be known as a "high-pressure" container and product, there was initiated an aggressive research program into the fields of containers, valves, and propellants, with a view to lowering the costs, making the products safer, and widening the scope of products that could be dispensed in a satisfactory way in aerosol form. Soon after World War II, a backlog of research and development made it possible to bring forth aerosol packages more in keeping with civilian needs, easier to carry and use, and requiring little or no mechanical aptitude on the part of the consumer.

Except for the perfume of Moore and the reference to cosmetics in the patent of Rotheim, the emphasis of the aerosol technology was not yet in the direction of cosmetics. Several efforts to market a shampoo had been made, one of which, manufactured by Charles Thoms, used nitrous oxide as the propellant.

From 1945 to 1948, Fiero of Daggett and Ramsdell conducted extensive developmental work on aerosol cosmetics, as indicated by patents in Argentina (17) and Panama (18). These patents describe colognes, perfumes, baby oil, hair makeup, hair lacquer, hand lotion, and foam shaving cream. The only product commercialized was an aerosol perfume, called "Gay Manhattan," which was first marketed in 1946.

Two years later a foam product was marketed by Gebauer; it had a relatively small sale and was recommended for shampoo. The following year, Estignard-Bluard (19) obtained a Belgian patent for pressurized foam products made by incorporating an aqueous soap solution or a liquid detergent in a liquefied gas.

Several hundred patents have since issued on valves, actuators, propellant systems, and formulations, but only three stand out as significant. The first patent on a plastic valve with a 1-in. mounting cup was issued to R. H. Abplanalp (20) in 1953. Another patent in 1953 issued to Spitzer, Reich, and Fine (21) covered shaving cream, using a fluorocarbon propellant. This patent was upheld in the courts, which resulted in the use of hydrocarbons for shaving creams to avoid litigation or royalties. A third patent was issued to M. Spiegel (22) in 1959 covering the use of polyvinylpyrrolidone in aerosol hair sprays. This patent was litigated in the courts and declared invalid.

Principle and Mechanism

The principle on which the aerosol package is built is relatively simple. All liquid materials exert a pressure on the walls of any container in which they are placed. This pressure increases with rising temperature. When a liquefied gas is placed in a container, and sealed in, part of the material vaporizes, and part of it remains in the liquid state, reaching an equilibrium. The vapor phase rises to the top, and the liquid phase goes to the bottom, while the concentrate is dissolved or dispersed in this lower or liquid layer. The gaseous phase is exerting its vapor pressure, not only against the walls of the container, but on the liquid, forcing this liquid through the dip tube and up to the point of the opening of the valve; therefore, when the valve is released by pressing the actuating button, the liquid phase, pressing against this orifice, comes out and continues to be propelled forward so long as the valve orifice remains open and sufficient propellant remains inside the container for vaporization.

Coming forth from the valve opening is a solution, in liquid form, of propellant and concentrate. The propellant, which has a boiling point usually far below room temperature, vaporizes almost instantly, thus leaving in the air a spray or a cloud of liquid particles (called "aerosols") and fluorinated hydrocarbon gases. If this spray or cloud is directed upon the body, whether

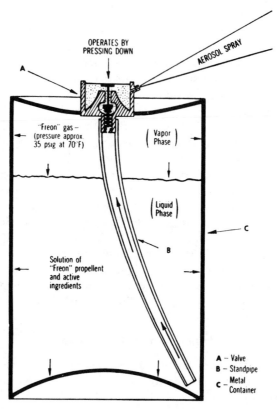

Fig. 5.   Cross-section of a typical space or surface-coating aerosol package. (Courtesy E. I. du Pont de Nemours and Co., Inc.)

the hair or skin, the propellant may vaporize before hitting the body or immediately upon hitting it, leaving behind the concentrate or active ingredient. In other instances, particularly in formulations of relatively high or medium viscosity dispensed by propellant mixtures exhibiting widely separated boiling points, varying amounts of high-boiling propellants will be entrapped for a shorter or longer period of time after the spray has been deposited on the skin or hair.

The same principle, with certain modifications, applies to the foam aerosols. The essential difference is that the foam product is usually an emulsion rather than a solution; furthermore, it is not a stable emulsion. For this reason some slight agitation of the container before use is necessary to produce the desired effect. In the case of foam products, the type of foam desired is also controlled by the valve, which may be constructed with or without a dip

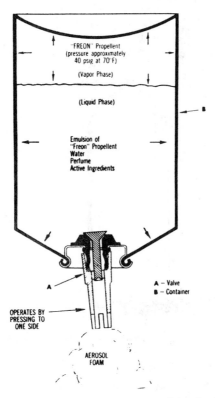

Fig. 6.   Cross-section of a typical aerosol package for foam product.

tube. The transformation from the emulsion to the foam takes place in the valve, where the emulsion is momentarily entrapped, and during this entrapment, the expansion of the propellant particles causes the formation of numerous small bubbles, giving a foam.

Finally, the aerosol principles outlined here are applied to powders, in which the solids are dispersed rather than dissolved in the propellant. The latter, upon expulsion, brings the powder with it, quickly volatilizing and leaving the powder on the surface of the skin or, in some instances, on the hair.

The principles of the aerosol package are illustrated in Figures 5 to 7.

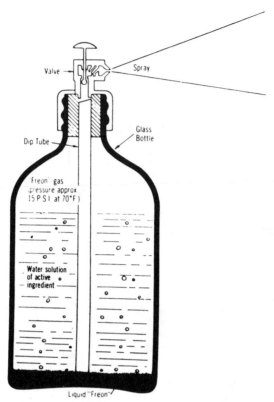

Fig. 7.   Typical three-phase glass aerosol. (Courtesy E. I. du Pont de Nemours and Co., Inc.)

The Package and Its Components

The aerosol cosmetic is a dispensing package and differs from other cosmetics in that it requires a container that will withstand pressures of up to 100 psig, a propellant, and a valve. In addition, the product must, in some way, accommodate the propellant. Only in the Sepro can or in the piston can is the propellant kept separate from the product. In all other aerosol products the propellant is either miscible, emulsified, dissolved, or in a layer adjacent to the product, generally referred to as the concentrate.

Container

Aerosol containers are manufactured in a wide range of sizes to meet the internal uses and needs of products packaged under pressure. The sizes range from $\frac{1}{2}$-oz to 24-oz capacity. They are made of tin plate, aluminum, stainless steel, coated and uncoated glass, and plastics.

Metals

Tin plate. By far the greatest number of containers used in aerosol packaging are three-piece tin plate. An indication of the importance of careful selection of aerosol container specifications is the fact that three-piece tin plate containers are now manufactured in over 5000 variables or combinations, consisting of fifteen sizes, types of plate, enamels, end contours, end compounds, coatings, and solders.

Dome-top three-piece cans are made in four different diameters and a variety of heights. Figure 8 presents the specifications for the different-size cans available and standardized by the Chemical Specialties Manufacturers Association. (*Note.* The last two digits of measurement designate 1/16 in. For example, 202 × 214 means 2 and 2/16 in. in diameter by 2 and 14/16 in. in height to the shoulder.)

The plate used for the three pieces of the can top, body, and bottom can be 0.25 electrolytic tin plate, 0.50, 1.00, 0.50/0.25, and 1.00/0.25. The last two are called differential coatings with a heavier coating of tin on one side than on the other. The tin coating designation refers to the amount of tin in lbs applied on a base box (112 sheets, size 14 × 20 in.) of steel or 31,360 in.[2] and is written lb./b.b.

Many types of linings are available to protect cosmetic products from reacting with the containers. These can be single or double coatings and can be either flat-applied or sprayed on for extra protection. If a product makes the iron anodic to the tin, coatings will not prevent perforation—in fact, they can even accelerate it, since corrosion is generally concentrated at a pinhole. The only purpose, therefore, of the coating is not to protect the container, but to protect the product. As little as 20 ppm of tin in a cosmetic product can adversely affect the perfume and as little as 10 ppm of iron can discolor some products, especially those containing hexachlorophene.

Many can linings are mixtures of copolymers of more than one basic resin type, blended to obtain certain performance characteristics. Aerosol can linings frequently contain a portion of "phenolic" to increase the chemical resistance of the film and some "epoxy" or "vinyl" to contribute to film flexibility or fabrication characteristics. For both economic and performance reasons, an enamel film is applied at the lowest film weight at which it will provide complete and continuous plate coverage.

One protection served by the top coat in a two-coat system is to repair imperfections and scratches present in the base coat, thereby achieving more complete coverage; also the top coat absorbs much of the enamel abuse

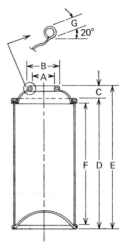

Fig. 8.   (part 1)

INDUSTRY SPECIFICATIONS FOR FABRICATED AEROSOL CANS

Presented and passed as modified in total at the 51st Annual Meeting of the Chemical Specialties Manufacturers Association in 1964, and revised at the 52nd Annual Meeting of CSMA in 1965.

Sales code diameter and height of all fabricated aerosol cans covered by industry specifications to date:

202 Diameter	207.5 Diameter	211 Diameter	300 Diameter
202 × 214	207.5 × 413	211 × 407.5	300 × 709
202 × 314	207.5 × 509	211 × 413	
202 × 406	207.5 × 605	211 × 510	
202 × 509	207.5 × 701	211 × 604	
		211 × 612	
		211 × 713	

Dimensions included in industry specifications to date:

A. Inside diameter of 1-in. cup opening.

B. Outside diameter of 1-in. cup opening.

C. Height of curl opening above double seam.

D. Height over double seam.

E. Overall height of container.

F. Height between double seams.

G. Thickness of curl around the 1-in. opening. (Gauges for measuring dimension G are available through CSMS.)

Nominal dimensions	A ±0.004	B ±0.010	C ±0.016	D ±0.031	E ±0.047	F Min.	G ±0.007
202 × 214	1.000	1.226	0.396				0.130
× 314	1.000	1.226	0.396				0.130
× 406	1.000	1.226	0.396				0.130
× 509	1.000	1.226	0.396				0.130
207.5 × 413	1.000	1.226	0.798				0.130
× 509	1.000	1.226	0.798	Sales code height—0.010 [see example]	Item C plus item D [see Example]	Sales code height—0.300 [see Example]	0.130
× 605	1.000	1.226	0.798				0.130
× 701	1.000	1.226	0.798				0.130
211 × 407.5	1.000	1.226	0.798				0.130
× 413	1.000	1.226	0.798				0.130
× 510	1.000	1.226	0.798				0.130
× 604	1.000	1.226	0.798				0.130
× 612	1.000	1.226	0.798				0.130
× 713	1.000	1.226	0.798				0.130
300 × 709	1.000	1.226	0.798				0.130

To determine items D, E, & F proceed as follows:

EXAMPLE

211 × 413

The sales code height is 413 or $4\frac{13}{16}$. [4.812].

$$D = 4.812 - 0.010$$
$$= 4.802$$
$$E = C + D$$
$$= 0.798 + 4.802$$
$$= 5.600$$
$$F = 4.812 - 0.300$$
$$= 4.512$$

Fig. 8. (part 2)

incurred in container fabrication, thereby protecting the base-coat enamel film.

An "Achilles' heel" as far as pinpoint corrosion or perforation is concerned is where the side seam meets the bottom seam. At this point the side stripe does not always cover the cut steel surface. This results in a large area of iron exposure relative to the tin which sets up an electrolytic couple in which the iron is anodic to the tin rather than vice versa, as may be the case in the other parts of the container. The inside side stripe can be applied by two methods. The postsolder application is a method in which the liquid stripe is sprayed on the hot side-seam area immediately after side-seam soldering. In a more recent presolder technique, liquid stripe is sprayed on the cold-formed side-seam area before soldering. Solder, being roll-applied from the outside surface of the side seam, flows through the side-seam structure to the inside area of the container. The presolder stripe film helps to restrict excessive solder flow, splashes, etc. With both methods of application, heat from the soldering operation covers the stripe material.

Spray-lined containers having inside metal exposure, substantially improved over that of conventional double-coat containers, are now utilized for certain aerosol antiperspirant products which heretofore could not be so packaged. In producing spray-lined containers, end units are double-coat enameled in the flat, while body units are single-coat enameled in the flat. Inside side striping is done at the time of body fabrication and then the body cylinder is sprayed inside overall prior to assembly.

Two types of bottom contour are available—the concave and the stacking; the stacking has a flat portion in the concave surface so that one can be placed on top of the other. The stacking feature is useful only when a small overcap, which fits only over the valve, is used. When cover caps, having the diameter of the container, are used, there is no need for the stacking-feature bottom. Tops and bottoms are available in silver, gold, and white enamel.

Each can company has developed its own end compounds for sealing the top and bottom of the can. These are generally designated by number only, since they are compounded rubbers. When the first deodorant products, which contained the same solvent system and propellant as hair spray, were marketed, there was considerable leakage through the bottom seal of the cans. It was found that the sealing compound which contained a resin and which would hold a product like hair spray, would not be suitable for a product like deodorant that contained no resin. The resin in the hair spray formula apparently was a factor that contributed to the sealing properties of the compound. New sealing compounds therefore had to be developed.

Solders for aerosol use are basically of two types—high tin solder, in which the metal is close to 100% tin in composition, and low tin solders, referred to as 98/2, in which the composition is 98% lead and 2% tin. The most commonly used solder in three-piece cans is 98% lead and 2% tin. High tin solder is used only when necessary, because of the high cost of tin metal. High tin solders are required for basically two reasons: (a) where product reacts with lead to form lead-insoluble compounds which deposit on the side-seam area, but may come off and result in valve plugging, or (b), where lead contributes to electrochemical action resulting in accelerated corrosion. High-strength versions of both high tin and low tin solders have been developed to prevent side-seam creep or, more technically, cold flow of the solder under the higher sustained pressures developed by nonliquefied compressed-gas-charged aerosol products. Because of the higher strength and container integrity provided by high-strength solders, these are now used exclusively by one can company. The high tin solder is called Duocom and consists of 98% tin and 2% antimony. High-strength lead solder is known as Tricom and consists of 98% lead, 1.5% tin, and 0.5% silver.

Another factor besides the solder in the strength of the side seam is the lap lock, referred to as the tab. The internal tab has greater strength than the outside tab and also is better aesthetically. Many marketers of aerosol cosmetics prefer the inside tab for its appearance; however, it has been found that inside-tab cans are more susceptible to corrosion than the outside-tab cans. This is especially true with water-base products. In one instance, with a water-base product, an outside tab can had a shelf life of at least 2 years, whereas the same product with an inside tab perforated in a matter of weeks.

A two-piece tin plate or black iron can called the Spratainer is also available from Crown Can Company in a 6-oz and a 12-oz size. These containers have a drawn body made from black plate or tin plate and have an internal protective coating. Since there is no side seam on this can, all-around external lithograph is possible. This container is suitable for pressure as high as 100 psig at 70°F and therefore is used for Propellant 12. It has been used for shaving creams and other emulsion cosmetic products. This container has not proved satisfactory for hair spray or other alcoholic products since, with this type of product, the interior coating would loosen and the free-floating lacquer film would clog the valve. The tin coating on a drawn can is not as continuous as in a three-piece can, since the can is plated before the final draw. It is this greater iron exposure in the drawn container that causes the coating to loosen with product that can permeate the coating. A new aerosol depilatory is being marketed in a Spratainer with no corrosion problem.

Aluminum.   Although there appears to be a greater interest recently in the utilization of aluminum as container material, there are many technical problems surrounding its use which still remain to be solved. The following advantages accompany the use of this metal:

1. Wide size range, particularly in the smaller sizes because of the lesser production problems as compared to tin plate.

2. More elegant appearance, with a greater degree of perfection in the art of lithography on aluminum, making for a more aesthetic container.

3. A belief that eventually aluminum containers can compete with tin plate.

The disadvantages of aluminum are these:

1. At present, it is more expensive.

2. Many products are corrosive to aluminum.

Two recently introduced cosmetic aerosols, the mouth spray and the feminine deodorant spray, are using an aluminum container. In the mouth spray, aluminum provides a small container not available in tin plate, and in the case of the deodorant spray, the aluminum container can be used at the high pressure necessary for this product.

Propellant 11 cannot be used in uncoated aluminum containers. Coatings used for other metals are generally not sufficient protection for aluminum, and therefore anodized aluminum containers must be used for hair sprays containing Propellant 11 as part of the formula.

European countries, especially Italy, have utilized aluminum cans for aerosol production to a far greater extent than the United States. In some European countries there is no tin plate can production, and aluminum cans produced locally are competitive with imported tin plate.

One-piece aluminum cans are available in diameters of $\frac{5}{8}$, $\frac{7}{8}$, 1, $1\frac{1}{4}$, $1\frac{3}{8}$, and $1\frac{1}{2}$ in. and in lengths from $2\frac{1}{2}$ to $6\frac{1}{2}$ in. The capacities of these tubes range from $\frac{1}{4}$ to 6 oz. Larger aluminum containers have been available in one-piece and two-piece types from European sources, as well as a two-piece type in the United States, and later one-piece and two-piece aluminum cans in 12-oz size were offered in the United States. Heretofore there had been a very limited availability of aluminum cans in sizes over 6 oz.

Because of corrosion problems, most pharmaceutical aerosol products use either aluminum or glass containers. In recent years, more cosmetic aerosol products have adopted aluminum containers. This trend will undoubtedly continue when and if more larger sizes in aluminum become available and the economics permit.

Glass

There are two types of aerosol glass bottles in use, the coated and the uncoated. The coated bottle is made by a dipping process and is available from Owens-Illinois and Wheaton Plastic-Cote Corporation. The coating is a polyvinyl chloride plus additives. From an aesthetic standpoint, the aerosol glass package rates best, and therefore nearly all aerosol colognes and perfumes are packaged in this container. Pressure in the uncoated glass bottle is limited to 15 psig at 70°F, whereas pressure in the coated bottle can go to 30 psig at 70°F. Pharmaceutical aerosols also use glass because of its relative inertness.

Because of the rapid growth of aerosol perfumes and colognes (120 million in 1969, having doubled in five years), the aerosol glass bottle has become much more readily available. An aerosol antiperspirant has been introduced in a coated glass bottle.

Where there are particularly difficult corrosion problems (for example, an aerosol depilatory) glass has solved the problem.

It would appear that because the cost of the glass aerosol will be higher than the metal equivalent, this package will be restricted to those products in which either appearance or corrosion resistance is a determining factor.

Tables I and II show the various plastic-coated bottles available and the capacities. Table III shows the uncoated bottles available in the standard 20-mm finish and also a 13-mm and a 15-mm finish. All the plastic-coated bottles are available only in the 20-mm finish.

Plastics

Perhaps the one area of aerosol packaging that has met with the most frustration and failure has been the plastic container. There have been at least ten companies which have tried to promote a plastic aerosol container, with little or no success. Plastics which have been tried and found wanting for one reason or another include: Nylon®, Melamine®, Zytel®, polypropylene, high-density polyethylene, polyvinylchloride, Delrin®, and Celcon®.

Although plastic containers have made strong inroads on glass containers for cosmetics, it is not likely to happen for some time with aerosols. Until a plastic material is developed that approaches the impermeability of metal, the inertness of glass, and the cost of either one, there seems little possibility for plastic aerosol packages. Attempts to reduce permeability of plastics with coatings and vacuum metalizing have not been successful.

Valves

The valve mechanism used in conjunction with the aerosol package is frequently considered the most critical part of the package. This is due to the high degree of precision and control needed in the manufacture of its components. There are now about ten manufacturers of valves, and each valve part has many variables. This results in a very large selection of valve variables for any given product. Figure 9 shows a cross-section of a valve in the open and closed position. The basic components of the valve are as follows:

1. *Dip tube.*   The dip tube, usually made of polyethylene, is used primarily as a means of delivering the material from the inside of the container to the valve mechanism. The inside diameter of the tube can be down as low as 0.030 in. or as high as 0.25 ($\frac{1}{8}$) in. Dip tubes from about 0.080 in. and down are referred to as "capillary" and tend to reduce spray rate. Normal-size dip tubes are about $\frac{1}{8}$ in. inside diameter. Oversize dip tubes—$\frac{1}{4}$-in. and over—are utilized as a storage chamber to allow usage upside down. The amount of material in the dip tube allows the aerosol to be used in the upside-down position for the period of time required to empty the tube.

2. *Housing.*   The housing or body (usually made of Nylon®, Delrin®, polyethylene, or stainless steel) holds all the valve parts together and is

TABLE I

Mold no.	Description	Overflow	Liquefied propellant[b]	Compressed Gas[b]	Approx. weight (oz)	Approx. height (in.)	Approx. diameter (in.)
AC-10621	Boston round	13.5 cc	10 cc	9 cc	$\frac{5}{64}$	$2\frac{9}{64}$	$1\frac{3}{64}$
AC-10369	Boston round	$1\frac{1}{4}$ oz	1 oz	$\frac{27}{32}$ oz	$2\frac{3}{64}$	$3\frac{1}{16}$	$1\frac{29}{64}$
AC-10370	Boston round	$2\frac{3}{4}$	$2\frac{1}{4}$	$1\frac{7}{8}$	$4\frac{9}{32}$	$3\frac{5}{8}$	$1\frac{29}{32}$
AC-10556	Boston round	$4\frac{23}{32}$	4	$3\frac{1}{4}$	$5\frac{1}{4}$	$4\frac{41}{64}$	$2\frac{1}{8}$
AC-10475	Boston round	$6\frac{1}{2}$	$5\frac{1}{2}$a	$4\frac{1}{2}$	$7\frac{29}{32}$	$4\frac{11}{16}$	$2\frac{31}{64}$
AC-10474	Tapered round	$6\frac{5}{8}$	$5\frac{3}{4}$a	$5\frac{5}{8}$	$8\frac{13}{32}$	$4\frac{53}{64}$	$2\frac{35}{64}$
AC-10491	Tapered round	$9\frac{29}{32}$	$8\frac{5}{8}$a	$6\frac{7}{8}$	$10\frac{61}{64}$	$5\frac{7}{8}$	$2\frac{25}{32}$
AC-10544	Double band	$3\frac{3}{64}$	$2\frac{5}{8}$	$2\frac{1}{8}$	$5\frac{1}{8}$	$4\frac{49}{64}$	$2\frac{13}{64}$
AC-10320	Double band	$4\frac{31}{64}$	$3\frac{7}{8}$	$3\frac{1}{8}$	$5\frac{47}{64}$	$5\frac{19}{64}$	$2\frac{7}{16}$
AC-10533	Double band	$7\frac{5}{32}$	$6\frac{15}{64}$a	5	$9\frac{39}{64}$	$6\frac{3}{64}$	$2\frac{25}{32}$
AC-10543	Pine tree	$2\frac{1}{8}$	$1\frac{3}{4}$	$1\frac{1}{2}$	$3\frac{55}{64}$	$4\frac{17}{64}$	$2\frac{1}{64}$
AC-10321	Pine tree	$4\frac{1}{2}$	$3\frac{7}{8}$	$3\frac{3}{8}$	$5\frac{63}{64}$	$5\frac{19}{64}$	$2\frac{31}{64}$

TABLE I—*Continued*

Mold no.	Description	Overflow (oz)	Liquefied propellant[b] (oz)	Compressed gas[b] (oz)	Approx. weight (oz)	Approx. height (in.)	Approx. diameter (in.)	Approx. width (in.)	Approx. thickness (in.)
AC-10590	Pinch tapered round	$2\frac{15}{16}$	$2\frac{1}{2}$	$2\frac{3}{64}$	$5\frac{1}{32}$	$5\frac{29}{64}$	$2\frac{1}{64}$		
AC-10526	Pinch tapered round	$4\frac{3}{8}$	$3\frac{3}{4}$	$3\frac{1}{32}$	$6\frac{5}{64}$	$6\frac{21}{64}$	$2\frac{19}{64}$		
AC-10557	Tall pinch	$4\frac{1}{32}$	$3\frac{1}{2}$	$2\frac{13}{16}$	$6\frac{27}{32}$	$6\frac{37}{64}$	$2\frac{21}{64}$		
AC-10394	Tall taper	$4\frac{1}{16}$	4	$3\frac{1}{4}$	$6\frac{47}{64}$	$6\frac{19}{64}$	$2\frac{5}{8}$		
AC-10938	S/S cylinder round	$\frac{5}{16}$	$\frac{1}{4}$	$\frac{13}{64}$	$\frac{5}{8}$	$2\frac{21}{64}$	$\frac{7}{8}$		
AC-10850	S/S cylinder round	$3\frac{7}{64}$	$1\frac{5}{32}$	$2\frac{5}{64}$	$1\frac{29}{64}$	$2\frac{31}{32}$	$1\frac{3}{64}$		
AC-10820	S/S cylinder round	$1\frac{9}{32}$	1	$\frac{7}{8}$	$1\frac{31}{32}$	$3\frac{15}{16}$	$1\frac{15}{64}$		
AC-10698	S/S cylinder round	$2\frac{3}{8}$	2	$1\frac{5}{8}$	$4\frac{7}{32}$	$4\frac{37}{64}$	$1\frac{39}{64}$		
AC-10733	Pinch ovals	$1\frac{13}{64}$	1	$\frac{3}{4}$	$3\frac{7}{64}$	$2\frac{31}{32}$		$1\frac{59}{64}$	$1\frac{3}{16}$
AC-10687	Pinch ovals	$2\frac{3}{8}$	2	$1\frac{5}{8}$	$5\frac{3}{32}$	$4\frac{3}{4}$		$2\frac{3}{8}$	$1\frac{27}{64}$
AC-10685	Pinch ovals	$4\frac{23}{32}$	4	$3\frac{1}{4}$	$7\frac{3}{8}$	$5\frac{53}{64}$		$2\frac{15}{16}$	$1\frac{27}{32}$

[a] Low percentage of LP—product dispensed as foam.

[b] Suggested fills—based on approximately 15% headspace for LP and 43% headspace for CG.

TABLE II

Mold no.	Capacity, overflow (oz)	Capacity, liquefied propellant	Capacity, compressed gas	Weight (oz)	Height (in.)	Diameter (in.)
AC-10491	$9\frac{29}{32}$	$8\frac{1}{2}$ oz	$6\frac{7}{8}$ oz	9	$5\frac{25}{32}$	$2\frac{39}{64}$
AC-10474	$6\frac{39}{64}$	$5\frac{3}{4}$ oz	$4\frac{5}{8}$ oz	$6\frac{1}{2}$	$4\frac{3}{4}$	$2\frac{13}{32}$
AC-10487	$\frac{31}{64}$	12 cc	10 cc	$1\frac{1}{4}$	$2\frac{5}{32}$	$1\frac{5}{32}$
AC-10369	$1\frac{1}{4}$	1 oz	$\frac{7}{8}$ oz	2	3	$1\frac{11}{32}$
AC-10370	$2\frac{3}{4}$	$2\frac{1}{4}$ oz	$1\frac{7}{8}$ oz	$3\frac{1}{2}$	$3\frac{9}{16}$	$1\frac{13}{16}$
AC-10556	$4\frac{23}{32}$	4 oz	$3\frac{1}{4}$ oz	4	$4\frac{9}{16}$	$1\frac{61}{64}$
AC-10475	$6\frac{1}{2}$	$5\frac{1}{2}$ oz	$4\frac{1}{2}$ oz	$6\frac{1}{2}$	$4\frac{39}{64}$	$2\frac{11}{32}$
AC-10533	$7\frac{5}{32}$	$6\frac{1}{8}$ oz	5 oz	8	$5\frac{21}{32}$	$2\frac{39}{64}$
AC-10320	$4\frac{31}{64}$	$3\frac{7}{8}$ oz	$3\frac{1}{8}$ oz	$4\frac{1}{2}$	$5\frac{7}{32}$	$2\frac{17}{64}$
AC-10544	$3\frac{3}{64}$	$2\frac{5}{8}$ oz	$2\frac{1}{8}$ oz	$3\frac{7}{8}$	$4\frac{11}{16}$	$2\frac{3}{64}$
AC-10543	$2\frac{1}{8}$	$1\frac{3}{4}$ oz	$1\frac{7}{16}$ oz	3	$4\frac{13}{64}$	$1\frac{55}{64}$
AC-10321	$4\frac{1}{2}$	$3\frac{7}{8}$ oz	$3\frac{1}{8}$ oz	$4\frac{1}{2}$	$5\frac{7}{32}$	$2\frac{21}{64}$
AC-10526	$4\frac{1}{2}$	$3\frac{7}{8}$ oz	$3\frac{1}{8}$ oz	5	$6\frac{1}{4}$	$2\frac{1}{8}$
AC-10557	$4\frac{1}{32}$	$3\frac{1}{2}$ oz	$2\frac{13}{16}$ oz	$5\frac{1}{2}$	$6\frac{1}{2}$	$2\frac{11}{64}$

itself secured in the mounting cup. The dip tube fits onto the housing, and the orifice in the housing can be wide open or down to 0.013 in., in which case it is called a "restricted housing." The housing may have a hole, and in this instance, it will act as a vapor tap. This vapor tap will result in the gaseous phase of the propellant mixing with the liquid phase as the valve is opened. The vapor tap also allows the container to be used in the upside-down

TABLE III.   Stock Aerosol Containers

	Rounds, 13-mm finish			
Mold no.	Practical fill	Overflow cap	Height	Diameter
S-775C	1.0 cc	1.8 cc	33 mm	15 mm
S-633	1.5	1.9	36	14
S-329	2.0	2.6	40	15
S-111B	3.5	5.0	45	18
S-12A	5.0	8.5	53	19
S-7A	10.0	12.7	60	23
S-791C	12.0	17.0	59	26
S-275	15.0	20.0	65	27
	Rounds, 15-mm finish			
S-2053F1	7 cc	8.3 cc	65 mm	17 mm

TABLE III—*Continued*

Rounds, 20-mm finish

Mold no.	Practical fill	Overflow cap	Height	Diameter
S-1744F1	8 cc	10.5 cc	59 mm	21 mm
S-2029F1	10 cc	13	54	24
S-1844F1	10 cc	13	64	21
S-204F1	15 cc	23	59	31
S-1409F1	20 cc	26	70	30
S-1388F1	1 oz	36	76	33
S-1527F1	1 oz	36	99	28
S-1665F1	1½ oz	55	80	39
S-1983F1	2 oz	75	97	40
S-2001F1	2 oz	69	113	35
S-2024F1	2 oz	83	90	44
S-1746F1	2 oz	67	107	36
S-1X85F1	2 oz	75	114	36
S-2138F1	3 oz	105	115	44
S-1743F1	3 oz	106	128	40

Ascot, 20-mm finish

S-1982F1	1 oz	35 cc	102 mm	37 mm
S-1843F1	2	73	128	46
S-1766F1	3	118	147	53

Fancies, 15-mm finish

Style	Mold no.	Practical fill	Overflow cap	Height	Width	Depth
Classic	S-2052F1	7 cc	8.3 cc	65 mm	17 mm	17 mm
Capri		1 oz	34	91	26	26
Capri	RS-897F	2 oz	68	114	41	41
Monte Carlo		1 oz	34	102	31	31
Monte Carlo		2 oz	68	135	42	42
Flamingo	RS-893F	1 oz	34	76	38	38
Flamingo		2 oz	68	93	46	46

Fancies, 20-mm finish

Gem	S-2004F1	3/8 oz	11.9 cc	58 mm	25 mm	25 mm
Ming	S-1440F1	2/3	21	68	28	28
Hapsberg	S-1234F1	2	69	101	50	33
Coronet	S-1356F1	2	67	99	52	52
Coronet	S-1421F1	3-1/2	123	117	63	63
Royal	S-1411F1	2	69	98	53	35
Nobility	S-1112F1	2-1/3	84	97	57	39
Seville	S-1300F1	2	68	101	51	35
Imperial	RS-131F1	3	114	117	55	55
Starlite	S-2094F1	3	105	142	49	46
Capri		3	102	128	46	46
Monte Carlo	RS-892F1	3	102	154	47	47
Flamingo		3	102	104	52	52

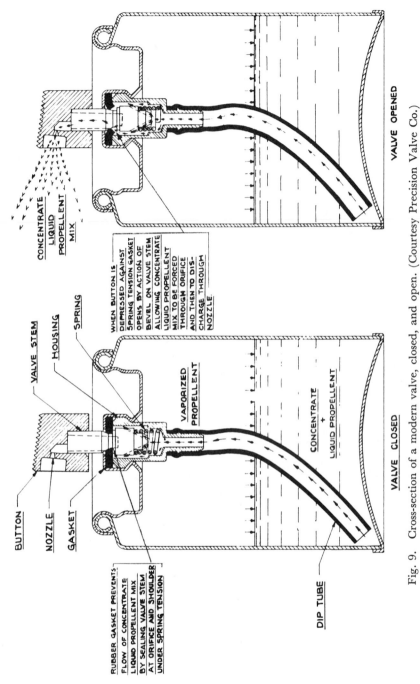

BUTTON

NOZZLE

GASKET

RUBBER GASKET PREVENTS
FLOW OF CONCENTRATE
LIQUID PROPELLENT MIX
BY SEALING VALVE STEM
AT ORIFICE AND SHOULDER
UNDER SPRING TENSION

VALVE STEM

HOUSING

SPRING

WHEN BUTTON IS
DEPRESSED AGAINST
SPRING TENSION GASKET
OPENS BY ACTION OF
BEVEL ON VALVE STEM
ALLOWING CONCENTRATE
LIQUID PROPELLENT
MIX TO BE FORCED
THROUGH ORIFICE
AND THEN TO DIS-
CHARGE THROUGH
NOZZLE

CONCENTRATE
LIQUID
PROPELLENT
MIX

VAPORIZED
PROPELLENT

CONCENTRATE
+
LIQUID PROPELLENT

DIP TUBE

VALVE CLOSED

VALVE OPENED

Fig. 9.   Cross-section of a modern valve, closed, and open. (Courtesy Precision Valve Co.)

440

position, which is desirable with products such as under-arm antiperspirants and deodorants.

3. *Stem.* The stem, usually made of Nylon®, Delrin®, brass, or stainless steel, transmits the product from the housing into the actuator and serves as a restricting device and an expansion chamber. The stem may have one or more orifices, and these are referred to as "metering orifices."

4. *Gasket.* This seals the stem orifice when the valve is closed and also cushions the housing against the cup, creating a seal. The gasket, usually Buna® N or Neoprene®, is a very critical part of the valve construction in that it must be made of a material that is unaffected by the product, maintains its tensile strength over a long period of time, and will not swell or deteriorate dimensionally so as to cause any leakage or malfunctioning of the valve.

5. *Spring.* This returns the stem to close the valve when the actuator is released. Most springs for cosmetic purposes are of stainless steel, other materials being easily subject to corrosion. At least one valve is available where an elastomer is used in place of a spring to return the valve to the closed position.

6. *Actuator.* The design of the actuating button (and its orifice) is dependent on the end use intended for the product; e.g., in the case of a spray type, a relatively small external orifice (0.018-in. diameter) is used, as compared with a foam-type orifice (0.300-in. diameter). There are various configurations of the orifice in the actuator which affect the spray pattern. The "straight-taper" exit orifice has approximately parallel walls and gives a fairly narrow spray cone. With a "reverse-type" exit orifice, the sides form a cone toward the outside end of the orifice, which results in a wide spray pattern. The mechanical break-up actuator has a whirling chamber which utilizes the centrifugal energy to get break-up of the particles. The energy of the droplets throws them out in a wide, hollow cone, if propellant is not emitted with the product, or in a narrower, solid cone, if there is a proportion of product miscible propellant.

7. *Valve Mounting Cup and Ferrules.* The functional parts of the valve are mounted in a metal component called the cup or ferrule, which is capable of being sealed onto the container.

By far the most common mounting is the 1-in. mounting cup, which fits all 1-in.-internal-diameter neck cans. The cup is made of tin plate and can be coated on one or both sides. A flowed-in rubber gasket serves as the seal when the cup is crimped.

Ferrules for glass bottles and for aluminum and stainless steel cans are generally made of aluminum. The basic dimension for glass containers is the 20-mm finish—i.e., the diameter of the lip on the bottle. Spinning or crimping techniques are used for working the ferrule under the lip of the container.

Valve Specialties

Actuator Cover Cap

The actuator button can be made a part of the cover cap, or the actuator can protrude through the cover cap.

Powder or Paint Valves

These are special valves which can be readily cleaned. In these valves, the stem and actuator are an integral molding.

Foam Valves

There are a number of specially designed actuators to deliver foam products. Some of these are designed to operate in an inverted position so that no dip tube is required.

Spray Anyway Valves

Most valves are designed to operate when the container is held in an upright position. It is desirable with some products, such as foot sprays and under-arm sprays, that the container be held upside down as well as right side up. There are four valves that allow operation in this manner:

1. Vapor-tap valve.
2. Double-tube valve with ball check.
3. Large-size dip tube.
4. Vapor tap with sliding collar (Gulf S.A.).

Metering Valves

These are designed to deliver a fixed quantity of product on each operation of the valve, and have been used for purse-size perfumes, oral-inhalation products, and mouth sprays.

Special Applicators

When it is necessary to deliver product into positions not normally accessible, special applicators have been designed:

1. Dental applicator—for use in mouth.
2. Oral applicator—for inhalation therapy.
3. Nasal applicator.
4. Vaginal applicator for contraceptives or sanitizing.

Codispensing Valves

A new development in valves is the codispensing valve which makes a two-compartment container out of the aerosol package. The valve is equipped with a pouch made of blown polyethylene or Mylar® laminated polyethylene, which can be filled with material other than that in the container. When the valve is opened, product from the pouch and container is mixed in a predetermined ratio and ejected as a foam or spray. This allows the packing in a pressure package of two materials which, if mixed, give an exothermic reaction, e.g., hydrogen peroxide and sodium sulphite.

There are already six cosmetic products using the codispensing principle on the market:

1. Hot shaving cream.
2. Hot hair conditioner.
3. Hot facial cleanser.
4. Hot facial moisturizer.
5. Hot facial oil treatment.
6. Hair dye.

The codispensing principle should result in a tremendously increased volume of cosmetic products in pressurized packages, since, like the first aerosol hair spray, it allows for products that heretofore were impossible to make.

There is considerable literature on aerosol valves (23–29).

Propellants

The propellant is that part of the contents of an aerosol package which provides the force that makes the product what it is, namely, self-dispensing; however, the propellant may also have auxiliary functions, as for instance, to act as a solvent or a diluent. Not only is the propellant the means for ejecting the contents from the package, but in many instances, it determines (or at least partially determines) numerous characteristics and properties of the ejected material. Propellants can be classified as follows.

1. Fluorinated hydrocarbons, which are now manufactured by several companies under the trade names of Freon®, Genetron®, Isotron®, Ucon® and Racon®. The use of chemical nomenclature for the fluorinated propellants is quite cumbersome because the names are long and easily confused. Since the Freon propellants were originally used as refrigerants, the numbering system used by the refrigeration industry was adopted by the aerosol industry in classifying the fluorinated propellants; also, because of the number

of trade names now available, rather than using "Freon-12"®, we use the designation "Propellant 12."

2. Hydrocarbons, such as propane, butane, and isobutane, are used, especially in water-base products. Up to 10% of these may be used in alcohol-base products, such as hair sprays.

3. Gases, such as nitrogen, nitrous oxide, and carbon dioxide are used in food products, but only to a very limited degree in cosmetic products.

4. Chlorinated compounds, such as methylene chloride and methyl chloroform, are used in cosmetic products to replace Propellant 11 because of cost considerations or because of solvency characteristics.

The rules for the numbering system used to identify fluorocarbon propellants are as follows.

1. The first digit on the right is the number of fluorine atoms in the molecule.

2. The second digit from the right is one more than the number of hydrogen atoms in the molecule.

3. The third digit from the right is one less than the number of carbon atoms in the molecule. When this digit is zero, it is omitted from the number.

4. The number of chlorine atoms in the compound is found by subtracting the number of fluorine and hydrogen atoms from the total number of atoms which can be connected to the carbon atoms, i.e., the valence of the carbon atoms not involved in carbon to carbon bonds.

5. In the case of isomers, each has the same number. The most symmetrical one is indicated by the number without any letter following it. As the isomers become more unsymmetrical, the letters, "a," "b," "c," etc., are appended. Symmetry is determined by adding the atomic weights of the groups attached to each carbon atom and subtracting one from the other. The smaller the difference, the more symmetrical the compound. For example, $CClF_2CClF_2$ is Propellant 114 and CCl_2FCF_3 is Propellant 114a.

6. The fourth digit from the right is the number of unsaturated linkages in the molecule. For example, the first 1 in Propellant 1141 represents the double bond in vinyl fluoride, $CH_2=CHF$.

7. If a molecule is cyclic, the number is preceded by a "C" as in perfluorocyclobutane, Propellant C-318.

The use of propellants in aerosol products, with special emphasis on cosmetics, has been studied by Reed (29,30). Table IV gives the fluorinated hydrocarbons of greatest current interest in cosmetics, their trade names, boiling points, liquid densities, and other physical properties. The boiling points and vapor pressures are the properties of prime importance to the users of these substances, because they determine the character of the spray or foam as well as the other properties of the finished product. It will be

TABLE IV. Physical Properties of Fluorinated Hydrocarbon Propellants

Number	11	12	21	22	113	114	152a	142b	1141	C-318
Formula	CCl_3F	CCl_2F_2	$CHCl_2F$	$CHClF_2$	CCl_2FCClF_2	$CClF_2CClF_2$	CH_3CHF_2	CH_3CClF_2	$CH_2=CHF$	$CF_2CF_2CF_2CF_2$
Molecular weight	137.4	120.9	102.9	86.5	187.4	170.9	66.1	100.5	46.0	200.0
Boiling point, °F	74.8	−21.6	48.1	−41.4	117.6	38.4	−11.2	15.1	−97.5	21.1
Freezing point, °F	−168	−252	−211	−256	−31	−137	−179	−204	−257	−42.5
Vapour pressure, psig										
at 70°F	13.4	70.2	8.4	122.5	5.5	12.9	61.7	29.1	355	25.4
at 130°F	24.3	181.0	50.5	300	3.4	58.8	176	97.2	745	92
Liquid density, g/cc										
at 70°F	1.485	1.325	1.323	1.209	1.574	1.468	0.911	1.119	0.638	1.513
at 130°F	1.403	1.191	1.193	1.064	1.493	1.360	0.813	1.028	0.350	
Vapour density at b.p., g/l	5.86	6.26	4.57	4.83	7.33	7.83	—	—	—	9.30
Heat of vaporization at b.p., Btu/lb	78.31	71.94	104.2	100.7	63.1	59.0	141	96.0	—	49.8
Liquid viscosity at 70°F cP	0.439	0.262	0.351	0.238	0.697	0.038	0.243	0.330	—	0.455
Surface tension at 77°F dynes/cm	19	9	19	9	19	13	—	—	—	8.3†
Critical temperature, °F	388	234	353	205	417	294	236	279	130.5	240
Critical pressure, psig	620	582	735	701	480	459	637	583	745	378
Flammability limit, vol.% in air	None	None	None	None	None	None	5.1–19.1	9.0–14.8	2.6–21.7	None

† At 70°F.

445

noted that they cover a wide range of boiling points and vapor pressures, thus making it possible to choose the propellant, or combination of propellants, with those properties most desired.

Although some aerosol cosmetics are made with a single propellant, it is customary to use a mixture or a solution of propellants. This is so because there are no individual fluorinated hydrocarbons available with a sufficiently wide range of boiling points to furnish the pressures and other properties required for each product; therefore, by using mixtures, boiling points and

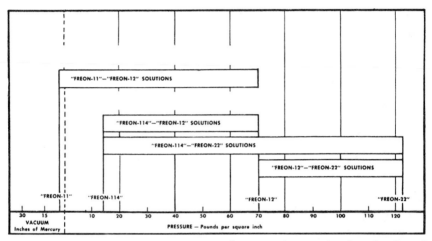

Fig. 10.   Range of pressures obtainable at 21°C with mixtures of various fluorinated hydrocarbon propellants. (Courtesy of E. I. du Pont de Nemours and Co., Inc.)

pressures are "tailor-made." Mixtures of fluorinated hydrocarbons are true solutions, having properties which are not merely the additive properties of the individual components; these solutions behave in a predictable manner.

Reed suggests that one consider Propellant 12 as the basic propellant, to be diluted with Propellant 11 or Propellant 114 to reduce its pressure to the range permitted by ICC regulations.

In Figure 10 the pressure ranges of these solutions are given. As may be seen, the vapor pressure of Propellant 22, for example, is higher than that of the other propellants; it has excellent solvent properties, so that in solution with the other propellants, it finds valuable use.

The temperature-pressure relationships of solutions of these propellants are most important in cosmetic aerosol formulation. Studies of such relationships have been published and presented graphically for the most frequently used solutions; these graphs are reproduced here as Figures 11 to 14. By a study of these charts, one can select the percentage of each fluorinated compound in the solution that would give the desired pressure at the proper

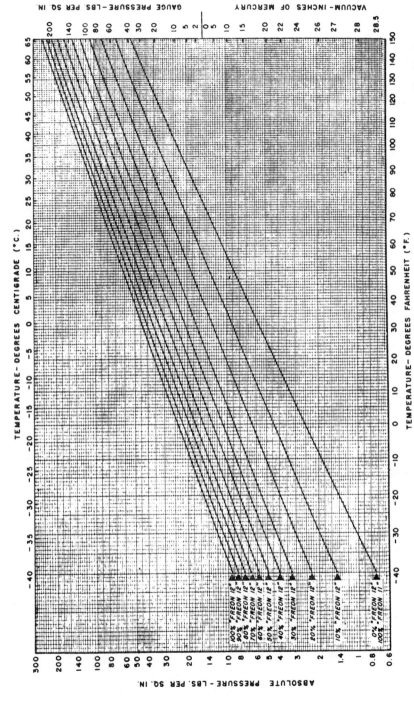

Fig. 11.　Vapor pressure of solutions of Propellants 11 and 12. Composition given is percentage by weight in the liquid phase.

447

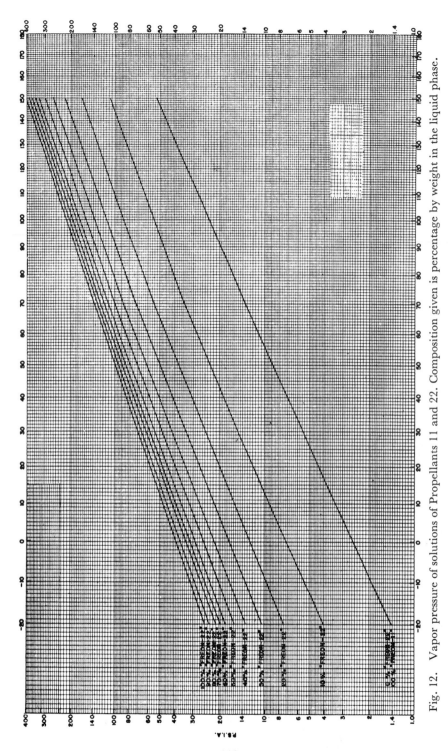

Fig. 12. Vapor pressure of solutions of Propellants 11 and 22. Composition given is percentage by weight in the liquid phase.

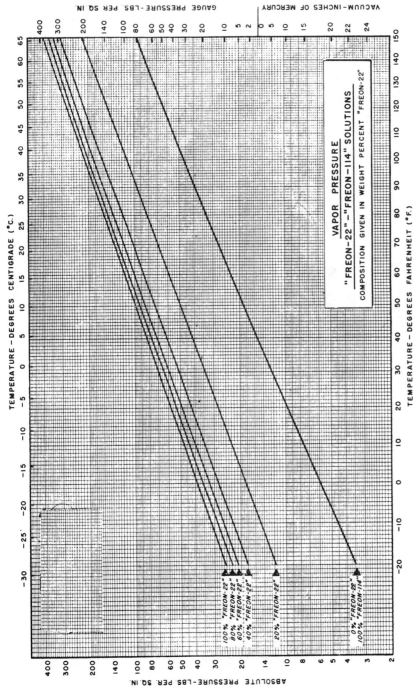

Fig. 13. Vapor pressure of solutions of Propellants 22 and 114. Composition given is percentage by weight in the liquid phase.

449

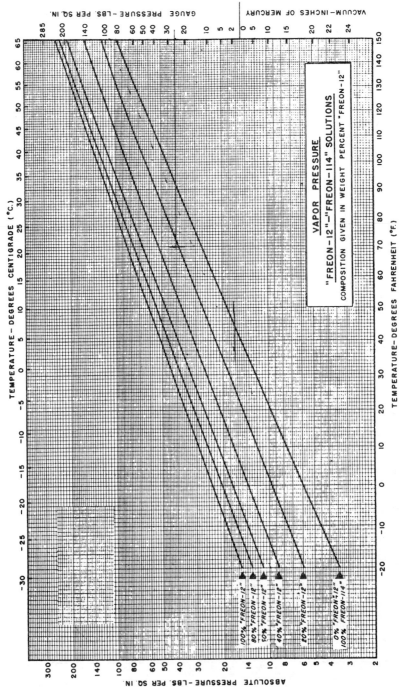

Fig. 14. Vapor pressure of solutions of Propellants 12 and 114. Composition given is percentage by weight in the liquid phase.

450

temperature under study. For example, Propellant 11, which has virtually no pressure at 21°C, and Propellant 12, which has a pressure of about 70 psig at that temperature, can be mixed in proportions to give any desired pressure between 0 and 70 psig at 21°C.* To obtain a pressure of 40 psig at 21°C, one would use about a 60/40 mixture of Propellants 12 and 11. Addition of active ingredients will, of course, lower this pressure.

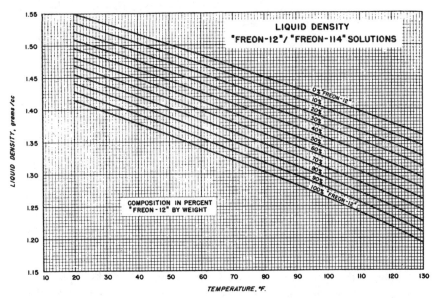

Fig. 15. Liquid density of solutions of Propellants 12 and 114. Composition given is percentage by weight in the liquid phase.

In addition to other properties, the liquid densities of the propellants are important; they are relatively high for organic compounds. Densities of solutions of various combinations of propellants have also been published, one of which (for Propellants 12/114) is shown in Figure 15. At 21°C for example, a solution of Propellant 12 and Propellant 114 with a liquid density of 1.35 would consist of 80% of the former and 20% of the latter.

Among other properties of the fluorinated hydrocarbons that have been studied are inhalation toxicities. These materials have a relatively low order of toxicity; this is particularly true of Propellants 12 and 114. A recent

* Although the entire aerosol literature has reported its data in degrees Fahrenheit (as is shown in several of the charts used here), Centigrade is being used in the text of this chapter to conform both with the remainder of this book and with the practice in American scientific publications.

brochure from du Pont (31) discusses fluorocarbons and safety. Because of the presence of other components, or because of the extremely fine particle size, toxicity studies must be made on each completed product, a matter that has been considered in greater detail by Gee and Fiero (32).

The fluorinated hydrocarbons themselves are nonflammable, although all aerosol cosmetics do not answer this description. This is particularly true of those products which contain alcohol (such as most hair lacquers and spray colognes). As a general rule, Reed (29,30) points out, when "the alcohol concentration of an aerosol product exceeds about 40 wt %, the spray can be ignited with a match flame and will continue to burn after the match flame has been removed."

The stability of the fluorinated hydrocarbon propellants has likewise been studied; in Reed's opinion, "most aerosol products present very few stability problems that are not encountered by the same products in conventional packaging methods." However, the fact that the propellant remains stable and unchanged does not necessarily mean that other component parts will not be affected by the propellant. This is particularly true of the packaging materials. Reed states that "the liquefied propellants under discussion have no effect, for all practical purposes, on metals such as steel, tin and aluminum." This need not be true, however, when the complete aerosol formulation is studied. In alkaline systems and in the presence of water, Propellants 11 and 22 seem to offer some difficulty; this may result in corrosion of the metal similar to that which occurred when synthetic-detergent shampoos were first marketed in aerosol form.

In view of the wide interest in packaging aerosol products in plastic containers, and in the use of plastics for valve parts, the effects of the propellants on plastics have also been investigated. Studies on the swelling of plastics and of elastomers in the liquefied propellants have been made. It was not possible to form any general rules as to the stability of the propellants in plastics; Reed concluded that it is necessary to test each plastic in order to determine its stability in any specific aerosol formulation.

Solubilities of the various propellants have been studied, and it is generally found that the fluorinated hydrocarbons are poor solvents for aqueous solutions or emulsions. This has necessitated the use of cosolvents, reformulation, or the marketing of products which separate on standing and which therefore must carry a legend instructing the user to shake the container before using. Among the best cosolvents is ethyl alcohol, particularly anhydrous grade, which is used in hair sprays and aerosol colognes. A series of charts showing the best solubilities and spray patterns has been worked out; several typical systems are reproduced in Figures 16 to 22.

Before discussing the factors involved in the choice of propellants for a given formulation, a word about the hydrocarbon liquefied gases should be

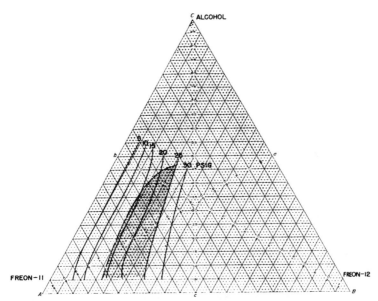

Fig. 16. Properties of solutions of Propellant 12, Propellant 11, and 100% ethyl alcohol. Temperature, 22°C; composition given is percentage by weight; solid lines indicate vapor pressure contours; shaded area indicates region of satisfactory spray character.

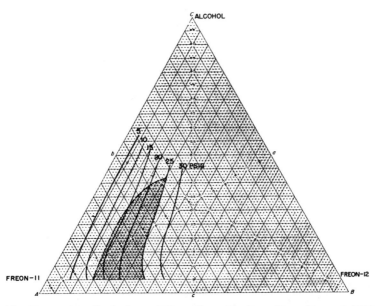

Fig. 17. Properties of solutions of Propellant 12, Propellant 11, and 95% (by volume) ethyl alcohol. Conditions same as given in Figure 16.

453

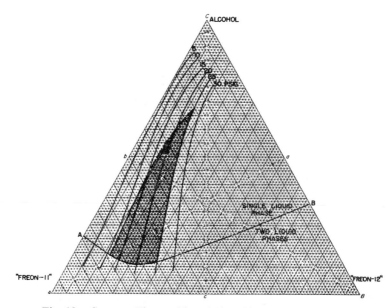

Fig. 18. Same as Figures 16 and 17, with 90% ethyl alcohol.

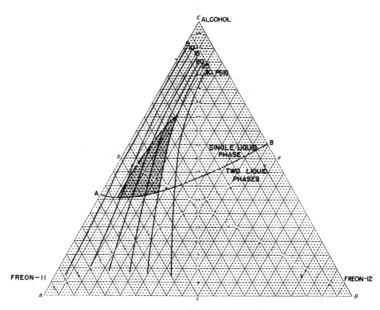

Fig. 19. Same as Figures 16 and 17, with 85% ethyl alcohol.

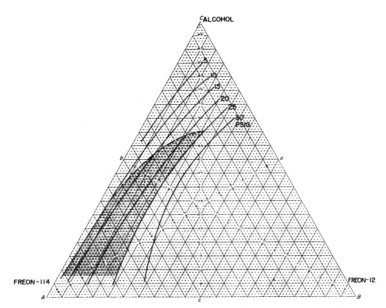

Fig. 20. Properties of solutions of Propellant 12, Propellant 114, and 100% ethyl alcohol. Conditions same as given in Figure 16.

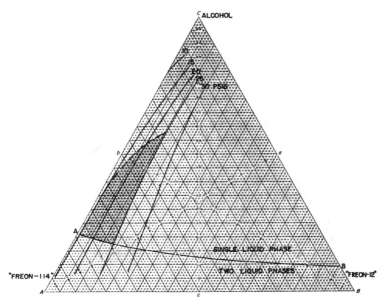

Fig. 21. Same as Figure 20, with 95% (by volume) ethyl alcohol.

455

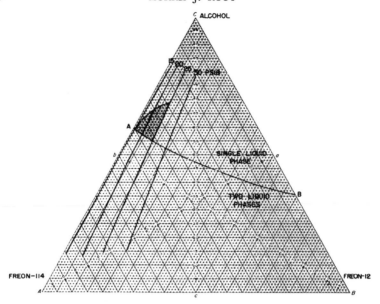

Fig. 22. Same as Figure 20, with 90% (by volume) ethyl alcohol. (Courtesy du Pont de Nemours and Co., Inc.)

added. These gases have advantages: low cost, a new range of vapor pressures and temperature-pressure relationships offer new possibilities not obtainable with the fluorinated hydrocarbons and solubilities for materials not readily soluble in the fluorocarbon propellants. Vapor pressures that can be obtained with butane, isobutane, and propane and mixtures thereof, are shown in Figure 23. Thermodynamic properties of butane, isobutane, and propane are given in Tables V, VI, and VII. They offer certain disadvantages in handling in that they are flammable and explosive. These dangers (particularly to the consumer) are minimal in foam products. Aerosol foams (specifically shaving

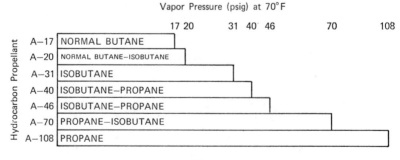

Fig. 23.

creams), in which a mixture of propane and isobutane is the propellant, have almost replaced the fluorocarbon-propelled products.

In choosing a propellant or a solution of propellants, the chemist must consider spray pattern, pressure, solubility, stability, and cost.

In general, there are three types of spray patterns. They can be described as "wet spray," "fine spray," and "dry spray," although they are not always distinct from each other. As a rule, the higher percentages of propellant give a finer spray, and conversely, the lower percentages of propellant give a wetter spray, although the upper and lower concentrations of propellant must be determined above all else by type of container, government regulations, and pressures obtained. Propellant solutions containing greater percentages of higher boiling materials give wetter sprays, and conversely, those solutions with lower percentages of higher boiling propellants give drier sprays. By uniting these two principles, the chemist is usually able to obtain the character of spray desired.

Where flammability is a problem, the higher the percentage of fluorinated hydrocarbons, the lesser is the tendency toward flammability.

So far as cost is concerned, the propellant is the most expensive ingredient (except in the case of aerosol colognes and other fragrance products, where the perfume oils are usually more costly). Because the propellants themselves vary considerably in price, one is sometimes able to choose such combinations of propellants as will be economically feasible, and to use sufficiently small proportions of such propellant mixtures to keep costs down. In foam products, as will be seen in the section on formulation, the percentage of propellant is usually sufficiently small to keep the cost low.

For reasons already mentioned, the chemist often chooses to formulate the product with a mixture of propellants (usually two), rather than with a single propellant. In the preliminary development of a product, it is advisable to restrict oneself to a commercially available mixture; later the formula may be modified and refined, and in this case a specially prepared mixture might be required. In making a new mixture of gases on which previous data are not available, it is necessary to study the vapor pressures and liquid densities of the mixture, and not to assume that they will fall into a predictable place between the properties of the constituent gases. Certain of these mixtures are azeotropic; i.e., the mixture has a constant boiling point that may be higher or lower than the boiling point of the individual components.

In summary, the following should serve as a guide to the use of propellants in an aerosol cosmetic:

Propellant 11 or trichloromonofluoromethane is a low-pressure propellant. It is used almost invariably in combination with other propellants, particularly Propellant 12. Its low pressure makes Propellant 11 usable in glass; however, the odor which it contributes makes it less attractive for products

TABLE V. Thermodynamic Properties of Normal Saturated Butane

(Enthalpies and entropies are referred to saturated liquid at −200°F where the values are zero.)

Temp. (°F) t	Pressure[a] (psi)		Specific volume[a] (ft³/lb)		Density[a] (lb/ft³)		Enthalpy[c] (btu/lb)		Latent heat (btu/lb) L^c	Entropy[c] (btu/lb, °F)		Temp (°F) t
	Absolute p	Gauge $g.p.$	Liquid v	Vapor V	Liquid l/v	Vapor l/V	Liquid hf	Vapor hg		Liquid sf	Vapor sg	
0	7.3	15.0[b]	0.02591	11.1	38.59	0.0901	103.8	275.3	171.5	0.226	0.572	0
5	8.2	13.2[b]	0.02603	9.98	38.41	0.100	106.2	276.9	170.7	0.231	0.572	5
10	9.2	11.1[b]	0.02615	8.95	38.24	0.112	108.8	278.5	169.7	0.236	0.574	10
15	10.4	8.8[b]	0.02627	8.05	38.07	0.124	111.5	280.0	168.5	0.242	0.571	15
20	11.6	6.3[b]	0.02639	7.23	37.89	0.138	114.0	281.6	167.6	0.248	0.571	20
25	13.0	3.6[b]	0.02651	6.55	37.72	0.153	116.7	283.1	166.4	0.254	0.571	25
30	14.4	0.6[b]	0.02664	5.90	37.54	0.169	119.2	284.9	165.7	0.260	0.571	30
35	16.0	1.3	0.02676	5.37	37.37	0.186	121.8	286.4	164.6	0.264	0.571	35
40	17.7	3.0	0.02689	4.88	37.19	0.205	124.2	288.0	163.8	0.270	0.571	40
45	19.6	4.9	0.02703	4.47	37.00	0.224	126.8	289.5	162.7	0.276	0.571	45
50	21.6	6.9	0.02716	4.07	36.82	0.246	129.6	291.2	161.6	0.282	0.571	50
55	23.8	9.1	0.02730	3.73	36.63	0.268	132.1	292.7	160.6	0.287	0.570	55
60	26.3	11.6	0.02743	3.40	36.45	0.294	134.8	294.2	159.4	0.293	0.570	60
65	28.9	14.2	0.02759	3.12	36.24	0.321	137.3	295.9	158.6	0.298	0.570	65
70	31.6	16.9	0.02773	2.88	36.06	0.347	140.1	297.5	157.4	0.304	0.570	70
75	34.5	19.8	0.02789	2.65	35.86	0.377	142.6	298.9	156.3	0.310	0.570	75
80	37.6	22.9	0.02805	2.46	35.65	0.407	145.0	300.3	155.3	0.315	0.570	80
85	40.9	26.2	0.02821	2.28	35.45	0.439	147.8	302.0	154.2	0.322	0.570	85
90	44.5	29.8	0.02838	2.10	35.24	0.476	150.5	303.5	153.0	0.326	0.571	90
95	48.2	33.5	0.02854	1.96	35.04	0.510	153.1	305.0	151.9	0.332	0.571	95

458

100	52.2	37.5	0.02870	1.81	34.84	0.552	156.2	306.7	150.5	0.340	0.571
105	56.4	41.7	0.02889	1.70	34.62	0.588	159.1	308.1	149.0	0.346	0.572
110	60.8	46.1	0.02906	1.58	34.41	0.633	161.9	309.5	147.6	0.352	0.572
115	65.6	50.9	0.02925	1.48	34.19	0.676	165.0	311.1	146.1	0.359	0.572
120	70.8	56.1	0.02945	1.37	33.96	0.730	167.8	312.7	144.9	0.365	0.572
125	76.0	61.3	0.02962	1.28	33.77	0.783	171.0	314.0	143.0	0.372	0.573
130	81.4	66.7	0.02980	1.19	33.56	0.840	174.0	315.5	141.5	0.378	0.573
135	87.0	72.3	0.03000	1.11	33.34	0.900	177.1	317.0	139.9	0.385	0.573
140	92.6	77.9	0.03020	1.04	33.14	0.965	179.9	318.2	138.3	0.391	0.574
145	100.0	85.3	0.03040	0.966	32.92	1.035	183.1	319.5	136.4	0.398	0.574
150	108.0	93.7	0.03060	0.897	32.70	1.115	186.5	321.0	134.5	0.405	0.574
155	115.0	100.3	0.03084	0.840	32.43	1.190	189.3	322.3	133.0	0.411	0.574
160	122.0	107.3	0.03112	0.785	32.15	1.275	192.7	323.8	131.1	0.418	0.575
165	130.0	115.3	0.03140	0.733	31.90	1.365	195.8	325.0	129.2	0.425	0.575
170	140.0	125.3	0.03165	0.687	31.62	1.455	199.2	326.1	126.9	0.433	0.575
175	150.0	135.3	0.03193	0.643	31.36	1.557	202.1	327.2	125.1	0.439	0.575
180	160.0	145.3	0.03218	0.602	31.10	1.660	205.0	328.4	123.4	0.445	0.575

a Based on material from Dana, Jenkins, Burdick, and Timm, published originally in *Refrig. Eng.*, **12**(12): 402 (June 1926).

b Inches of mercury below one standard atmosphere (29.92 in.).

c From *Mollier diagrams for butane*, W. C. Edmister, Standard Oil Co. (Indiana).

TABLE VI. Thermodynamic Properties of Saturated Isobutane
(Enthalpies and entropies are referred to saturated liquid at 0°F where the values are zero.)

Temp. (°F) t	Pressure[a] (psi)		Specific volume[b] (ft³/lb)		Density[b] (lb/ft³)		Enthalpy (btu/lb)		Latent heat (btu/lb)	Entropy (btu/lb,°F)		Temp. (°F) t
	Absolute p	Gauge $g.p.$	Liquid v	Vapor V	Liquid l/v	Vapor l/V	Liquid hf	Vapor g		Liquid sf	Vapor sg	
−20	7.50	14.6[b]	0.02610	11.0	38.35	0.0952	−9.0	156.5	165.0	−0.020	0.356	−20
−15	8.30	13.0[b]	0.02620	9.90	38.15	0.101	−7.0	157.0	164.0	−0.015	0.354	−15
−10	9.28	11.0[b]	0.02635	8.91	37.95	0.112	−4.5	158.5	163.0	−0.010	0.353	−10
−5	10.4	8.8[b]	0.02645	7.99	37.80	0.125	−2.5	159.5	162.0	−0.005	0.351	−5
0	11.6	6.3[b]	0.02660	7.17	37.60	0.139	0	160.5	160.5	0.000	0.350	0
+5	13.1	3.3[b]	0.02675	6.41	37.40	0.156	+2.5	162.0	159.5	0.005	0.348	+5
10	14.6	0.2[b]	0.02690	5.75	37.20	0.174	5.0	163.5	158.5	0.011	0.348	10
15	16.3	1.6	0.02705	5.18	37.00	0.193	7.5	164.5	157.0	0.016	0.347	15
20	18.2	3.5	0.02715	4.68	36.80	0.214	10.0	166.0	156.0	0.021	0.346	20
25	20.2	5.5	0.02730	4.24	36.60	0.236	13.0	167.5	154.5	0.027	0.346	25
30	22.3	7.6	0.02745	3.86	36.40	0.259	15.5	169.0	153.5	0.032	0.346	30
35	24.6	9.9	0.02760	3.52	36.20	0.284	18.0	170.5	152.5	0.038	0.346	35
40	26.9	12.2	0.02780	3.22	36.00	0.311	21.0	172.0	151.0	0.044	0.346	40
45	29.5	14.8	0.02795	2.96	35.80	0.338	24.0	174.0	150.0	0.049	0.346	45
50	32.5	17.8	0.02810	2.71	35.60	0.369	27.0	175.5	148.5	0.055	0.346	50
55	35.5	20.8	0.02825	2.49	35.40	0.402	30.0	177.5	147.5	0.061	0.347	55
60	38.7	24.0	0.02840	2.28	35.20	0.439	33.0	179.0	146.0	0.067	0.348	60
65	42.2	27.5	0.02855	2.10	35.00	0.476	36.5	181.0	144.5	0.073	0.349	65
70	45.8	31.1	0.02875	1.94	34.80	0.515	39.5	183.0	143.5	0.079	0.350	70
75	49.7	35.0	0.02890	1.79	34.60	0.559	43.0	185.0	142.0	0.086	0.351	75

460

80	53.9	39.2	0.02910	1.66	34.35	0.602	46.5	187.0	140.5	0.092	0.352	80
85	58.6	43.9	0.02930	1.54	34.10	0.649	50.0	189.0	139.0	0.098	0.353	85
90	63.3	48.6	0.02950	1.42	33.90	0.704	53.5	191.0	137.5	0.105	0.356	90
95	68.4	53.7	0.02965	1.32	33.70	0.758	57.5	193.5	136.0	0.112	0.358	95
100	73.7	59.0	0.02990	1.23	33.45	0.813	61.0	195.0	134.5	0.118	0.359	100
105	79.3	64.6	0.03005	1.14	33.25	0.877	65.0	198.0	133.0	0.125	0.360	105
110	85.1	70.4	0.03030	1.07	33.00	0.935	69.0	200.0	131.0	0.132	0.362	110
115	91.4	76.7	0.03050	0.990	32.80	1.01	73.0	202.5	129.5	0.139	0.364	115
120	98.0	83.3	0.03075	0.926	32.50	1.08	77.0	204.5	127.5	0.147	0.367	120
125	104.8	90.1	0.03095	0.867	32.30	1.15	81.5	207.5	126.0	0.154	0.369	125
130	112.0	97.3	0.03125	0.811	32.00	1.23	86.0	209.0	123.0	0.161	0.371	130
135	119.3	104.6	0.03145	0.760	31.80	1.32	90.5	211.0	120.5	0.169	0.375	135
140	126.8	112.1	0.03175	0.710	31.50	1.41	95.0	213.2	118.2	0.176	0.377	140
145	136.0	121.3	0.03195	0.662	31.30	1.51	99.6	215.2	115.6	0.183	0.379	145
150	145.0	130.3	0.03225	0.620	31.03	1.61	104.5	217.5	113.0	0.190	0.382	150
155	155.0	140.3	0.03255	0.580	30.73	1.72	109.3	219.7	110.4	0.197	0.384	155
160	165.0	150.3	0.03285	0.542	30.43	1.84	114.2	221.7	107.5	0.204	0.387	160
165	175.0	160.3	0.03320	0.505	30.14	1.97	119.3	224.0	104.7	0.212	0.389	165
170	186.0	171.3	0.03350	0.475	29.85	2.10	124.7	225.7	101.0	0.219	0.392	170
175	198.0	183.3	0.03385	0.448	29.54	2.23	129.9	228.0	98.1	0.226	0.395	175
180	210.0	195.3	0.03420	0.420	29.23	2.38	135.0	230.0	95.0	0.233	0.397	180

a Based on material from Dana, Jenkins, Burdick, and Timm, published originally in *Refrig. Eng.*, **12** (12): 402 (June 1926).
b Inches of mercury below one standard atmosphere (29.92 in.).
c Specific volume and density beyond 135°F, B. H. Sage and W. N. Lacey, *Ind. Eng. Chem.*, June 1938.

TABLE VII. Thermodynamic Properties of Saturated Propane

(Enthalpies and entropies are referred to saturated liquid at −200°F where the values are zero.)

Temp. (°F) t	Pressure[a] (psi)		Specific volume[a] (ft³/lb)		Density[a] (lb/ft³)		Enthalpy[c] (btu/lb)		Latent[c] heat (btu/lb)	Entropy[c] (btu/lb, °F)		Temp. (°F) t
	Absolute p	Gauge g.p.	Liquid v	Vapor V	Liquid l/v	Vapor l/V	Liquid hf	Vapor hg	L	Liquid sf	Vapor sg	
−75	6.37	17.0[b]	0.02660	14.5	37.59	0.0690	65.5	255.5	190.0	0.142	0.624	−75
−70	7.37	14.9[b]	0.02674	12.9	37.40	0.0775	68.0	257.0	189.0	0.148	0.623	−70
−65	8.48	12.7[b]	0.02688	11.3	37.20	0.0885	71.5	258.0	186.5	0.154	0.621	−65
−60	9.72	10.1[b]	0.02703	9.93	37.00	0.111	74.0	259.5	185.5	0.160	0.620	−60
−55	11.1	7.3[b]	0.02717	8.70	36.80	0.115	77.0	261.0	184.0	0.167	0.618	−55
−50	12.6	4.3[b]	0.02732	7.74	36.60	0.129	79.5	263.7	183.2	0.173	0.617	−50
−45	14.4	0.6[b]	0.02748	6.89	36.39	0.145	82.3	264.2	181.9	0.179	0.615	−45
−40	16.2	1.5	0.02763	6.13	36.19	0.163	85.0	265.8	180.8	0.185	0.614	−40
−35	18.1	3.4	0.02779	5.51	35.99	0.181	87.5	267.2	179.7	0.190	0.613	−35
−30	20.3	5.6	0.02795	4.93	35.78	0.203	90.2	268.9	178.7	0.196	0.612	−30
−25	22.7	8.0	0.02811	4.46	35.58	0.224	92.8	270.3	177.5	0.202	0.610	−25
−20	25.4	10.7	0.02827	4.00	35.37	0.250	95.6	271.8	176.2	0.208	0.608	−20
−15	28.3	13.6	0.02844	3.60	35.16	0.278	98.3	273.2	174.9	0.214	0.607	−15
−10	31.4	16.7	0.02860	3.26	34.96	0.307	101.0	274.9	173.9	0.220	0.607	−10
−5	34.7	20.0	0.02878	2.97	34.75	0.337	103.8	276.2	172.4	0.226	0.606	−5
0	38.2	23.5	0.02895	2.71	34.54	0.369	106.2	277.7	171.5	0.231	0.605	0
+5	41.9	27.2	0.02913	2.48	34.33	0.403	108.8	279.0	170.2	0.236	0.604	+5
10	46.0	31.3	0.02931	2.27	34.12	0.441	11.3	280.5	167.2	0.246	0.603	10
15	50.6	35.9	0.02950	2.07	33.90	0.483	114.0	281.8	167.8	0.248	0.602	15
20	55.5	40.8	0.02970	1.90	33.67	0.526	116.8	283.1	166.3	0.254	0.601	20
25	60.9	46.2	0.02991	1.74	33.43	0.575	119.7	284.4	164.7	0.260	0.600	25
30	66.3	51.6	0.03012	1.60	33.20	0.625	122.3	285.7	163.4	0.266	0.599	30

35	0.598	0.272	162.0	287.0	125.0	0.676	32.97	1.48	0.03033	57.3	72.0	35
40	0.597	0.278	160.3	288.3	128.0	0.730	32.73	1.37	0.03055	63.3	78.0	40
45	0.596	0.285	158.4	289.5	131.1	0.787	32.49	1.27	0.03078	69.9	84.6	45
50	0.596	0.292	156.5	290.7	134.2	0.847	32.24	1.18	0.03102	77.1	91.8	50
55	0.596	0.298	154.8	292.0	137.2	0.909	32.00	1.10	0.03125	84.6	99.3	55
60	0.595	0.306	152.6	293.2	140.6	0·990	31.75	1.01	0.03150	92.4	107.1	60
65	0.594	0.313	150.7	294.5	143.8	1.06	31.50	0.945	0.03174	100.7	115.4	65
70	0.594	0.321	148.3	295.8	147.5	1.13	31.24	0.883	0.03201	109.3	124.0	70
75	0.594	0.327	146.6	296.9	150.3	1.21	30.97	0.825	0.03229	118.5	133.2	75
80	0.593	0.335	145.1	299.1	154.0	1.30	30.70	0.770	0.03257	128.1	142.8	80
85	0.593	0.342	142.2	299.2	157.0	1.39	30.42	0.722	0.03287	138.4	153.1	85
90	0.593	0.349	140.0	300.3	160.3	1.49	30.15	0.673	0.03317	149.0	164.0	90
95	0.592	0.356	137.9	301.3	163.4	1.58	29.87	0.632	0.03348	160.0	175.0	95
100	0.592	0.363	135.6	302.4	166.8	1.69	29.58	0·591	0.03381	172.0	187.0	100
105	0.592	0.370	133.4	303.2	169.8	1.81	29.27	0.553	0.03416	185.0	200.0	105
110	0.591	0.376	131.2	304.0	172.8	1.92	28.96	0.520	0.03453	197.0	212.0	110
115	0.590	0.383	128.5	304.7	176.2	2.05	28.63	0.488	0.03493	211.0	226.0	115
120	0.589	0.391	125.4	305.2	179.8	2.18	28.30	0.459	0.03534	225.0	240.0	120
125	0.588	0.399	122.3	305.8	183.5	2.31	27.97	0.432	0.03575	239.0	254.0	125
130	0.587	0.406	119.3	306.1	186.8	2.48	27.64	0.404	0.03618	257.3	272.0	130
135	0.586	0.413	116.3	306.3	190.0	2.62	27.32	0.382	0.03662	273.3	288.0	135
140	0.585	0.422	112.5	306.5	194.0	2.78	27.00	0.360	0.03707	290.3	305.0	140

a Based on material from Dana, Jenkins, Burdick, and Timm, published originally in *Refrig. Eng.*, **12** (12): 403 (June 1926).

b Inches of mercury below one standard atmosphere (29.92 in.).

c From *Mollier diagrams for propane*, W. C. Edmister, Standard Oil Co. (Indiana).

such as toilet waters and perfumes, which have thus far been the most widely used glass-packaged cosmetic aerosols. In hair lacquers, and other products containing alcohol, the latter should be anhydrous grade when Propellant 11 is used; otherwise, free hydrochloric acid may form, causing corrosion of the container.

Propellant 12 or dichlorodifluoromethane is a high-pressure product, which, by itself, is used in high-pressure aerosols as well as in foam products. In combination with other propellants (particularly 11 and 114), it becomes useful for cosmetic sprays. Combinations can be prepared which will give almost any desired pressures or spray patterns. Propellant 12 is not as good a solubilizer for most cosmetic products as is Propellant 11; however, it is not as likely to cause corrosion, and it is particularly satisfactory for perfume aerosols where it contributes practically no discernible odor of its own. If it is used in foam without another propellant, a small proportion of mineral oil is often incorporated in the product to provide adequate vapor-pressure-depressant effect to the maximum 60 psig limitation for the container.

Propellant 114 or dichlorotetrafluoroethane is sold under the trade names of Freon®-114 and Genetron®-114a, which are stereoisomeric compounds. They are relatively low-pressure propellants, each with pressure sufficient to permit use by itself in glass bottles, particularly if a sufficient proportion of the total contents consists of the propellant, so that there is enough force to expel the material from the bottles. However, neither is used by itself, except for perfumes and colognes; and in such instances the cost is usually high. They have relatively good solubility for perfume oils and particularly for perfume-alcohol mixtures, and affect the fragrance very little, if at all. Most frequently, one of the isomers is used as a diluent or solvent for Propellant 12, and such mixtures offer a wide range of possible pressures, with lower pressures for glass bottles when the 114 (or 114a) is predominating, and with medium pressures for "beer can" aerosols when the proportion of Propellant 12 exceeds that of 114 (or 114a).

Propellant 22, or monochlorodifluoromethane, exerts pressures which are too high for ordinary use; however, in mixtures with 114, particularly where the latter is dominant, 22 offers interesting formulation possibilities. In alkaline solutions, its hydrolysis rate is relatively higher than those of Propellants 11, 12, and 114.

Propellant 142b, or 1,1,1-chlorodifluoroethane, is one of the more recently developed fluorinated hydrocarbons, and has been recommended for use in glass, plastic, and aluminum containers. It may be incorporated either *per se* or in mixtures with 114, to provide a favorable range of low vapor pressures. Its low liquid density and high vapor volume permit more product to be packaged with less propellant.

In evaluating a propellant for its various spray characteristics, the initial work should always be done at a constant temperature; 21 to 23°C is suggested. Then, before a formulation is decided upon, the product should be tested at higher and at lower temperatures, in order to determine the spray patterns, rate of flow, and other characteristics under conditions of actual use; for this purpose, it is felt that the tests should be run at 15 and 33°C. It is emphasized that these are temperatures of potential use, not temperatures of storage. For evaluating the shelf life under storage conditions, the finished products are kept at −5 and +50°C for periods of at least 1 month, and then allowed to remain at room temperatures again for 1 day, before they are further studied.

Finally, a third evaluation of the propellant takes place when the container is full, and again when it has been considerably emptied, that is, at about one-fourth full. This is particularly important for foam products, because the amount of propellant used for them is so small that there is danger of inability to expel the contents of the container when the latter is virtually empty. Such an evaluation of the container should be conducted with the following in mind:

1. This method of test is not to be divorced from the previous ones; in other words, the test under various temperatures of use and after considerable shelf storage at high and low temperatures must be concluded by a study of the test containers when full and partially emptied.

2. The study of the test container when partially empty should not be conducted by pressing the actuating button until most of the contents have been expelled. The expulsion should be over a period of time, so that any valve-clogging can be determined at the same time.

3. By recording the tare of the bottle or can, it is possible for the chemist to know at any time how much still remains in the container. A container in which more than 2 % of the contents of a spray product or more than 7 % of the contents of a foam product cannot be expelled is not considered satisfactory.

Concentrate

The last component of the product, and in many respects the most important, is the "concentrate." This can be defined as the active and inert ingredients (usually liquid, but sometimes solid) placed in the container to be expelled therefrom by the propellant. Usually it is found that one cannot take existing cosmetic formulations and convert them into aerosol cosmetics merely by adding a chosen propellant in a given proportion. Such adaptation and reformulation require flexibility and a desire on the part of the formulator to veer away from the previously accepted and to explore in new directions. For example, the usual shaving cream is far too viscous for aerating; it would not soften the beard sufficiently if dispensed from an aerosol container; and it shows incompatibilities with the available propellants. By reformulation, however, a new material, an aerosol-type shaving cream concentrate, can be produced.

This entire matter of "concentrate" will be dealt with later, in the section on the individual products.

Production

The production of aerosols requires a different approach from that usually taken in the manufacture of cosmetics. Aerosol production requires in many instances a thorough training in chemical and mechanical engineering, the basic problem being that of economic handling of liquefied gases under pressure or at low temperatures.

Improper handling of liquefied gases can result in great losses (in addition to other difficulties), because the gases are, in the ordinary sense, nondetectable. Improper handling can further result in substandard production, either because the containers are deficient in propellant, or because the propellants are not present in the correct proportions.

An aerosol plant, or the aerosol section of a plant, usually has special filling equipment and storage tanks, as well as other standard equipment, made of specially selected metals. In many cases such a plant will have large refrigeration units and specially designed heat exchangers, in order to cope with the wide variety of aerosol products.

Basically, there are three methods of putting the propellant into an aerosol product: they are cold-fill (also known as refrigeration-fill), pressure-fill, and under-the-cup methods. A well-equipped aerosol production plant should have production facilities available for all three types, and should also have adequate equipment for the following purposes:

> Propellant storage and handling.
> Manufacture and handling of concentrate.
> Container handling.
> Coding.
> Product filling.
> Water bath.
> Drying.
> Check weighing.
> Valve inserter.
> Overcap placer.
> Labeling.
> Cartoning.
> Finished goods handling.

Cold Filling

In the cold-fill method, the concentrate is first manufactured in the same manner that is used for conventional cosmetics. In certain products a high-boiling propellant might be added to the concentrate as a diluent or solvent in order to prevent precipitation during cooling. If a propellant is added to the concentrate prior to its being pumped into the heat exchange system, then that concentrate must be handled in storage as if it were a liquefied gas; i.e., it must be stored in pressurized vessels, both for reasons of safety and to prevent dissipation of the propellant.

The concentrate is usually refrigerated to a temperature of 10 to 20° above the temperature at which the propellant mixture is to be added. Either the propellant is purchased in a premixed form (if it is a mixture), or it can be premixed in the aerosol production plant by means of proportioning equipment, in a bulk receiver, and then put into the propellant heat exchangers. It is then refrigerated to a temperature at which the vapor pressure of the propellant is about 10 psia. This is done to take care of the heat pickup from the air, from the cans, and from the concentrate during the filling operation.

The temperature at which one can fill must be predetermined by viscosity measurements. The lower limit is the lowest temperature at which one has a free-flowing liquid and at which the various components of the concentrate are not adversely affected. If it is found that there is a critical viscosity or a critical temperature with respect to precipitation, then the refrigeration system can be adjusted to reduce the temperature of the propellant below that ordinarily required, in order to anticipate the additional heat pickup from a less cold concentrate. If this should cause precipitation in the container, then it must be predetermined whether this is a reversible phenomenon; i.e., whether the precipitate will go back into solution during storage.

Generally speaking, cold filling calls for adding the concentrate first, although there are instances where this is not true. In such cases (which, incidentally, are relatively rare), there may be poor solubility of certain ingredients even at moderately low temperatures, or there may be increased viscosity at such temperatures, to the point where the concentrate cannot be fed through the filling lines into the containers. In such instances it has become necessary to add part of the concentrate in solid form, that is, as a "pill" or a "powder," prior to adding the remainder of the concentrate and the propellant.

The concentrate and the propellants may be fed from the filling machines to the containers simultaneously, or in sequence. In the latter case, the concentrate is added first, followed by the propellant.

Ordinarily, the addition of the propellant during the cold-fill process results in sufficient flash vaporization to displace the air in the container;

however, excessive pressures due to air in the headspace may develop from three sources: (a) dissolved air in the propellant, acquired over a period of time in the bulk-handling system; (b) the propellant being added to the container at such a low temperature as to preclude flashing; or (c) air entrapped in the concentrate in which it is dissolved, particularly in aqueous foam products.

If the product being filled is a nonaqueous system, it is essential that in all handling prior to filling, dehydrators be used in the production line to prevent any condensation and addition of moisture which might cause corrosion or other ill effects. Moisture elimination is necessary also to prevent accumulation of any ice in the heat exchangers or in the lines themselves.

After the propellant and concentrate have been added to the containers, the valve is inserted and then crimped, spun, or seamed onto the container. This is a most critical operation. It should be speedy, not only for economy, but in order to prevent sufficient pickup of heat on the line to cause volatilization of any propellant. The crimping or sealing operation requires constant control and careful supervision in order to assure proper closure of every container. Improper closure results in leaking and valve clogging.

Following the crimping, the container is tested, in conformity with ICC regulations, by heating to 55°C or 130°F, usually in a water bath. During this immersion, inspectors can easily detect any leaks. Removed from the hot water bath, the container is sometimes subjected to an individual spray-testing operation. If the can has not been previously lithographed or silk-screened, the label is added at this point, and then the container is ready for packing. The protective dome is placed over the valve, other auxiliary components are added, and the package is complete.

The cold-filling method has as its main advantage the speed of operation, which is translated into cost of production; it is more economical from a labor viewpoint. The cold-filling method, furthermore, offers less opportunity for entrapment of air. Cold filling offers certain disadvantages, however. First, there is the possibility of condensation of moisture in the product. For anhydrous products, the exclusion of moisture may be essential. Second, cold filling requires a high capital investment for refrigeration equipment. Third, the cold-filling process presents more limited formulation possibilities, because of the necessity to refrigerate the concentrate prior to filling; therefore the concentrate must be so formulated that any precipitation at very low temperatures is reversible.

Pressure Filling

In pressure filling, the concentrate is added in the same manner as in ordinary cosmetic manufacture, except that here too a high-boiling propellant could be added to the concentrate. In this case also, the concentrate is handled as a liquefied gas.

In pressure filling, the concentrate is almost invariably added first, at room temperature; however, there are certain pressure-filling operations where the container is evacuated first, either by vacuum or by propellant displacement of air, prior to addition of the concentrate. In this case, the valve is put on first, and the concentrate is actually introduced through the valve mechanism. More frequently, the concentrate is added to the container as the first operation in pressure filling, and air is displaced by one or two of three methods:

1. By the introduction of a small amount of liquefied gas, which flashes off, displacing the air.
2. By flushing the headspace over the concentrate with a fluorinated hydrocarbon, particularly Propellant 12.
3. By creating a vacuum in the headspace prior to valve closure. The valve is inserted and crimped or spun on following air displacement, and the container moves to the propellant-injection stage, where a predetermined amount of the propellant is introduced into the package through the valve.

Pressure filling can be carried out with the actuating button off the valve mechanism or with the actuating button in place. When the filling is done with the actuator button off, an additional operation for replacing the actuator is required.

Following the filling and sealing of the can, the container goes through exactly the same operation as when cold-filled: inspection in a hot water bath and spray testing.

Pressure filling has as its main disadvantage the lesser speed with which the line runs; hence, from the viewpoint of labor and overhead, it is a more expensive process. With newer equipment, however, especially of the multiple-head variety, this disadvantage can be overcome. Pressure filling has as its second disadvantage the greater possibility for the entrapment of air, although this danger can be reduced to a minimum and, in fact, completely eliminated by proper and careful evacuation and flushing. Air entrapment is extremely important, for if the air is not evacuated, there is a danger of build-up of excessive internal pressures with consequent risk of explosion, at worst, or of acceleration of the rate of corrosion.

Pressure filling offers certain distinct advantages, however. The headspace can be evacuated or flushed, thus counteracting the high pressures built up inadvertently by entrapped air. Pressure filling offers far greater possibilities for flexibility of formulation and of production. It permits changing from one product to another with greater ease and without arduous and time-consuming cleaning of the lines. For all practical purposes, it excludes the possibility of contaminating a product with the material from a previous fill. There is little or no danger of moisture condensation. The equipment for pressure filling is less expensive than cold filling, and the required capital investment is smaller.

Under-the-Cup Filling

Combining the advantages of the older cold-filling and the later pressure-filling methods is the latest method of under-the-cup filling. In this process, the product or concentrate is added to the container with a customary filling machine at room temperature, much as in pressure filling. The propellant is added through the 1-in. cup after the can has been evacuated. The valve is then crimped. The sequence is, therefore, (a) evacuation of the can, (b) lifting of the valve cup, (c) volumetric filling of the propellant, and (d), crimping of the valve. With a 9-head rotary undercap filling machine, speeds of up to 180 cans/min are attainable. The under-the-cup unit is available as a single-head, 6-, 9-, or 18-head machine. The under-the-cup filler allows as fast filling as the cold-fill and has the advantage of pressure filling, in that no refrigeration is required. The disadvantages of the under-the-cup method are high initial cost, highly skilled maintenance required, and high propellant losses. Under-cup equipment is more costly than a pressure filler or a cold filler because of the complexity required. Maintenance costs are high because of gasket replacements and need for skilled personnel. Propellant that is left in the valve cup (approximately 3 cc) is lost, and therefore propellant economy is not good. A propellant recovery system is available, but is not too satisfactory for all propellant blends.

With each product, it is possible to evaluate which of the three methods is best suited. The following considerations determine this:

1. Amount of propellant fill.
2. Cost of propellant.
3. Size of container.
4. Production speed required.
5. Propellant fill accuracy required.
6. Elimination of moisture in product.
7. Low air in head space.
8. Aqueous or nonaqueous product.

In order to handle all aerosol products, cold filling, pressure filling, and under-cup filling must be available.

Formulation

In the formulation of a pressure-propelled cosmetic, it is nearly always necessary to make changes from the original nonaerosol formulation; in other words, one must formulate for pressure since the propellant becomes a part of the product. In addition, since this is a dynamic package, consideration must be given to the physical form and temperature of the emitted product. The following are, therefore, factors which are normally not encountered in standard cosmetic formulation:

1. *Spray patterns.* The nature of the spray, its wetness or dryness, is adjusted by the percentage and proportion of propellants used, the construction of the valve, and the characteristics of the product. The spray pattern can be varied by any one of these factors, or by all three. Some products—for example, high-molecular-weight resins—are extremely difficult to spray even at high pressures. This is probably due to intramolecular forces which make the shearing processes of the particles difficult. The CSMA *Aerosol Guide*, 5th ed., March 1966, has a method for the analysis of spray patterns.

2. *Foam characteristics.* The foam pattern is determined by the propellant, concentrate, and the type of valve used. The amount of air present in the can may be critical in changing a wet foam to a dry, stiff foam. Foam products are also very sensitive to the amount of propellant. Downing and Young (34) reported that "even though the propellants are present in the same concentration, those with higher pressures tend to give a stiffer and more elastic foam than those with lower pressures." There are three main foam characteristic types: (*a*) stable foam, (*b*) quick-breaking foam, (*c*) spray foam. These have been further characterized by Paul Sanders, who has done a considerable amount of work in this area (33–37, 39, 40). Richman and Shangraw studied the rheology and theories of formation of foam aerosol products (38, 41–43). A greater amount of research work has been done on the aerosol foam products than on other aerosols.

3. *Temperature of spray.* Not only is the storage temperature of an aerosol product important, but the temperature of the product as emitted, especially for skin application, must be considered. Low storage temperatures will affect the pressure in the container and therefore the spray and foam characteristics. Propellant evaporation as the product is emitted absorbs heat and therefore tends to cool the product. If the propellant hits the skin before evaporation has taken place, there will be a chilling sensation. Products that have the lowest boiling points are the warmest on the skin because evaporation has taken place before it hits the skin. This, of course, assumes that the aerosol container has been held at least 5 in. away.

4. *Reactivity.* Possibilities of reaction between the various ingredients have been investigated by Root (44). Those of greatest concern involve the various constituents of the product with the component parts of the valve, the can, and sometimes the propellant.

5. *Solubility.* The solubilities of various cosmetic materials in the different propellants have been reported (45,46). Materials which are slightly

soluble or which become insoluble below room temperature should be avoided, since if they come out of solution, clogging of the valve orifices can result.

6. *Corrosion.* Corrosion is probably the most important factor to be considered in an aerosol cosmetic packaged in a metal container. This can be critical not only to the package but to the product as well. Odor, color, and pH changes can be due to corrosion, not only of the container, but the metal parts, such as the spring of the valve. Even the plastic parts of the valve are subject to attack by some products. A very comprehensive bulletin has been published by Allied Chemical Corporation on the subject of corrosion (47).

7. *Fragrances.* Even more critical than in the standard cosmetic product is fragrance in aerosols. Because the product is sprayed in many cases, this is the first contact with characteristics of the product. The fragrance can be affected by the container, the valve gaskets, and the propellant. Aerosol colognes and perfumes never use Propellant 11 because of its odor. Propellant breakdown by hydrolysis can also cause changes in odor.

Hair Products

The most successful aerosol cosmetic product—or *any* aerosol product—has been hair spray. At the time that this is being written, more than half a billion cans are being produced in the United States annually. Other hair products include the following:

1. Hairsets and conditioners.
2. Hairdressings.
3. Hair color rinses and sprays.
4. Wave lotions.
5. Shampoos.

Hair Sprays

Hair sprays contain from 1.5 to 2.7% solids. These solids include the resin, plasticizers, modifiers, perfume oil, and neutralizer when it is used.

At various times during the development of hair sprays, several resins have taken the lead and have then been replaced. The order goes in this fashion: Shellac, PVP, PVP/VA, National Starch 28-1310, and, most recently, Gantrez. It seems that there are fashions in hair spray resins almost like there are in clothes. Other resins, such as Ciba 325, Vem, and Vicryl have played minor roles. There are two types of resins that are being used, those that require neutralization for water solubilization, and those that do not.

Resins with carboxyl group requiring neutralization	Resins that do not require neutralization
Shellac	PVP
National starch 28-1310	PVP/VA
National starch 28-2930	Ciba 325
Vem 640	
Vicryl XOR 63	
Gantrez N-119 ester	

Neutralizing agents used include the following:

1. AMP—aminomethylpropanol.
2. AMPD—aminomethylpropanediol.
3. AEPD—aminoethylpropanediol.
4. TIPA—triisopropanolamine.
5. DDA—dodecylamine.
6. NH$_4$OH—ammonium hydroxide.

Modifiers and plasticizers can be used when necessary. The following are being used in hair spray formulations:

1. Castor oil.
2. Dimethyl phthalate.
3. Ethoxylated lanolin.
4. Acetyl triethyl citrate.
5. Glycol monooleate.
6. Silicones.
7. Dioctyl phthalate.

The solvent in all hair sprays is SDA 40 anhydrous ethyl alcohol. Propellants used are 11 and 12, with additions of isobutane and methylene chloride for economy reasons. The isobutane content of the propellant is generally limited to 10% because of flammability. Methylene chloride is limited to 10% of the total formulation because of possible attack of can-sealing compounds and valve gaskets. Formulas 1 to 8 are for hair sprays.

Formula 1

Shellac-Type Hair Spray

Dewaxed unbleached shellac	1.25%
Castor oil	0.10
Lanolin ester	0.15
Perfume oil	0.15
Ethyl alcohol, anhydrous, SDA 40	28.35
Propellants 12/11, 35:65	70.00

PVP-Type Hair Spray

PVP K-30	1.25%
Lanolin derivative	0.10
Silicone	0.10
Perfume	0.15
Ethyl alcohol, anhydrous, SDA 40	28.40
Propellants 12/11, 35:65	70.00

Formula 3

PVP/VA Hair Spray

PVP/VA, 70/30	1.50%
Lanolin derivative	0.05
Silicone	0.08
Perfume oil	0.15
Ethyl alcohol, anhydrous, SDA 40	28.22
Propellants 12/11 35:65	70.00

Formula 4

Resin Hair Spray

Resin 28-1310	3.45%
AMPD (2-amino-2-methyl-1,3-propanediol)	0.38
Alcohol-soluble lanolin	0.90
Isopropyl myristate	0.40
Dipropylene glycol	0.10
Perfume oil	0.35
Ethyl alcohol, anhydrous, SDA 40	94.42
Above concentrate	35%
Propellants 11/12, 65:35	65

Formula 5

Resin Hair Spray

Resin 28-2930	1.50%
AMPD	0.11
Silicone fluid	0.15
Perfume oil	0.15
Ethyl alcohol, anhydrous, SDA 40	33.09
Propellants 11/12, 65:35	65.00

Formula 6

Gantrez ES-425	3.96%
Triisopropanolamine	0.07
Perfume oil	0.95
Ethyl alcohol, anhydrous, SDA 40	25.02
Propellants 11/12, 65:35	70.00

Formula 7

Vicryl XOR 63	3.50%
AEPD	0.19
Silicone 556	0.05
Hodag CSA 80	0.17
Perfume oil	0.10
Ethyl alcohol, anhydrous, SDA 40	30.99
Propellants 11/12, 65:35	65.00

Formula 8

Vem 640, 50%	2.23%
AEPD	0.12
Silicone 556	0.13
Hodag CSA 80	0.11
Perfume oil	0.06
Ethyl alcohol, anhydrous, SDA 40	46.45
Methylene chloride	6.00
Propellant 12	16.38
Propellant 11	23.41
Isobutane	5.21

Hairsets and Conditioners

There has been a recent interest in water-base hair sprays, which are used as setting agents and conditioners. These products are based on carboxylated resins which are neutralized to make them water-soluble. Although water-based hair sprays were marketed some years ago, they made no impact on the market. Hair sprays are used to hold the coiffure or hair style, and must be anhydrous. Water in this type of product destroys the temporary set in the hair. The new water-based sprays are put on wet hair; the hair is then wound on curlers and dried. The resins are neutralized with a volatile alkali so that the drying converts the resin to the water-insoluble acid form of the resin. These wave sets are superior to previous products based on gums, carbopol gels, and water-soluble resins.

There are already several such products on the market, and it appears that this aerosol hair product will show the greatest growth in the coming years.

Hair sets and conditioners have also been marketed as men's products. Formulations, except for perfume, are very similar to those that are marketed for women. Men do not like to have a product that stiffens the hair—on the other hand, neither do they like a product that leaves the hair oily or sticky. There have been a few anhydrous hair sprays marketed for men. In these products the percentage of resin is lower and a suitable fragrance is used.

Hairdressing

Hairdressings containing no resins have been growing in popularity for both men and women. These are fast-drying products which contain fatty materials, such as isopropyl palmitate, lanolin, and mineral oil. The percentage of propellant is high—usually 90% or over.

Color Rinse and Sprays

Metallic hair sprays have been marketed in the United States, but with only small success, since they are what might be called special-occasion products. Hair sprays with certified dyes were also marketed, but have met with little acceptance, mainly because of the difficulty of preventing the spray from getting on the skin and clothes. There have been some color shampoo aerosols, but these, too, have not been popular.

A paraphenylenediamine hair dye with hydrogen peroxide in the pouch of the codispensing valve has come on the market. There are still some problems with this product, since the metering of the peroxide and dye is not as accurate as it should be; moreover, because of permeation problems, the product contains no ammonia and, therefore does not lighten hair, as many users might desire. With the solution of these problems, the new pressure-package hair dye could replace the current two-bottle peroxide-dye package.

Wave Lotions

There has been a tremendous amount of work done on aerosol cold-wave products. As of today, only one quick-breaking, cold-waving lotion based on thioglycolate is marketed.

Shampoos

Although aerosol shampoos came on the market as early as 1950 with quite an impact, they were removed shortly thereafter because of leaking cans. This unfortunate first experience has served to keep similar products off the retail shelves since then.

Aerosol shampoos have been sold in the professional field and in the direct home sale market. The technical problem of the container has been largely overcome; this shampoo has been sold to beauty shops in both tin cans and aluminum. A return to the retail market is anticipated.

The advantages of an aerosol shampoo are the following:

1. Dispensing package.
2. Highly concentrated.
3. Cannot be contaminated.
4. Can be promoted as a family package.
5. Can be placed in the shower and actuated with one finger.
6. No cap to take off and put on.

On the other hand, it is necessary to sell a smaller package at what seems like a higher price. The product, however, is economically sound, as evidenced by its repeat sales to the professional buyers of shampoos—the beauty operators.

Skin Products

Deodorants and Antiperspirants

The most dramatic growth of an aerosol has taken place with aerosol deodorants and antiperspirants. In a short period of four years, the products made great commercial headway in replacing roll-ons, pads, creams, squeeze bottles, and sticks. Eventually they may replace hair sprays and become the largest-selling aerosol product.

The first successful deodorant product was based on hexachlorophene and aluminum sulphocarbolate, and was promoted for men; the first antiperspirant was a powdered product which contained micronized aluminum chlorhydroxide, talc, isopropyl palmitate, and propellant.

At present there are four different types of aerosol antiperspirants on the market:

1. Powdered antiperspirant.

2. Solution antiperspirant. In this product, the aluminum chlorhydroxide is in solution. A product of this type is marketed in a tin can and plastic-coated glass bottle.

3. Powdered dispersion. In this product, a micronized aluminum chlorhydroxide is dispersed in an oil, such as isopropyl myristate, squalene, or other oil. Propellants 11 and 12 make up 90% of this product.

4. Solution and dispersion. In this product, aluminum chlorhydroxide is present in solution and also as a dispersion. The product has the advantage of the solution in that it becomes effective immediately and has reserve

aluminum salt available like the dispersion, which goes into solution when sweating occurs.

Fragrances

A major portion of the colognes are now marketed in aerosol bottles, both coated and uncoated. A perfume aerosol is a built-in, nonspillable perfume atomizer, which does not have the disadvantages of clogging and fragrance deterioration that beset the bulb atomizers.

Aerosol fragrances have also been marketed in an emulsified foam, known as a foam cologne. These products contain a high percentage of perfume oil.

Sunscreen

Although aerosol sunscreen products had sold well for many years in Europe, especially Italy, it was not until much later that these products began to be popular in the United States. The reason was the development of the new cocoa butter aerosol spray product, which clings to the skin and whose spray is not easily blown away from the skin by the wind. Previous foam products were not very satisfactory, and the spray products were too fine and were generally dispersed by the wind before they reached the skin; also, when they did make contact with the skin, there was insufficient material to be rubbed over a large area of skin. An example of a cocoa butter formulation is shown in Formula 9.

Formula 9

Mineral oil	56.50%
Propylparaben	0.25
Paraffin	10.50
Cocoa butter	24.50
Sunscreen	7.75
Perfume oil	0.50
	100.00
Concentrate	50.00%
Propellants 11/12, 50:50	50.00

Shaving Cream

Aerosol shave creams have continued their growth, but because of the phenomenal growth of deodorants and antiperspirants, they dropped from the second to the third position in aerosol cosmetic sales in the United States. Because of the Spitzer-Reich-Fine patent (21), which covered fluorocarbon-propelled shaving creams, many people went to hydrocarbon-propelled products. Although this patent expired in 1970, this writer has little expectation of a reversion on a large scale to the fluorocarbon propellant for shaving lather, because of cost factors and the development of the hydrocarbon propellant technology. A hot shaving cream with the codispensing valve, based on an exothermic reaction, has attracted some interest, but seems at this point unlikely to replace the older formulations.

Shaving Accessories

An aerosol after-shave foam is being marketed for men who use electric shavers, as well as those who use a razor. These products are quick-breaking alcoholic foams. An example of such a product is shown in Formula 10.

Formula 10

After Shave Collapsible Foam

Polychol 40	1.5%
Cetyl alcohol	1.5
SDA 40 alcohol (190 proof)	58.0
Hyamine 1622	0.1
Menthol	0.1
Perfume	1.0
Water, distilled	37.8
Concentrate	92.0%
Propellants 12/114, 40:60	8.0

Feminine Deodorant Spray

A newcomer to the aerosol feminine products is a deodorant spray for the vaginal area. This product was first introduced in Switzerland, and within the last two years it has grown to a 17 million can-per-year product in the United States. The product is marketed in an aluminum can because the propellant used has a pressure of 70 psig at 70°F. and could give difficulties with a side-seam can. A typical product is shown in Formula 11.

Formula 11

Isopropyl myristate	4.0%
Hexachlorophene	0.3
Perfume oil	1.7
Propellant 12	94.0

A vapor tap valve is used with this product so that it can be used right side up or upside down. Because Propellant 12 has a very low boiling point ($-21.6°F$), it evaporates before it touches the skin, and therefore the spray is relatively warm. The chilling effects of various propellants have been studied by T. F. Dunne (48,49) and G. F. Broderick and Lloyd Flanner (50).

Nail Preparations

There have been many attempts to make an aerosol nail lacquer, but because of the application problem, no such product is yet available. The only nail preparation in aerosol form thus far marketed has been a nail dryer. The product is a fluorinated hydrocarbon with the addition of an oil such as peanut or persic oil, plus perfume. The oil gives gloss to the polish and the propellant hastens the drying process.

Powders

There have been talc products for the body marketed in aerosol form. Since the maximum amount of powder is about 5 % in an aerosol to prevent clogging, this turns out to be a rather expensive package for talc. There is the further objection that the talc cannot be sprayed on the body very readily without getting it all over the room.

There have been at least two aerosol powder shampoos marketed. These products contain starch, perfume, and propellant. The starch product must be brushed out of the hair. It appears that this product has a very limited market, since it is only desirable when, for one reason or another, the hair cannot be shampooed in the regular manner.

Face Creams and Lotions

There have been a number of aerosol products for the face in the form of foam, such as moisturizing, cleansing, emollient, and foundation products. These have been specialty products and have not reached the mass market. A spray product for fixing the makeup has met with some mass-marketing possibilities. This product consists of a propellant, PVP, and perfume oil. It has a chilling effect on the skin and fixes the makeup by putting a thin layer of resin over the face. One difficulty with the use of this product is that the eyes have to be closed during application.

Oral Products

The only oral cosmetic product to be introduced in aerosol form since the appearance of toothpaste in pressurized containers has been the mouth freshener. This is a one-half-ounce aluminum aerosol with a metered spray. The product contains propellant, flavoring, and a sweetener.

There is little doubt at this time that aerosols will continue to play a very significant part in the growth of the cosmetic industry. New products will depend on the developments made by container, propellant, valve, and chemical specialty manufacturers, on those of the aerosol "fillers" or manufacturers, on those of the finished product houses (the marketers), and on developments emanating from manufacturers of other raw materials, such as emulsifiers and fragrances.

Appendix *

Glossary of Terms Used in the Aerosol Industry (Tentative)

Active ingredient. Component of an aerosol formulation that produces the specific effect for which the formulation is designed.

Aerosol. A suspension of fine solid or liquid particles in air or gas, as smoke, fog, or mist. As defined by the Department of Agriculture, 100% of the particles in an insecticidal aerosol spray must have a diameter less than 50 μ and 80% of the particles must have a diameter less than 30 μ.

Aerosol insecticides storage test. Tentative official method (sponsored by CSMA) for determining storage characteristics of aerosol insecticides.

"Aerosol" product. Self-contained sprayable product in which the propellant force is supplied by a liquefied gas. Includes space, residual, surface coating, foam, and various other types of products but does not include gas-pressurized products such as whipping cream. The term "aerosol" as used here is not confined to the scientific definition.

Aerosol test method for flying insects. Official bio-assay method (sponsored by CSMA) using houseflies and OTA.

Auxiliary solvent. Liquid material used in addition to the primary solvent. Generally used to replace part of the primary solvent to produce some specific effect or as a matter of economics.

Chemical attack. Chemical reaction or solvent effect, causing failure or deterioration of plastic and rubber parts, organic coatings, metals, or lithography involved in the completed package.

Compatibility. Broad term meaning that the various components of an aerosol formulation can be used together without undesirable physical or chemical results.

Concentrate. A basic ingredient or mixture of ingredients to which other ingredients, active or inactive, are added.

Container. Metal, glass, or plastic shell in which an aerosol formulation is packaged.

Corrosion. Chemical alteration of the metal parts of container or valve. May lead to package failure and/or product deterioration.

Cosolvent. Solvent used to improve the mutual solubility of other ingredients.

Crimp. One operation by which the valve may be permanently seated in some aerosol containers.

* This glossary was drawn up by the Chemical Specialties Manufacturers Association and is reproduced by permission of that organization.

Density.   Weight of a given volume of material at a specified temperature.

Delivery rate.   Weight of mixture discharged from dispenser per unit of time at a specified temperature. Usually expressed as g/sec at 80°F.

Dispenser.   Metal, glass, or plastic shell with valve from which an aerosol or pressurized formulation is dispensed.

Eductor tube.   Tubing connecting the lower portion of container or dispenser with valve. Sometimes miscalled "syphon tube" or "dip tube."

Foam product.   Aerosol formulation containing a solution or emulsion which is dispensed in a highly expanded fluffy form by a liquefied gas propellant.

Head space.   Volume in upper portion of dispensor not filled with liquid contents. Usually expressed as percent of total volume of dispenser at a specified temperature.

High volatile ingredients.   See Volatile Ingredients.

Inert (or inactive) ingredient.   Component of an aerosol formulation that does not contribute to the specific effect of the formulation. In some cases, may be quite arbitrarily defined. For example, with insecticides, only the propellants are considered as inert ingredients.

Low volatile ingredients.   See Nonvolatile Ingredients.

Metering valve.   Valve that delivers a definite, limited amount of aerosol formulation each time the valve mechanism is operated.

Nonvolatile ingredients.   Components of an aerosol formulation with a vapor pressure less than atmospheric pressure (< 14.7 psia) at temperature of 105°F. Sometimes called *low volatile* components.

Official test aerosol, or OTA.   A standard insecticide dispenser and formulation prepared by CSMA for use in Official Aerosol Test Method for Flying Insects.

Particle size.   Diameter of solid or liquid particles expressed in μ (thousandths of a millimeter).

Pressure.   Internal force per unit area exerted by any material. Since the pressure is directly dependent on the temperature, the latter must be specified. The pressure may be reported in either of two ways:

1. Absolute pressure—the total pressure with zero as a reference point. Usually expressed as pounds per square inch absolute (psia).

2. Gauge pressure—the pressure in excess of atmospheric pressure. Under standard conditions at sea level, the numerical value of the absolute pressure is 14.7 higher than that of the gauge pressure. The gauge pressure is usually expressed as pounds per square inch gauge (psig).

Product deterioration.   Chemical reaction or physical change within or between components considered compatible in original formulation. May be due to time or temperature of storage or other factors.

Product formulation.   Specific formulation of completed product, including propellant(s). Usually expressed as weight/weight (w/w) percent.

Propellant.   Liquefied gas with a vapor pressure greater than atmospheric pressure (>14.7 psia) at a temperature of 105°F.

Solubility.   The extent to which one material will dissolve in another. Generally expressed as percent by weight. May also be expressed as percent by volume or parts per 100 parts of solvent by weight or volume. The temperature should be specified.

Solvent.   Liquid part of an aerosol formulation used to dissolve solid or other liquid parts.

Spray.   The dispersed discharge from an aerosol-type dispenser in the form of small droplets or particles. Does not include foam-type discharge.

Spray coating.   Aerosol spray product for surface application, which leaves a residual clear or pigmented finish for protective or decorative purposes.

Stability.   Ability of a product to maintain its original characteristics over extended storage periods, under normal variations in temperature conditions.

Synergist.   An auxiliary material that has the property of increasing the effect of the active ingredient even though it may have little specific activity itself.

NOTE: In the case of insecticides, synergists are considered as active ingredients.

Valve.   Mechanism for discharging products from aerosol-type dispensers.

Viscosity.   Internal resistance to flow of a solid (powder), liquid, or gas at a specified temperature. A definite measurement for the consistency of a material.

Volatile ingredients.   Components of an aerosol formulation with a vapor pressure greater than atmospheric pressure (>14.7 psia) at a temperature of 105°F. Sometimes called *high volatile* components.

REFERENCES

1.   Whytlaw-Gray, R. W., and Patterson, H. S.: *Smoke,* E. Arnold, London, 1932.

2.   Sinclair, D.: *Handbook on aerosols,* Washington, D.C., 1950, p. 64.

3.   Avy, A. P.: *Les aérosols,* Dunod, Paris, 1956, p. 2.

4.   Chemical Specialties Manufacturers Association: Glossary of terms used in the aerosol industry, 5th ed., Mar. 1966.   (This is here reprinted as an appendix to this chapter).

5.   Lynde, N.: U. S. Pat. 34,894 (1862).

6.   Helbing, H., and Pertsch, G.: U. S. Pat. 628,463 (1899).

7.   Gebauer, C. L.: U. S. Pat. 668,815 (1901).

8.   Gebauer, C. L.: U. S. Pat. 711,045 (1902).

9.   Moore, R. W.: U. S. Pat. 1,378,866 (1903).

10.   Mobley, L. K.: U. S. Pat. 1,378,481 (1921).

11.   Rotheim, E.: U. S. Pat. 1,800,156 (1931).

12.   Rotheim, E.: U. S. Pat. 1,892,750 (1933).

13.   Midgley, T., Jr., Henne, A. L., and McNary, R. R.: U. S. Pat. 1,926,396 (1933).

14.   Bichowsky, F. R.: U. S. Pat. 2,021,981 (1935).

15.   Goodhue, L. D.: Insecticidal aerosol production: spraying solutions in liquefied gases, *Ind. Eng. Chem.,* **34:** 1456 (1942).

16.   Goodhue, L. D., and Sullivan, W. N.: U. S. Pat. 2,321,023 (1943).

17.   Fiero, G. W.: Arg. Pat. 59,548 (1947).

18.   Fiero, G. W.: Panama Pat. 157 (1948).

19.   Estignard-Bluard, J.: Belg. Pat. 487,623 (1949).

20.   Abplanalp, R. H.: U. S. Pat. 2,631,814 (1953).

21.   Spitzer, J. G., Reich, I., and Fine, N.: U. S. Pat. 2,655,480 (1953).

22. Spiegel, M.: U. S. Pat. 2,871,161 (1959).

23. Beard, W. C., in Shepherd, H. R. (ed.): *Aerosols—science and technology*, Interscience Publishers, New York, 1961, Chapter 5, pp. 119–211.

24. Herzka, A., and Pickthall, J.: *Pressurized packaging*, 2nd ed., Academic Press, New York, 1961, Chapter IV, pp. 102–113.

25. Herzka, A., Harris, R. C., and Platt, N. E.: *International encyclopedia of pressurized packaging (aerosols)*, 1966, Chapter 6, pp. 78–118.

26. Beard, W. C.: Aerosol valves, up-to-date, *Proc. CSMA*, 1965, pp. 60–68.

27. Greenebaum, James E.: What's new in aerosol valves, *Proc. CSMA*, 1968, pp. 37–42.

28. Devera, A. T.: Specialty valves, fitments and applicators, *Proc. CSMA*, 1968, pp. 42–43.

29. Reed, F. T.: The propellent in aerosol products, *JSCC*, **7**: 137 (1956).

30. Reed, F. T.: Propellents for low pressure cosmetic aerosols, *Proc. CSMA*, 1954, p. 31.

31. E. I. du Pont de Nemours Company: Freon compounds and safety, Freon Products Div., Wilmington, Delaware, 1969.

32. Gee, A. H., and Fiero, G. W.: Biological testing of aerosols for safety, *Proc. CSMA*, 1954, p. 35.

33. Sanders, P. A.: Molecular complex formation in aerosol emulsions and foams, *JSCC*, **17**: 801 (1966).

34. Downing, R. C., and Young, E.: Formulation of aerosol products, *Soap*, **24**: 142 (Sept. 1953).

35. Sanders, P. A.: Aqueous alcohol aerosol foams, *DCI*, **99**: 56 (1966).

36. Sanders, P. A.: New developments in aerosol foam technology, *Am. Perf.*, **81**: 31 (February 1966).

37. Sanders, P. A.: Non-aqueous aerosol foams, *Aerosol Age*, **11**: 33 (November 1960).

38. Richman, M. D., and Shangraw, R. F.: Foams—a review of theory of formation, formulation, and methods of evaluation, *Aerosol Age*, **11**: 36 (May 1966) and **11**: 30 (June 1966).

39. Sanders, P. A.: Aerosol emulsion systems, *JSCC*, **9**: 274 (1958).

40. Sanders, P. A.: Surfactants for aerosol foams, *Soap*, **77**: 85 (October 1962).

41. Richman, M. D., and Shangraw, R. F.: Rheological evaluation of pressurized foams, *Aerosol Age*, **11**: 28 (July 1966) and **11**: 39 (August 1966).

42. Richman, M. D., and Shangraw, R. F.: A study of pressurized foams with emphasis on rheological evaluation, *Aerosol Age*, **11**: 45 (September 1966).

43. Richman, M. D., and Shangraw, R. F.: Pressurized foams with emphasis on rheological evaluation, *Aerosol Age*, **11**: 32 (October 1966) and **11**: 28 (November (1966).

44. Root, M. J.: Formulating for pressure, *JSCC*, **7**: 149 (1956).

45. Parmelee, H. M.: Water solubility of Freon refrigerants. Part I. Compounds boiling below 32°F. *Refrig. Eng.*, **63**: 1341 (1953).

46. Reed, F. T.: The behavior of certain typical emulsifying agents in Freon water systems, Kinetic Technical Memorandum No. 4, E. I. du Pont de Nemours Co., Inc., 1950.

47. Allied Chemical Corporation, Morristown, N. J.: Genetron product information bulletin 7-60, 1960.

48. Dunne, T. F.: Determination of possible chilling effects of propellants, *Aerosol Age*, **4:** 36 (May 1959).

49. Dunne, T. F.: An Improved instrument for the determination of aerosol chilling effects, *Aerosol Age*, **4:** 52 (December 1959).

50. Broderick, G. F. and Flanner, L.: Chilling effects of aerosol sprays, *Soap*, **41:** 98 (June 1965).

Bibliographical Note. The reader is referred to the rich literature published in the annual *Proceedings of the Chemical Specialties Manufacturers Association.*

Chapter 27

AEROSOL HAIR PRODUCTS

Stephen Shernov

In 1947 and 1948 two patents were granted to Fiero in Argentina and Panama respectively (1), describing among other products aerosol dressings and hair lacquers.

In the past decades one product described by Fiero, hair-holding sprays, earlier known as hair lacquers, has become the outstanding aerosol cosmetic. Hair-holding sprays, referred to nowadays as hair sprays, have grown to such a leading position that for several years the volume has more than doubled that of any other aerosol and is one-fourth of all aerosols produced.

The prognosticators of the cosmetic industry believe that sales of hair sprays will continue to grow, as evidenced by their estimates for 1975—a total volume of 824 million units, representing an increase of 97 % over a nine-year span. The amount of hair spray used in the coming years will depend on the increased acceptance of hair spray by young girls and older women, and also on hair styles, of which no prediction can be made.

Another segment of the hair spray market has been developing; that of sprays for men. According to one source (2), this market was initiated when men borrowed cans of hair spray from their wives, sisters, or girl friends. Men started to use hair spray and found that it held their hair in place without a "plastered-down look." Furthermore, with the longer hair styles for men, they are finding that hair spray helps to prevent a "fly-away look," a problem common to women. Men have been attracted by the ease of application, so that hair spray aimed directly at males is expected to have an impact in the near future.

Some of the other aerosolized hair products are shampoos, which, when first produced, proved to be corrosive in cans and had to be withdrawn. Several years later, noncorrosive shampoos were formulated and merchandised, but have only a small part of the total shampoo market.

Aerosol hair conditioners, setting lotions, cold waves, and men's hair-dressing products have not yet enjoyed any significant part of the aerosol hair products market. No real volume can be published at this time on the amount of units produced, but as the public is made more aware of their function and availability, these products may make a considerable contribution to the aerosol hair products market.

Hair Sprays

The pressurized hair spray, as a rule, is an alcoholic solution of film-producing resins, together with an appropriate propellant, usually packaged in a tin can. When sprayed on the hair, the product forms droplets of resin, which, when dry, imparts support and stiffening properties to the individual hair fibers. Brookins (3) has shown that these droplets work primarily by forming junctions between adjacent or intersecting hair fibers, to yield a rigid network.

There are several different types of film formers or resins. One of the most widely used is polyvinylpyrrolidone, better known as PVP (4). The patent covering this product relates to a film-forming composition, and more particularly to a sprayable film-forming composition, which is nonflammable and nonalcoholic, and which is suitable for therapeutic purposes.

PVP is available in several viscosity grades, ranging from very low to high molecular weights, and is soluble in aqueous and organic solvent systems. The resin can also be utilized with vinyl acetate as a copolymer series, known as PVP/VA (5). This series ranges from the E-335 to the E-735 viscosity. The higher the E series, the more viscous the material, because the higher the E series the greater the PVP/vinyl acetate ratio. The PVP resins are hydrophilic, tending to take up moisture and becoming dull and tacky. Hydrophobic properties can be imparted by the addition of shellac, ethyl cellulose, and certain other synthetic resins. (Not more than 10% of shellac can be added to PVP without causing instability.) When attempting to impart water repellency to the PVP film by the use of shellac or ethyl cellulose, it is useful to include a coupling agent. The best agent used is benzyl alcohol, but the amount is critical; too much leads to a tacky film, and too little to an opaque whiteness.

Formula 1 represents a hair spray with considerable hydrophobic properties. It employs PVP in association with a small amount of shellac. Formula 2 makes use of the PVP/VA copolymer, and Formula 3 of PVP itself, without shellac or copolymer.

Shellac

Shellac, as already indicated in Formula 1, is a resin used in hair sprays. It is an excellent fixative but is water-insoluble. Its use, therefore, presents a

Formula 1

PVP, K-30	4.0%
Shellac, dewaxed	0.4
Benzyl alcohol	0.1
Perfume oil	0.2
Ethyl alcohol, anhydrous	95.3
Above concentrate	40.0%
Propellants 12/11, 50:50	60.0

Formula 2

PVP/VA E735	2.0%
Citroflex A-2	0.2
Perfume oil	0.2
Ethyl alcohol, anhydrous	97.6
Above concentrate	40.0%
Propellants 12/11, 60:40	60.0

Formula 3

PVP, K-30	3.0%
Ethoxylan 100	0.3
Perfume oil	0.2
Ethyl alcohol, anhydrous	96.5
Above concentrate	45.0%
Propellants 12/11, 65:35	55.0

problem in its removal from the hair. Since shellac is nonhygroscopic over a large pH range, shampooing will not solubilize it. This difficulty existed for many years, but was accepted by women until other resins, like PVP, came into vogue.

"Shellac" is a word used in industry to refer to all forms of processed lac. Lac is a unique natural resin, the only commercial resin of animal origin (6). It is an organic amorphous solid. It is the secretion of the female insect, *Laccifer Lacca* (family *Coccidea*), which is parasitic on certain trees and bushes in India, Burma, and Thailand. The secreted resin functions as a protective covering for the female insect's offspring.

The shellacs used by the industry can be broken down into two main product types—orange shellac and bleached shellac. Orange shellac is manufactured from seed lac, the resinous secretion of the lac insect scraped from the host trees. In producing the orange shellac, seed lac is purified by various heat or solvent-extraction processes to yield shellacs in a range of colors from natural orange to very yellow. Bleached shellac is manufactured from either seed lac or orange shellac, by chemical bleaching processes, to yield resins of light color.

Both orange and bleached shellac are available in two main grades: wax, containing 5 % natural shellac wax, and wax-free, filtered to remove all traces of wax.

In aerosol hair sprays, it is advisable to use a shellac having somewhat different properties from those usually found in bleached shellac. It has been reported that tests supported by actual use for many years have shown that there are special grades of imported orange dewaxed shellac best suited for aerosol. This dewaxed or "superblond" product has stability properties that cannot be obtained in a chemically bleached shellac. It is decolorized by activated carbon, and is free of chlorine, a substance which has caused perfume breakdown and can corrosion. Another stability factor is low moisture content, which would aid shelf life in alcoholic solutions. Formulas 4 and 5 indicate aerosol hair sprays based on dewaxed shellac.

Formula 4

Shellac, dewaxed	5.0%
Castor oil	2.0
Propylene glycol	1.5
Perfume oil	0.2
Ethyl alcohol, anhydrous	91.3
Above concentrate	40.0%
Propellants 12/11, 50:50	60.0

Formula 5

Shellac, dewaxed	8.0%
I.P.M.	1.5
Cetyl alcohol	0.5
Alcohol 40, anhydrous	90.0
Above concentrate	45.0%
Propellants 12/11, 60:40	55.0%

"Hard-to-Hold," "Superhold," and "Extra Firm" Hair Sprays

In the past several years a new type of hair spray product has been merchandised. This product is sold as a companion to the standard PVP hair spray. Often the same trade name is used as the standard PVP spray, except that the words "superhold" or "extra firm" are imprinted on the container. These words do not imply that the product is initially harder-feeling than the standard hair spray on the hair, but rather that it is less affected by high humidity.

A procedure to determine "hard-to-hold" properties is based on a method of determining percent curl retention at high humidity. The hair swatch sample is prepared by weighing out 3.0 ± 0.1 g of a 14-in. hair tress and binding it tightly at one end with a string. The swatch is shampooed before each test, rinsed, and damp-dried with paper towels. Drying is completed with a hair dryer from a distance of 12 in. for 15 min and then in an oven at 40°C for 15 min. The swatch is then combed to smooth out tangled hair and labeled.

The swatch is sprayed for 10 sec with the aerosol hair spray from a distance of 12 in. by moving the unit up and down the length and on both sides of the swatch. The swatch is combed twice, rolled on an open mesh hair roller, and secured with the roller clip. The curled swatch is dried with a hair dryer from a distance of 6 in. for 15 min.

After drying, the hair is removed from the curler and the distance from the tie to the bottom of the curl (L_1) is immediately recorded. The suspended swatch is placed in the constant humidity chamber at 96 to 98% relative humidity, 78°F, for 10 min. It is then removed and the distance from the tie to the bottom of the curl (L_2) is recorded. Percent curl retention is calculated as follows:

$$\% \text{ curl retention} = 100 \frac{L_2 - L_1 \times 100}{L_1}$$

The swatch should also be evaluated in terms of the following for all types of hair sprays:

Combing, judged by the manner in which a comb passes through the tress, the absence or presence of snarls and frizzing.
Feel, evaluated by touch and described as soft and silky to coarse and dry.
Sheen and luster, determined by comparison with a control.
Flaking, judged by the amount of resin flaking deposited during two combings.
Static charge, determined by combing the swatch several times with a clean and dry comb to see if the individual hairs have a tendency to repel each other.

Specific "Hard to Hold" Resins

Carboxylated vinyl acetate resins.   One resin series used in "hard-to-hold" formulas is the carboxylated vinyl acetate copolymers, soluble in anhydrous ethyl alcohol (7), and available under the trade name of Resyn 28-1310. By neutralizing the resin, a wider range of water solubility can be achieved. The usual neutralizer used in an amino hydroxy compound, such as AMPD (2-amino-2-methyl-1,3-propanediol). The copolymer is readily modified by reaction with AMPD, to form a resin which can be water-dispersible or water-soluble. Changes in the film properties of this resin, such as hygroscopicity and water solubility, can be controlled by varying the amount of amino hydroxy compound. A method for the determination of the percentage of neutralization is shown below:

$$\text{parts by weight (g) of base (AMPD) required} = \frac{A \times 1.16 \times B \times C}{1000}$$

where

A = parts by weight (g) of Resyn 28-1310 used
B = equivalent weight of base used
C = neutralization required (decimal)

The carboxylated resin is readily compatible with shellac and proteinaceous compounds such as Nibin and Polypeptide AAS, but is not compatible with the PVP resins. Together they will form aggregates and precipitate. A product based on Resyn 28-1310 is shown in Formula 6.

Formula 6

Resyn 28-1310	3.51%
AMPD	0.32
Perfume oil	0.40
Ethyl alcohol, anhydrous	95.77
Above concentrate	40.00%
Propellants 12/11, 50:50	60.00

Alkyl monoester resins. Another resin used in "superhold" products is the alkyl monoesters of poly(methyl vinyl ether/maleic acid), sold under the trade name of Gantrez. The structure is

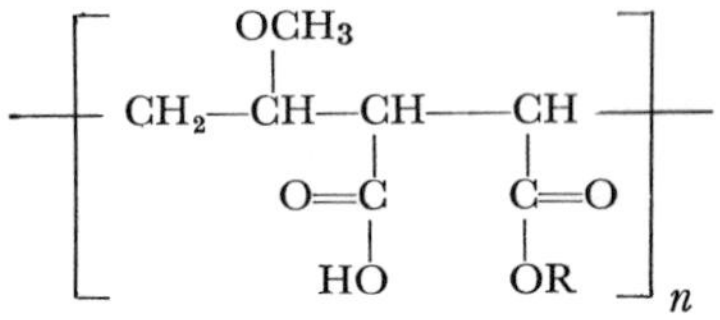

where R is an alkyl group, such as ethyl, isopropyl, or butyl.

The resins of this series are supplied in alcoholic solutions. The polymers are readily miscible with esters, ketones, glycol ethers, and alcohol. They are readily incorporated in aqueous and organic solvents and propellant systems. The resins are supplied in two forms, ethyl ester in ethyl alcohol or isopropanol, and butyl ester in the same alcohols. Just as the carboxylated resins are neutralized for greater water solubility or dispersibility, so are the monoester resins. The neutralizing agents can be 2-amino-2-methylpropanol (AMP), 2-amino-2-methyl-1,3-propanediol (AMPD), or triisopropanol-amine (TIPA), which neutralize the unesterified carboxylic acid moiety of the monoester resin (8–10).

A method for determining the stoichiometric neutralization for the mono-ester resins follows:

$$\text{(g neut. agent/g ester solids)} = \frac{(1\% \text{ stoich. neut.}) \times (\text{eq. wt. neut. agent})}{100 \times (\text{eq. wt. ester})}$$

Eq. wt. neut. agent	Eq. wt. ester
TIPA = 191	ES-225 = 202
AMPD = 105	ES-335-1 = 216
AMP = 89	ES-425 (ES-435) = 230
NH_4OH = 35	

A product based on a monoester resin is shown in Formula 7.

Formula 7

Monoester resin	3.0%
TIPA	0.2
Perfume oil	0.2
Citroflex A-2	0.3
Ethyl alcohol, anhydrous	96.3
Above concentrate	45%
Propellants 12/11, 60:40	55

Acrylic copolymer resins. The acrylic copolymer resins known under the trade name of the Dicrylans are good hair fixatives (11–14). They do not become tacky under high humidity conditions, nor do they lose their gloss. There is good curl retention under high humidity; yet the film is completely removed with water, making this resin desirable for extra-firm-holding hair spray formulations. Furthermore, this resin offers the advantage of not being subject to neutralization. The copolymer resin is supplied in 50% alcoholic solutions, in two forms, one recommended for aqueous alcoholic solutions, the other for nonaqueous. An example of the anhydrous product is given in Formula 8.

Formula 8

Copolymer resin, 50%	7.000%
Acetulan	0.335
Perfume oil	0.200
Ethyl alcohol, anhydrous	92.465
Above concentrate	40.000%
Propellants 12/11, 50:50	60.000

Plasticizers

Plasticizers (or plasterizers, as they are sometimes called) are liquids that, when added to resins, impart flexibility or resiliency to the resin film. Without plasticizers the resin is more likely to crack because of the movement of the hair, giving the appearance of flaking dandruff. Plasticizers are usually nonvolatile liquids solubilized in the resin phase and coating the resin molecules, thus increasing the elasticity and decreasing the brittleness.

The plasticizer most commonly used in earlier hair spray formulations was dibutyl phthalate. Today there are many more plasticizing chemicals available. Among those used with PVP, shellac, and the carboxylated resins are acetyl triethyl citrate, acetyl tributyl citrate, and tributyl citrate, as well as some of the alcohols, such as dipropylene glycol and benzyl alcohol.

Some aerosol formulators use the nonvolatile components of the perfume oil to soften or plasticize the resin. Plasticizers give the formulator the opportunity to adjust the particular resin from a stiff to a soft film. A plasticizer may or may not be incorporated into a hair spray formulation, depending on the characteristics of the hair spray desired.

Federal Government Regulations on Labeling

The following are the three government requirements for labeling hair sprays, as completely described in the Chemical Specialty Manufacturing Aerosol Guide Book. They are based on testing procedures for determining flammability of self-pressurized containers.

1. Flame extension
   a. Nonflammable. Flame projection under 8 in. at full valve opening.
   b. Combustible. Flame projection over 8 in. but under 18 in. at valve opening.
   c. Flammable. Flame projection of over 18 in. or a full flashback of the flame to the actuator when container is at full valve opening.
2. Tap open-cup test
   a. Nonflammable. The contents of the aerosol having a flash point above 80°F.
   b. Flammable. The contents of the aerosol having a flash point between 20 and 80°F.
   c. Extremely flammable. The contents of the aerosol having a flash point below 20°F.
3. Closed drum test. The standards for classification of this test are that the product is rated flammable if, when sprayed through each specified opening in the drum (the other two openings being closed), an explosion

or rapid burning of the vapor air mixture present is sufficient to cause the hinged cover to move in a time of 60 sec or less. It is also required that handling, storage and disposal information be printed on the label.

Nonfluorinated Propellants

To lower the cost of the propellant mixture, hydrocarbons such as iso-butane, *n*-butane, *n*-pentane, and *n*-hexane have been mixed with the fluorinated propellants. Even though the hydrocarbons are flammable, by mixing the hydrocarbons with the fluorinated propellant, nonflammable blends can be achieved. One such propellant mixture being used is Propellant A, a blend of Propellant 12, isobutane, and Propellant 11 (45:10:45). This mixture will pass open-cup tests for flammability because the nonflammable molecules of P-12 and P-11 surround the flammable molecules of the iso-butane, producing what is called a head-and-tail mixture; the head is the P-12 coming off first, and then come the isobutane and finally the P-11, producing the tail. Other blends can be produced to pass flammability tests by using this principle. A product utilizing this mixture is made with 40% hair spray concentrate and 60% of Propellant A.

Nonfluorinated halogenated compounds, such as vinyl chloride, methylene chloride, and methyl chloroform are also used. Vinyl chloride is usually mixed with P-12 and P-11, producing the same type of mixture as Propellant A. It also can be mixed with P-12 alone, at a ratio of 80% by weight of P-12 and 20% by weight of vinyl chloride, to produce a nonflammable blend. Methylene chloride and methyl chloroform are nonflammable and have been used in place of P-11 in P-12/11 mixtures. They act as vapor depressants, as does P-11, but cost about $0.06/lb less than P-11.

Methylene chloride and methyl chloroform are also used in combination with nitrous oxide, with or without P-12. When formulating with a compressed gas, enough gas must be placed in the container to prevent an excessive pressure drop, in order to avoid producing wet, streamy sprays when 50% or more of the can is empty. Formulas 9 to 11 illustrate the use of these various propellants.

Formula 9

Hair spray concentrate	40%
P-12/P-11/Vinyl chloride, 20:45:35	60

Formula 10

Methylene chloride	15%
Methyl chloroform	10
Hair spray concentrate	50
Propellant 12	25

Formula 11

Methylene chloride	16%
Methyl chloroform	30
Hair spray concentrate	50
Nitrous oxide (80 psig)	4

Water-Based Hair Sprays

Another solvent for the resin in a hair spray is water, instead of the conventional alcoholic solutions. One such system, covered by a patent, is a three-phase aerosol (15). The hydrocarbon propellant produces a stratified layer on the top of the water concentrate. This type of system has certain inherent problems. The product cannot be shaken because this would produce an uneven dispersion of propellant in the concentrate, resulting in an irregular spray pattern. In certain instances all the contents will not empty. Also a vapor tap valve must be used to ensure that the stratified layer of propellant will be emptied at the same time as the concentrate. If drying time is important, ethyl alcohol can be mixed with water.

Another method of using water employs a temporary emulsion formed by the addition of surfactants and shaking the hydrocarbon propellant into the aqueous concentrate. This method is better than the three-phase system because more uniform spray patterns can be achieved. There is no necessity for a vapor tap valve under these conditions. This is illustrated in Formula 12, an emulsion type.

Formula 12

Ethyl alcohol 40	20.0%
Water, deionized	42.8
Hair spray resin	1.5
Emcol 14	0.5
Perfume oil	0.2
Isobutane	35.0

Using an extremely flammable propellant such'as dimethyl ether, two-phase systems can be attained with ratios as high as 35 % by weight of the ether and 65 % by weight of water. The spray emitted is nonflammable and nonfoaming. This type of product is covered by a United States patent (16). It is described in Formula 13.

Hair Spray Additives

Additives have been added to hair spray because of their apparent beneficial effect on hair. Such materials as proteins and lanolin and its derivatives

Formula 13

PVP/VA E735	2.0%
Dimethyl phthalate	0.2
Silicone (G.E. SF 1066)	0.1
Water, deionized	97.7
Above concentrate	70%
Dimethyl ether	30

have been used. Proteins in the form of polypeptides are available in anhydrous alcohol. It has been claimed that proteins aid damaged hair.

Lanolin and lanolin derivatives have been placed in hair spray formulations to give added sheen to the hair.

Hair Conditioners

When hair is overteased or bleached, a conditioner is needed to bring back its soft feel and luster. Conditioners used in aerosols are long-chain fatty esters and lanolin derivatives. The conditioning agent can be carried in a spray or a foam-producing formulation.

One such foam-producing formula (17) relates to a self-propelling homogeneous liquid formulation which, when dispensed in a suitable aerosol container, produces foams of limited stability. This formulation is made of a saturated aliphatic monohydric alcohol containing one to three carbon atoms, preferably ethyl alcohol, water, a surfactant, and a propellant in certain critical proportions. The product is shown in Formula 14, a foam, whereas a spray hair conditioner is shown in Formula 15.

Formula 14

Part A

Surfactant	2.0%
Solulan 16	3.0
Ucon 50 HB-5100	1.0
Diisopropyl sebacate	1.3
Ethyl alcohol, anhydrous	50.0

Part B

Water deionized	42.4

Part C

Perfume oil	0.3

Procedure: Heat part A until solids are dissolved. Heat part B and add to part A with moderate stirring. Add part C after the A and B mixture is at ambient temperature with stirring.

Above concentrate	92%
Propellants 12/114, 30:70	8

Formula 15

Acetulan	25.0%
Isopropyl myristate	46.5
Dipropylene glycol	28.2
Perfume oil	0.3
Above concentrate	10%
Propellants 12/11, 40:60	90

Aerosol Shampoos

Besides hair spray, one of the early aerosol products was shampoos. These aerosol shampoos had poor shelf life. The cans corroded within months after production.

A patent relating to a composition of matter suitable for use as a shampoo and, more particularly, to a pressurized liquid shampoo for employment in an aerosol, has been described (18). The formulation comprises the ammonium salts of sulfated monoglycerides of higher fatty acids as the detergent, a liquefied gaseous organic propellant, and a minor proportion of a suitable saturated higher fatty alcohol in an aqueous, lower alcoholic medium. The product is shown in Formula 16.

Formula 16

Ammonium salt of monoglyceride sulfate of coconut oil	35.0%
Perfume oil	0.7
Lanolin	0.8
Cetyl alcohol	6.0
Isopropanol	17.0
Water, deionized	37.0
Ammonium sulfate	3.5

Procedure: Heat the ammonium salt together with the cetyl alcohol, lanolin, and isopropanol. Add the ammonium sulfate to the water. Mix until dissolved. Heat the water and sulfate mixture. Add the coconut fatty mixture to the water mixture with stirring. Cool and add the perfume oil.

Above concentrate	88%
Propellants 12/114, 60:40	12

Dry Shampoos

Recently several companies have marketed dry shampoo products. Many of these products contain "vulca," a cross-linked corn starch material. The material is micronized to 5 μ and is suspended in a propellant blend similar to a body talc formulation. Isopropyl myristate is added for valve lubrication. The product is sprayed on the hair, leaving a deposit of vulca powder, and is brushed out, removing any hair spray and dirt. A typical product is shown in Formula 17.

Formula 17

Vulca 90	10.0
Isopropyl myristate	0.5
Propellants 12/11, 50:50	89.5

Cold Waves

An aerosol cold wave has been patented, with claims for an improved method for the cold permanent waving of the hair packaged in pressurized containers (19). The cold waving is accomplished by coating the hair with some chemical composition which will soften and plasticize the hair at room temperature. Such chemicals reduce the keratin structure of the hair, thereby breaking disulfide linkages in the hair fiber. The patent also relates to attempts that have been made in eliminating the contact of the hands with the wave lotion, which is alkaline with a pH above 9. Cold-wave solutions may cause irritation when repeatedly brought into direct contact with the skin of sensitive individuals.

The volatility and susceptibility to oxidation of the active ingredients in many of the conventional hair-waving lotions make the ordinary methods of commercial packaging inappropriate for a container of lotion which can be used and stored repeatedly without loss of activity. The aerosol wave lotion is hermetically sealed, giving the product better stability by lessening oxidation. A product is described in Formula 18.

Formula 18

Thioglycolic acid (as monoethanol-ammonium salt)	5.0%
Dithiodiglycolic acid (as monoethanol-ammonium salt)	2.0
Perfume oil (stable at above pH 9)	0.5
Ethyl alcohol	60.0
Monoethanolamine (adjust to pH 9.3)	2.2
Water, deionized	30.3

Procedure: Add the thioglycolic and dithiodiglycolic acids to water. Mix until uniform. Add the perfume oil to the ethyl alcohol. Mix until uniform. Add the ethyl alcohol mix to the water with stirring. Add the monoethanolamine to the mixture three with stirring.

Above concentrate	90%
Propellants 12/114, 15:85	10

Brilliantines

The brilliantine-type formulas have been marketed mainly for men. The product places a thin layer of oil on the hair. Aerosol brilliantines containing mineral oil have been removed from the market because there have been cases of pneumonia attributed to the inhalation of the mineral oil spray. Dipropylene glycol and isopropylan can be used in place of mineral oil, as shown in Formula 19.

Formula 19

Dipropylene glycol	44.85%
Isopropylan 50	44.85
Perfume oil	0.30
Ethyl alcohol, anhydrous	10.00
Above concentrate	10%
Propellants 12/11, 40:60	90

Hair-Setting Lotions

Aerosol hair-setting lotions contain a large amount of hair spray resin, usually in an aqueous alcoholic system, for the express purpose of forming a hard film on the hair before it is combed out. The setting lotion differs from the ordinary hair spray in that the hair spray holds the final style and the setting lotion adds body to the hair before styling. It is usually applied after shampooing, either as a wet spray or a pressure-sensitive-type foam. Other materials can be added to form dual-purpose products, such as cationic material to produce a rinse as an aid in combability to remove snarls.

If a foam is desired, it could be patterned after the previously mentioned Klausner patent (17), which would contain a surfactant usually nonionic, but some anionic agents have been used with good results in an alcohol-and-water mixture. The ratio of alcohol to water would depend on how fast or slow one wishes the foam to break, the more alcohol the faster, and the more water the slower.

The active ingredients are also carried with 4% by weight of a propellant.

Spray and foam products are shown in Formulas 20 and 21, respectively, and a rinse-and-set product in Formula 22.

Formula 20

PVP/K30	4.0%
Water, deionized	15.0
Ethyl alcohol 40	80.7
Perfume oil	0.3
Above concentrate	65%
Propellant 12	35

Formula 21

Surfactant (nonionic)	2.0%
Water, deionized	38.0
Ethyl alcohol 40	52.8
PVP/VA 735	7.0
Perfume oil	0.2

Procedure: Mix surfactant with water. Stir. Mix until dissolved. Mix alcohol, perfume, and resin together with stirring. Mix until uniform. Add the water mixture to the alcohol mixture with stirring. Mix until uniform.

Above concentrate	96%
Propellant 152a	4

Formula 22

Surfactant	1.5%
Water, deionized	40.0
Cationic agent	1.0
Ethyl alcohol	49.3
Resin	8.0
Perfume oil	0.2

Procedure: Mix together water, cationic agent, and surfactant with stirring until uniform. Mix alcohol, perfume, and resin together, with stirring, until uniform. Add water mixture to alcohol mixture with stirring and mix until uniform.

Above concentrate	96%
Propellant 152a	4

Hair Coloring

Aerosol hair-coloring products have appeared on the market in two forms, as an insoluble metallic coating and as a solubilized certified color carried in a pressure-sensitive foam.

This type of product is described in a patent (20) that relates to a composition capable of forming a relatively stable foam with a composition suitable as a carrier for the dyes. The foam leaves little residue, reducing the necessity for rinsing the hair following use.

Since contact with skin or scalp of the user is undesirable, use of a pressure-sensitive foam makes application easier and lessens staining to the skin. Formula 23 illustrates such a product.

Formula 23

Sulfated ethanolamide of coconut	2.00%
Fatty acids	
Sodium lauryl sulfate	1.00
Sodium stearate	1.90
Fast Black BB (D&C Black No. 1,	
Color Index #307)	0.01
Water, deionized	89.09
Propellant 12	6.00

Procedure: Mix together water, sodium stearate, and sulfate with stirring until dissolved. Heat the water mixture to 65°C. Heat the coconut fatty acid to 65°C. Add the fatty acid to the heated water mixture with moderate stirring. Add the dye material after the mixture has cooled to 40°C. Mix until uniform.

REFERENCES

1. Fiero, G. W.: Arg. Pat. 59,548 (1947); Panama Pat. 157 (1948).
2. A Consumer survey on hair spray. Freon Products Division, E. I. duPont de Nemours and Co., Inc.
3. Brookins, M. G.: The action of hair sprays on hair, *JSCC*, **16**: 309 (1965).
4. Stoner, G. G.: U. S. Pat. 3,073,794 (1963).
5. Witmer, D. B.: U. S. Pat. 3,171,784 (1965).
6. Martin, J. W.: Shellac in hair sprays, *DCI*, **94**: 828 (1964).
7. Reiter, R. W., and Horning, R. G.: U. S. Pat. 2,996,471 (1961).
8. Coons, R. J.: Can. Pat. 694,866 (1964).
9. Johnson, J. H.: U. S. Pat. 2,913,437 (1959).
10. Mills, C. L.: U. S. Pat. 2,957,838 (1960).
11. Root, M. J., and Bohac, S.: A new terpolymer for hair sprays, *Proc Sci. Sect. TGA*, **42**: 32 (1964).
12. Maeder, A.: U. S. Pat. 2,897,172 (1959).
13. Maeder, A.: U. S. Pat. 3,025,219 (1962).
14. Maeder, A.: U. S. Pat. 3,257,281 (1966).
15. Clapp, P.: U. S. Pat. 2,995,278 (1962).
16. Presant, F., and Carrion, C., Jr.: U. S. Pat. 3,207,386 (1965).
17. Klausner, K.: U. S. Pat. 3,131,152 (1964).
18. Allen, J. M., and Scott, G. V.: U. S. Pat. 2,878,231 (1959).
19. Shepard, W. W., Romney, M. E., and Moore, R. S.: U. S. Pat. 3,103,468 (1963).
20. Horn, R. H.: U. S. Pat. 3,092,555 (1963).

Chapter 28

BATH PREPARATIONS

Robert E. Sauté

The origin of the bathtub can be traced back to the ancient Egyptians, who were building baths as we know them in 1700 B.C. These "lustral" chambers, as they were called, were used for anointing the body as a part of religious rites as well as for bathing (1). The Greeks, and later the Romans, used bath preparations such as perfumed oils and unguents during and after bathing to give the skin a smooth, even color and a soothing feeling. The Greeks are credited with developing the shower in their quest for cleanliness. The Romans built luxurious baths and developed a technique for bathing. Upon entering the bath, the bather was anointed with oil and then went into a room, or court, where he indulged in violent exercise. After exercising, he proceeded to the hot room and then to the steam room. At this point he was probably scraped of the accumulation of sweat and oil with curved metal strigils. The bather then went to the warm room before going into the cold bath or swimming pool. The body was again anointed after the bath or swim, and this completed the bath ritual (2).

Wine baths were used by the ladies of the court during the reign of Mary, Queen of Scots. The wine baths were relatively expensive, and the younger ladies of the court had to resort to taking milk baths, which have enjoyed popularity at various times throughout history. Mineral water, or medicinal, baths are popular and found in most countries. Mediums other than water and steam have been used for baths. Examples of these are mud, sand, peat, and solutions of aromatic herbs.

Modern bath preparations are classified in two main groups—bathing products used in the bath water and afterbath products used on either the wet or dry skin immediately after washing. Bath products are usually formulated to serve one or more of the following purposes:

1. To clean the body by removing dirt and odor.
2. To help soften hard water.
3. To give an aesthetic effect to the bath by adding fragrance and/or color to the water.
4. To give the user a relaxed or refreshed feeling.
5. To impart an emollient effect as well as fragrance on the skin.
6. To prevent a ring from forming around the bathtub.

Soap is probably the most commonly used product in the bath today, but it is not within the scope of this chapter because it is not considered a cosmetic by various legal definitions. The technology of its manufacture is quite dissimilar from other bath preparations and is covered in another chapter of this text.

Bubble Bath

Bubble bath preparations have enjoyed a tremendous increase in popularity in the past ten years, not only in the adult market but in the children's area, with novelty packaging that is very appealing. They are designed to fill the tub with foam and fragrance, as well as to prevent hard water rings from forming around the tub. Many of the children's products claim to float the dirt away and eliminate harsh scrubbing. In any case, as long as children enjoy using them and mothers insist on children taking baths, there should be a market for novelty bubble baths.

Bubble baths are found in various physical forms: powders, granules or beads, liquids, gels, tablets, capsules, crystals, and cakes (3–5). The three most popular forms are the liquid, powder, and granule types. A good bubble bath should possess the following qualities.

1. The product should produce a copious foam in low concentrations without the use of excessive water pressure.
2. The foam should remain stable in the presence of hard or soft water, soap, soil, and a relatively wide range in temperature.
3. When the product is used as directed, it should not be irritating to the skin or mucous membrane.
4. The product should prevent the formation of a lime soap ring around the bathtub.

There are many foaming agents available on the market today that produce good foams. One of the main complaints concerning bubble bath is the instability of the foam, which is due primarily to the use of soap in the belief that the bubble baths do not cleanse the skin properly. The use of soap at the end of the bath may be advisable in that it will facilitate the removal of the foam from the tub after the bath.

Various items have been incorporated into bubble baths with claims which are not usually substantiated by convincing experimental data. Some of these are vitamins such as A, E, F, H, and D, chlorophyll, and medicinal herb extract. Others contain lanolin or lanolin derivatives which are said to decrease irritation and lubricate the skin. Most of the oil additives tend to cut down the amount of foam produced; consequently a lower active level might produce the same results in respect to irritation.

Synthetic gums, natural gums, and alkanolamides are used in bubble bath formulations to help stabilize foam and to thicken the less viscous liquids and give them more "body." Sodium carboxymethyl cellulose has been used to keep dispersed soil from redepositing on the skin or in the tub because of its ability to suspend soil. Large amounts are used to form the gel types of bubble baths.

Powdered bubble baths have become very popular because they are readily available in a finished form at a low price. The inexpensive products are spray-dried alkyl aryl sulfonates which can be obtained in various densities to fill large-size, attractive packages with hollow, spherical beads of bubble bath. These beads can be colored and/or perfumed easily, dissolve quickly, and produce good foam. The bead is 40% alkyl aryl sulfonate, and the balance is made up mainly of sodium sulfate or sodium chloride. Sodium lauryl sulfate is a good foaming agent which can be used in most areas with good results, but its effectiveness decreases in proportion to the hardness of the water. A technical grade which contains about 45% active is available in a low density powder. The sodium lauryl sulfate surfactant is more irritating than the alkyl aryl sulfonate, and this should be kept in mind when formulating the product (6).

When a luxury line of bath products is being formulated which can absorb the additional cost, sodium lauryl sulfoacetate might be used. It is also supplied as a low-density flake at a 70% active level. A series of anionic surfactants consisting of fatty acid esters of sodium isethionate are available in powder form. These are prepared from the fatty acids of coconut oil and are more compatible with soaps.

Fillers and water softeners are incorporated in powdered bubble bath to add bulk or act as carriers. Sodium sesquicarbonate, sodium hexametaphosphate, and tetrasodium pyrophosphate are used as water softeners. One of the main problems in formulating a powdered bubble bath is to keep it free-flowing and prevent it from caking. This can be done by adding tricalcium phosphate or sodium silica aluminate. The perfume oil can be added to bentonite, starch, or some other suitable absorbent to disperse it through the product. Color can also be added by premixing with one of the fillers. (See Formula 1 for spray-dried bead-type bubble baths).

The perfume can be atomized on the bead as it leaves the vacuum drier,

Formula 1. Bead Bubble Bath

Alkyl aryl sulfonate, 40% active	96%
Starch, low-moisture	2
Sodium silica aluminate	1
Perfume	1
Color	Trace

or it can be mixed with the sodium silica aluminate and starch. Color can be added to the starch if used. The color must be water-soluble. The mixer used to blend the ingredients should have a gentle action which will not break down the beads while mixing. (See Formulas 2, 3, and 4 for examples of powdered bubble baths that will give desirable results).

Formulas for Bubble Bath

	2	3	4
Sodium lauryl sulfoacetate	25%	—	30%
Sodium lauryl sulfate	—	25%	—
Sodium hexametaphosphate	7	—	—
Sodium chloride	—	70	—
Sodium carboxymethyl cellulose, low-viscosity	—	2	1
Sodium sesquicarbonate	65	—	—
Sodium tripolyphosphate	—	—	66
Perfume	3	3	3
Color	Trace	Trace	Trace

Most of the bubble bath powders can be compressed into tablets or cakes, but this does have a tendency to slow down the solubilization and, consequently, the formation of foam.

Lubricants such as starch and talc might be necessary to free the cakes or tablets from the punch and dies used to compress the cakes. It is most important to work in a controlled humidity to prevent excess sticking.

Bubble bath solutions or liquid bubble baths are very popular and require the use of only a few ingredients. The base is a concentrated solution of a high-foaming detergent. In the past, bubble bath formulas were based primarily on liquid soap and phosphates with other sequestering agents such as citrates added to keep insoluble soaps from forming in the bath. Phosphates such as those used in the powdered bubble baths can be used in the liquids as well (e.g., sodium hexametaphosphate).

It is difficult to incorporate inorganic water softeners into liquid bubble baths in that they tend to salt out the organic surfactants and do affect the stability of the fragrance oils.

Modern bubble baths are composed chiefly of synthetic surfactants, especially the good foamers, and do not necessarily have to be good detergents. Many of the good foams have a detergent action, but if they do not clean the skin as well as desired, a detergent can be added to the formulation. The objective in the development of a bubble bath is not usually to produce a cleansing preparation but merely to fill the tub with a voluminous amount of fragrant bubbles.

Some of the newer sequestering agents such as the tetrasodium salt of ethylenediamine tetraacetic acid are excellent for the complexing of calcium and magnesium ions in bath water to prevent "killing" the foam. They also have a tendency to keep solutions clear in the bottle and help to stop discoloration due to trace-metal contaminants reacting with the fragrance oils.

Triethanolamine lauryl sulfate is the detergent most often used in liquid bubble baths. It is usually available in approximately a 40 % solution. It has a low cloud point of about 0°C and a clarification point of 14°C. Sodium lauryl sulfate is not very soluble in water and starts to crystallize out of solutions at a few degrees below normal room temperature. Ammonium lauryl sulfate also has a relatively high cloud point (14°C) and does not lose its opacity until it reaches 24°C. The magnesium salt of sulfated lauryl alcohol is as soluble as the triethanolamine salt and is much more stable to light. It facilitates the use of lighter colors and maintains lighter shades better if colored solutions are to be made. Sodium lauryl ether sulfate is much more soluble in cold water than sodium lauryl sulfate and is more resistant to hard water because of the higher solubility of its magnesium and calcium salts.

The alkyl benzene sulfonate and lauryl sulfates have a tendency to be poor foamers in hard water. The insoluble calcium and magnesium soaps formed by bath soaps plus hard water kill the foam quickly, and bath soap even in soft water also causes the foam to collapse. Sulfated derivatives of alkylphenol polyglycol ethers and the sodium salt of N-methyl tauride of coconut fatty acids give good foam characteristics in hard water. One of the most popular low-cost detergents for liquid bubble baths is triethanolamine dodecyl benzene sulfonate. It is available in a 60 % active solution and has a viscosity of approximately 1500 cps. It can be diluted with up to twice its volume of water and remain quite viscous. The sodium salt is more soluble than the triethanolamine salts and lighter in color. It has one disadvantage in that it tends to form a sediment on standing in solution, but this can be overcome to some extent by adding alcohol to the preparation.

The alkylolamides are used as lime soap dispersers and viscosity builders. The diethanolamine condensates of lauric acid and coconut fatty acids are good foamers and stabilizers of foam produced by anionic detergents in the presence of soap. They do give a darker color to light-colored detergent solutions.

The concentration of a surfactant in a liquid bubble bath can range between 15 and 35% depending on the amount of perfume to be solubilized and the quantity of foam desired. Viscosity can be built up by adding sodium chloride or sodium sulfate, which reduces the solubility of the organic surfactant. Viscosity is built by varying the charge on the surface. A small amount of salt will increase the viscosity, but the further addition of salt will result in a reverse effect, which will cause total loss of viscosity owing to the common ion effect. These salts also affect the clarity of the solution as well. The judicious choice of very soluble detergents over a wide range of temperature is necessary to produce clear bubble bath solutions. If the bubble bath solution does become cloudy, it is often desirable to add a water-miscible organic solvent such as ethyl alcohol, isopropyl alcohol, or hexylene glycol to act as a cosolvent. Perfume may be added at a level of 1 to 10% depending on the fragrance oil used and results expected. (See Formulas 5 to 7 for several simple liquid bubble bath formulas.) They are easy to make in that they require the use of only a vessel and a stirrer.

Formulas for Liquid Bubble Bath

	5	6	7
Triethanolamine lauryl sulfate, 45%	18%	—	50%
Lauric diethanolamide	12	—	5
Triethanolamine dodecyl benzene sulfonate, 60%	—	15%	—
Fatty acid amine condensate	—	32	—
Coconut monoethanolamide ethoxylate	—	—	5
Propylene glycol	—	—	5
Perfume	1	2	4
Water-soluble dye	Trace	Trace	Trace
Tetrasodium ethylenediamine tetraacetic acid	—	—	Trace
Water q.s.	100	100	100

A quick and easy way to evaluate a bubble-bath or foam-producing surfactant is to place 1 mm of liquid or 1 g of powdered product in a glass-stoppered, 1-liter cylindrical graduate. Add 200 ml of 100°F water with a hardness of 300 ppm. One gram of soap chips can be added at this point or dissolved in the water before placing it in the graduate. The cylinder is then inverted 30 times in 1 min, and the foam immediately measured and recorded. The readings are taken again after 2-, 5-, and 10-min intervals to evaluate foam stability. The longer the foam lasts, the more apt the customer is to be pleased with the product. When the foam has broken, no scum should be apparent on the surface of the water.

Bath Oils

Bath oils have become very acceptable as functional products as well as for their cosmetic appeal. These products are known by many names such as beauty bath for dry skin, bath perfume, beauty oil for bath, etc. (7). They are available in sprays, liquids, capsules, towelettes, and even in dry powders or bead form. In order to facilitate the understanding of these product types, we shall define them as emollient or functional types and nonemollient or fragrance products.

The cosmetic or fragrance bath oils are usually very simple mixtures of perfume oil and isopropyl myristate or some other suitable oil which will form a clear product when mixed with the fragrance oils. This product is not intended to be used as an emollient, and directions usually recommend the use of just a few drops in a tub of water to give the bath water and bather a pleasant odor. The fragrance oil level in these products ranges from about 35 to 65%.

Soluble bath oils have been available for many years and are quite popular. These products contain large quantities of surfactants to solubilize the high fragrance oil concentrations and to dissolve or disperse the oils readily in the bath water. They leave no residue in the tub and no emollient effect on the body. Foaming soluble bath oils can be made by adding a foaming agent and a foam stabilizer. These products could be classified as bubble baths with high fragrance oil levels. (See Formulas 8 to 10 for the various types of nonemollient bath oils.)

Formulas for Nonemollient Bath Oils

	8	9	10
	Perfume Bath Oil	*Soluble Bath Oil*	*Foaming Bath Oil*
POE 20 sorbitan monopalmitate	—	5.00%	22.2%
Fragrance oil	35%	5.00	4.4
Isopropyl myristate	65	—	—
Methyl *p*-hydroxybenzoate	—	0.18	—
Propyl *p*-hydroxybenzoate	—	0.02	—
Sodium lauryl sulfate	—	—	10.0
Ninol AA-63	—	—	1.0
Sorbic acid	—	—	0.2
Water q.s.	—	100.00	100.0

Emollient bath oils make up the largest segment of the bath oil market, because of their therapeutic effect on dry skin. There are many pharmaceutical products as well as many pleasantly scented cosmetic products which are effective relievers of dry, itchy skin (8). The effectiveness of these products has been studied by several authors such as Taylor (9) and Knox (10). Taylor reported on a quantitative method for determining the amount of oil deposited on the skin when the arm is placed in a cylindrical bath. He concluded that mineral oil products adhere better on the skin than vegetable oil formulations, and oilated oatmeal preparations (colloidal oatmeal combined with mineral oil and lanolin) leave little residual oil deposited on the skin. Knox and Ogura (11) used pulverized keratin and incubated it in several different bath oil formulations. The adherence of the oils to keratin was determined by an ether extraction technique. Knox also pointed out that mineral oil preparations adhere better to keratin than vegetable oil preparations.

Another method of measuring the deposition of oils on the skin was published by Stolar (12). His method consists of dispersing 0.200 g of oil containing 4000 USP units of vitamin A palmitate in 1 liter of tap water with a Lightening mixer for 5 min. The hand is immersed in the bath, kept at 40°C, and at the end of 15 min allowed to air-dry. A picture is then taken under ultraviolet light and the continuity and film deposition are observed. The oil deposited is then eluted from the skin with mineral oil and assayed spectrophotometrically at 325 μ to determine quantitatively the amount of oil which adhered to the skin.

Becher and Courtney (13) have published on the physicochemical approach to formulating a floating bath oil. The authors point out that the spreading coefficient, surface tension, and interfacial tension all play an important role in the formulation of an effective, acceptable product. When a bath oil is placed in the bath water, it either spreads over the surface, covering a large area (W_a = work of adhesion; that aspect of the surface-free energy which maximizes the interface between two immiscible liquids), or it forms lenslike globules on the surface (W_c = work of cohesion; that aspect of the surface free energy which causes an immiscible liquid to take on a shape of minimal surface area). The difference between the two supplies the driving force for spreading or the spreading coefficient S. Spreading occurs when $S > 0$, and nonspreading when $S < 0$.

Becher and Courtney give the following equation for calculating the spreading coefficient of a bath oil:

$$S = W_a - W_c = Y_w - Y_o - Y_{ow}$$

S = spreading coefficient
Y_w = surface tension of aqueous phase
Y_0 = surface tension of oil phase
Y_{ow} = interfacial tension between the two phases

Ideally, a floating bath oil should cover the surface of the water in a monomolecular layer and be deposited on the skin in a very thin film, covering as much of the skin surface as possible. This has never been attainable in any of the current formulations on the market. A lubricating bath oil should not be deposited in a heavy, greasy layer which is nonaesthetic to the user. Nor should it leave a heavy, oily film which is hard to remove from the tub. The velocity of spreading should be considered for aesthetic reasons, because a quick spreading film will be more appealing to the consumer. The speed of spreading is directly proportional to the spreading coefficient divided by the viscosity of the liquid on which the oil is spread. Since the substrate in this case is water, it is apparent that a spreading coefficient of approximately +45 would be optimum. A relatively simple method can be used to observe spreadability and velocity of bath oils. A pan about 12 × 12 in. containing water at 50°C is dusted with starch. Ten drops of bath oil are dropped into the center of the pan. The oil will move the starch away from the center to the edge of the oil surface, showing both the area and velocity quite clearly.

The spreading coefficient increases in magnitude with increasing HLB (10,14,15). It would appear that the use of the highest HLB would be the best choice of surfactant for a floating bath oil, but there are other factors to be considered, such as the solubility and compatibility. For example, if the surfactant were too highly water-soluble, it might be removed from the bath oil and cause a well-dispersed film to become a nonaesthetic group of oil globules by the end of the bath. The temperature of the bath also plays a role in the spreadability of the floating bath oils. A temperature change of 20°C can change a spreading bath oil to a nonspreading bath oil. Soap in the water will affect the spreading coefficient by lowering it.

There are many oils used to formulate the emollient-type bath oil, such as high- and low-viscosity mineral oils, vegetable oils such as olive oil, cottonseed oil, corn oil, peanut oil, safflower oil, peach kernel oil, etc. Lanolin and lanolin derivatives as well as fatty acids, fatty alcohols, and their esters are used to give better penetration of the oils into or on the skin and give an emollient effect without producing a greasy feeling on the skin (16). The two most commonly found materials in bath oils are mineral oil and isopropyl myristate. The bath oils should be formulated to produce a clear liquid, stable to light and pleasantly scented. (See Formulas 11 and 12 for formulas that will produce good bath oils which will be ideal from a spreading coefficient standpoint. They can be prepared easily by simply mixing the ingredients together.)

Another type of emollient bath oil is called the "instant bloom" bath oil. This product has a rich milklike appearance when added to the bath water. Instant bloom types are sometimes preferred to the floating type because they form a uniformly dispersed emollient and fragrance throughout the

Formulas for Floating Bath Oils

	11	12
Mineral oil	50%	40%
Isopropyl myristate	48	50
Liquid lanolin	—	5
Arlatone T	1	1
Perfume	1	4
Color	Trace	Trace

bath water. The uniform dispersion is said to give a thorough contact with the bather's body. These oils also give the bather a visual sense of rich, high emolliency when the oil is poured into the water, and they leave little to no oily ring in the tub. The oil-soluble, nonionic emulsifiers are surfactants of choice for these dispersible bath oils.

Polyoxyethylene-(2)-oleyl ether is an oil-soluble surfactant, having an HLB value of 4.9. It is usually selected for making water-in-oil emulsions as opposed to oil-in-water emulsions. When used in a bath oil where small amounts of water are used, it forms a W/O-type emulsion, and when added to the bath where large amounts of water are present, it inverts to an O/W emulsion. Just the agitation caused by the incoming water from the faucet and the movements of the bather will form a relatively stable O/W emulsion. (See Formula 13 for a typical formulation.) A little heat is needed to facilitate the simple mixing necessary to prepare this formulation.

Formula 13.   *Instant Bloom Bath Oil*

Mineral oil	72%
Isopropyl myristate	20
Brij 93	5
Perfume	3

Procedure: Mix ingredients, allow to stand, and filter.

A more novel type of emollient bath oil is the foaming type, which does not currently enjoy the sales volume of the two previous types. This type differs from the floating and instant bloom types in that it forms a very stable bloom in the bath water while yielding an acceptable amount of foam. The same emollient oils can be used for this formula as those used in the other emollient types. (See Formula 14 for a typical formulation.) A little heat is needed to facilitate the simple mixing necessary to prepare this formulation.

The selection of a fragrance for bath oils is very important. The fragrance oils should be soluble in the particular system intended for use. A high-quality fragrance is well worth the extra cost, for many products are sold

Formula 14. Foaming Emollient Bath Oil

Solar 25	20%
Mineral oil, white	60
Brij 93	13
Ninol AA-63 Extra	2
Perfume	5

because of their pleasant fragrance as well as their emollient properties. Medicinal products are usually not as highly scented, and frequently simple hypoallergenic fragrances are used.

Other emollient bath products are similar in formulation to the bath oils but dispensed or applied in a different way. Some emollient bath oils are applied to the face cloth after a shower and applied directly to the skin. Others are used to saturate paper towels, and these are packaged in foil packs for individual use. A new form of bath oil application is an attachment to the shower head containing the oil. Water mixes with a small amount of oil in the shower head and is sprayed on the bather.

Bath oils packaged in aerosol containers have been promoted for use after showers. After-shower foams dispensed from aerosol containers produce slow- or quick-breaking foams which impart various degrees of emollient effects when applied after the bath. (Examples of aerosolized bath oils are given in Formulas 15 and 16.)

Formula 15. Bath Oil Spray

Concentrate	
Isopropyl myristate	4.0%
Mineral oil, light	4.0
Solulan 16	1.0
Perfume	0.5
Ethyl alcohol, denatured,	
anhydrous	90.5
Aerosol formulation	
Concentrate	60.0%
Propellant 12	40.0

Procedure: Mix the concentrate ingredients thoroughly in the alcohol. Heat slightly if necessary.

The powdered emollient bath products are usually made by mixing various oils with oatmeal. Other suitable carriers might be used, but there is a residual in the tub, and a problem might arise from blocked drains if one is not careful. Bath oils are sometimes encapsulated in gelatin capsules. The capsules have a tendency to dissolve slowly and usually do not dissolve completely during a normal bath. The temperature of the water, the thickness of

Formula 16.   Quick-Breaking Foam Bath Oil

Concentrate
Polawax	1.5%
Lantrol	1.0
Water, deionized	37.0
Perfume	0.5
Ethyl alcohol, denatured	60.0

Procedure: Heat the Polawax and Lantrol in the alcohol to dissolve the solids. Heat the water and add to the alcohol mixture with constant stirring until uniform.

Aerosol formulation
Concentrate	92.0%
Propellants 12/114	8.0

the capsule wall, and the hardness of the gelatin have an effect on the solubility of the capsule shell. Special equipment is needed to make these soft gelatin capsules, and the bath oil must be anhydrous not to dissolve the gelatin wall while in the package.

Bath Powders

Bath powders have been in existence for some time and are usually sold as talcs, beauty dusts, emollient bath powders, etc., usually in elaborate containers with equally elaborate puffs or mitts. Aerosolized bath powders are a relatively new entry into the field. Bath powders were sold primarily for women, but in today's market there are several products being sold in boxes with puffs for men. Bath powders serve two functions: to apply a fragrance to the skin, and to supply a lubricant which will absorb some moisture, facilitate dressing, and prevent chafe.

The chief ingredient in bath powders today is talc. Many other ingredients have been used, such as corn starch, rice starch, boric acid, chalk, etc. The addition of materials such as magnesium carbonate and kaolin can be used to absorb the perfume oil prior to mixing it with the talc. They will also help hold the fragrance in the finished product. Zinc stearate is used to give slip to the finished product. Zinc oxide and chalk are used to make heavy-density powders, whereas Cabosil (Cabot Corporation), a fumed silica, may be used to give the finished product a lighter density.

Talc is used in a high concentration, and it is important to use a good grade of talc which has a low odor level. Frequently the odor is bad enough to warrant the use of a small amount of musklike fixatives to cover the earthy odor prior to using the talc in a finished powder. A small amount of titanium dioxide can be used to give the powder some opacity. If color is desired, the earth oxide colors should be used.

Cosmetic-grade iron oxide will give a light pink color and ochre, a light tan. Various mixtures of these two iron oxide colors will produce various flesh-colored shades. The proper way to incorporate these colors is to make color extenders with them. This is done by using about 25 % of the dry colors with approximately 75 % talc and grinding the ingredients together until they produce a uniform color and do not develop a more intense shade after additional grinding or micropulverizing. A small amount can be ground with a mortar and pestle or in a Waring-type blender, then compared to the remainder of the extender to determine whether or not the color is fully developed. These extenders are then added to large batches of powders and are dispersed quickly with a minimum mixing time with better color reproducibility from batch to batch. (See Formulas 17 and 18 for a heavy and a light bath powder. See Formulas 19 and 20 for a lubricating and an emollient bath powder, respectively.)

Bath Powder Formulas

	17 Heavy	18 Light
Talc	70%	90%
Magnesium carbonate	5	4
Kaolin	7	—
Zinc stearate	4	4
Zinc oxide	2	—
Chalk, heavy	11	—
Cabosil	—	1
Perfume	1	1
Color and extender	q.s.	q.s.

	19 Lubricating	20 Emollient
Talc	85%	86%
Calcium carbonate	5	5
Zinc stearate	9	6
Perfume	1	1
Isopropyl myristate	—	2

Procedure: A general procedure for mixing various bath powders: Place all the dry ingredients, except the magnesium carbonate, in a powder mixer and mix until uniform. If a fixative is to be added to overcome the base odor, it can be added at this time, dissolved in alcohol. It can be sprayed on or mixed with the powders until uniform. The oils can be dissolved in alcohol or sprayed on in a fine mist with constant mixing until the liquids are well dispersed. The fragrance oil may also be mixed with the magnesium carbonate in a separate container and sifted into the batch while mixing until the perfume is mixed in thoroughly. Color can be added at this point and again mixed until the color is completely developed. The thoroughly mixed powder can be micropulverized or passed through a bolting cloth prior to packaging.

A new entry into the bath powder market is the aerosolized powder. Like conventional bath powders, it is used as a lubricant and means of applying fragrance. These products must be formulated to prevent excess dusting, that is, to prevent powder from being deposited all over the bathroom, and yet must not cake on the skin when applied. The particle size of the ingredients must be such that they will pass through the valve and actuator button without clogging the aerosol valve system. Any soluble materials used in the product must not crystallize out in the orifices and cause a blockage. (See Formula 21 for a typical formulation for an aerosol bath powder.) It can be packaged in an aerosol can using a special powder valve specifically designed to dispense powders.

Formula 21. Aerosol Bath Powder

Concentrate	
Talc	16.6%
Magnesium stearate	2.2
Isopropyl myristate	0.5
Arlacel 83	0.5
Perfume	0.5
Propellant 11	79.7
Aerosol formulation	
Concentrate	55.0%
Propellant 11	45.0

Procedure: Mix the isopropyl myristate, Arlacel 83, perfume, and propellant. Add the previously mixed talc and magnesium stearate and stir constantly, making a slurry. Keep agitated while filling. If powder-filling equipment is available, the powders may be added separately.

Bath Salts

Bath salts are designed to give fragrance, color, and water-softening properties to the bath. Any or all of these functions can be formulated into an elegant bath product known as bath salts, or bath crystals. Occasionally they are found in capsules and pellets. Some are in the form of large crystals and others in a powder form or small crystals.

Among the simplest and oldest forms of bath salts are fragranced and colored rock salt crystals. Because of their inert character, rock salt crystals may be scented with any desired perfume and colored with any certified color. This type of bath salt does not soften the water or have any cleansing function. One of the disadvantages of this product is that the crystals do not dissolve quickly and could take 5 to 10 min to dissolve completely, depending on the size of the crystals and the temperature of the water.

Epsom salt and Glauber's salt are sometimes used as a base for bath salts. They are neutral salts such as rock salt and do dissolve in bath water quickly. There are several other neutral salts that could be used, but they show few if any advantages and are usually too expensive.

The procedure used to manufacture these salts is relatively simple. The salt crystals of the desired size are placed in a ribbon mixer, and a solution of the perfume oil, alcohol, and the desired color is added while mixing. Between 0.5 and 1.0% of a perfume oil is used for these products. After the product has been thoroughly mixed so that the perfume and color, if it is used, are evenly deposited, the crystals are spread out on trays to allow the alcohol to evaporate.

The water-softening types of bath salts are composed chiefly of sodium phosphates or sodium sesquicarbonate. Trisodium phosphate is the most commonly used phosphate salt, but is seldom used alone, because of its high alkalinity. Borax or sodium sesquicarbonate is usually mixed with the trisodium phosphate to buffer its high alkalinity. This type of bath salt has the advantage of softening the water and in turn aiding in the cleansing action of soap. The surface tension of the water is lowered and, consequently, makes the skin wet easier. People who have a sensitivity to alkaline products may respond with a skin irritation. Perfume oils must be formulated to withstand the alkalinity of these formulations, and if color is used, it too will have to be alkaline-resistant.

Other phosphates may be used in place of the trisodium phosphate, such as sodium hexametaphosphate, tetrasodium pyrophosphate, sodium tripolyphosphate, etc. Some of these would be better sequestering agents and water softeners. Different shapes and sizes may be used to give various effects to finished products. (See Formulas 22 to 25 for some typical formulas for bath salts.)

The bath salts may be colored by dissolving the certified colors in water and then adding as much alcohol as the dye will permit without precipitating from the solution. The alcohol prevents the alkali from dissolving, keeps the

Bath Salt Formulas

	22	23	24	25
Rock salt or sodium chloride crystals	98%	48%	—	—
Trisodium phosphate	—	50	50%	—
Sodium sesquicarbonate	—	—	48	90%
Borax, powdered	—	—	—	8
Perfume	2	2	2	2
Color	Trace	Trace	Trace	Trace

Procedure: Mix the dry ingredients before the perfume and color are added.

crystals free flowing, and does not let them affect the color. All color systems should be chosen with care and properly evaluated for stability in the system being used. Fading, darkening, and changes in hue are frequently seen in bath salts.

Effervescent bath salts are not very common because of the inherent problems associated with these products. They must be kept free from moisture, and this is almost impossible in the bathroom where they are used. Individual foil (polyethylene laminate, heat-sealed) wraps appear to be the best way to package these products. They can be prepared in tablet, granule, or powder form. Sodium bicarbonate and citric acid are the most commonly used ingredients. Granules can be made by mixing the dry ingredients with alcohol and forcing the wetted mass through a screen of proper size to form the desired finished product. The wet granules are then dried in a hot oven and immediately packaged under humidity-controlled conditions. The presence of moisture will cause the effervescent reaction to take place. It has been recommended that a hot steam oven be used to dry the wet granulation, which causes a slight reaction to take place that produces a formation of a coating of neutral salt, which prevents the granulation from reacting further with the moisture of the air. Anhydrous alcohol should be used to granulate the salts. Alcohol-soluble gum binders are used to give the granules more strength, but only in very small quantities. The bath salts can be formed into tablets by compressing them into bath-size disks. Suitable lubricants such as starch or talc can be used in about a 2% concentration. The compression should be carried out in humidity-controlled rooms.

Powdered foaming agents can be used in the formulations as well as chelating agents and other water-softening agents such as the phosphates. (See Formula 26 for a typical formula for effervescent bath salts.)

Formula 26.   Effervescent Bath Salts

Sodium bicarbonate	45%
Citric acid	38
Sodium hexametaphosphate	9
Carboxymethyl cellulose	2
Sodium lauryl sulfate	5
Perfume	1
Color	Trace

Procedure: Combine the dry ingredients in powdered form and mix thoroughly. Dissolve the perfume and the color in enough alcohol to moisten the batch (about 2 lb/100 lb) and add to powders. Mix the mass until perfumed uniformly, then granulate through a screen and dry. The finished granules can be compressed by adding a lubricant. The powder or tablets must be packaged in airtight containers. The product should be processed in humidity-controlled facilities.

There are several other products associated with the bath or advertised as after-bath products, such as skin softeners, body lotions, and after-bath colognes, that are not discussed in this chapter. Most of these products are covered in other chapters under skin lotions and creams; the friction lotions and after-bath colognes are discussed under colognes and toilet water.

Bath products, like all cosmetics, should be evaluated for stability under high (110 to 120°F), low (35 to 40°F), and alternating (35° to 40°F for 24 hr; 110 to 120°F for 24 hr) temperature conditions. Liquid products as well as powders should be evaluated for their irritating effect on skin. Any product which may be splashed into the eye should pass the eye test. Sprays such as the aerosol products should be subjected to inhalation studies. Many states now require the use of biodegradable detergent systems. It is apparent that there are many factors other than the formulation of products which must be considered during and after the formulation of bath products. All of these should be considered before a product is marketed.

REFERENCES

1. *Encyclopaedia Britannica*, 11th ed., Vol. 3.
2. *Encyclopaedia Britannica*, 14th ed., Vol. 3.
3. Alexander, P.: Bubble bath, *Mfg. Chemist*, **36**: 41 (1965).
4. Bath preparations. II. Bubble bath solutions, *Schimmel Briefs*, No. 343, 1963.
5. Argiriadi, A.: Bubble bath formulation, *DCI*, **99**: 49 (1966).
6. Tronnier, H., Schuster, G., and Madde, H.: Zusammenhange zwischen Wascheffekt und Hautvertraglichkeit anionaktive Tenside, *Arch. Klin. Exp. Dermatol.*, **221**: 232 (1965).
7. Bath preparations: an expanding market, *Givaudanian*, September 1966.
8. Reisch, M., and Marciano, M. R.: A bacteriostatic emollient: its use as a cleanser and bath additive, *Ind. Med. Surg.*, **32**: 430 (1963).
9. Taylor, E. A.: Oil adsorption: a method for determining the affinity of skin to adsorb oil from aqueous dispersions of water-dispensable oil preparations, *J. Invest. Dermatol.*, **37**: 69 (1961).
10. Knox, J. M., Everett, M. A., and Curtis, A. C.: The oil bath, *Arch. Dermatol.*, **78**: 642 (1958).
11. Knox, J. M., and Ogura, R.: Adherence of bath oil to keratin, *Brit. Med. J.*, **2** (5416): 1048 (1964).
12. Stolar, M. E.: Evaluation of certain factors influencing oil deposition on skin after immersion in an oil bath, *JSCC*, **17**: 607 (1966).
13. Becher, P., and Courtney, D. L.: A physico-chemical approach to floating bath oil, *JSCC*, **17**: 607 (1966).
14. Ross, S., Chen, E. S., Becher, P., and Ranauto, H. J.: Spreading coefficients and hydrophile-lipophile balance of aqueous solutions of emulsifying agents, *J. Phys. Chem.*, **63**: 1681 (1959).
15. Becher, P.: Spreading, HLB, and emulsion stability, *JSCC*, **11**: 325 (1960).
16. Dispersible bath oils, *Schimmel Briefs*, No. 259, 1965.

Chapter 29

NAIL LACQUERS AND REMOVERS

William C. Doviak

A nail lacquer or enamel in order to be successful must contain a film former whose characteristics are ease of brushing application, fast drying, and hardening, and whose formulations should enhance adhesion to the nail without losing its resistance to chipping and abrasion. These characteristics can be accomplished by the proper formulation of the necessary constituents of a nail enamel (1). These are the following:

1. Primary film former.
2. Secondary film-forming resin.
3. Plasticizers.
4. Solvents.
5. Colorants.
6. Specialty fillers.

Conventional nail lacquers can contain one of many film-forming substances as the base or primary resin. One can choose from (1) cellulose acetate, (2) cellulose acetate-butyrate, (3) ethyl cellulose, (4) polymers of the methacrylates, (5) various vinyl polymers, (6) and, recently, (7) sucrose acetate isobutyrate. As a primary film former, none of the above is as outstanding as nitrocellulose, because of its hardness, toughness, resistance to abrasion, and excellent solvent release. It is interesting to note that nitrocellulose is also the oldest known man-made substance among the film-forming agents, which dry solely by evaporation and without subsequent oxidation or polymerization. Nitrocellulose, however, has several disadvantages: its tendency to shrink and become brittle, its low order of gloss, and its only fair adhesion to most surfaces. It is necessary, therefore, to modify the nitrocellulose with secondary resins and plasticizers to improve its gloss and adhesion, impart flexibility, and minimize shrinkage. It is necessary to utilize solvents, in order to

521

place these nonvolatile components in a liquid, workable form; the compounder can then add pigments to give opacity and shading.

Raw Materials

In evaluating any of the raw materials used in nail lacquer, it is of primary importance to determine its harmlessness to the nail, nail bed, and surrounding tissue. The finished product must also be checked to determine, under actual use, its harmlessness and nonirritating properties.

Primary Film Formers

Nitrocellulose is the commercial name for cellulose nitrate, a fluffy white fibrous or cubed product resulting from the reaction of cellulose with nitric acid. Cellulose, chiefly obtained from wood pulp, can be considered as a trihydric alcohol, the empirical formula for which would be written as $[C_6H_7O_2(OH)_3]_n$. Cellulose is thus composed of a large number of β-anhydroglucose units joined together by acetal linkages. The average number of anhydroglucose units in the molecule is several thousand for most celluloses; it ranges from 500 to 2500 for chemically purified celluloses. Under suitable, controlled conditions, cellulose reacts with nitric acid to form the ester. In theory, it should be possible to replace the three hydroxyl groups in each anhydroglucose unit with nitrate groups to give cellulose trinitrate. Such a product, having a degree of substitution of three, would contain 14.14% nitrogen.

In practice, it is not possible to achieve cellulose trinitrate, the practical upper limit in substituting nitrate for hydroxyl groups being about 2.9, corresponding to a nitrogen content of 13.8%. Furthermore, it has been shown that to provide properties most valuable for industrial uses, nitrocellulose should have an average degree of substitution between 1.8 and 2.3. The number of anhydroglucose units in the molecule of nitrocellulose is referred to as degree of polymerization. This, in turn, is related to viscosity of the solution formed when the nitrocellulose is dissolved at a given concentration in a particular solvent. The product from the nitration is then washed repeatedly with water in order to remove all traces of the acid. Upon completion of the purification treatments, the nitrocellulose in water is pumped into bins where the major part of the water is allowed to drain out. Hydraulic presses apply further pressure to force out more water. The nitrocellulose for shipment is packed wet, with either water or alcohol, the alcohols being denatured ethyl, isopropyl, or n-butyl alcohol. The volatile content is reduced to something under 30% by application of further pressure. The compressed block of nitrocellulose, wet with either one of the alcohols, is then put through a block breaker. This so loosens the fibers that they will be readily soluble in the solvents used in compounding lacquer. Nitrocellulose, for use in lacquers, can be compounded with such nitrocellulose designated as RS, AS, and SS. RS type has an average nitrogen content of 12 (11.8 to 12.2) %. The RS type is completely soluble in esters, ketones, ether-alcohol mixtures, and glycol ethers. It also has excellent tolerance for aromatic hydrocarbon diluents, such as toluene, and is compatible with many resins. The AS type has an average nitrogen content of 11.5 (11.3 to 11.7) %. Although it is soluble in the same types of solvents as RS type, it can also be dissolved in solvent systems containing high proportions of low-molecular-weight anhydrous alcohols. The SS type has an average nitrogen content of 11 (10.9 to 11.2) %. It is considerably more soluble in alcohols than is the RS type.

RS nitrocellulose is the type generally utilized for nail lacquer formulations and is available in a number of viscosity grades from 10 cps to 5000 sec. These viscosities are expressed in terms of the time required for a metal ball to fall through a standard solution of a given sample, as defined by a nitrocellulose producer (5):

> The falling ball method consists of measuring the time required for a ball of specified metal and size to fall a definite, measured distance through the liquid being tested while the liquid is contained in a glass tube of a uniform and definitely prescribed diameter, and while its temperature is held within strictly prescribed limits.

The stability of nitrocellulose indicates the relative rate at which the material breaks down chemically. This stability is most important in the formulation of nail lacquers, and it can be measured in the laboratory by determining the time required for the evolution of a given amount of oxides of nitrogen, while the sample is heated at an elevated temperature. The formation of nitrous oxides, owing to any instability, may cause severe discoloration of the user's fingernails. It is most important that the compounding chemist ascertain the stability of the nitrocellulose utilized in his formulations. The chemist should also be cautioned against the use of any materials, such as solvents and plasticizers among others, that will also adversely affect the stability of the nitrocellulose formulation. Following is the description of a method for testing the stability of nitrocellulose, using methyl violet (6):

> Glass tubes of pyrex, 300 mm long × 17 mm O.D. and 15 mm I.D., with one end closed and rounded off like a test tube, are required. The nitrocellulose should be thoroughly dry, as the presence of alcohol affects the end point; presence of alcohol would also lower the amount of nitrocellulose weighed out. A weighed out sample of 2.5 g of dried nitrocellulose is shaken or pressed down in the bottom of the tube so that it occupies the lower 2 in. of the tube. None of the nitrocellulose should be left clinging to the tube walls above the sample mass. The methyl violet test paper is creased for one-half its length and then is inserted into the tube, with the creased end being down, until the lower edge of the paper is 1 in. above the top of the nitrocellulose sample. The paper must remain in this relative position throughout the test.

The tube is stoppered with a cork having a notch filed in one side from top to bottom to allow for unrestricted expansion and contraction of the air in the tube. The sample tube is placed in a constant temperature bath, which is already up to its proper temperature of 135.5 $\pm$ 0.5°C. After the sample has been heated for 10 min, the tube is lifted and the test paper is inspected. Thereafter, successive inspections are made at the end of each 5-min interval until the end point is reached. The end point is the time, in multiples of 5 min, required for the methyl violet paper to lose completely its violet color. If, for instance, the color is only partly changed in 20 min but has completely changed in 25 min, the stability of the sample is recorded as 25 min. Properly stabilized nitrocellulose should show a stability of at least 20 min.

The above test should be repeated for each lot of nitrocellulose that will be compounded, because the degree of stability can vary from lot to lot owing to its manufacturing conditions.

Secondary Resins

Resins are used in conjunction with nitrocellulose compositions to improve the degree of film build and also to promote better depth, gloss, and adhesion. It has heretofore been stated that there are many resins which can be used in combination with nitrocellulose, but few have been found useful in nail lacquers. Natural resins were the first resinous modifiers to be used in combination with nitrocellulose. Some examples of this type of resin are dammar, elemi, shellac, sandarac, and pontianak. Of these, dammar was probably the most satisfactory, because when properly dewaxed it has a high degree of compatibility with nitrocellulose, and it imparted to the lacquer good gloss and adhesion. Most of these natural resins have no film-forming properties. In other words, they do not provide films with good tensile strengths, flexibility, or elasticity. Since the advent of synthetic resins, those properties lacking in the natural resins can be had. Some of the synthetic resins utilized in nail lacquer formulations are the drying and nondrying alkyd type, polyvinyl acetate, the many butyrates, and aryl sulfonamide-formaldehyde resins. The sulfonamide resins are known to provide excellent depth, gloss, flow, adhesion, and good water resistance to films. Solutions of the resin alone have low viscosities, and therefore relatively large amounts may be added to nitrocellulose without materially affecting the workability of the lacquer, or the hardness or flexibility of the coating (7).

Formaldehyde and p-toluene sulfonamide, employed in equimolecular proportions, condensed to form a viscose mass, which on heating to 110°C yields a hard, colorless resin (8).

The brilliant clarity, the near water-white color, and the excellent properties it imparts to lacquers are the reasons this resin assumes the predominant position in the production of nail lacquers.

Plasticizers

Plasticizers for nitrocellulose are classified variously as solvent or non-solvent, and as polymeric or monomeric. The plasticizer is invariably lower in molecular weight than the nitrocellulose. The function of a plasticizer is to give control to flexibility and elongation of the film. The ideal plasticizer would have a combination of properties, including the following: It should have permanence (zero volatility) so its effect would persist throughout the life of the film, should be free from color, odor, and taste. It should be non-toxic and have high efficiency; in other words, provide maximum flexibility with minimum loss in strength. It should have no effect on the chemical stability of the nitrocellulose; a stabilizing effect would be desirable. It should impart uniform plasticity to the nail enamel over a wide range of temperatures.

Since no plasticizer meets all these desirable properties, plus others which might be wanted, every choice of a plasticizer must be a compromise, depending on the properties essential to the finished nail enamel.

Several plasticizers commonly used in nail lacquer formulations are tricresyl phosphate, dibutyl phthalate, butyl phthalyl butyl glycolate, dioctyl phthalate, triphenyl phosphate, dibutoxy ethyl phthalate, camphor, and castor oil. Of these, dioctyl phthalate, dibutyl phthalate, and tricresyl phosphate are rather widely used. These plasticizers are examples of the solvent type. They act as nonvolatile solvents which remain in the dried film. Solvent-type plasticizers may be defined as those which exhibit complete miscibility with nitrocellulose at all proportions. Their presence in a nail lacquer increases the dilution ratio for diluents, such as toluene and xylene. Also, they increase resin compatibility and reduce or eliminate entirely the spewing of nonsolvent oil plasticizers. The use of castor oil and/or camphor as a nitrocellulose plasticizer has, today, no useful purpose except to impart an odor that can be related to nail lacquers, which may be desired by some users. Even though many of the above plasticizers have been used from time to time, they should be examined and tested to determine that their presence is harmless.

Solvents

The solvents or volatile portion of a nitrocellulose nail enamel formulation provides the means for dispersing the film-forming and nonvolatile portion, so that a uniform mixture of these components may be obtained. Solvents for use in nitrocellulose nail lacquer formulations must be considered in the three general interrelated categories: active solvents, couplers or latent solvents, and diluents.

Active solvents are those liquids that dissolve nitrocellulose; they include ketones, esters, amides, glycol-ethers, and nitroparaffins. Their rates of evaporation are classified as fast, medium, and slow evaporating.

Couplers are generally alcohols. In themselves they are not solvents for nitrocellulose, but when used in conjunction with active solvents, they increase the strength of the latter. Because the alcohol couples with the ester solvent, synergism takes place. A solution of nitrocellulose in an active solvent alone will have a greater viscosity than those similar solutions containing mixtures of the active solvents and the alcohol. Depending on the percentage of alcohol utilized, the flow of the nail enamel can be improved.

Diluents are nonsolvents for nitrocellulose. They are used to stabilize the viscosity in nail lacquers in order to reduce the number of applications of a base coat, help to carry resins into solution, and lower the overall cost of the lacquer formulations. Aromatic hydrocarbons, notably toluene and xylene, are the most common diluents found in nail lacquers. A term frequently used in formulating nail lacquers is "dilution ratio." This term denotes the total amount of any given diluent which can be added to a solution of nitrocellulose in a particular solvent without causing gelation or precipitation of the nitrocellulose. It is not normal to formulate solvent mixtures too closely approaching the dilution ratio. Such a formulation will cause the lacquer to be so high in viscosity as to prevent a smooth deposit of the film.

The properties of the most frequently used nail lacquer solvents are given in Table I.

Colorants

The coloring of nail enamels is twofold in purpose: (a) to impart to the clear lacquer an acceptable shade for cosmetic use, and (b) to opacify the nail lacquer film so that even the most delicate of shades will cover the nail. All coloring materials used in nail lacquers must be certified by the FDA for the particular use. There are many colorants included in this list that can be used in the formulating of nail lacquer shades, but from the practical standpoint, the choice is limited to those which have good permanence and are insoluble in the lacquer solvents in order to avoid staining and discoloration of the nails, as well as to avoid reaction with the nail lacquer vehicle itself. The organic red pigments which are most widely used in the shading of nail enamel are the lithol rubines (D&C Reds No. 6 and 7), the lithols (D&C Reds No. 10, 11, 12, and 13), and Tob-Bon Maroon (D&C Red No. 34). Of the yellow pigments, external D&C Yellows No. 5 and 6 Lake meet the necessary requirements.

Among the inorganic pigments that can be utilized in nail lacquers are the cosmetic grades of yellow and red oxides, iron blue, iron black, carbon

TABLE I. Properties of Lacquer Solvents

Solvent	Mol. wt.	Sp. gr. (20/20°C)	Boiling range (°C)	Vapor pressure mm Hg 20°C	Evaporation rate n-butyl acetate=1	Dilution rate with toluene
Amyl acetate	88.10	0.902	76 to 79	65.00	4.1	3.4
n-Butyl acetate	116.16	0.876	115 to 130	8.70	1.0	2.9
Butyl Cellosolve acetate	160.12	0.943	188 to 192	<1.0	<0.10	2.1
Cellosolve acetate	132.16	0.973	145 to 165	1.20	0.21	2.5
Methyl Cellosolve acetate	118.13	1.005	132 to 152	3.30	0.31	2.3
Acetone	58.08	0.791	55 to 57	186.20	7.70	4.4
Methyl ethyl ketone	72.10	0.806	79 to 81	71.20	4.60	4.3
Methyl isobutyl ketone	100.16	0.802	114 to 117	15.20	1.60	3.6
Butyl Cellosolve	118.17	0.903	166 to 173	0.60	0.10	3.5
Cellosolve	90.12	0.930	132 to 137	3.80	0.32	4.9
Methyl Cellosolve	76.09	0.965	122 to 126	6.20	0.47	4.0
Ethyl alcohol, 95%	46.07	0.818[a]	75 to 80	44.00	2.40	—
Isopropyl alcohol, 99%	60.09	0.790[a]	81 to 83	31.50	2.20	—
n-Butyl alcohol	74.12	0.811	116 to 119	5.50	0.50	—
Toluene	92.13	0.867	109 to 111	22.40	2.10	—
Xylene	106.16	0.870	137 to 143	9.50	0.70	—

[a] Specific gravities of ethyl and isopropyl alcohols at 60/60°F.

black, and purified titanium dioxide. In order to obtain opacity for different shades, one should use titanium dioxide because of its extreme opacifying characteristics and its chemical inertness. Because of the high specific gravity of the inorganic pigments, they have a definite disadvantage. These pigments, as well as the nacreous pigments, such as natural "pearl essence," titanated mica, and bismuth-coated mica, can be held in suspension; this method is fully described in the section on specialty fillers.

Soluble dyes should not be used in nail enamels or, if at all, they should be used sparingly, because of their staining effect on the skin and nails. They are generally used in tinting nail lacquer removers, rather than in nail enamels.

One can determine the possibility of any staining effect of the pigments by mixing a 5-g sample of the pigment with 100 g of solvent, and then filtering. The filtrate should remain clear and colorless. If found satisfactory in this test, the pigment is then incorporated in the nail lacquer in a proportion much greater than that contemplated for actual shading. The test enamel is then applied to the nails, and left undisturbed for a period of 7 days. Upon removal of the enamel, no staining or discoloration of the nail or the contiguous skin should be observed.

The pigments, in order to be utilized for incorporation into nail lacquers, should be dispersed, utilizing strong shearing force. This dispersion of the pigment particles can be accomplished either by the use of ball mills or on roller mills equipped for such a purpose. Because of the extreme shearing stresses which can be developed in the grinding, the two-roll mill is preferred for the dispersion of pigments for nail lacquer (9).

The grinding charge usually consists of the pigment, nitrocellulose, plasticizer, and sufficient solvent to form the dry ingredients into a plastic mass. It is most important that the formulator or manufacturer of the dispersions use only those plasticizers, solvents, and resins which are also utilized in the basic nail lacquer formulation. When the mass is completely solvated, it should be charged on to the rolls and passed through to form a sheet, which is then repeated by passing through the rolls as is necessary in order to obtain the desired degree of dispersion. During the process of dispersion, the heat generated by the friction of the milling process volatizes all of the solvent and some of the plasticizer. It is then allowed to cool so that when the sheet hardens, it can be broken into small pieces or chips, and stored for future use. In order to determine whether or not a proper degree of dispersion has taken place, a sample of the small chips of the sheet should be dissolved in a 50:50 mixture of n-butyl acetate and ethyl acetate. A film of 0.001 in. should be cast on clear glass and examined under a microscope for the presence of any undispersed aggregates. If, by careful examination, there are such undispersed aggregates, the process should be repeated until they do not exist.

Because of the hazards involved in preparing dispersions of this type, it is wise to leave this phase of lacquer work to a specialist in pigment dispersions.

Inasmuch as the pigment dispersions vary from lot to lot, the nail lacquer colorist should, on obtaining each batch, make up a solution by dissolving the dispersion chips in solvents that are the same as utilized in his base lacquer formulation. This base color solution should then be checked against a standard, so that all color formulations remain constant.

Specialty Fillers

There are many special materials, not used for coloring, which impart to the nail lacquer certain characteristic iridescence. One such product consists of a suspension of brilliant, reflective, transparent crystals obtained from the scale and skin of fish, such as many herrings and small ocean fish. Guanine (2-amino-6-hydroxy purine) as these platelike crystals are called, is scraped from the skin of the fish, levigated with water, and subjected to several washes of water and butyl acetate. A final wash-out with butyl acetate should remove all presence of water. This material is then added to a nitrocellulose solution to form a paste of sufficient viscosity to keep the crystals suspended. It is most important that the nitrocellulose solution and solvents utilized in the process be the same as formulated in the base nail lacquer.

Recently, other nacreous pigments have been used to impart iridensence. These include titanium dioxide-coated mica flakes (10) and bismuth oxychloride-coated mica flakes. From time to time, various colored aluminum chips, generally called glitter, have been utilized. Inasmuch as the nail lacquer bases commonly used with these nacreous pigments and natural pearl essence have refractive indices from approximately 1.48 to 1.72, the refractive index necessary for the nacreous pigments to afford the greatest iridescence should be at least 1.8. When these fillers are so utilized, they can be employed either alone or with colorants in the same manner as noniridescent shades.

A disadvantage resulting from the use of such nacreous pigments, in particular the coated micas and the aluminum chips, is that the particles are large enough so that they tend to settle to the bottom of the nail lacquer container, where they become immovable. There have been many unsuccessful attempts to avoid such settling by the addition of various thickening agents. Recently, with some degree of success, a method of obtaining nonsettling iridescent nail lacquers was utilized (9). This involves incorporating an organophilic tetra substituted ammonium cation-modified montmorillonite clay; such clays impart a thixotropic behavior to the nail lacquer. Such clays have been utilized since the mid-1950's. However, while under proper formulation a suspension nail lacquer can be formulated, such materials can also produce undesirable side effects, and require most rigid controls for storage and filling.

The suspension nail lacquers utilizing these various clays, which impart the thixotropic effect, have very narrow parameters of stability. The users and fillers of such suspension nail lacquers should take precautions to ensure the stability of this material as long as possible. After receipt of the material, filling should take place as soon as possible. If this cannot be accomplished, the containers must not be opened, but should be stored at temperatures between 65 and 75°F. Prior to filling, the bulk containers must be placed on a roller-type mixer for approximately 2 hr to allow the thixotropic gellation system to sufficiently return to a solution. If this is not done, the lacquer can become layered, resulting in heavy gelling in some filled bottles, with little gelling in others. The filling lines must be absolutely clean of any contaminants, such as oils, solvents, or other lacquers. Such contaminants, especially oils and incompatible solvents, will cause a breakdown of the system and an eventual settling of the pigments. It is also not advisable to store partial drums of suspension-type nail lacquer. Storing such partial drums, because of the evaporation of the solvent system, can also cause severe gelling and a breakdown of the suspension system. Furthermore, various components, such as brushes and even bottles, may contain or be coated with materials, such as vinyl silicone or others, that are incompatible with the suspension-nail lacquer.

A nail lacquer that does not effect a complete suspension, but has a high percentage of suspension quality, has been developed. It does not contain clay additives but, rather, is manufactured by a series of controlled compounding steps resulting in good wearing ability, strength, and smooth, even flow.

Formulation of Nail Lacquers

In considering the formulation of nail lacquer, one must strive to obtain gloss, durability or resistance to wear, and flexibility, all of which are a function of the film thickness of the lacquer. Therefore it is desirable to obtain as much film in a single application or coat as is consistent with the ease of application, speed of drying, and quick hardening of the lacquer. Viscosity requirements and practical limitations on the amount of nonvolatile matter may make the application of two or three coats necessary to obtain adequate film thickness and sufficient opacity. The normal viscosity range of nail lacquers lies between 300 and 450 cps. Lacquers falling below this range will be too thin to remain on the brush and will not deposit a film sufficient to cover the average nail. Enamels of over 450 cps will be too thick and cannot be brushed out to flow properly, resulting in a streaky rough application.

RS nitrocellulose is the most frequently used type in nail lacquer formulations. Of the many viscosity ranges available, RS $\frac{1}{4}$ sec, RS $\frac{1}{2}$ sec, and RS 5–6 sec should adequately serve all the needs of the nail lacquer formulator. The RS $\frac{1}{4}$ sec type is used where a high solids content is desired, with possibly some sacrifice in wear resistance. RS $\frac{1}{2}$ sec is used for better wear resistance and reasonably high nonvolatile content. RS 5–6 sec is similar in all respects to the RS $\frac{1}{2}$ sec type, except for its higher viscosity. In order to obtain a nail lacquer in the desired viscosity ranges, the formulator should not utilize more than 15 to 18% by weight solution of nitrocellulose RS $\frac{1}{2}$ sec type. One must then add, proportionately, the secondary resins and plasticizers. If the formulator selects the sulfonamide resin as his secondary resin, then a ratio of 2 parts of dry nitrocellulose to 1 part of resin will result in a film having better adhesion and luster and a higher degree of hardness than if nitrocellulose were used alone.

Films that are composed solely of nitrocellulose and a secondary resin are too brittle for practical use and will fail because of cracking and chipping.

The use of an appropriate plasticizer in the film will minimize this type of failure. A plasticizer in a proportion of 20 to 50% based on the dry weight of the nitrocellulose present, will give films possessing the requisite degree of flexibility.

Solvent mixtures for nail lacquers cover a wide range of compositions and may include several solvents, so as to obtain control of the evaporation rate, flow, viscosity, and drying of the enamel. Up to 50% of diluent in the solvent mixture has little or no adverse effect on the practical working viscosity of the lacquer (11); the remainder should consist of active lacquer solvents and coupler, in other words, 80% of butyl acetate and 20% of an alcohol. A typical solvent mixture for nail lacquer may be butyl or ethyl acetate 40%, alcohol 10%, and toluene 50%. The use of butyl alcohol in the suspension-type lacquers containing the various modified montmorillonite clays should be avoided. Controlled samples in the laboratory have shown that such lacquers containing butyl alcohol are incompatible and can cause adverse reactions owing to the hydroxyl radicals present. A typical nail lacquer is shown in Formula 1. Many variations in such a formula may be made, both in its total nonvolatile content and in the ratios of the nonvolatile components. The solvent balance may be modified to be slower- or faster drying, and the viscosity may be adjusted upward or downward to meet particular requirements.

Formula 1 may be pigmented by adding a sufficient quantity of concentrated dispersion of color to bring the pigment up to 3 to 5% of the total. Although some preparations of this type are being made with only 2% of pigment, or less, the higher amount can improve the wear resistance of the enamel. Due allowance should be made in the clear formula for the volatile and nonvolatile matter introduced by the addition of the color concentrate (12).

Formula 2 is a typical product for use as a base lacquer for iridescent material. Up to 10% by weight of natural pearl essence may be added to the base nail lacquer, depending on the desired effect. It may be further adapted for use as a suspension nail lacquer by the addition of 1 to 4% of the various montmorillonite-type clays, sold under the names of Bentone 27, 34, or 38.

Formula 1

Nitrocellulose RS $\frac{1}{2}$ sec, dry	15.00%
Sulfonamide resin (Santolite)	7.50
Dibutyl phthalate	3.75
Butyl acetate	29.35
Ethyl alcohol (from nitrocellulose)	6.40
Butyl alcohol	1.10
Toluene	36.90

Formula 2

Nitrocellulose RS $\frac{1}{2}$ sec, dry	14.00%
Nitrocellulose RS 5–6 sec, dry	2.00
Sulfonamide resin (Santolite)	7.00
Dibutyl phthalate	3.00
Tricresyl phosphate	0.75
Butyl acetate	16.25
Ethyl acetate	13.15
Ethyl or isopropyl alcohol (from nitrocellulose)	6.85
Ethyl or isopropyl alcohol	1.00
Toluene	36.00

Since nails vary from person to person in size, shape, smoothness, and even in composition, it is neither possible nor reasonable to expect that a single nail lacquer composition will show the same characteristics of application, adhesion, and wear resistance under the multiplicity of conditions to which it may be exposed. In order to prepare the nail for better acceptance of the enamel, to help in building film thickness, and to increase the adhesion of the enamel to the nail, base coats have been developed. Base coats often contain higher quantities of secondary resins to increase adhesion, and they are formulated to dry more rapidly and harder than enamels. A lower nonvolatile content and a lower viscosity are considered appropriate for base coats since, in this instance, the somewhat thinner film is likely to give more desirable results. A typical clear base coat may be prepared according to Formula 3. Base coats may be made either clear or slightly pigmented, although there is no reason why they cannot be pigmented as fully as regular nail lacquers.

For greater luster, depth of finish, increased film thickness, and resistance to abrasion and wear, top coats are applied over the nail. Top coats are formulated to impart toughness, extra gloss, and minimum disturbance of the underlying enamel. Their formulation is characterized by a higher nitrocellulose than plasticizer content, a lower secondary resin content,

Nail Lacquer Formulas

	3	4
Nitrocellulose RS $\frac{1}{2}$ sec, dry	10%	16%
Sulfonamide resin (Santolite)	10	3
Dibutyl phthalate	2	4
Tricresyl phosphate	—	1
Ethyl or isopropyl alcohol	5	10
Ethyl acetate	29	11
Butyl acetate	6	12
Toluene	38	43

and a relatively high diluent content; their tendency is to be fast-drying. A typical top coat may be prepared in accordance with Formula 4.

Formulations shown for the nail lacquer, base coat, and top coat shall contain as their nonvolatile components similar materials. Such a system allows for some correlation between the coefficients of expansion of the three coatings, thus minimizing the possibility of failure by peeling because of too great a disparity among the three elements of the coating system. The use of tricresyl phosphate in the top coat formulation will impart some degree of flame retardance and permanence owing to air, oil, and water loss.

Blushing Effect

Nitrocellulose lacquers containing solvents and diluents that evaporate rapidly may sometimes dry to a film with a white, chalky appearance. This is known as blushing, and is usually due to the precipitation by water of the nitrocellulose or other film-forming ingredients. Blushing is encountered most often during periods when the atmospheric humidity is high. When solvents evaporate rapidly, the air in contact with the lacquer film is cooled. As a result, when the humidity is high, the air may be cooled below the dewpoint, and moisture be absorbed from the air into the film. If this moisture is not carried out of the film by the evaporating solvents and remains behind in sufficient quantity to precipitate the film-forming ingredients, the result is a blushed coating. There are several remedies for blushing. If blushing is encountered in the use of lacquer, the difficulty may be overcome by adding thinners that evaporate more slowly than those normally used. Their use has the effect of retarding the drying rate and causing less cooling of the surrounding air.

There are two types of blushing other than that discussed in the preceding paragraph which may result from improperly formulated nail lacquer solutions: (a) Gum or resin blush may be produced by reducing a clear lacquer with an excess of an alcohol which is a nonsolvent for the resin present. Then the clear lacquer, upon drying, produces a blushed film because in the latter stages of solvent evaporation some of the resin is precipitated by the alcohol. (b) A diluent blush may be produced in the same manner as that caused by the alcohol. If a slow-evaporating aromatic hydrocarbon is used as the diluent with a fast-evaporating active nitrocellulose solvent, blushing can be caused in the latter stages of the solvent evaporation by precipitation of nitrocellulose.

Manufacture of Nail Enamels

The manufacture of nail lacquers should consist of two distinct operations: first, the manufacture and compounding of the base lacquers and, second, the coloring of such lacquers to give cosmetically acceptable shades. The mixing tanks and storage vessels should be constructed of aluminum or stainless steel to prevent discoloration of the lacquers caused by the contact with ferrous metals. The mixing tanks, depending on capacity, should be equipped with either stationary or portable mixing units with turbine or propeller-type agitators. In the manufacture of lacquer systems containing clays to effect the suspension, as heretofore described, it is necessary to employ mixing equipment that is capable of developing a high rate of sheer, such as a Hochmeyer disperser or other similar mixing devices. In manufacturing the clear lacquer bases, 75% of the solvent and diluent portions of the formula is first charged into the mixing tank. With the agitator running, nitrocellulose is added, then thoroughly mixed until completely in solution. The resin and plasticizer, along with the balance of the solvents, are then added in this order, and agitation continued for several hr, or until solution of all ingredients is complete. The clear lacquer viscosity is checked at this point, and adjusted if necessary. The clear lacquer is then pumped through a filter press, or run through a centrifuge. The lacquer is stored in suitable containers until required for further use. Filtration or centrifugation of lacquers is necessary to remove any foreign particles, materials, or insoluble particles which may be present. Such filtration or centrifugation also improves the brilliance and clarity of the lacquer solutions, and the film deposited from a brilliant solution has appreciably more gloss than from lacquer which has not been so treated. The manufacture of suspension-type systems can be accomplished by several methods, depending on the formulator's equipment and desired technique. He can have manufactured for him a clay dispersed in nitrocellulose similar to the color dispersions noted above or, depending on his equipment, incorporate the clays directly in accordance with the techniques recommended by the supplier of such clays.

Pigmented or color lacquers are made by charging an appropriate amount of clear lacquer into a mixing tank, while agitating, adding the requisite amount of color concentrate, or specialty fillers, such as pearl essence, to give the desired effect. If the manufacturer has taken all precautions of uniformity as to color concentrates and lacquer bases, he should be able to utilize fixed formulations for the desired shades. The color matcher can check his batch by comparing it with the master color standard. Perhaps the simplest method to accomplish this is the application of the standard and a batch sample to one's nails, along with a check of the standard and the batch sample in equivalent bottles. Another method which is quite reliable, and is in general use, is accomplished by pouring a standard and batch sample side by side on a glass plate. The pour should be approximately 1.5 in. in diameter, and the glass plate raised to a vertical position to allow the pours to flow down so that the adjacent edges just touch each other. An opal or milk-white glass should be used for cream shades, and a clear glass for the iridescent shades. The opacity and undertone should be checked and adjusted accordingly. Finally, a most critical method can be employed by observing the shade and the degree of opacity of equal film thicknesses when the standard and batch sample are drawn down on a chart of white paper, crossed with black $\frac{3}{8}$-in. lines or a chart which is 50% white and 50% black. The film thicknesses should be 0.006 in. as drawn down by a knife similar to a "Bird applicator" utilizing a vacuum plate (13).

Safety Considerations

Although lacquer manufacturing operations are potentially hazardous, experience has shown that the operation can be made safe so that accidents are rare and one avoids even minor damage to plant and materials, or injury to personnel. To achieve this low level of accident occurrence, a suitable location for the plant must be selected; proper layout of the plant in its external and internal features must be provided; safe operating procedures must be practiced; and good housekeeping must be maintained at all times.

An important point in plant layout is the desirability of segregation in the storage of both raw and finished materials from the manufacturing area. Tanks, for solvent storage, should also be in a separate area. Raw and finished products may be stored in a separate building apart from the building or buildings housing the manufacturing process. There are several advantages with such an arrangement. The fire risk is reduced and, in the event of fire, it is unlikely that all units would be involved—i.e., raw material storage, manufacturing buildings, and finished product storage.

The following are some considerations that should be given when building a new plant, modifying existing facilities, or operating a plant as it exists.

1. Storage of raw materials and finished products. Flammable solvents should usually be stored in tanks well isolated from the operating plant. The storage, handling, and use of flammable and combustible liquids should be in accordance with nationally recognized good practice. This includes consideration of requirements such as distances to lines of adjoining properties, spacing between tanks, diking, adequate vents, and the proper accessories such as flash arrestors, etc. The loading and unloading stations should be separate from the processing area and other buildings. The unloading site should be at a location reasonably well removed from the sources of ignition. Raw materials, such as pigments, plasticizers, and similar materials, should be stored in a separate room cut off from the processing areas by a wall or partition having at least a 2-hr fire-resistant rating, and openings should be equipped with approved fire doors. The finished products should be stored outside of buildings, in a separate building, or in a separate room, cut off from the processing areas by a wall or partition having at least a 2-hr fire-resistant rating, and openings should be equipped with approved fire doors (14).

2. *Process buildings.* Congestion should be avoided in planning the manufacturing operation. Raw-material and finished-goods storage areas preferably should not be in the same building housing manufacturing operations. If these facilities are in the same building, they should be cut off from the manufacturing operations by a wall or partition having a fire-resistant rating of at least 2 hr and openings should be equipped with approved fire doors. Buildings should be of fire-resistant or noncombustible construction without load-bearing walls. Floors should be constructed of nonsparking and nonconductive materials, and all metal pipes, fittings, tanks, etc. should be connected to a continuous ground cable which will afford readings less than 2 ohms. In addition, all mixing tanks, mixers, portable equipment, and drums should have means of discharging any build-up of static electricity afforded them. This can be accomplished by the use of portable grounding cables properly attached to the equipment being used.

Buildings which house solvent process operations should be explosion-vented so that, in case of a vapor-air explosion, structural damage and loss are minimized. An important factor in plant design is adequate ventilation to prevent the accumulation of vapors, which may be flammable and also toxic. A preferred method is installed ventilation close to the floor level to sweep out the vapors, since most of the vapors are heavier than air.

Tools, such as wrenches, scrapers, bars, etc., should be made of nonferrous metals, such as brass. Scoops for handling of nitrocellulose should be made of fiber material. Considering the nature of the business and character of the materials processed, sprinkler system protection should be installed in the various buildings. Equipment, such as mixers, tanks, and similar vessels, may be protected by foam, dry chemical, or carbon dioxide. This equipment should be so installed as to be capable of either automatic or manual operation. Of major importance is the training of all employees concerned in the use of all fire-fighting equipment. They must also be educated on the necessities of good housekeeping and the characteristics of all raw material being used, as well as means of escape from burning buildings as an effective measure to guard against accidents and loss of life (15).

Evaluation of Nail Lacquers

Completed nail lacquers, prior to release from the manufacturing plant, should be exposed to quality control and checked against the standards and specifications established for nonvolatile content, drying time, smoothness of flow, gloss, hardness, color, application, water resistance, abrasion resistance, adhesion, flexibility, and viscosity.

Nonvolatile Content

There are many methods for the determination of the nonvolatiles present in the lacquer. Generally accepted is the "dish method." Place 1 ± 0.2 g of the sample in a tared, flat dish about 8 cm in diameter. Spread the sample evenly with a tared wire and place in an oven at $105 \pm 2°C$. At intervals, break up any skins formed with the wire. After 1 hr, remove, cool, and weigh the dish. Reheat for 1 additional hr in the same manner and reweigh. Use the greater weight loss in calculating the nonvolatile content of the sample.

Drying Time

Apply a film of the sample with a 0.006-in. Bird applicator under controlled temperature and humidity conditions, at 25°C and 50% relative humidity, to a completely nonporous surface, such as a plate of glass or melamine-coated paper. Measure with a stopwatch the time required to form a dry-to-touch film. Dry-to-touch is the condition at which the film may be touched with a clean fingertip without the resultant transfer of any material to the finger.

Smoothness of Flow

Apply a pour of approximately 1.5 in. to a glass plate and raise vertically. Observe the film with a magnification glass of approximately 5 power. The film should reveal no presence of foreign matter or coarse particles, and should be free of a severe orange-peel effect.

Gloss

Comparisons against a standard should be made by using the flow pour described above. This is a general visual observation. More critical determination of gloss can be made on a Gardner or other similar instrument (13).

Hardness

Film surface hardness can be tested with a Sward rocker as follows: Films of 0.006 in. should be cast on a glass plate and dried for 48 hr at 25°C and an additional 2 hr at 71°C. The films then should be conditioned for an additional 48 hr at 25°C before testing. A simple comparative test is one in which a pour on a glass plate can be scratched with the thumbnail.

Color

A simple check of color and opacity can be made by a pour of both batch and standard onto a glass plate (opal for creams, clear for iridescent shades).

The plate should then be set vertically so that both pours flow down with edges touching each other. More definitive and critical checks can be made by utilizing a Bird film applicator of 0.006 in. and drawing down a film on a white and black panel, similar to a Morest type. Differences noted in either color or opacity by the operator can then be adjusted.

Application

The most reliable means of testing is the actual application to one's nails. One must check overall characteristics, such as evenness and smoothness of brushing, with particular attention to air bubbles in the film, and also streaking. Further examination of other properties already described can also be done.

Water Resistance

The method used should generally be the water immersion test. The operator should apply a 0.006-in. film on three glass plates and dry them in an oven at 25°C for 24 hr. The plates should be removed and placed in a desiccator for another 24 hr and then removed and weighed to the nearest 0.1 mg. The plates should then be immersed in a water bath containing distilled water at 37°C for 24 hr. The panels should then be removed, and dried by placing the plate between absorption paper and reweighed. They are again accurately weighed, and the loss in weight computed as percent of the original weight of the sample.

Abrasion Resistance

The standard method of measuring resistance to abrasion is by the use of a Taber or similar abrader. On such a device a film on a panel is subject to wear by abrasive wheels and specified revolutions. Loss in weight after a given number of these revolutions is the measure of comparable resistance to abrasion.

Adhesion

Two methods are generally employed. One is the use of a Hoffman scratch adhesion unit, which records the grams loading needed to scratch a film coating from a substrate glass plate. The other is the tape adhesion method, in which the operator will form a 0.006-in. film on a glass plate and allow it to dry for 24 hr at room temperature. A series of $\frac{1}{4}$-in. crosshatches are made on the film with a razor blade or knife, adhesive tape is applied to the film, and the percent of the crosshatched area removed by the tape is recorded.

Flexibility

Flexibility of the film can be measured on a mandril set in accordance with ASTM Method D-1737-62. Generally, this method determines the flexibility of films which have been applied to panels; the specified-size Mandril is withdrawn from the unit and placed in such a manner that the panel is on top of and bent over the mandril rod. The film coating is then examined for cracks over the area of the bend, and compared against a standard.

Viscosity

Several methods can be employed. Primarily, the efflux type, such as a Ford 4 cup is generally used for standard enamels. The rotational instruments, such as the Brookfield viscometer, are also employed. The suspension thixotropic type lacquer presents particular problems in this regard and careful measurements can be made on a Brookfield viscometer equipped with a helipath unit, so that the rotation is read in cps, while the spindle is being lowered into the sample.

Further comprehensive methods can be employed, as outlined by the further works of Mattiello (16), Gardner (13), and the specifications of the ASTM (17).

The wearability of nail lacquer on the nail itself by women chosen by the operator is the most informative from the actual field-use standpoint. The operator should pick subjects of varying types (housewife, typist, and model, for example). A controlled sampling should be done on alternate nails, such as complete with base and topcoats, one, two, and three coats, etc. The test period should be for the duration of 1 week, and the operator should reverse the applications to the subject's left or right hand for a second week. The test will produce results indicating what the ultimate consumer will experience.

Nail Lacquer Removers

The removers of nail lacquers and enamels are generally solvents of nitrocellulose combined with oils, emollients, or other agents designed to prevent or reduce the drying of the skin and nails, owing to the oil extraction by the solvents. Some formulations contain dissolved or emulsified oils and fats in the solvents. The most successful have been those containing stearates or lanolin derivatives. A typical oily-type remover is shown in Formulas 5 (18) and 6, the latter containing solvents similar to those used in a lacquer.

Nonsmearing removers have become popular in the last several years. This type, as the name implies, is intended to remove the lacquer or enamel without smudging or leaving a lacquer residue on the nails or adjoining

Formula 5

Methyl ethyl ketone	85%
Diethylene glycol monoethyl ether	10
Butyl stearate	5
Perfume and color	q.s.

Formula 6

Ethyl acetate	20%
Acetone	66
Butyl acetate	5
Water	8
Oil, white, or lanolin derivative	1
Perfume and color	q.s.

skin areas. Typical formulations utilizing fast solvents with water are shown in Formulas 7 and 8.

Nonsmearing remover	*Formula 7*	*Formula 8*
Acetone	80%	75%
Ethyl acetate	15	15
Butyl acetate	—	5
Water	5	5
Perfume and color	q.s.	q.s.

Cream nail polish removers are not generally popular; nevertheless they elicit some interest, and such a product is shown in Formula 9 (19).

Formula 9

Carnauba wax	15.0%
Toluene	15.0
Bentonite	0.5
Methyl ethyl ketone	69.5

Procedure: Heat the carnauba wax, toluene, and bentonite sufficiently to dissolve the wax. Remove from heat and, with vigorous agitation, add the methyl ethyl ketone. (*Note.* Because of flammability of the solvents, proper care and precaution should be taken in heating.)

The writer places great importance on the quality not only of the finished product, but of the raw materials required in manufacturing. One must continually check the physical and chemical specifications of raw materials that go into the product. Although the setting up of standard specifications for raw material is many times a difficult problem, it is one which should never be avoided. Ideally it should cover all essential characteristics. It is impossible to overestimate the importance of standardizing the methods of testing, so that they will give results of comparable precision and accuracy in the hands of different operators and, in many instances, different laboratories. They should also be designed to furnish sufficient but not excessive precision and accuracy; one should bear in mind that it is wasteful to use a method which is too fine for the purpose, and dangerous to use one which is not fine enough. The laboratory quality-control procedures should be supplemented by shelf-testing under various conditions of temperature, light exposure, etc. The ultimate objective of this type of testing is the correlation of the results of the laboratory tests with the shelf-life tests performed (20).

REFERENCES

1. Peirano, John: Nail lacquers and removers, in Edward Sagarin (Ed.): *Cosmetics: science and technology*, Interscience Publishers, New York, 1957.

2. Fuller, H. C.: U.S. Pat. 2,173,755 (1939).

3. Peter, R. C.: U.S. Pat. 2,195,971 (1940).

4. *Chem. Abst.*, **47**: 16 (August 25, 1953).

5. *Nitrocellulose handbook*, Hercules Corp., Wilmington, Del., 1969.

6. American Society for Testing Materials, ASTM-D-301-56.

7. Mattiello, J. J.: *Protective and decorative coatings*, Vol. I, Wiley, New York, 1941.

8. Ellis, C.: *Chemistry of synthetic resins*, Vol. I, Reinhold, New York, 1935.

9. Kuritzkes, A. M.: U.S. Pat. 3,422,185 (1969).

10. Jackson, J.: U.S. Pat. 3,342,617 (1967).

11. von Fischer, W.: *Paint and varnish technology*, Reinhold, New York, 1948.

12. Peirano: *op. cit.*, p. 686.

13. Gardner, H. A.: *Physical examination of paints, varnishes, lacquers and colors*, Gardner Laboratory, Inc., Bethesda, Md., 1956.

14. National Fire Protection Association, Bulletins 30, 35, 77, 85 (1968).

15. *Ibid.*, Bulletins 35, 101 (1968).

16. Mattiello: *op. cit.*, Vol. 4.

17. ASTM standards on paint, varnish, and lacquer.

18. Peirano: *op. cit.*, p. 691.

19. *Ibid.*, p. 692.

20. ASTM manual on quality control of material.

Chapter 30

FINGERNAIL ELONGATORS AND ACCESSORY NAIL PREPARATIONS

Leonard J. Viola

In addition to nail polish and polish remover, cosmetic preparations for the nails include cuticle removers, creams and lotions to combat brittleness, nail strengtheners, nicotine removers and nail bleaches, nail polish powders, nail whiteners, the recently introduced aerosol nail driers, and the newest innovation—fingernail elongators.

FINGERNAIL ELONGATORS

Perhaps the greatest advance in nail products since the introduction of enamel has been the plastic nail elongator, a synthetic material molded on the natural nail, where it remains firm and tight and grows with the wearer's nail, having a life that is claimed to be at least 4 weeks (1). In addition to transforming stubby fingernails into lengthy and attractive ones, the elongators are used to repair injured nails and to strengthen soft nails. Furthermore, they are a boon for the habitual nail biter, since the polymethacrylate plastic film is so hard that the teeth slip off the "nail," rather than bite into it.

Constituent Materials

Plastic nails were developed from formulations first encountered as fillings for teeth (2,3). The basic materials for such formulations consist of at least one of the following items: a monomer (methyl methacrylate), a finely divided polymer (polymethyl methacrylate), a catalyst, and a polymerization promoter. When the materials are admixed for use, they may be formulated to range from a viscous liquid to a thick dough. The admixture quickly hardens by polymerization to a tough, nonporous plastic.

In order to begin the process of elongating the nails, all previous polish and access cuticle must be removed. The nail surfaces are emeried off to remove excess oil, to shape out rough spots, and to increase adhesion for the plastic. To prevent the plastic formulation from polymerizing to the skin around the nail, a coating of "separator" is generally applied to the skin. The "separator" may be either a water-soluble resin, which is easily removed, or a material with little adhesion for the plastic.

A guide form made of stiff silvered paper or plastic (polystyrene or polyethylene), shaped to follow the contour of the fingertips, is fitted under the tip of the nail. The nails are then built over this guide.

The fingernail elongator consists of a powder and a liquid. The powder contains the polymer and the catalyst; the liquid consists of monomers, promoters, and possibly a plasticizer. The liquid may be admixed with the powder immediately prior to application, or the liquid and powder may be applied separately and mixed on the nail. In the first process, the mixture must be so formulated as not to solidify too rapidly. The gelation time of the admixed solution is 7 to 9 min. A constant consistency is obtained by adding small amounts of the liquid to keep the mixture fluid. However, once it is on the nails, hardening must take place in a reasonable time (usually 6 to 8 min).

In the other method of application, one brushes the liquid on the nails, rewets the brush to aid in picking up the powder, and then places the powder on the edge of the nail, at the outer tip. This process is repeated until the nail has been built up or "elongated." After the tip of the nail has been processed, the nail is built up from the center to the outer edge. When the nail has the shape and length desired, it is allowed to harden. In this process, the brush must be cleaned after each addition in order to avoid having the catalyst and the monomer come in contact with each other. The hardening stage may be expedited by placing the hands in water, thus excluding oxygen, as the latter inhibits polymerization. After the plastic has hardened sufficiently, a process that usually takes from 4 to 6 min, the new nails are trimmed and nail polish is applied.

Polymerization of Vinyl Compounds

As pointed out by Mark and Tobolsky (4), the chain character of vinyl polymerization was recognized in a series of fundamental papers by Chalmers (5), Staudinger and Frost (6), Mark and Raff (7), Flory (8), Gee and co-workers (9–12), and Melville (13), among others. The important principles of the chain-reaction mechanism in these polymerizations, both thermal and with benzoyl peroxide as catalyst, were clarified by Schulz, Dinglinger, and Husemann (14). They found that during most of the polymerization, the average molecular weight of the polymer formed was constant. The initiation of free radical-catalyzed polymerization involves dissociation of the catalyst into free radicals, which can then initiate polymerization by reacting with monomers. Polymerization by free radicals can be produced by heat, light, fast elementary particles (electrons, protons, α-particles, or neutrons), or free radical catalyst. The free radical catalyst should not strictly be called "catalyst" since it takes part permanently in the reaction and combines chemically with the resulting macromolecule. It is more correctly called an "initiator."

These methods of producing free radicals can be expressed by the following equations:

$$(a) \quad R-\underset{\overset{|}{H}}{C}=CH_2 + heat \longrightarrow R-\underset{\overset{|}{H}}{\overset{}{C}}-CH_2\cdot$$

$$(b) \quad R-\underset{\overset{|}{H}}{C}=CH_2 + (h\nu) \longrightarrow R-\underset{\overset{|}{H}}{\overset{}{C}}-CH_2\cdot$$

$$(c) \quad C_6H_5\overset{\overset{O}{\|}}{C}-O-O-\overset{\overset{O}{\|}}{C}-C_6H_5 \longrightarrow 2C_6H_5\overset{\overset{O}{\|}}{C}-C\cdot$$

The mechanism of thermal initiation, equation (a), and photochemical initiation, equation (b), is still somewhat obscure. Many workers have postulated that the absorption of energy by the double bond of the monomer leads to the excitation of the π-electrons into separate antibonding orbitals with the formation of a subsequent diradical (15,16). This diradical could by hydrogen atom transfer to, or from, a neighboring molecule, thus forming two monoradicals.

On the other hand, it is possible, particularly with high-energy irradiation, that scission of the absorbing molecule occurs to yield two monoradicals. However, there is little doubt in polymerization that propagation is carried out by a monoradical species.

The usual method of forming free radicals in bulk polymerization is by the use of a free radical catalyst such as benzoyl peroxide (17), equation (c). Splitting of this catalyst under heat, for example, or in the presence of a promoter at room temperature, results in two free radicals. These radicals are in a highly active state because of the presence of the free electron and can open the double bond of the monomer. Thus a covalent bond with the benzoyl radical is formed, and at the same time the energy is transferred to the other end of the adduct and reproduces the unpaired electron, equation (d). These two reactions, equations (c) and (d), are known as the initiation step:

$$(d) \quad C_6H_5-\overset{\overset{O}{\|}}{C}-O\cdot + H_2C=\underset{X}{CH} \longrightarrow C_6H_5\overset{\overset{O}{\|}}{C}-O-CH_2-\underset{X}{\overset{\overset{H}{|}}{C}}\cdot$$

This adduct reacts with a second monomer, then with a third monomer, and continues to react in this way, forming a head-to-tail polymer:

$$(e) \quad R(CH_2\overset{\overset{\displaystyle H}{|}}{\underset{\underset{\displaystyle X}{|}}{C}})_{n-1}CH_2\overset{\overset{\displaystyle H}{|}}{\underset{\underset{\displaystyle X}{|}}{C}}\cdot + CH_2{=}\overset{\overset{\displaystyle H}{|}}{\underset{\underset{\displaystyle X}{|}}{CH}} \longrightarrow R(CH_2CH)_n CH_2\overset{\overset{\displaystyle H}{|}}{\underset{\underset{\displaystyle X}{|}}{C}}\cdot$$

where R = benzoyl radical
X = methyl or alkyl group

This process is called the growth or propagation reaction.

The chain grows until the terminal radical on a growing chain is destroyed or otherwise rendered inactive. Two processes by which this termination step may occur are chain coupling,

$$(f) \quad R(CH_2\overset{H}{\underset{X}{C}})_n{-}CH_2{-}\overset{H}{\underset{X}{C}}\cdot + \cdot\overset{H}{\underset{X}{C}}{-}CH_2{-}(\overset{H}{\underset{X}{C}}{-}CH_2)_m R \longrightarrow$$

$$R(CH_2\overset{H}{\underset{X}{C}})_n{-}CH_2{-}\overset{H}{\underset{X}{C}}{-}\overset{H}{\underset{X}{C}}{-}CH_2{-}(\overset{H}{\underset{X}{C}}{-}CH_2)_m R$$

and/or disproportionation through the transfer of the hydrogen atom,

$$(g) \quad R(CH_2\overset{H}{\underset{X}{C}})_n{-}CH_2{-}\overset{H}{\underset{X}{C}}\cdot + \cdot\overset{H}{\underset{X}{C}}{-}CH_2{-}(\overset{H}{\underset{X}{C}}{-}CH_2)_m R \longrightarrow$$

$$R(CH_2\overset{}{\underset{X}{C}})_n CH_2\overset{H}{\underset{X}{C}}H + \overset{H}{\underset{X}{C}}{=}\overset{}{\underset{}{C}}{-}(\overset{H}{\underset{X}{C}}{-}CH_2)_m R$$

In methyl methacrylate polymerization, there is good evidence that coupling or combination is the dominant chain-terminating process and that disproportionation occurs, at most, to a minor extent (18).

Copolymerization. When two or more monomers are polymerized together in the same media, a copolymer is formed which has properties generally quite different from those of a mechanical mixture of the individual polymers. When two polymers copolymerize, there are two polymer radicals and two monomer radicals to consider, making four possible propagation reactions. The composition of the copolymer increments varies continuously throughout the polymerization. As the degree of conversion increases with monomers of widely differing reactivities, the faster reacting monomer will be entirely used up before 100% conversion, and thus the composition of the polymer changes.

Effect of oxygen. The inhibiting effect of oxygen on the polymerization of methyl methacrylate has been widely investigated (19). Barnes (20) explains this effect by assuming that peroxide formation occurs in preference to polymerization, thus destroying active centers or activated molecules which otherwise would have initiated chain growth. It has been shown (21) that a polymeric peroxide of low molecular weight is formed from methyl methacrylate. This peroxide decomposes to give a low polymer, methyl pyruvate, and formaldehyde. The probable reaction is

$$
\left[CH_3-\underset{\underset{COOCH_3}{|}}{\overset{\overset{CH_3}{|}}{C}}-O-O \right]_n \longrightarrow \text{low polymer} + \underset{COOCH_3}{\overset{CH_3}{C}}{=}O \quad + OCH_2
$$

Raw Materials

Monomer

The monomer most frequently used in dental restorations and in fingernail elongators has been methyl methacrylate. It is preferred because of its ease of polymerization, its availability, relatively low order of toxicity, and the outstanding properties of its polymer and copolymers. Vinyl compounds such as methyl methacrylate undergo additional polymerization by either a "free radical" or an "ionic" mechanism. Thermal polymerization, photopolymerization, and polymerization initiated by peroxides, persulfates, or azo compounds are examples of the free radical mechanism. Ionic mechanisms are acid- or base-catalyzed (carbonium ion or carbanion and are initiated by sulfuric acid, boron fluoride, aluminum chloride, sodium amide, sodium alkyls, or similar Lewis acids and bases. Because of the moisture sensitivity of ionic polymerization, free radical polymerization is the method generally used. In the presence of a free radical catalyst, methyl methacrylate is readily converted from a liquid into a solid polymer:

$$
H_2C{=}\underset{\underset{COOCH_3}{|}}{\overset{\overset{CH_3}{|}}{C}} \longrightarrow \left[-\underset{\underset{H}{|}}{\overset{\overset{H}{|}}{C}}-\underset{\underset{COOCH_3}{|}}{\overset{\overset{CH_3}{|}}{C}}- \right]_n
$$

A comparison of the rates of copolymerization of methyl methacrylate with various other monomers, using 0.1% benzoyl peroxide at 60°C, results in the following order of decreasing rates (22): methyl acrylate, acrylonitrile, vinyl acetate, methyl methacrylate, vinyl chloride, styrene, methacrylonitrile, allyl acetate. This order, however, has no direct relationship to the rates at which the same monomers copolymerize.

Because of their ease of polymerization, commercial methyl methacrylate and other vinyl monomers contain a polymerization inhibitor. This is usually hydroquinone, monomethyl ether of hydroquinone, or pyrogallol. Hydroquinone, the most widely used of these products, is not itself the actual inhibitor, since in the total absence of oxygen it does not inhibit polymerization. Benzoquinone, the oxidative product of hydroquinone, is the inhibitor; it inhibits polymerization by reacting with the chain radicals as they are formed (23,24), instead of allowing the addition of another monomer molecule. If an excess of free radicals is formed by the use of a promoter and an excess of catalyst, all of the inhibitor is used up and polymerization occurs.

The hydroquinone inhibitor may be removed by washing the monomer with a sodium hydroxide-salt mixture. Monomethyl ether of hydroquinone or pyrogallol is usually removed by distilling the monomer under vacuum.

Methyl methacrylate is soluble in most common organic solvents. It is insoluble in formamide and only slightly soluble in glycerol and ethylene glycol. It is miscible with most of the esters of acrylic and methacrylic acid. The monomer slowly dissolves methyl methacrylate polymers, except when the latter are cross-linked or of an extremely high degree of polymerization. The commercial polymer granulars must be constantly stirred while being dissolved in the monomer; otherwise a solid polymer plug will form which is very difficult to dissolve. A small fraction of the commercial polymer is not soluble in the monomer, but only swells. This fraction may be removed by filtration. In addition, methyl methacrylate is a solvent for other polymers and copolymers of acrylic and methacrylic esters.

Toxicity. The vapors of methyl methacrylate have been found to be less toxic to mice than those of ethyl acetate (25). Ethyl and butyl methacrylate were found to exhibit the same toxicological effects as methyl methacrylate, but to a slightly lower degree. Papers on the toxicity of methyl and ethyl acrylate show that the monomers are quite toxic (26,27). The recommended range of maximum vapor concentrations for humans is in the order of 50 to 75 ppm (28).

Both methyl and ethyl acrylate are lachrymatory and irritating to the skin. Inhalation of large amounts of methyl methacrylate can cause irritation of the mucous membrane, gastrointestinal disturbances, headaches, and possible irritation of the kidneys (29). These allergic symptoms disappear after treatment and removal of the personnel from the area where methacrylates are being used (28).

Polymerization properties. The methacrylate and acrylate esters are outstanding in their ability to copolymerize with many types of monomers varying widely in polarities (30). In fact, there are few if any cases of marked inhibition between monomer pairs (19). Some of the possible advantages of copolymerization are to improve the odor, to decrease the toxicity, and to increase the rate of polymerization of the monomer. In addition, copolymerization can improve the polymer by increasing toughness, improving adhesion, and internally plasticizing the polymer (30).

The acrylic and methacrylic esters copolymerize readily with each other, except where one attempts to copolymerize large proportions of long-chain alkyl esters (e.g., stearyl methacrylate or dodecyl acrylate) with lower alkyl esters.

There are many monomers which are unable to copolymerize, but in the presence of methyl methacrylate as the third monomer, a terpolymer is formed. One example is the formation of a polymer with methyl methacrylate, vinyl acetate, and styrene (31).

Methyl methacrylate may be copolymerized with a cross-linking monomer. The effect of the cross-linking agent is to form a harder and more brittle thermoplastic polymer, and to decrease the solubility of the polymer formed. The hardness and brittleness of these cross-linked polymers are dependent on the amount of cross-linking agent and the length of the chain separating the olefinic double bonds of the cross-linking agent; i.e., a polymer cross-linked with ethylene dimethacrylate will generally be harder and more brittle than one cross-linked with tetraethylene glycol dimethacrylate. The cross-linking agent is generally present in the proportion of 1 to 10% of the monomer used.

Polymer

In order to accelerate the rate of polymerization, reduce shrinkage, and reduce odor, a monomer-polymer dough or slurry, generally of methyl methacrylate, is used. It has been known for some years in the plastics industry that if polymethyl methacrylate is added to the monomer to form a syrup, the polymerization rate is increased. The increase in the rate of polymerization is attributed not to the greater number of growing free radical centers, but to a diminution in the rate of chain termination resulting from the increased viscosity of the medium (32). This greater viscosity evidently retards considerably the mobility of the growing macroradical, thus retarding the possibility of two macroradicals terminating. On the other hand, the monomer, because of its size, still retains a relatively high mobility and can thus add to the chain. Polymers of highest molecular weights are obtained when the monomer is a poor solvent for the polymer formed. In confirmation of the viscosity-chain termination theory, Trommsdorff and co-workers (33) showed that even dissolving cellulose tripropionate in methyl methacrylate accelerated the rate of polymerization. Other extraneous polymers which, when added to methyl methacrylate, also increase the rate of polymerization are polystyrene, cellulose acetate, and polyisobutylene (33).

Because of the outstanding properties of methyl methacrylate, and because of its considerable use in dentistry, it was the logical polymer to be chosen for fingernail elongators. Commercial polymethyl methacrylate is a hard, optically clear plastic, with good dimensional stability, water resistance, and resistance to aging. The properties of the unplasticized plastic are given in Table I. The polymer displays good resistance to weak alkalis, weak acids, and aliphatic compounds; however, it is not resistant to esters, ketones, aromatic compounds, chlorinated hydrocarbons, and concentrated acids. When the polymer is free of cross-linking, it is soluble in acetone, ethylene dichloride, carbon tetrachloride, toluene, and acetic acid. Of the lower alkyl methacrylate polymers, only polymethyl methacrylate is soluble in formic acid at 25°C. One serious limitation to polymethyl methacrylate is its poor abrasion and mar resistance. However, this is of little importance in fingernail elongators.

TABLE I.   Properties of Methyl Methacrylate Polymer (34)

Specific gravity	1.18 psi	Hardness, Rockwell	800H; 105M
Tensile strength	10,000 psi	Heat-distortion temp. (ASTM D648-427)	199°F
Flexural strength	18,000 psi	Index of refraction	1.49
Impact strength, Izod	0.3 ft-lb/in.	Mar resistance (ASTM D673-42T)	60%

The polymer of methyl methacrylate is supplied under a variety of trade names and with some variations in properties. Plexiglas Y-100 is a bead polymer of nonuniform size spheres, all the particles passing through an 18-mesh screen. Plexiglas DC is a special screening of Y-100, all the particles passing through a 40-mesh screen (35). One grade consists of a polymer of the molecular weight of approximately 250,000, or a degree of polymerization of about 2500. Granular polyethyl methacrylate is also available.

When comparing the high polymers of the various methacrylate esters, it is found that the physical properties depend on the length and degree of branching of the alcohol radical. Hardness and softening points are increased by branching, especially when the branching is at the second carbon atom of the alcohol radical.

The methacrylate polymers from methyl to n-octyl are all higher-softening, harder, and higher in brittle points than the corresponding methyl to n-octyl acrylate polymers (19). In comparison, the polymer of methyl methacrylate at room temperature is a hard material; the ethyl methacrylate is slightly softer and, therefore, not used widely in dentifrices, unless cross-linked, since it distorts somewhat at the temperature in the mouth. The polymer of butyl methacrylate exhibits solubility in a wider range of solvents than methyl methacrylate and gives a considerably softer and more extensible film, but is still free of tack. This trend toward tackier, softer polymer, as shown by brittle points, continues through the n-lauryl esters.

Promoter

The role of the promoter is to accelerate or promote polymerization or copolymerization of methyl methacrylate and other vinyl monomers, when in the presence of peroxide or persulfate catalyst. The activation energy to open the molecule of the catalyst, which is usually —O—O—, —C—$\overset{|}{\mathrm{N}}$—, or a —C—O— bond, amounts to about 15 to 30 kg cal (16). Consequently there are catalysts which act at relatively low temperatures, such as hydroperoxides and aliphatic azodinitriles (40 to 50°C), whereas others, such as peroxides or aliphatic azodiesters, may require temperatures as high as 150 to 175°C to decompose into free radicals. Therefore the promoter itself is not the polymerization catalyst, but it lowers the activation energy and causes the catalyst to release free radicals and thus initiate polymerization at much lower temperatures, e.g., at room temperature.

Knock and Glenn (36) report that the most effective nitrogen-containing compounds which promote methyl methacrylate polymerization or copolymerization (in the presence of a peroxide or persulfate catalyst) are *m*-tolyl diethanolamine, phenyl diethanolamine, *β*-hydroxyethyl ethyl aniline, *p*-diethyl amino chlorobenzene, and *m*-dimethyl sulfonamido phenyl diethanolamine. These compounds have approximately equal activity as promoters. Among others which they mention that have a somewhat lower promoter activity are *N*-phenyl glycine, phthalamide, and creatinine. Other tertiary aryl amines, such as *N,N*-dimethyl aniline and *N,N*-diethyl *p*-toluidine, are also effective promoters of benzoyl peroxide-catalyzed vinyl-type polymerization (37–40).

Catalyst

The function of the catalyst is to form free radicals which in turn initiate polymerization. The types of free radical catalysts which are generally used to initiate polymerization of vinyl compounds are the following:

1. Organic peroxides (benzoyl peroxide) and hydroperoxides (cumene hydroperoxide).
2. Hydrogen peroxide and persulfate catalyst (hydrogen peroxide formers), including ammonium, potassium, and magnesium persulfates.
3. Azo catalysts, including *a,a'*-azodiisobutyronitrile (41,42), *p*-bromobenzenediazohydroxide, triphenylmethyl azobenzene, and *a,a'*-azobis (*a,a'*-dimethyl valeronitrile) (43).

Because of their solubility and availability, the organic peroxides are generally used in bulk polymerization.

In fingernail elongators, it is desirable to have a solid peroxide, since it can be packaged with the polymer. If a liquid peroxide is used, it must be packaged in a separate container, or it must be mixed with and be absorbed (or adsorbed) by the polymer. This latter method may produce some complications, especially in peroxide stability.

The preferred catalysts are benzoyl and lauroyl peroxides, in that order. The former is a white crystalline compound, melting at 103°C, which decomposes explosively at 106°C. Lauroyl peroxide is a softer, granular compound, melting at 48 to 50°C.

The persulfates, which are hydrogen peroxide precursors, are excellent catalysts for polymerizing vinyl monomers; because of their lack of solubility in the methacrylate and acrylate monomers, they are not used for bulk polymerization. These initiators (and hydrogen peroxide) are used, however, with excellent results in aqueous emulsion polymerizations, both as the sole initiator or in a redox system.

Although the azo catalysts have not been used to any great degree, the recently developed low-temperature aliphatic compounds may find wider application. These catalysts have the advantage of initiating by a first-order reaction, without any side reactions; however, they must be carefully studied from a toxicological point of view.

Methyl methacrylate and other vinyl monomers may also be polymerized by irradiation with ultraviolet light. Benzoin, diacetyl, or other diketones which absorb ultraviolet light and are free of polymerization-inhibiting groups are effective photosensitizers for methacrylic esters (44–46).

Formulation

Fingernail elongators must consist of materials that will polymerize at room temperature. The reaction rates are accelerated by higher temperature, and in many cases by increasing the amounts of catalyst, promoter, or both. However, the greater the catalyst concentration, the lower is the molecular weight of the polymer formed. A substantial decrease in molecular weight diminishes the moisture and solvent resistance of the polymer. Generally, 1 to 4 % of the catalyst and the promoter, based on the weight of the monomer, is sufficient. Sometimes, a combination of promoters or of catalysts is more effective than each of these compounds alone.

The liquid component contains one or more monomers with the inhibitor, the promoter, and possibly a plasticizer, pigment, or polymer, but should never contain the catalyst. The powder contains the polymer, catalyst, and possibly a pigment, opacifier, or filler. A simple product is described in Formula 1.

To the powdered or solid component, one may add a filler, such as silica, talc, mica, metal oxides, or inorganic insoluble salts. A product based on the use of aluminum silicate as the filler is described in Formula 2. The liquid component of Formula 2 may have too low a viscosity, and therefore it may

Formula 1*

Solid component
Methyl methacrylate poly-
 mer, granular   96.5%
Benzoyl peroxide   3.0
Pigments and opacifiers (if
 desired)   0.5
Liquid component
Methyl methacrylate mono-
 mer, with hydroquinone   99.0%
p-Diethylaminochlorobenzene   1.0

Instructions for use: Mix 1.8 parts by weight of the powder with 1 part by weight of the liquid; polymerization to a hard plastic results at room temperature in 15 min, or at 37°C in 10 min. Note: the pigment may be Iron Oxide and the opacifier may be titanium dioxide or barium sulfate.

be desirable to thicken this solution. This may be done by placing a portion of the polymer in the liquid component, as shown in Formula 3.

Formula 2*

Solid component
Methyl methacrylate polymer,
 granular   75%
Aluminum silicate   23
Benzoyl peroxide   2
Liquid component
Methyl methacrylate mono-
 mer, with hydroquinone   99%
Phenyl diethanolamine   1

Instructions for use: Mix 2 parts of the powder with 1 part of the liquid to form a dough, polymerization to a hard solid taking place in 6 min.

Formula 3*

Solid component
Same as Formula 2
Liquid component
Methyl methacrylate mono-
 mer, with hydroquinone   80%
Methyl methacrylate polymer,
 granular   19
Phenyl diethanolamine   1

Procedure and instructions: Since all of the commercial polymethyl methacrylate may not go into solution, filtration may be necessary. Mix $1\frac{1}{2}$ parts of the powder with 1 part by weight of the liquid, and polymerization to a solid plastic follows in 6 min at 37°C.

* Ref. 36.

Thus far, benzoyl peroxide has been used as the catalyst. In Formula 4 the catalyst is changed to lauroyl peroxide, and instead of a single promoter, a combination of promoters is used.

Formula 4*

Solid component
  Methyl methacrylate polymer,
    granular   97.90%
  Lauroyl peroxide   2.00
  Pigment and opacifier (if desired)   0.10
Liquid component
  Methyl methacrylate monomer,
    with hydroquinone   99.00%
  Phenyl diethanolamine   0.50 ⎫
  p-Diethylamino diphenyl   0.25 ⎬ Promoters
  p-Diethylaminochlorobenzene   0.25 ⎭

Instructions for use: Mix 1.8 parts by weight of the powder with 1 part by weight of the liquid; this polymerizes to a hard, strong plastic at room temperature in 20 min, and at 37°C in 15 min.

  * Ref. 36.

To obtain a more flexible plastic, one may add a plasticizer to the liquid component, in a proportion of about 1 to 10%. Plasticizers suitable for this purpose include triphenyl phosphate, tricresyl phosphate, and dimethyl, diethyl, or dibutyl phthalate. Improved flexibility may also be obtained by forming a copolymer with a small concentration (5 to 20%) of the monomer of an acrylate or methacrylate ester of higher molecular weight, as shown in Formula 5.

Formula 5

Solid component
  Same as Formula 2
Liquid component
  Methyl methacrylate mono-
    mer, with hydroquinone   83%
  Lauryl methacrylate   15
  Diethyl aniline   2

To improve the solvent resistance and dimensional stability of the polymer, one may add a cross-linking agent, in a concentration of 2 to 10% as shown in Formula 6.

Several liquid compositions which polymerize at room temperature, when admixed with a solid polymer, require removal of the inhibitor before use

Formula 6

Solid component	
Same as Formula 2	
Liquid component	
Methyl methacrylate mono-	
mer, with hydroquinone	93%
Ethylene dimethacrylate	5
Phenyl diethanolamine	2

(46). These formulations are based on the copolymerization of dichloro-styrene with other monomers. Commercial dichlorostyrene usually contains mixtures of 2,5-, 2,4-, and 2,6-isomers. The 2,5-dichlorostyrene is slightly more reactive than the 2,4-isomer, and the 2,6-dichlorostyrene is relatively sluggish in polymerization. Therefore the percentage of the 2,6-isomer should be kept at a minimum, i.e., less than 5 %. The use of dichlorostyrene in finger-nail elongators is illustrated in Formulas 7 and 8. These products may polymerize at room temperature without removing the inhibitor, if a pro-moter is added.

	Formula 7*	*Formula 8**
Solid component		
Methyl methacrylate polymer,		
low molecular weight	7.20 parts	6.80 parts
Benzoyl peroxide	0.08	0.08
Liquid component		
Dichlorostyrene	3.00	3.00
Methyl methacrylate monomer	1.00	—
Ethyl butadiene-2-carboxylate	—	1.00

* Ref. 46.

Although polyethyl methacrylate distorts in hot water, it may be copoly-merized and/or cross-linked to overcome this difficulty. A product, originally designed for use in dentistry, is shown in Formula 9 (47); it may be used as such or readily adapted for fingernail elongators.

A Belgian patent (48) employs a copolymer of equal parts of ethyl acrylate and ethyl methacrylate as the polymer. This preparation, Formula 10, is used for replacing or repairing broken or missing fingernails.

A recent French Patent (49) for a fingernail elongator and strengthener claims that a nonperoxide redox catalyst such as dimethyl paratoluidine is effective in polymerizing methyl methacrylate without the aid of a peroxide catalyst. However, polymerization appears to be slow since it is suggested that a peroxide catalyst, such as benzoyl peroxide, can be added to speed

Formula 9*

Solid component
 Ethyl methacrylate polymer   16.00 cc
 Benzoyl peroxide   0.04
Liquid component
 Ethyl methacrylate monomer   1.00
 Diallyl itaconate   1.00
 Styrene   2.00

Instructions: When 4 parts by volume of the solid powder is added to 1 part by volume of the liquid, and heated with boiling water for $1\frac{1}{2}$ hr, a hard copolymer is formed. With increased amounts of catalyst and a promoter, polymerization proceeds at room temperature.

 * Ref. 47.

Formula 10

Solid component
 Copolymer of ethyl acrylate
  and ethyl methacrylate,
  1:1   96%
 Benzoyl peroxide   4
Liquid component
 Methyl methacrylate   97 to 98 %
 Tri-*p*-tolylamine (tertiary
  amine)   2 to 3%

Instructions: When 1.3 parts by volume of the solid powder is mixed with 1 part by volume of the liquid, a paste forms. This can be applied with a brush to the nails. It hardens rapidly.

up the rate of polymerization. The curable paste contains a polymer, poly-(methyl methacrylate), a monomer such as methyl methacrylate, and an adhesive such as polyvinyl chloride and the nonperoxide "redox" compound. It appears to the author that the dimethyl paratoluidine acts as a promoter. Thus it probably decomposes the peroxide formed from methyl methacrylate and oxygen (air) to release free radicals and initiate the polymerization.

The advantages of the higher methacrylate esters, from the viewpoint of less odor and diminished toxicity, may be obtained in a monomer-polymer system if the monomer is polymerized by heating, preferably under an inert gas, until a gel or heavy syrup is formed. This polymer after cooling can be dissolved in the monomer and a promoter and inhibitor are added. When ready for use, the catalyst is added to the system.

All materials used for the formulation of fingernail elongators should be tested for irritation of the skin, and discoloration and other deleterious effects on the nail; these tests should be conducted both on materials singly and in the finished formulation. Particular attention is directed toward discoloration that may be caused by certain promoters (e.g., dimethyl aniline). An insufficiently tested formulation may cause very severe damage, and in extreme cases may result in falling off of the nail.

However, fingernail elongators and dental preparations containing methyl methacrylate and benzoyl peroxide have a fairly clean record (50). Canizarous (51) reported a case of contact dermatitis from a manicurist who was constantly exposed to artificial nail preparations. He found the sensitizing agent to be methyl methacrylate monomer. Schwartz and Peck (52) carried out patch tests on affected dental technicians engaged in molding dentures and found methyl methacrylate monomer and also benzoyl peroxide catalyst to be the allergens.

Topical preparations containing benzoyl peroxide have recently come into wide use for treating acne. Pace (53), in clinical studies on benzoyl peroxide preparations for acne, reported an incidence of contact sensitivity of 1 % when using a petrolatum-lanolin base. In an oil-in-water emulsion base the incidence was 2.5 %. Pace reported that all nine reactions to the cream base preparation (O/W emulsion) were females and all but one had sensitive skin. Using a controlled positive patch test, Eaglstein (54) applied 5 % benzoyl peroxide in a petrolatum base on the backs of 41 subjects. After 48 hr one control patient had a positive patch test to benzoyl peroxide. However, this subject also had a positive reaction to materials from his boots which had caused dermatitis for several years.

Manufacture and Handling

Metallic salts in the range of 0.1 to 100 ppm catalyze the polymerization of vinyl monomers. These compounds include cupric, ferric, chromic, and silver salts. However, increased quantities may inhibit the rate of polymerization. Because of the sensitivity of salts on the rate of polymerization, all equipment should be thoroughly cleaned with distilled or deionized water. All tanks used for manufacture should be of 316 stainless steel; 304 stainless steel tends to rust after long use with these materials. For gaskets and valve packing, Teflon or Turil sheet metal is excellent; Anchorite No. 450 is satisfactory where the color of the monomer is not important.

Drums containing the lower monomeric acrylic esters should be protected from direct sunlight during storage. The temperature of the drum should be maintained below 25°C. Because of the possibility of spillage and dissemination of vapors, the lower monomeric acrylic esters in drums should be handled entirely outside of the building. The monomer may then be pumped into a closed measuring tank, using an antistatic hose, preferably made of polyethylene.

Benzoyl peroxide is stable at room temperature but is flammable and is capable of exploding under certain conditions (55,56). It can be detonated by heating to 100°C and by shock. Therefore benzoyl peroxide must be protected from heat by open flame, and must not be ground or milled. Exposure to dust and direct sunlight may be hazardous (57).

The mixing kettle should be vented to the outside of the building to reduce vapors. The usual precautions with a flammable material, such as use of sparkproof motors and switches, should be taken.

CUTICLE REMOVERS

Partially covering the lunula of the nail is a special membrane called the eponychium, better known to the layman as the cuticle (58). This membrane consists chiefly of the epidermal stratum corneum skin layer. In the case of the nail, this keratinized skin collects around the nail grooves and, together with the natural fatty secretions, forms an irregular appendage which mars the beauty of an otherwise well-kept nail (59). In order to form a neat-appearing nail, the cuticle is either pushed back off the nail or is completely removed from the skin, depending on the amount of the cuticle, its toughness, and the degree of removal sought. Purely mechanical methods of removing the cuticle have been largely superseded by chemical methods, because of the difficulty of manipulation and the time consumed in the mechanical approach. Chemical removal involves either plasticizing or degrading the keratin of the stratum corneum and thus aids in pushing back or removing the cuticle.

The nail cuticle, consisting mostly of dead cells of the skin, is part of the skin epidermis or cuticle. The epidermis is a simple tissue composed entirely of cells, the outer layer of which consists of dead cells and an epidermal fatty lipoid material. This cornified layer constantly sheds its cells; the nail cuticle largely consists of this layer. However, since no separation of layers exists, and subdivisions only represent changes in cell structures as they move toward the surface, such layers often merging imperceptibly into the next (60), the subject of cuticle removal must therefore take into account the chemistry of the whole epidermis.

Following the studies of the importance of cystine in the hair, the main outline of the similar situation in the epidermis was established by Giroud and co-workers (61,62). Using the classical nitroprusside test, they found the stratum malpighi to stratum corneum to be high in free —SH groups, while the stratum corneum gave a negative reaction, or at most only a few isolated regions of —SH groups. However, by blocking free —SH groups in the epidermis and applying a reducing agent, it was shown that the stratum corneum was the only layer which produced the intense Prussian blue reaction. These observations demonstrated that the stratum corneum contains —S—S— bonds (63).

Further work showed that the —SH and —S—S— bonds played important roles in the physical properties of the epidermis, as they do in wool and hair. Rudall (64) has shown that the mammalian (including human) epidermis exhibits long-range elasticity, and the same reversible intramolecular, alpha-beta transformation (65) occurs as in hair keratin (66) and myosin (67). In addition, this transformation of the epidermis varies at different levels of the tissue structure.

Experiments using more modern histochemical sulfhydryl reagents do not fully support the older concept that the sulfhydryl groups of the malpighian layer are converted to the disulfide bridge of the keratin molecule (68–70). Instead, the findings support the theory that epidermal keratinization starts in the depths of the malpighian layer. A principal obstacle to the complete understanding of the mechanism of keratin formation is the lack of correlation between the morphological and biochemical knowledge of the process (71,72).

Chemical Basis of Cuticle Removers

Alkalis have been the chief agents used as cuticle softeners and removers for about 60 years. Although they are very harsh, they are both safe and effective. This mechanism by which alkalis attack keratin has been widely studied. One of the main reactions in alkali degradation of keratin is the breakdown of the disulfide bridge by the hydroxyl ions. Harris (73) found that dilute solutions of caustic soda were able to attack 50 % of the total cystine in wool. Horn and co-workers (74) isolated lanthionine from alkali-treated wool. Schöberl and Hornung (75), on the basis of experiments with simple cystine derivatives, suggest that an aldehyde is formed as an intermediate in the synthesis of lanthionine (76), the postulated mechanism being as follows:

$$RCH_2SSCH_2R + H_2O \longrightarrow RCH_2SH + RCH_2SOH$$
$$RCH_2SOH \longrightarrow RCHO + H_2S$$

Rosenthal and Oster (77), also working with model compounds, believe that no sulfenic acid is formed as an intermediate product, but that cleavage of the disulfide bond takes place by a β-elimination reaction, with formation of an aldehyde. However, Cuthbertson, and Phillips (78) were unable to detect aldehyde groups in alkali-treated wool, and advanced the theory that α-amino acrylic acid formed, which thus added to the thiol to form lanthionine. Further data supporting this mechanism can be found in the work of Bergmann and Stather (79), who ascertained that α-amino acrylic acid, and not an aldehyde, is produced on treating certain cystine peptides with alkali. This work and other substantiating data would indicate that peptides and keratin may react in an entirely different way from the disulfide bonds in simple organic compounds (80). In any event, it is generally agreed that an unsymmetrical molecule results from the alkali cleavage of disulfide groups in keratin, one of the products being a mercaptan.

In addition, alkali solutions react rapidly with many other groups in keratin and thus a general degradation results, although the disulfide and acid amide groups are probably the first to react. Alkalis are more destructive and less selective to peptide bonds than acids; therefore smaller peptides are formed on alkaline hydrolysis. In addition to cystine, some of the other amino acids decomposed are arginine, histidine, and serine (81).

Formulation

Until recently, the most widely used cuticle removers were based on potassium hydroxide or sodium hydroxide, in a concentration range of 1 to 5 %. Potassium hydroxide is usually preferred, since it is slightly slower and less harsh in action. Although both alkalis are used, they leave much to be desired. They may cause some burning of the skin, and for that reason irritation test of the finished product is most important. Furthermore, on long standing in glass bottles, the alkali attacks the glass, forming an unsightly precipitate (represented as $Na_4CaSi_6O_{15}$) which causes etching of the glass. In order to increase viscosity, counteract irritation, and retard evaporation, glycerol, propylene glycol, or sorbitol is added to the formulation. Formulas 11 to 13 use one of these materials, in combination with either sodium or potassium hydroxide. To counteract the harshness of the solution due to the high pH, and in order to retard the etching of the glass container, various alkali polybasic salts are used as cuticle removers. Such products are illustrated by Formulas 14 to 16.

Formulas for Cuticle Removers

	11*	12†	13‡
Sodium hydroxide	5.0%	—	2.8%
Potassium hydroxide	—	2.5%	—
Ethyl alcohol SD 39-B	37.0	—	—
Propylene glycol	10.5	—	—
Glycerol	—	19.0	11.5
Sorbitol	—	—	8.2
Perfume	0.5	0.5	—
Water	47.0	78.0	77.5

* Ref. 82.
† Ref. 83.
‡ Ref. 84.

Formulas for Cuticle Removers

	14	15*	16
Trisodium phosphate	—	8%	—
Tetrasodium pyrophosphate	—	—	8.0%
Sodium carbonate	6%	—	—
Glycerol	12	20	12.0
Perfume	q.s.	—	0.5
Rose water	—	50	—
Water	82	22	79.5

* Ref. 83.

Still milder results are obtained by using organic bases in concentrations of 5 to 10%. In the case of triethanolamine, higher concentrations may be used. Triethanolamine and other organic bases are used in Formulas 17 to 19.

Formulas for Cuticle Removers

	17	18†	19‡
Triethanolamine	12%	—	—
Isopropanolamine	—	10%	—
Monoethanolamine	—	—	6.0%
2-Amino-2-methyl-1-propanol	—	—	2.5
Glycerol	10	10	11.5
Perfume	q.s.	q.s.	q.s.
Water	78	80	80.0

† Ref. 59.
‡ Ref. 84.

These milder solutions, however, are not so effective as the more caustic alkali. Furthermore, on long standing, they too may produce a precipitate. Therefore the preparations listed thus far should be packed in an opaque glass container, and preferably fitted with a rubber, plastic, or other alkali-resistant stopper (59).

In order to prevent flaking of the glass container and to permit packing in a clear bottle, a coconut oil (85) or oleate soap can be used, with an excess amount of alkali. This approach is illustrated in Formulas 20 and 21.

Recently one of the manufacturers of ethylenediamine tetraacetic acid (EDTA) advertised that this compound added to a cuticle remover formula prevented flaking of the glass bottles. Alkalies such as sodium and potassium hydroxide usually start attacking soda-lime glass at pHs above 9 by breaking $Si-O$ bonds; the rate increases with the increasing pH and temperature. Chelating agents such as EDTA and the pyrophosphates chelate with the silicous flakes, thus preventing a precipitate, although the rate of attack of the alkali on the glass may be increased (86).

Formulas for Cuticle Removers

	20§	21§
Coconut fatty acid	10%	—
Oleic acid	—	7.0%
Tetrapotassium pyrophosphate	4	—
Potassium hydroxide	3	2.5
Triethanolamine	8	6.0
Water	75	84.5

§ Based on ref. 85.

It has been shown that small amounts of potassium silicate and chlorides or sulfates of certain bivalent and trivalent metals, such as calcium, barium, zinc, and aluminum, greatly reduce alkaline attack under certain conditions. Beryllium has a powerful inhibiting effect on alkaline attack of glass (86). There are alkaline-resistant glasses, such as Corning glass 7280, an alkali-zirconium silicate, which is about 10 times more resistant to alkali attack than soda-lime glass (87).

In addition to glass, plastic containers can be used. The polyolefins, such as polyethylene and polypropylene, are resistant to alkaline attack.

Cuticle softeners and removers in cream form have the advantages of not spilling, either out of the package or all over the hands, and of being adaptable for packaging in a tube. Two creams are illustrated in Formulas 22 and 23. Both are rather mild in their action, and would be classified as softeners rather than removers of the cuticle.

For a stronger product, which would act as a cuticle remover, one would turn to stearate soaps, glycerol, and a regulated excess of alkali; or, by altering a vanishing cream formulation, one may incorporate an excess of alkali or of organic amine, or both (59). An example of a cuticle remover using polyoxyethylene ether emulsifiers, which are not prone to hydrolysis under strongly alkaline conditions as esters, is illustrated in Formula 24. The Veegum is used for thickening. This lotion can readily be made into a cream by adjusting the emulsifiers.

Quaternary ammonium salts have been used as cuticle softeners (89). The quaternaries are used primarily for their bactericidal effects, rather than for any cuticle-removing properties. However, they give a soft, smooth feel to the cuticle because of their affinity toward protein. Steinhardt and Zaiser (90) found that for a series of alkyl quaternary ammonium compounds, the affinity of the cation for protein increased with increasing molecular size. Among organic cations of the same molecular weight, asymmetric ions had the greater affinity toward proteins. Quaternary ammonium compounds not only have a greater affinity for proteins than alkalis, but they also attack disulfide bonds at a lower pH.

Part A	*Formula 22**	*Formula 23†*
Protegin X	—	28%
Cholesterin absorption base	20.0%	—
Petrolatum, short-fiber	24.0	10
Cetyl alcohol	5.0	—
Stearyl alcohol	—	8
Cocoa butter	5.0	—
Isopropyl palmitate	—	4
Beeswax, white	12.0	—
Part B		
Water	23.0	47
Potassium hydroxide	—	2
Duponol C	—	1
Borax	0.5	—
Perfume	0.4	q.s.
Preservative	0.1	—
Tincture of benzoin	10.0	—

Procedure: Melt part A at about 80°C. Heat part B to about the same temperature, and slowly add B to A, while stirring. Allow to cool to 65°C, and then add perfume, preservative, and tincture of benzoin.

* Ref. 82.
‡ Ref. 84.

*Cuticle Remover Lotion, Formula 24**

Part A	
Veegum	1.00%
Water	74.75
Part B	
Vancide BN	0.25%
Triethanolamine	11.00
Glycerol	10.00
Part C	
Brij 30	1.20
Brij 35	1.80

Procedure: Add the Veegum to the water slowly, agitating continually until smooth. Add B to A, and heat to 70°C. Heat C to 75°C, add, and mix to 40°C.

* Ref. 88.

The quaternary ammonium hydroxides and salts are generally used in the concentration of 0.1 to 1.0 %, the former level being highly bactericidal to gram-positive and gram-negative bacteria. For cuticle-softening, it is possible to make either a clear solution or a cream. A thickening agent should be added, since the quaternaries, in low concentrations, have very little viscosity. Nonionic thickening agents, such as methocels, polyvinyl alcohols, polyacrylamide, dextran, PVP, vinyl pyrrolidone-vinyl acetate copolymers, cetyl alcohol, soap-free fatty amides, and propylene glycol alginate, may be used.

Cuticle softeners based on quaternary ammonium salts are shown in Formulas 25 to 27. In Formulas 25 and 26 the urea is used as a swelling agent

Formulas for Quaternary Ammonium Cuticle Softeners

	25	26	27
Cetyl trimethyl ammonium bromide	0.3%	—	0.5%
Cetyl trimethyl ammonium hydroxide	—	0.25%	—
Urea	5.0	5.00	—
Lanolin, anhydrous	—	—	16.5
Sorbitol	5.0	5.00	—
Cetyl alcohol	—	—	16.5
Isopropyl myristate	—	—	16.5
Witch hazel extract	89.7	—	—
Water	—	89.75	50.0

for keratin, thus softening the cuticle. In addition to the cetyl trimethyl ammonium bromide and hydroxide used in these three formulas, one may use several other quaternaries, e.g., cetyl dimethyl benzyl ammonium chloride, cetyl dimethyl ethyl ammonium bromide, lauryl dimethyl benzyl ammonium chloride, dimethyl didocenyl ammonium chloride, laurylisoquinolinium bromide, N-soya N-ethyl morpholinium sulfate, cetyl trimethyl ammonium stearate, lauryl dimethyl benzyl dimethyl ammonium chloride, octadecenyl-9-dimethyl ethyl ammonium bromide, diisobutyl cresoxy ethoxy ethyldimethylbenzyl ammonium chloride, diisobutylphenoxyethoxy ethyldimethylbenzyl ammonium chloride, octadecyl dimethyl ethyl ammonium bromide, stearyl dimethyl benzyl ammonium chloride, cetyl pyridinium chloride, and cetyl pyridinium bromide, among others.

Other compounds which have been used or suggested as cuticle removers include salicylic acid, sodium salicylate, formamide, and acetamide. Sodium salicylate is used in products presently being marketed. Formamide, suggested in concentrations of 40 to 60%, slowly hydrolyzes in water, and therefore offers a stability problem. In a homologous series, the amides decrease in effectiveness as the molecular weight increases; thus acetamide is next in effectiveness after formamide (91).

NAIL CREAMS AND LOTIONS

Nail creams and lotions are used primarily to combat brittleness of nails or fragilitas unguium, a common cause of complaint among women. The causes of this condition are manifold; slight systemic disorders or mild infections affect the rate of growth as well as the brittleness of the nail (92). The fragility of the nail may be influenced by the condition of the nail bed, which in turn is dependent on the blood supply (60). Systemic disorders which have been reported to cause the nails to become dry and tend to split easily include, among others, hypothyroidism, hyperthyroidism, and gout (93).

According to Low (94), there is also a condition known as "eggshell nail," in which the nail plate is soft, and semitransparent, bends easily, and splits at the end. This condition has been associated with arthritis, peripheral neuritis, leprosy, and hemiplegia. This affliction is somewhat similar to onychia syphilitica sicca, a dry atrophic splitting condition of the nails which may be a visible sign of late syphilis (95). Fox (58) pointed out that the systemic diseases, such as anemias, avitaminoses, and endocrine disturbances, may present important changes in the nail. Furthermore, the nails frequently become reedy in old age (93).

Brittle nails may also be a congenital condition. Such defects are commonly associated with abnormalities of the skin, hair, and teeth. For example, individuals with dry, squamous, atrophic skin, in whom the hair is usually scanty, dry, and brittle, and the teeth undeveloped and separated, often have nails that are thin, friable, and atrophic or dystrophic (96).

Many causes of brittle or reedy nails are reported to be due to external factors. Achten (97) claims a great deal of the trouble can be laid at the door of the detergents, alkalis, acids, naphtha, toluene, xylol, and alcohols used in everyday domestic work. These substances remove the lipids from the nail; they dehydrate the nail, taking out all the water-soluble substances between the keratin lattice, and even modify the keratin structure by acting on the scleroprotein side chains (disulfide bridges, electrovalent bonds, and hydrogen bonding). Repeated chemical attack of this kind alters the very fabric of the nails, which then break and split.

Cases of scaling of the nail plate allegedly caused by nail lacquers and manicuring aids have been reported (98). Silver and Chiego (92) call attention to the possible dehydrative action of solvents used in nail lacquers and in lacquer removers. It is this dehydrative action, rather than defatting, that is believed to be the cause of nail brittleness since the nail shows a low fat content, 0.15 to 0.76%, compared to the high water content, 7 to 12% (99). However, Silver and Chiego report that in a survey in which brittleness was reported to be common among housewives, the highest incidence occurred among those not using nail lacquers. These authors (100) considered alkali degradation of the keratin in the use of washing and cleansing preparations to be a primary factor in causing brittleness of the nails. This conclusion, based on experiments using 0.1 N caustic for testing, generally is thought to be too severe to represent actual, everyday use of commercial preparations. However, one must not overlook the catalytic effect that the anions of anionic surfactants have on hydrolysis of amide and peptide bonds in proteins (101).

Any local irritation of the skin in the vicinity of the nail and nail beds which may reach the nail matrix can distort the nails and change their consistency, making them less flexible (102). Prolonged immersion of the nails in water may produce maceration and infection of the nail folds with bacteria or fungi. Such conditions are common among bartenders, cooks, dishwashers, fishermen, laundry workers, and others whose occupations bring their hands into frequent contact with water (58).

Cosmetics designed for the relief of brittle nails are based on suitable emollient oils or creams. Pardo Castelló (103) advises bathing the affected nails in warm water, followed by massage with olive oil; several weeks are usually required before the nails plates regain normal consistency and elasticity. Emollient creams are said to be useful when applied several hr before the use of nail lacquers (85) or after the nails have been enameled (83). Such applications are designed to prevent the lacquer from spoiling, resulting in a loss of luster and adhesion. Harry (60) recommends application of the cream after soaking the hands in warm, soapy water and thoroughly drying them before retiring for the night. The procedure is to be carried out at least once and preferably two or three times per week (on alternate nights), the number of applications depending on the severity of the condition.

Essential constituents used in creams and lotions for maintaining elasticity of the nails are lanolin, lanolin absorption base, cocoa butter, beeswax, paraffin wax, short-fiber petroleum jelly, mineral oil, vegetable oils, cetyl alcohol, cholesterin, lecithin, and linseed oil derivatives (isolinoleates). Preparations based on these ingredients are shown in Formulas 28 to 33. Formulas 28 to 30 are beeswax-borax emulsions, whereas Formulas 31 to 33 utilize a lanolin absorption base. Attention is called to the turtle oil used in Formula 30; this ingredient had lost much of its former popularity, perhaps because of exaggerated claims: Recently, however, it has regained some of its former popularity. Although no preservative is specifically indicated in Formulas 31 and 33 it is recommended that all creams contain at least one preservative, and preferably two; the preservatives of choice in creams of this type would seem to be methyl p-hydroxybenzoate for the water phase, and propyl p-hydroxybenzoate for the oil phase.

A product containing a quaternary ammonium compound which is

Formulas for Nail Creams

	28*	29†	30‡
Beeswax	12.0%	12%	22.0 g
Lanolin, anhydrous	—	15	15.0
Petrolatum, short-fiber	24.0	—	—
Cocoa butter	5.0	8	—
Avocado pear oil	—	—	15.0
Cod liver oil	—	—	15.0
Turtle oil	—	—	15.0
Cetyl alcohol	5.0	3	—
Cholesterol	—	1	0.3
Cholesterin absorption base	20.0	—	—
Mineral oil	—	30	—
Water	23.0	30	30.0
Borax	0.5	1	1.5
Preservative	0.1	q.s.	q.s.
Tincture of benzoin	10.0	—	—
Perfume	0.4	q.s.	—
			113.8 g

* Ref. 82.
† Ref. 60.
‡ Ref. 59.

Formulas for Nail Creams

	31*	32†	33‡
Lanolin absorption base	25.0%	20.0 g	25%
Lanolin, anhydrous	—	7.0	10
Mineral oil	16.0	15.0	19
Cocoa butter	—	3.0	—
Beeswax	3.7	—	—
Lecithin (soya)	—	0.3	—
Cetyl alcohol	—	1.2	—
Water, distilled	55.0	35.5	45
Methyl *p*-hydroxybenzoate	—	0.2	—
Perfume	0.3	0.3	1
		82.5 g	

* Ref. 83.
† Ref. 104.
‡ Ref. 82.

claimed to relieve brittleness, dryness, and splitting and provides a good base coat for the application of nail polish is shown in Formula 34 (105).

Formula 34*

Stearyl trimethyl ammonium chloride	3.0%
Nonyl phenol ethoxylated (10 M)	1.5
Monoethanolamine stearate	0.5
Water	95.0

* Ref. 105.

Two interesting formulas based on *S*-carboxymethyl cysteine

$$\left(\begin{array}{c} NH_2 \\ | \\ HOOC\text{—}CH\text{—}CH_2\text{—}S\text{—}CH_2COOH \end{array} \right)$$

are shown in Formulas 35 and 36.

Formulas for Nail Creams

	35*	36*
S-Carboxymethyl cysteine	2.00%	2.0%
Cholesterol palmitate	10.00	10.0
Cholesterol	2.00	2.0
Thyroxine	0.02	—
Vitamin A palmitate	50,000 units	—
Glycerol monostearate	10.00	10.0
Lanolin hydrogenate	13.00	13.0
Sorbitol	3.00	5.0
Tween 80	4.00	4.0
Sorbic acid	—	0.2
Water, distilled	q.s. 100.00	q.s. 100
		q.s. 100.0

* Ref. 106.

A treatment that claims to prevent nail dryness and splitting nails, and to reduce brittleness, utilizes an amphoteric amino fatty acid, such as *N*-lauryl-β-amino propionic acid (Deriphat 170C.) in an organic solvent (107). The solvents are ones commonly used in the lacquer industry, such as acetone, methyl ethyl ketone, butyl acetate, and the like. The optimum results are obtained by using 80 to 70% by weight of ethyl alcohol, and from 20 to 30% by weight of acetone, as the solvent. The amino-fatty acid concentration used is from 0.75 to 4.50% by weight.

Protein and protein derivatives of the nails have been found to help in the treatment of nails. Studies have shown that proteins are sorbed by keratin in hair (108–110) and nails (111), using the technique of analyzing for the retained hydroxyproline, a compound found only in collagen, or by using tritium-labeled polypeptide (112). Proteins applied to the nail generally condition the nail, give body, and increase the resistance of the nails to cracking and splitting. In addition they can protect against deleterious substances.

Nail creams and lotions have been prepared from many different types of proteins. In one patent nail creams and lotions are prepared from proteins hydrolyzed with alkali (113). The hydrolyzed proteins can be prepared from collagens, such as ossein, and elastins or fibroins such as silk, lyssus, or the like; they use keratin such as in horns, hair, nails, feathers, or scales. Also hydrolyzed products of scleroprotein may be used, such as gelatin which is obtained by hydrolyzing collagens. These hydrolyzed proteins are passed through a cationic exchange resin to remove the sodium ions before use.

A product using one of these proteins is shown in Formula 37.

Formula for Nail Cream 37*

Feather scleroprotein (dry)	5.0 g
tert-Ester of *o*-phosphoric acid and cetyl tetraglycolic ether	15.0
Cetyl alcohol	4.5
Isopropyl myristate	25.0
Methyl *p*-hydroxybenzoate	0.3
Perfume	0.2 cc
Ethyl alcohol	5.0 cc
Water, distilled	45.0 cc

* Ref. 113.

In another patent (114), enzymatically hydrolyzed plankton obtained from hot sulfur springs is hydrolyzed to form a mixture of keratin-type amino acids and short-chain peptides. A recommended nail lotion consists of 15 parts of hydrolyzed plankton, 5 parts glycerol, and 80 parts water. The plankton contained the following amino acids: cysteine 1%, aspartic acid 3%, glutamic acid 4%, glycine 5%, proline 10%, valine 11%, phenylalanine 12%, and leucine 10%.

Another French patent (115) utilizes a hydrolyzed gelatin or collagen, molecular weight 500 to 1000. This polypeptide is then acetylated and made into a lacquer containing the acetylated polypeptide, nitrocellulose, and an organic solvent.

Richter (116) has patented an aqueous solution containing a relatively high amount (above 3.5%) of enzymatically hydrolyzed gelatin, disodium cupric ethylenediamine tetraacetate, and cystine or keratin. The preferred composition of the nail product is shown in Formula 38.

*Formula of Nail Lotion 38**

Gelatin	6.00 to 8.00%
Hydrolyzing agent for gelatin, e.g.,	
trypsin	0.04 to 0.06
Copper (EDTA) chelate	0.20 to 0.40
Hydrogen peroxide	0.50 to 2.00
Cystine	0.30 to 0.70
Water	100.0

Procedure: The gelatin is dissolved and partially hydrolyzed by heating in boiling water. After cooling to about 40°F, trypsin is added to further hydrolyze the gelatin. The copper chelating agent is added and then the hydrogen peroxide, 35%. Peroxide increases the penetrating ability of the solution (probably by decomposing some amino acids). The solution is then heated to 60°C to complete the reaction of the peroxide. The solution is allowed to cool to room temperature. Cystine which is dissolved in 28% ammonia is then added. The *p*H is adjusted to about 6 with an acid, such as tartaric acid.

* Ref. 116.

There are available commercial hydrolyzed collagen and polypeptide derivatives which are currently being used as fingernail strengtheners and conditioners or are currently being evaluated for their use.

The commercial hydrolyzed collagens or polypeptides are usually made by either alkali hydrolysis, e.g., sodium hydroxide, or by enzyme hydrolysis, e.g., papain. An advantage of the enzyme method of hydrolysis is that a low-salt polypeptide product is formed. However, low-salt or salt-free products are also available, using the alkali hydrolyzed method, such as salt-free polypeptide SF. The alkali hydrolyzed polypeptides have an average molecular weight of about 2000 and are available as the ammonium salt (Polypeptide LSN and SF) or as the sodium salt Polypeptide 37. In the enzymatic hydrolysis (papain) a polypeptide is available with an average molecular weight of 1000, WSP × 250, and with an average molecular weight of 10,000, WSP × 1000. The polypeptide derivatives, however, seem to have the greatest potential for nail use since they are available with a variety of properties. Other than molecular weight differences, they can be cationic, anionic or nonionic, and have a hydrophile-lipophile balance range so as to be soluble in oil, alcohol, and/or water. Among these derivatives are the following: Polypeptide AAS cationic polypeptide (some ammonium quaternary groups) formed from a polypeptide with an average molecular weight of 1000, having water solubility and/or alcohol solubility. Tritium tagged studies (112) show this derivative to be absorbed about 2.5 times more ($2800\ \mu g/g$) on keratin (virgin hair) than the polypeptide LSN at a 15% concentration.

Acyl derivatives of polypeptides with free carboxylic acid groups, such as oleyl polypeptide acid and WSPA 2000, a fatty acid amide polypeptide acid, show weak cationic properties below their isoelectric points and also have alcohol solubility. These derivatives are a condensation of fatty acid chloride with a polypeptide, where the fatty acid can be lauric, palmitic, oleic, or stearic acid.

A new series of proteins has recently been prepared with a hydrophile-lipophile balance ranging from 18 to 2 (117,118). These are represented by the following types of derivatives.

Super-Pro No. 5A is a water-soluble collagen-derived polypeptide, molecular weight average 500, condensed with a coconut fatty acid and partially cross-linked to high-molecular-weight polyols (above molecular weight 300). Any remaining free carboxyl groups are adjusted to neutrality, using a blend of potassium hydroxide and triethanolamine.

Super-Pro No. 11 is a complex formed by the interaction of sterols (lanolin alcohols) with an acylated low-molecular-weight polypeptide (average molecular weight 350). This is represented by the formula:

$$R'\overset{O}{\overset{\|}{C}}-\overset{H}{\overset{|}{N}}-\left[\overset{R}{\underset{H}{\overset{|}{\underset{|}{C}}}}-\overset{O}{\overset{\|}{C}}-\overset{H}{\overset{|}{N}}-\overset{R}{\underset{H}{\overset{|}{\underset{|}{C}}}}\right]_n-\overset{O}{\overset{\|}{C}}-OR''$$

where R^1 = long-chain hydrocarbon
R'' = a sterol group
R = various groups found in amino acids

This compound is completely water-insoluble and has an hydrophile-lipophile balance of about 2.

Super-Pro No. 25 is a complex which has been treated to form a phosphate with the lipoprotein, and has an HLB of about 6, intermediate between Super-Pro No. 5A and Super-Pro No. 11, on the HLB scale.

Super-Pro No. 35 is a linear polypeptide with an HLB of 3 to 4. It is more lipophylic than Super-Pro No. 25.

Proteins must be preserved. The usual preservative is formaldehyde (0.2% formalin). However, methyl and propyl or butyl p-hydroxybenzoate in combination with Roccal (benzalkonium chloride) or Bronpol (2-bromo-2-nitropropane-1,3-diol) are also effective preservatives. It should be noted that benzalkonium chloride loses activity with anionic surfactants. Dioxin 0.2% (6-acetoxy-2,4-dimetyl-m-dioxane) is also a suitable preservative which may, however, darken with aging and should be stability-tested. Also potassium sorbate 0.2% or sorbic acid, in combination with the parahydroxybenzoates, is effective at times, particularly at a pH less than six. Some protein derivatives, such as polypeptide AAS, may not require a preservative.

In addition to creams and lotions to make nails more pliable and flexible, there are preparations to strengthen and harden soft and devitalized nails. These preparations usually contain ingredients that chemically react with the keratin in the nail, either by covalent bond formation with the amino group, or by salt formation with metals having a valence number greater than 1.

Nail hardeners containing formaldehyde or formaldehyde donors caused great excitement when introduced (50). These reportedly effective preparations, however, allegedly have sensitizing capabilities, with resulting lesions and infections of the soft tissue, which includes the cuticle surrounding the nail. It may be possible to have a safe product, using up to 4% concentration of formaldehyde, if a means could be devised for shielding the surrounding soft tissue during application. The Food and Drug Administration, however, desires proof of safety for this type of product by requiring a New Drug Application.

Some attempts have been made to lessen the skin irritation potentials of formaldehyde. In a recent patent (119) 5 to 15% formaldehyde is designed to be applied once a week to the part of the nail that extends beyond the finger or toe. In another recent patent, dimethylol thiourea (made by reacting thiourea with formaldehyde) is used (120). The patentee claims to have overcome the toxic undesirable side effects of formaldehyde. Dimethylol thiourea should impart to native keratin higher elasticity and strength than formaldehyde. As with formaldehyde, dimethylol thiourea reacts with the free amino groups in keratin, forming cross-links. However, since dimethylene thiourea is a longer-chain bridge than the formaldehyde methylene bridge, it imparts greater elasticity to the nail. Furthermore, dimethylol thiourea, after reacting with the keratin, is capable of polymerizing with other dimethylol molecules, thereby forming a film which is anchored to the keratin by covalent bonds.

In addition to methylol thiourea for strengthening nails, Joos has patented two other methylol derivatives (121). One consists of a methylol carboxylic amide of the type

$$H-O-CH_2-NH-C-(X)-Y-C-X-N-H-CH_2-OH;$$

where Y is an alkylene group and X is O, S, or an NH group. The other compound is a cyclic dimethylol derivative. Aqueous solutions of methylol derivatives usually contain some free formaldehyde, and therefore the same sensitizing potentials may exist as with formaldehyde.

A nail hardener containing aldehydes other than formaldehyde is also patented (122). Aldehydes such as gluteraldehyde, dialdehyde sucrose, or dialdehyde starch, with either a water-soluble film former such as PVP,

PVP/VA copolymer, or a solvent-soluble film former such as nitrocellulose, polyvinyl acetate, or polyvinyl acetobutyrate with a plasticizer, can also be used.

Since metals with a valence number greater than 1 may form cross-linking salts between adjacent carboxy groups in keratin, various di- and trivalent metals have been recommended as fingernail strengtheners. It is claimed that an aqueous solution of 1 to 5% of the chloride, sulfate, or acetate of aluminum, zirconium, zinc, or strontium hardens or toughens the keratin in fingernails and toenails, thus increasing resistance to cracking and splitting (123). Alum has been recommended for many years as a hardener and strengthener for weak nails. A typical product is shown in Formula 39.

Formula 39*

Alum	7%
Glycerol	23
Water	70

        * Ref. 124

Other chemicals which have been used or recommended for this type of product are tannic acid (125,126), sodium iodide (127), 3,5-diiodothyronine, and tyrosine derivatives (128).

A novel new method of strengthening and repairing fingernails is one in which solutions, composed of 1% nylon fibers (3 deniers, $\frac{1}{8}$-in. long) and 99% clear nail lacquer or nail polish, is applied (129). Rayon fibers and other fibers may also be used (130,131).

NAIL DRIERS

Nail driers are aerosol preparations formulated on the principle of depositing a fine film of oil over a freshly enameled nail. Aerosol preparations being discussed elsewhere in this volume, this will be but a brief mention of the product.

The adhesive film deposited by the nail drier reduces the tackiness of the wet enamel and thus prevents it from being marred if accidentally touched against an object. Furthermore, it guards against dulling the film if it is immersed in water before the enamel is completely dry.

In addition to the propellant, the preparation usually contains an oil, such as mineral oil, refined peanut oil, olive oil, white sesame oil, a silicone, and frequently dipropylene glycol. A bactericidal agent, usually a quaternary ammonium salt, may be added to help prevent infections of the cuticle or of the skin in the vicinity of the nail. Finally, it may be necessary to use a solvent for the oil.

Nicotine Removers and Nail Bleaches

Lotions, creams, and other preparations can be formulated for effective removal of ink, tobacco, vegetable, and other stains from the fingers and nails. These products were more popular before the advent of nail lacquers and, after the lacquers were introduced, when the fashion did not call for extending the lacquer to the tip of the nail.

Nail bleaches consist in the main of oxidizing agents, such as peroxides, perborates, and hypochlorites. As shown in Formula 40, reducing agents, such as sodium sulfite, may also be used either alone or alternatively with oxidizing agents.

Formula 40*

Sodium sulfite	20%
Water	75
* Ref. 129.	

Various organic acids, such as citric, tartaric, oxalic, and acetic, are also used. In Formula 41 a simple solution of citric acid in water is shown; Formula 42 employs zinc peroxide as the oxidizing agent and titanium dioxide as the whitener; Formula 43 uses hydrogen peroxide. Since zinc peroxide decomposes on contact with water, care must be taken to eliminate moisture while preparing Formula 42.

Formulas for Nicotine Removers and Nail Bleaches

	41*	42†	43*
Citric acid	8%	—	—
Titanium dioxide	—	20.0%	—
Hydrogen peroxide (20 vol)	—	—	40%
Talc	—	20.0	—
Zinc peroxide	—	7.5	—
Petrolatum	—	26.0	—
Mineral oil	—	26.5	—
Water	92	—	60
Perfume	q.s.	q.s.	q.s.
Preservative	q.s.	—	—
* Ref. 59.			
† Ref. 82.			

An effective product may be devised by using a combination of sodium perborate monohydrate, boric or tartaric acid, china clay, and diatomaceous earth. The resultant powder is made into a paste by mixing with water just before application (104). Other formulas may consist of a scouring agent, such as powdered pumice or marble dust, in a cream base.

Nail Whiteners

The function of the nail whitener is to whiten the unenameled ends of the fingernails. The whitener is applied by means of an orange stick or directly from a pencil, to the underside of the protruding ends of the nail, and the nails are then rubbed until a uniform film is formed. The preparation lost its importance with the modern trend to apply lacquer to the entire surface of the nail. However, there is still a small market among women who have not followed the modern fashion in nail lacquers and who desire white tips (132).

The nail whitener can be a cream or a pencil, the latter being a harder, stiffer product. The product usually consists of a white pigment and a vehicle, the latter varying according to the degree of hardness desired. Suitable pigments include titanium dioxide, zinc oxide, lithopone, zirconium hydroxide, and extenders such as kaolin (white grade), talc, and chalk (83). The pigment is usually present in a proportion of 20 to 30%. For a cream product, a soft ointment base can be used; for sticks and pencils, one uses such materials as stearic acid, beeswax, and ozokerite. Liquid nail whiteners can be made by suspending the pigments in a water-soluble resin or gum.

Nail Polish Powders

Today little more than an historical curiosity, the nail polish powders or creams were used before the modern nail lacquers, serving as buffing or polishing agent for the nail. A typical product is illustrated by Formula 44.

Formula 44

Tin oxide	90%
Titanium dioxide	8
Oleic acid	2
* Ref. 82.	

REFERENCES

1. False length for nails, *Life*, **37**: 87 (October 25, 1954).
2. Acrylics for the cuticle, *Chem. Week*, **75**: 74 (November 13, 1954).
3. Schildknecht, C. E.: *Vinyl and related polymers*, Wiley, New York, 1952.
4. Mark, H. F., and Tobolsky, A. V.: *Physical chemistry of high polymeric systems*, Interscience Publishers, New York-London, 1950, p. 401.
5. Chalmers, W.: The mechanism of macropolymerization reactions, *Can. J. Research*, **7**: 113 (1932); *ibid.*: *J. Am. Chem. Soc.*, **56**: 912 (1934).

6. Staudinger, H., and Frost, W.: Über hochpolymere Verbindungen, 129. Mitteil. Über die Polymerisation als Kettenreaction, *Ber.*, **68**: 2351 (1935).

7. Mark, H., and Raff, R.: Die Kinetik der thermischen Polymerisation von Styrol, *Z. physik. Chem.*, **B31**: 275 (1936).

8. Flory, P. J.: The mechanism of vinyl polymerizations, *J. Am. Chem. Soc.*, **59**: 241 (1937).

9. Gee, G., and Rideal, E. K.: The kinetics of polymerisation processes, *Trans. Faraday Soc.*, **31**: 969 (1935).

10. Gee, G.: The kinetics of polymerisation processes. Part II, *Trans. Faraday Soc.*, **32**: 656 (1936).

11. Gee, G., and Rideal, E. K.: The kinetics of polymerisation processes. Part III. The effects of catalysts and inhibitors, *Trans. Faraday Soc.*, **32**: 666 (1936).

12. Cuthbertson, A. C., Gee, G., and Rideal, E. K.: On the polymerization of vinyl acetate, *Proc. Roy. Soc. London*, **A170**: 300 (1939).

13. Melville, H. W.: The photochemical polymerization of methyl methacrylate vapour, *Proc. Roy. Soc. London*, **A163**: 511 (1937).

14. Schulz, G. V., Dinglinger, A., and Husemann, E.: Über die Kinetik der Kettenpolymerisationen. VII. Die thermische Polymerisation von Styrol in verschiedenen Lösungsmitteln, *Z. physik. Chem.*, **B43**: 385 (1939).

15. North, A. N.: *The international encyclopedia of physical chemistry and chemical physics*, Pergamon Press, New York, Topic 17:37, 1966.

16. Kirk, R., and Othmer, D.: *Encyclopedia of chemical technology*, Interscience Publishers, New York, 1953, Vol. 10, p. 962.

17. Billmeyer, F. W.: *Textbook of polymer science*, Interscience Publishers, New York-London, 1964.

18. Flory, P. J.: *Principles of polymer chemistry*, Cornell University Press, Ithaca, N.Y., 1953.

19. Schildknecht, C. E.: *Vinyl and related polymers, their preparation, properties, and applications in rubbers, plastics, fibers, and in medical and industrial arts*, John Wiley, New York, and Chapman & Hall, London, 1952.

20. Barnes, C. E.: Mechanism of vinyl polymerization. I. Role of oxygen, *J. Am. Chem. Soc.*, **67**: 217 (1945).

21. Barnes, C. E., Elofson, R. M., and Jones, G. D.: Role of oxygen in vinyl polymerization. II. Isolation and structure of the peroxides of vinyl compounds, *J. Am. Chem. Soc.*, **72**: 210 (1950).

22. Mayo, F. R., Lewis, F. M., and Walling, C.: Copolymerization. VIII. The relation between structure and reactivity of monomers in copolymerization, *J. Am. Chem. Soc.*, **70**: 1529 (1948).

23. Breitenbach, J. W., Springer, A., and Horeischy, K.: Die stabilisierende Wirkung des Hydrochinons auf die Wärmepolymerisation des Styrols, *Ber.*, **71**: 1438 (1938).

24. *Ibid.*: Zur Kenntnis der Versögerung der Wärmepolymerisation des Styrols durch *p*-Benzochinon, *Ber.*, **74**: 1386 (1941).

25. Spealman, C. R., Main, R. J., Haag, H. B., and Larson, P. S.: Monomeric methyl methacrylate—studies on toxicity, *Ind. Med.*, **14**: 292 (1945).

26. Pozzani, U. C., Weil, C. S., and Carpenter, C. P.: Subacute vapor toxicity and range-finding data for ethyl acrylate, *J. Ind. Hyg. Toxicol.*, **31**: 311 (1949).

27. Treon, J. T., Sigmon, H., Wright, H., and Kitzmuller, K. V.: The toxicity of methyl and ethyl acrylate, *J. Ind. Hyg. Toxicol.*, **31**: 317 (1949).

28. Riddle, E. H.: *Monomeric acrylic esters*, Reinhold, New York, 1954.

29. "Lucite" acrylic resin: molding of monomer-polymer doughs, Information Bulletin No. X-27c, E. I. du Pont de Nemours & Co., Inc., Polychemicals Department, 1954.

30. Copolymerization of monomeric acrylic esters, Rohm & Haas, 1954.

31. I. G. Farbenindustrie Aktiengesellschaft: Br. Pat. 498,464 (1939).

32. Norrish, R. G. W., and Brookman, E. F.: The mechanism of polymerization reactions. I. The polymerization of styrene and methyl methacrylate, *Proc. Roy. Soc. London*, **A171**: 147 (1939).

33. Trommsdorff, E., Köhle, H., and Lagally, P.: Zur Polymerisation des Methacrylsäuremethylesters, *Makromol. Chem.*, **1**: 169 (1948).

34. "Lucite" acrylic resin: embedment of specimens, Information Bulletin No. X-28b. E. I. du Pont de Nemours & Co., Inc., Polychemicals Department.

35. *Modern Plastics*, encyclopedia issue, 1955.

36. Knock, F. E., and Glenn, J. F.: U. S. Pat. 2,558,139 (1951).

37. Horner, L., and Scherf, K.: Über den Einfluss der Substituenten auf die chemische Reaktivität. I. Der Zerfall von Dibenzylperoxyd durch substituierte tertiäre Amine, *Ann.*, **573**: 35 (1951).

38. Horner, L., and Scherf, K.: *Ibid.* II. Über die Ursachren der "sterischen Hinderung" bei tertiären Aminen, *Ann.*, **574**: 202 (1951).

39. Horner, L., and Scherf, K.: *Ibid.* III. Versuche zum "elektroduktilen" Charakter ungesättigter Systeme, *Ann.*, **574**: 212 (1951).

40. Horner, L., and Betzel, C.: Studien zum Ablauf der Substitution. V. Beitrag zum experimentellen Nachweis von Durchgangsradikalen, *Ann.*, **579**: 175 (1953).

41. Hunt, M.: U. S. Pat. 2,471,959 (1949).

42. Burk, R. E.: U. S. Pat. 2,500,023 (1950).

43. Arnett, L. M.: Kinetic of the polymerization of methyl methacrylate with aliphatic azobisnitriles as initiators, *J. Am. Chem. Soc.*, **74**: 2027 (1952).

44. Agre, C. L.: U. S. Pat. 2,367,660 (1945).

45. Christ, R. E.: U. S. Pat. 2,367,670 (1945).

46. Knock, F. E.: U. S. Pat. 2,569,767 (1951).

47. Feigan, R. C., and Prange, C. H.: U. S. Pat. 2,318,845 (1943).

48. Bidoul, A.: Belg. Pat. 630,562 (1963).

49. Inter-Taylor A. G.: Fr. Pat. 1,435,000 (1966); Ger. Appl. Jan. 29, 1964.

50. Wells, F. V., and Lubowe, I. I.: *Cosmetics and the skin*, Reinhold, New York, 1964, p. 293.

51. Canizarous, O.: Contact dermatitis due to the acrylic materials used in artificial nails, *Arch. Dermatol.*, **74**: 141 (1956).

52. Schwartz, L., and Peck, S. M.: *Cosmetics and dermatitis*, Hoeber, New York, 1946.

53. Pace, W. E.: Benzoyl peroxide-sulfur cream for acne vulgaris, *Can. Med. Assoc. J.*, **93**: 252 (1965).

54. Eaglstein, W. H.: Allergic contact dermatitis to benzoyl peroxide, *Arch. Dermatol.*, **97**: 527 (1968).

55. Davies, A. G.: *Organic peroxide*, Butterworth, London, 1961, p. 197.

56.   Manly, T. D.: Organic peroxy compounds in industry, *Ind. Chem.*, **32**: 319 (1956).

57.   Lappin, F. R.: Safety warnings concerning benzoyl peroxide, dewar flasks, *Chem. Eng. News*, **26**: 3518 (1948).

58.   Lesser, M. A.: The nails, *DCI*, **49**: 511 (1941).

59.   Harry, R. G.: Manicure preparations, *Mfg. Perf.*, **4**: 108 (1939).

60.   Harry, R. G.: *Modern cosmeticology*, 4th ed., Leonard Hill, London, 1955, 1968.

61.   Giroud, A., Bulliard, H., and Leblong, C. P.: Les deux types foundamentaux de kératinisation, *Bull. histol. appl. tech. microsc.*, **11**: 129 (1934).

62.   Giroud, A., and Bulliard, H.: Les substances à fonction sulfhydryle dans l'épiderme, *Arch. anat. microsc.*, **31**: 271 (1935).

63.   Rudall, K. M.: The proteins of the mammalian epidermis, in Anson, M.L., Bailey, K., and Edsall, J. T.: *Advances in protein chemistry*, Academic Press, New York, 1952.

64.   Rudall, K. M.: Symposium on fibrous proteins, *J. Soc. Dyers Colourists*, 1946, p. 15.

65.   Gershon, S. D., Goldberg, M. A., and Rieger, M. M.: Permanent waving, in this book.

66.   Astbury, W. T., and Woods, H. J.: X-ray studies of hair, wool, and related fibres. II. The molecular structure and elastic properties of hair keratin, *Philos. Trans. Roy. Soc. London*, **232A**: 333 (1933).

67.   Astbury, W. T., and Dickinson, S.: X-ray studies of the molecular structure of myosin, *Proc. Roy. Soc. London*, **B129**: 307 (1940).

68.   Van Scott, E. J., and Flesch, P.: Sulfhydryl and disulfide in keratinization, *Science*, **119**: 70 (1954).

69.   Mescon, H., and Flesch, P.: Modification of Bennett's method for the histo-chemical demonstration of free sulfhydryl groups in skin, *J. Invest. Dermatol.*, **18**: 261 (1952).

70.   Barrnett, R. J.: The histochemical distribution of protein-bound sulfhydryl groups, *J. Natl. Cancer Inst.*, **13**: 905 (1953).

71.   Norman, I., and Porter, G.: Trapped atoms and radicals in a glass "cage," *Nature*, **174**: 509 (1954).

72.   Butcher, F. O., and Segunaes, R. F.: *Fundamentals of keratization*, American Association for Advancement of Science, Washington, D.C., 1962.

73.   Harris, M.: Effect of alkalies on wool, *J. Res. Natl. Bur. Stand.*, **15**: 63 (1935).

74.   Horn, M.J., Jones, J. B., and Ringel, S. J.: Isolation of a new sulfur-containing amino acid (lanthionine) from sodium carbonate-treated wool, *J. Biol. Chem.*, **138**: 141 (1941).

75.   Schöberl, A., and Hornung, T.: Über die Einwirkung von Alkalien auf Cystin und Cystinderivate: ein Betrag zur Frage des labilen Schwefels in Eiwesstoffen, *Ann.*, **534**: 210 (1938).

76.   Schöberl, A., and Hornung, T.: *op. cit.*, through Alexander, P., and Hudson, R. F.: *Wool: its chemistry and physics*, Reinhold, New York, 1954, p. 255.

77.   Rosenthal, N. A., and Oster, G.: Recent progress in the chemistry of disulfides, *JSCC*, **5**: 286 (1954).

78. Cuthbertson, W. R., and Phillips, H.: The action of alkalis on wool. I. The subdivision of the combined cystine into two fractions differing in their rate and mode of reaction with alkalis, *Biochem. J.*, **39**: 7 (1945).

79. Bergmann, M., and Stather, F.: Umlagerungen peptidähnlicher Stoff. 7. Umwandlung eines cystinhaltigen Diketopiperazins, *Hoppe-Seylers Z. physiol. Chem.*, **152**: 189 (1926).

80. Alexander, P., and Hudson, R. F.: *op. cit.* (ref. 76), p. 257.

81. Warner, R. C., and Cannan, R. K.: The formation of ammonia from proteins in alkaline solution, *J. Biol. Chem.*, **142**: 725 (1942).

82. Thomssen, E. G.: *Modern cosmetics*, Drug & Cosmetic Industry, New York, 1947.

83. Jannaway, S. P.: Manicure preparations, *PEOR*, **29**: 472 (1938).

84. Keithler, W. R.: *The formulation of cosmetics and cosmetic specialties*, Drug & Cosmetic Industry, New York, 1956.

85. deNavarre, M. G.: *The chemistry and manufacture of cosmetics*, Van Nostrand, New York, 1941.

86. Bacon, F. R.: The chemical durability of silicate glass, *Glass Ind.*, **49**: 1 (1968).

87. Kirk, R., and Othmer, D.: *Encyclopedia of chemical technology*, Vol. 10, 2nd ed., Interscience-Wiley, 1966, p. 580.

88. R. T. Vanderbilt Co., Technical Bulletin No. 56.

89. Hilfer, H.: Manicure preparations, lacquer removers, *DCI*, **64**: 432 (1949).

90. Steinhardt, J., and Zaiser, E. M.: Combination of wool protein with cations and hydroxyl ions, *J. Biol. Chem.*, **183**: 789 (1950).

91. Thuesen, D.: U. S. Pat. 2,041,158 (1936).

92. Silver, H., and Chiego, B.: Nails and nail changes: brittleness of nails (fragilitas unguium), *J. Invest. Dermatol.*, **3**: 357 (1940).

93. Fungous infection of nails (queries and minor notes), *J. Am. Med. Assoc.*, **115**, 1216 (1940).

94. Low, R. C.: Changes in the nail as an aid to diagnosis and prognosis, *Practitioner*, **142**: 627 (1939).

95. Fox, E. C.: Diseases of the nails: report of cases of onycholysis, *Arch. Dermatol. Syphilol.*, **41**: 98 (1940).

96. Schwartz, L., and Tulipan, L.: *Text-book of occupational diseases of the skin*, Lea & Febiger, Philadelphia, 1939.

97. Achten, G.: The normal nail, *Am. Perf.*, **79**: 23 (September 1964).

98. Hollander, L.: Dermatitis produced by cosmetics (la gerardine), *J. Am. Med. Assoc.*, **101**: 259 (1933).

99. Silver, H., and Chiego, B.: Nails and nail changes: modern concepts of anatomy and biochemistry of nails, *J. Invest. Dermatol.*, **3**: 133 (1940).

100. Chiego, B., and Silver, H.: Effect of alkalis on stability of keratins, *J. Invest. Dermatol.*, **5**: 95 (1942).

101. Steinhardt, J., and Fugitt, C. H.: Catalyzed hydrolysis of amide and peptide bonds in proteins, *J. Res. Natl. Bur. Stand.*, **29**: 315 (1942).

102. Brittle, splitting nails (queries and minor notes), *J. Am. Med. Assoc.*, **113**: 444 (1939).

103. Pardo Castelló, V.: *Diseases of the nails*, Charles C. Thomas, Springfield, Ill., 1936.

104. Vallance, J. M.: Cosmetic specialties. 3. Manicure and bathroom preparations, *Mfg. Perf.*, **4**: 283 (1939).

105. Drake, R., and Whitley, L.: U. S. Pat. 3,034,965 (1962).

106. Joillie, M., Laurre, M., Maillard, G., and Muller, P.: Fr. Pat. M-3111 (1965) and Belg. Pat. 650,426 (1965).

107. McKissick, R. W., and Eberhard, J. F.: U. S. Pat. 3,441,64S (1969).

108. Bouthilet, R., and Karler, A.: Cosmetic effects of substantive proteins, *Proc. Sci. Sec. TGA*, **44**: 27 (1965).

109. Karjala, S. A., Williamson, J. E., and Karler, A.: Studies on the substantivity of collagen-derived polypeptides to human hair, *JSCC*, **17**: 513 (1966).

110. Karjala, S. A., Johnsen, V. L., and Chiostri, R. F.: Substantive proteins in cosmetics, *Am. Perf.*, **82**: October 1967.

111. Johnsen, V. L.: Private communication.

112. Riso, R.: Private communication.

113. Bonadeo, J.: Br. Pat. 1,111,934 (1968).

114. Benad, J.: Fr. Pat. 1,536,017 (1967).

115. Rutkin, P., and Laford, L.: Fr. Pat. 1,572,598 (1969).

116. Richter, A.: U. S. Pat. 3,257,280 (1966).

117. Riso, R.: Protein derivatives, *Aerosol Age*, **14**: 30 (December 1969).

118. Riso, R.: Proteins: then and now, *Am. Perf.*, **85**: 129 (September 1970).

119. Knudson, M.: U. S. Pat. 3,382,151 (1968).

120. Joos, B.: U. S. Pat. 3,349,000 (1967).

121. Joos, B.: Fr. Pat. 1,556,612 (1969).

122. Zvick, C., Bouvet, R., and Hittner, S.: Fr. Pat. 1,521,072 (1968).

123. Williams, E., and Williams, D.: U. S. Pat. 3,034,966 (1962).

124. Winter, F.: Manicure preparations, *SPC*, **20**: 44 (January 1947).

125. Bersis, M.: Fr. Pat. 1,411,389 (1965).

126. Balsiger, G. F.: U. S. Pat. 3,510,554 (1970).

127. Achten, G.: The normal nail, *Am. Perf.*, **79**: 23 (September 1964).

128. Hellbaum, A.: Br. Pat. 859,546 (1961).

129. Jewel, P. W.: U. S. Pat. 3,301,760 (1967).

130. Weisman, M.: Fr. Pat. 1,453,089 (1966).

131. Max Factor: Fr. Pat. 1,529,329 (1968).

132. Hilfer, H.: Manicure preparations, *DCI*, **64**: 556 (1949).

Chapter 31

EYE LOTIONS

John D. Mullins

Products for use in the eye are classified as ophthalmic preparations and are defined officially in the United States Pharmacopeia as "sterile solutions free from foreign particles and suitably compounded for instillation into the eye" (1). Products of this type are also called eyedrops, eyewashes, and collyria.

It seems unnecessary to define cosmetic eye lotions as significantly different from other ophthalmic products. Intended use, pharmacologic activity, labeling, etc. will certainly bear on the final product classification as a drug or as a cosmetic. Certainly neither classification implies any diminution of specifications, manufacturing requirements, or product characteristics. The considerations general to ophthalmic products apply equally to cosmetic or pharmaceutical compositions.

The definition of an ophthalmic product as a sterile solution is a rather recent event. The United States Pharmacopeia XV (1955) was the first to include sterility as a requirement (2). The Food and Drug Administration took the position in 1953 that a nonsterile ophthalmic solution was adulterated (3).

Prior to World War II, one finds only a scattering of prepared ophthalmic solutions in either the cosmetic or the pharmaceutical categories. In most cases such products were prepared in the retail pharmacy and were intended for immediate use. This is reflected by the literature on subjects such as product stability, buffers, etc. Even in the 1940's and early 1950's stability is discussed in terms of immediate use (4).

In an historical sense, eye lotions can be traced to antiquity. Collyria are mentioned in the Evers papyrus and in the Hippocratic codex (5). The drug belladonna derives its name from the Italian *bella* (beautiful) *donna*

581

(lady), and reflects the use of the juice of this plant to dilate the pupils of such ladies during the Middle Ages (6).

General Considerations

The human eye is a particularly sensitive organ. It reacts quickly to any change in environment. For this reason, solutions for use in the eye must be prepared with meticulous care. Requirements which must be considered in the preparation of ophthalmic products are as follows:

Sterility
Clarity
Buffer and pH
Tonicity
Preservatives
Additives
Safety and irritation
Packaging and stability
Manufacturing
Products

Many, if not all, of these considerations are interrelated and cannot be regarded as isolated factors. Stability is of course related to pH and buffer and to packaging. The buffer system must be considered with tonicity and comfort in mind, and so on.

Sterility

The sterilization of eye products is a major factor in preventing serious eye infections and must be considered both the most important and most exacting procedure in the preparation of products for use in the eye. Failure to achieve sterility may cause serious eye injury, including the possibility of complete loss of vision (7).

The means of achieving sterility are varied and dependent on the physical and chemical characteristics of the product and its components, manufacturing scale, equipment, and product packaging. General procedures include the following:

Dry heat
Steam under pressure
Sterilizing solutions
Gas, i.e., ethylene oxide
Ultraviolet light
Filtration
Irradiation, i.e., Cobalt[60], etc.

The details of each of these procedures are given in the United States Pharmacopeia XVII, pp. 830 and following (8). In general, however, a manufactured eye lotion would include several sterilization procedures rather than a single method. Equipment, packaging, raw materials, and finished product must be fitted to the most appropriate methods in order to achieve total product sterility.

Dry heat sterilization is normally carried out at 160 to 170°C for 2 to 4 hr. Time and temperature schedules must be determined for the specific material to be treated. It is particularly important that warm-up time, circulation, and other variables be compensated for in order to assure exposure of the material itself to the required temperature and time cycle.

Steam sterilization is carried out in an autoclave and is based on the sterilizing action of steam under pressure. Temperature is usually 121°C, measured at the steam discharge line. To assure proper heat distribution, recording thermocouples sealed into representative containers are often recommended. Both the temperature and the time at a specific temperature should be recorded.

Heat-labile materials in solution may be sterilized by filtration through a suitable sterilizing medium. Filtration media vary considerably in the means by which sterilization is effected as well as in composition. Sterile filtration by adsorption or by a physical sieving mechanism should yield a sterile, particulate-free filtrate. The filter medium should not react with or remove components of the solution, i.e., preservatives, buffer salts, etc.

Sterile filtration may be accomplished by the use of suitable porosity membrane filters, filter pads, microporous porcelain, or specially sintered materials. A variety of each of these materials is available commercially, together with data on typical uses. Reactivity or a lack of reactivity of filter media should never be presumed. Specific products should be evaluated as a part of the process of selecting a suitable filtration medium.

Gas sterilization of heat-sensitive substances may be carried out by exposure to either propylene oxide or ethylene oxide. The latter is more commonly used and requires specialized, although not elaborate, equipment usually described as a gas autoclave.

Ethylene oxide for sterilization purposes is commercially available, diluted either with carbon dioxide or with halogenated hydrocarbons. Ethylene oxide is a colorless flammable gas at standard temperature and pressure. The gas liquefies below 12°C. Ethylene oxide is highly irritating to the eyes and to mucous membranes. Chemical structure is given as

$$CH_2\text{------}CH_2$$
$$\diagdown \quad \diagup$$
$$O$$

The use of ethylene oxide as a sterilizing agent was at one time severely limited because of its flammable and explosive properties. The use of carbon dioxide as a diluent, however, greatly reduces this danger. The commercial mixture known as Carboxide® contains 10% ethylene oxide and 90% carbon dioxide. A list of the commercially available ethylene oxide mixtures is given in Table I.

TABLE I. Ethylene Oxide Mixtures Available Commercially

Trade name	Mixture	Manufacturer
Cry-Oxide®	11% Ethylene oxide 79% Trichloromonofluoromethane 10% Dichlorodifluoromethane	Ben Venue Laboratories, Bedford, Ohio
Benvicide®	11% Ethylene oxide 54% Trichloromonofluoromethane 35% Dichlorodifluoromethane	The Matheson Co., East Rutherford, N.J.
Pennoxide®	12% Ethylene oxide 88% Dichlorodifluoromethane	Pennsylvania Eng. Co. Philadelphia, Pa.
Steroxide-12®	12% Ethylene oxide 88% Dichlorodifluoromethane	Castle Ritter Pfaudler Corp., Rochester, N.Y.
Carboxide®	10% Ethylene oxide 90% Carbon dioxide	Union Carbide Corp., Linde Division, New York, N.Y.
Oxyfume Sterilant-20®	20% Ethylene oxide 80% Carbon dioxide	Union Carbide Corp., Linde Division, New York, N.Y.
Steroxide-20®	20% Ethylene oxide 80% Carbon dioxide	Castle Ritter Pfaudler Corp., Rochester, N.Y.

Sterilization by the use of ethylene oxide demands careful consideration of conditions required to effect sterility. Temperature and pressure requirements are somewhat low in contrast to wet or dry methods; however, a careful control of exposure time, ethylene oxide concentration, and humidity is essential to achieve sterility.

Temperature usually ranges from 50 to 60°C. Ethylene oxide concentration may vary somewhat depending on humidity and exposure period; however, the usual gas concentration ranges from 500 to 1000 mg/liter with a 40 to 80% relative humidity and a 3 to 6 hr exposure cycle (9). More recently Aspery (10) reports sporicidal results using 90% humidity, 714 mg/liter ethylene oxide, and an exposure period of 3 to $3\frac{1}{2}$ hr.

Ethylene oxide exerts its bactericidal activity by alkylation of protein molecules. With the increased use of this method of sterilization, the potential for human toxicity has become obvious. In order to avoid harmful contact an aeration period should be included as a part of the total ethylene oxide sterilization cycle. The length of time necessary to dissipate possibly toxic residues depends on the material being sterilized, the packaging, sterilization cycle, and the end use of the article.

In any consideration of possible toxic residues from gas sterilization, both ethylene oxide and its reaction products must be included. Ethylene glycol is formed from the reaction of ethylene oxide with water. Ethylene halohydrins are formed if halogen ions are available for reaction. Polyvinyl chloride, for example, may yield ethylene chlorohydrin by reaction with ethylene oxide. Because of conflicting results and varying procedures, there is as yet no agreement on the specific toxicity levels obtained with ethylene oxide or its reaction products. Cleary (11) has reviewed this situation. His discussion reflects the concern over the possible toxicity of sterilant residues and also summarizes attempts to undertake comprehensive analytical and toxicologic studies of these residues.

Clarity

As officially defined, ophthalmic products must be free from foreign particles. Clarity is usually achieved by filtration. The degree of clarity can be estimated by the use of light-scattering devices such as the Coleman Nepho-Colorimeter. Nephelos units are numerical values on a 100 scale and are used to define the region between absolute clarity and visible turbidity. The nephelos unit then is a measure of departure from clarity (12).

Solution clarity is normally obtained by proper filtration either as a part of sterile filtration or by the use of a clarifying filter. Nonfibrous filters such as clarifying membranes or in-line sintered glass are often effective in clarifying solutions. It must be realized, of course, that a clarified solution must be filled into a clean container. This implies both particle-free rinse water and a particle-free filling area.

Buffer and *p*H

Product *p*H and the buffer required to establish *p*H are a part of product design and are usual essentials for product stability. The *p*H of the tear fluid is generally accepted to be approximately 7.4, although this value may vary. The active ingredients commonly used in eye lotions are usually acid salts of weak bases; as such, these are most stable at an acid *p*H.

This, then, is the usual dilemma in the formulation of eye products. Good product stability requires an acid *p*H. Product comfort, on the other hand, should ideally be based on a *p*H of approximately 7.4. It is not possible to compromise product stability. On the other hand, some formulation latitude is permissible in choice of buffer salts and buffer capacity.

It is generally accepted that a low (acid) *p*H *per se* will not necessarily cause stinging or discomfort during use. Difficulties generally arise as a result of buffer capacity at a specific *p*H rather than the *p*H value itself.

Solution pH is, of course, a measure of acidity or alkalinity; it is defined as the negative log of the hydrogen ion concentration. "Buffer capacity" is a term used to indicate the capacity or ability of a buffer system to resist change. A properly formulated eye product should include a buffer with a capacity sufficient to maintain product pH during the proposed shelf life. At the same time, the capacity should be minimized in order to permit the tear fluid to readjust to pH 7.4 following instillation of the product into the eye. Discomfort is caused by the inability of the tear fluid to overcome the buffer and adjust pH.

Many buffer solutions have been recommended for use in eye products. The U.S. Pharmacopeia (13) lists several together with suggested pH. It should be emphasized that if product stability is considered a function of pH, then stability should be verified by specific tests. Stability should never be accepted by reference.

Tonicity

Tonicity refers to the osmotic pressure exerted by salts in solution. A solution is generally considered isotonic with another solution when the magnitudes of the colligative properties of the solutions are equal. As a convenience, a solution is said to be isotonic when its tonicity is equal to that exerted by an 0.9% aqueous solution of sodium chloride. Since isotonicity implies equivalent colligative properties, freezing-point depression is usually selected as the basis for determining sodium chloride equivalency.

As a general rule, eye lotions should be isotonic in order to minimize discomfort. Normally this is not difficult to achieve. As a practical matter, exact isotonicity is not a strict prerequisite for product comfort. The eye can usually tolerate solutions equivalent to a range of 0.5 to 1.8% sodium chloride (14).

Solution tonicity has been investigated extensively over the years. Since osmotic pressure is a function of particles in solution, it is in turn dependent on the use of electrolytes and nonelectrolytes, and on the degree of ionization in addition to concentration. One result of prior investigations has been the accumulation of a large number of sodium chloride equivalents. These values have been computed for many of the common buffer salts, preservatives, and the like, and are published in various reference texts. The *Merck Index* (15) is a typical excellent source of such data.

Pharmaceutical tests such as Husa (16) and *Remington's Practice of Pharmacy* (17) discuss isotonic solutions in detail including several calculation procedures. These are reviewed and analyzed by Martin (18), who also includes isotonic values for a wide variety of substances.

Preservatives

With the single exception of unit dose products, sterile ophthalmic solutions must contain a suitable antimicrobial preservative to prevent the growth of microorganisms inadvertently introduced during use. The need for adequate preservation of ophthalmic solutions was recognized and discussed during the 1930's (19). The possibility of contamination was realized at that time; however, means of preservation ranged from the use of preservatives such as chlorobutanol to daily sterilization by boiling. The early use of preservatives was, in fact, recommended for economic reasons and to obviate the "tiresome" daily boiling.

The selection of an adequate chemical preservative for ophthalmic solutions is by no means a simple procedure. Ideally a preservative substance should meet the following criteria:

Broad spectrum antibacterial activity to include efficacy against both Gram-positive and Gram-negative organisms.
Satisfactory chemical and physical stability over a wide pH range.
Compatibility with chemical components and with packaging materials.
Safe and nonirritating during use.

Preservative substances must be evaluated for suitability as a part of the total product. *In vitro* bacteriological studies must be carried out against representative organisms present in adequate concentrations for specific time periods. Activity against several strains of *Pseudomonas aeruginosa* should be included in any preservative evaluation. This organism is recognized as a most dangerous eye pathogen. Detailed procedures for investigating the antimicrobial efficacy of preservatives are given in the United States Pharmacopeia XVIII (20).

An evaluation of the stability of preservatives should include an indication of chemical stability. A lack of stability could result in the formation of decomposition products which may be irritating or which may be inactive as preservatives. Chlorobutanol typifies this situation. Although this substance has been used as a preservative for many years, its chemical stability is highly dependent on pH. Stability decreases sharply as the pH is increased above the pH 5 to 6 range (21).

Many preservative substances may be inactivated by binding with other formulation additives, in particular with macromolecular compounds. These effects have been investigated by Kostenbauder and others (22–28).

Results indicate a selective inactivation which must be recognized by the formulator. It must also be recognized that generalizations may be overly limiting. For example, chlorobutanol and benzyl alcohol are inactivated by polysorbate 80 and polyvinylpyrrolidone, but not by methylcellulose.

Cetylpyridinium chloride is inactivated by methylcellulose; benzalkonium is not. This has been reviewed extensively by Lachman (29).

Preservative safety testing is normally carried out as a part of the general toxicologic assessment of the ophthalmic product. It should be recognized, however, that preservatives are potent substances and can easily cause eye irritation or discomfort even in comparatively low concentrations.

There are relatively few classes of chemical compounds which are used as preservatives in ophthalmic solutions. In no class do we have the ideal preservative substance. The general classes together with specific examples are reviewed in the following.

Quaternary Ammonium Compounds

These substances are represented chemically as

$$\left[R-\overset{\overset{\displaystyle R_2}{|}}{\underset{\underset{\displaystyle R_4}{|}}{N}}-R_3 \right]^+ \quad Y^-$$

The nitrogen is pentavalent with the R–N linkage nonionic. The N–Y valence is ionic; typically Y is a halogen. R' is nonpolar, generally a long-chain aliphatic; R_2, R_3, and R_4 may include short-chain aliphatics, an aromatic, or a heterocyclic group. Benzalkonium Chloride, USP, is a mixture of alkyl dimethyl benzyl ammonium chlorides of the general formula

$$[C_6H_5CH_2N(CH_3)_2R]Cl,$$

in which R represents a mixture of the alkyls from C_8H_{17} to $C_{18}H_{37}$ (30).

Quaternary ammonium preservatives typically are strongly cationic in reaction. They will react with and be inactivated by soaps and anionic materials. They are generally incompatible with salicylates and nitrates and may adsorb strongly on rubber or plastic surfaces.

This class of compounds is best exemplified by benzalkonium chloride, USP. This substance is the single most commonly used preservative in ophthalmic solutions, although it is by no means the ideal preservative.

Benzalkonium is commonly used in concentrations of 0.004 to 0.01 % (1:25,000 to 1:10,000) although 0.02 % (1:5,000) is used in some ophthalmic products. Preservative efficacy is generally satisfactory against a broad spectrum of organisms, although it is not uniformly effective against *Ps. aeruginosa* strains.

Benzalkonium is generally compatible with other preservatives and is occasionally combined with mercurials or aromatic compounds such as phenyl ethyl alcohol. The mercurial combination seems particularly logical as a means of obtaining a wide spectrum including superior antifungal activity.

Organic Mercurials

This class of compounds includes phenylmercuric salts, and thimerosal. Structures are given as

Phenylmercuric nitrate

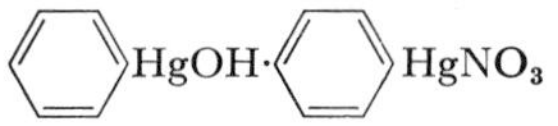

Thimerosal (sodium ethylmercurithiosalicylate)

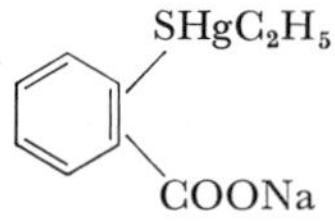

As a class, these compounds are useful preservatives; taken as a whole, their frequency of use approaches that of benzalkonium chloride. The usual concentration varies from 0.01 to 0.001 %.

Phenylmercuric acetate is sometimes combined with quaternary preservatives to obtain a broadened spectrum of activity as well as to increase bactericidal activity. The tenfold concentration range found in the use of these mercurials may indicate some uncertainty in proper use. Foster (31) remarks that the higher concentrations may be intended to control spores.

Esters of p-*Hydroxybenzoic Acid*

Typically this class of compounds is represented by methyl- and propyl-parabens. These preservatives are less frequently used in ophthalmics; their action is probably limited to a static effect, particularly in the concentrations used. Solutions of the parabens in concentrations greater than 0.1 % tend to cause discomfort when used in the eye (32).

Chlorobutanol

This compound perhaps deserves the term "time-honored." It has been repeatedly recommended for use as an ophthalmic preservative from the early 1950's onward (33).

Chlorobutanol is undoubtedly a good preservative; however, it is far from ideal. As noted previously, solution pH must be approximately 5.0 for adequate chemical stability. Solutions are, however, somewhat autostabilizing in that hydrochloric acid is produced as one of the hydrolysis products. This reduces pH and, in turn, limits decomposition. Chlorobutanol may also be lost when contained in low-density polyethylene packaging. Regardless of pH, chlorobutanol diffuses through the plastic wall.

For maximum activity, chlorobutanol should be used at a concentration of approximately 0.5 % (34,35). This is close to the solubility maximum of about 0.75 %.

Aromatic Alcohols

Benzyl alcohol and phenyl ethyl alcohol are less common but useful antibacterial agents, particularly the latter. Phenyl ethyl alcohol at 0.5 to 0.7 % is said to be particularly effective against Gram-negative organisms (36). The compound is compatible with most ophthalmic substances and is comfortable when used in the eye (37). Phenyl ethyl alcohol will tend to be lost when used as a preservative for products contained in low-density polyethylene. It has been used in combination both with organic mercurials and with quaternaries.

Miscellaneous

References to chlorinated compounds such as chlorocresol and chlorhexidine appear on occasion, particularly in the English literature. Neither compound has been used commercially to any real extent in this country. Chlorhexidine in particular has been recommended with some enthusiasm by Australian investigators (38). Unfortunately chlorhexidine has several disadvantages, including incompatibilities with such common substances as bicarbonates, borates, phosphates, and sulfates.

Additives

Various additives are permitted in ophthalmic solutions, limited of course by safety. Macromolecules such as polysorbate 80 may be used as solubilizing or clarifying agents; methyl cellulose or hydroxypropyl methyl cellulose may be added to increase viscosity; aromatics such as camphor, eucalyptol, etc. may be included. In each case, the effect of such additives on the product and the package must be investigated. Adsorption, loss by migration through the plastic container, and interaction with product components must be considered together with consumer acceptance.

Safety and Irritation

The degree and extent of toxicologic studies carried out on an ophthalmic product will be dependent largely on the intended use and duration of use of the finished product. A product for chronic use may well require more extended testing than a product intended for acute use.

Acute toxicity tests such as an LD/50 determination can be carried out using rodents; however, the typical ophthalmic toxicity evaluation is usually carried out using the albino rabbit as the test animal. The product to be tested is applied to one eye of the animal, using the fellow eye as the control. Application may be several times daily for a period of at least 21 days. It is usually helpful to administer the proposed product together with at least one multiple of the concentration of the active ingredient(s) in order to estimate, if possible, a toxic response.

Ocular response to dosage is evaluated both with the unaided eye as well as with a slit lamp. Observations should be made at intervals as established by the study protocol. In addition to ocular response, the weight gain (or loss) and other physical parameters should be evaluated throughout the test period.

The ocular evaluation should be based on detailed observations of the cornea, the iris, and the bulbar and palpebral conjunctiva. Numerical scores should be assigned to the lesions observed using the scoring technique described by Draize (39,40).

Following completion of the 21-day (or more) multiple exposure test for ophthalmic safety, the test animals should be autopsied. The eyes and adnexa should be studied in particular detail.

It should be borne in mind that animal safety is not directly equivalent to product comfort and consumer acceptance. The Draize scoring procedure will detect irritation but not necessarily the relatively minor reactions such as stinging or burning. These must be elucidated by careful testing of the final product on a test panel, consumer group, or patients, depending on intended final product use.

Packaging and Stability

Products for use in the eye have traditionally been packaged in a manner consistent with ease of use and technology available. The glass bottle and dropper and the eye cup have, for the most part, given way to the low-density polyethylene dropper bottle.

The glass bottle and dropper are still in use; however, the combination is cumbersome and increases both packaging and sterility problems. The use of glass packaging is usually dictated by stability considerations in which the glass container and screw cap minimize air and light exposure. The dropper assembly is generally packaged separately, sealed within a cellophane or blister package which permits sterilization of the dropper.

Low-density polyethylene packaging offers the great advantages of patient convenience plus reduced chance of inadvertent contamination during use. This type of packaging does away with the cumbersome duality of bottle and dropper by combining the ophthalmic container and delivery system into a single delivery unit.

Although the low density polyethylene package represents a real consumer advantage, its use as a container is not without disadvantages. The material can by no means be considered a flexible form of glass. It may well have an effect on, or may be effected by, the finished product.

Low-density polyethylene, i.e., polyethylene polymers having a density in the range of 0.91 to 0.92, is flexible and has a low incidence of stress cracking. At the same time, permeability and moisture/vapor transfer is at a maximum. Product storage in low-density polyethylene may result in the rapid loss of low-molecular-weight additives such as preservatives, moisture loss, or comparatively rapid oxidation of active ingredients (41).

Product stability in plastic packaging can only be based on experimental data; it cannot be estimated on the basis of prior experience or product performance in glass. It should be borne in mind that physical and chemical testing at elevated temperatures in plastic will serve to emphasize the plastic/product interactions. These interactions will not always be emphasized in a linear or predictable fashion; thus great care should be exercised in the interpretation of accumulated stability test data.

Manufacturing

The large-scale manufacture of ophthalmic products is a difficult and exacting task. It represents a distillation of all of the foregoing considerations to produce, on a commercial scale, a sterile solution free from foreign particles and suitably compounded for instillation into the eye.

In addition to the considerations discussed previously, manufacturing requires the application of sterile procedures to large-scale equipment. Makeup tanks, transfer lines, and other heat-resistant equipment may be sterilized by clean steam. Other procedures are applied according to product requirements.

Manufacturing, particularly the filling operation, must be carried out in a manner and in an area designed to produce a particle-free product as well as a sterile product. This can be accomplished in part by the use of laminar flow air control. Laminar air control, either horizontal or vertical, under proper conditions will aid in the manufacture of a superior quality product. Laminar flow per se will not assure a sterile product. Normally the standards of a class-100 clean room should apply to the manufacture of ophthalmic products (42).

Laminar air flow was originally developed by aerospace research groups. The process can be defined as air flow in which the total body of air within a confined area moves along parallel lines at a uniform velocity. Direction of flow may be either horizontal or vertical.

A reasonable general guide to proper manufacturing procedures is the good manufacturing practice proposals listed by the Food and Drug Administration (43). These, combined with a sound grasp of fundamentals together with insights of previous experience as compiled in the pertinent literature, form the basis for the development and manufacture of products suitable for ophthalmic use.

Products

Eye lotions are products for ophthalmic use and, as such, conform to the definition and requirements of ophthalmic solutions. Products are typically drops, washes, and solutions or suspensions of various types.

Pharmacologic action may not equal that obtained by the ladies of ancient Italy with their belladonna drops, but would include decongestant and astringent activities and others as indicated by labeling and intended use.

Worthy of mention also are products which might be termed indirect ophthalmic solutions. These include the products and solutions which are needed for the maintenance and use of contact lenses. Such products may or may not be used directly in the eye; however, their formulation must be carried out with the restrictions and requirements of ophthalmic solutions in mind.

Contact lenses are worn because of their cosmetic qualities and because of vision requirements. It is estimated that some 700,000 individuals are fitted with contact lenses each year. The majority of these are less than 25 years of age. Contact lens wearers require solutions to clean and to wet lenses in addition to solutions used for storing lenses.

Nearly all contact lenses in current use are made of polymethyl methacrylate, although experimental lenses made of silicone and also hydroxyethyl methacrylate are being investigated. Polymethyl methacrylate represents a nearly ideal material for lens fabrication. The polymer has good optical clarity, is relatively hard, yet easy to grind and polish. Lenses can be made in a variety of colors and sizes. A lack of heat stability and the hydrophobic nature of its surface are disadvantages of polymethyl methacrylate (44).

Because of its hydrophobic surface, the direct insertion of a contact lens may cause a considerable amount of discomfort. The tear fluid itself is a reasonably good wetting solution and in time will wet the lens. In the interim, however, the wearer undergoes unnecessary discomfort and excessive tearing. Preconditioning the lens with a wetting agent prevents or minimizes this discomfort.

A *wetting solution* by definition, therefore, should have many of the properties of the tear fluid, but in addition should be able rapidly to convert the hydrophobic lens surface to a hydrophilic surface. This is generally accomplished by the addition of a wetting agent such as polyvinyl alcohol or polysorbate 80. Viscosity-imparting agents are usually added to wetting solutions in order to aid in cushioning and positioning the lens on the cornea (45).

It is important to recall the general considerations of ophthalmic products in the formulation of products such as wetting solutions which are used indirectly in the eye. Too high a *p*H, for example, may cause hydrolysis of the partially acetylated polyvinyl alcohol if this has been used as the wetting agent. In all cases, wetting must be adequate and product comfort must be superior. Any trace of product discomfort would be magnified if the product is used in conjunction with a contact lens. Preservatives used in contact lens products are those discussed previously; however, particular care should be given to preservative activity and concentration because of the increased level of sensitivity of contact lens wearers.

The overnight storage of contact lenses was, at one time, the subject of considerable debate. Several investigators recommended that the lenses be stored dry. An equal number recommended wet storage. Over a period of time, the latter view prevailed (46). Wet storage is now the accepted practice. Lenses are usually stored in an antimicrobial solution which, during normal storage, exerts a bactericidal action against usual contaminants.

Since the soaking solution is usually rinsed from the lens before use, a preservative solution somewhat stronger than normal can be used. Although one must guard against inadvertent use in the eye, storage solutions may reasonably contain antibacterial preservatives in a maximum concentration, i.e., a maximum safe concentration. For organic mercurials this concentration is about 1:10,000; for quaternaries about 1:5,000. Disodium edetate in concentrations up to 0.25% has been used to enhance the antimicrobial activity of benzalkonium chloride in solutions of this type. It should be noted that soaking solutions, depending on specific labeling claims and statements, may come within the legal definition of an economic poison and thus be subject to registration by the United States Department of Agriculture (47).

After removal and prior to storage, contact lenses are cleaned to remove accumulated lipids, proteins, and the like. Cleaning is usually accomplished with the aid of a cleaning solution consisting (generally) of a solution of nonionic detergents which reduce surface tension, emulsify lipids, and aid in the removal of protein or other insoluble debris. After cleaning, the lens is rinsed in water and is ready for storage in the soaking solution.

As a matter of convenience, lens cleaning and storage are often carried out in the same receptacle. Such cleaning and storage kits are usually fabricated of plastic—generally softer than the polymethyl methacrylate lens material. Such cleaning and storage units should meet the following general criteria:

The plastic should be nonreactive with preservative materials.
The unit should be easily and completely cleaned.
Lenses should be submerged completely in the storage solution.
Right and left lens storage compartments should be color- and shape-coded.
The plastic used should obviate lens scratching.

The compositions shown in Formulas 1 to 5 are representative of typical nonprescription eye products, i.e., washes, tears, blanching agents, and contact lens products. Such products are discussed in detail by Lofholm (48). The compositions given do not necessarily imply adequate stability, packaging, and other important considerations. Such parameters must be determined during specific product design and development, as discussed earlier in this chapter. Formula 1 is a sterile ophthalmic irrigating solution. The product must be sterile, and is commonly packaged in 1- to 4-oz low-density polyethylene containers. Formula 2 is a sterile product, known as artificial tears, and recommended for use in dry eyes, or to provide additional lubrication for contact lens wearers. Formula 3 is an ophthalmic decongestant,

Formula 1

Boric acid ⎱	adjust to pH 7
Sodium carbonate ⎰	
Sodium chloride	q.s. isotonic
Benzalkonium chloride	0.004 to 0.01%
Water, purified	q.s. 100.00%

Formula 2

Sodium chloride	0.800%
Potassium chloride	0.150
Disodium edetate	0.050
Benzalkonium chloride	0.004 to 0.10%
Polyvinyl alcohol	1.00 to 2.00
Methyl cellulose, 4000 cps	0.25 to 0.50
Water, purified	q.s. 100

Formula 3

Phenylephrine hydrochloride	0.120%
Benzalkonium chloride	0.004
Water, purified	q.s. 100.000

typically used as a vasoconstrictor. The visual result is a whitening or blanching effect in the eye. It should be noted that vasoconstrictor products should not be used by individuals with glaucoma. Sterility, pH, stability, and proper packaging must be considered.

Formula 4 is a typical contact lens wetting solution, containing a wetting agent in a somewhat viscous, isotonic vehicle. Packaging is normally in low-density polyethylene. Formula 5 is a sterile antibacterial solution, used to store and soak contact lenses. This is also typically packaged in low-density polyethylene.

Formula 4

Sodium chloride	to make isotonic
Polyvinyl alcohol	1.5000%
Methyl cellulose, 4000 cps	0.5000
Benzalkonium chloride	0.004 to 0.010
Disodium edetate	0.010
Water, purified	q.s. 100.000

Formula 5

Benzalkonium chloride	0.004 to 0.010%
Disodium edetate	0.010
Water, purified	q.s. 100.000

REFERENCES

1. *The pharmacopeia of the United States of America*, 17th revision, Mack Printing Co., Easton, Pa., 1965, p. 790.

2. *Ibid.*, 15th revision, 1955, p. 814.

3. Martin, E. W. (Ed.): *Husa's pharmaceutical dispensing*, 5th ed., Mack Publishing Co., Easton, Pa., 1959, p. 256.

4. The stability and stability testing of pharmaceuticals: an annotated bibliography, 1939–1963, Committee on Stability Testing, Quality Control Section, Pharmaceutical Manufacturers Assoc., Washington, D.C., 1964.

5. Kremers, E., and Urdang, G.: *History of pharmacy*, Lippincott, Philadelphia, 1964, p. 11.

6. Youngken, H. W.: *A textbook of pharmacognosy*, 5th ed., Blakiston, Philadelphia, 1947, p. 755.

7. Ehrenstein, E.: *Remington's practice of pharmacy*, ed. Martin and Cook, 12th ed., Mack Publishing Co., Easton, Pa., 1961, p. 359.

8. *The pharmacopeia of the United States of America*, 17th revision, Mack Printing Co., Easton, Pa., 1965, p. 810.

9. Ehrlen, I. R.: Advances in sterilization techniques, *Sven. Farm. Tidskr.* **68:** 818 (1964).

10. Aspery, G. M., Jr.: Gas sterilizer evaluation with sporicidal testing, *Bull. Par. Drug Assoc.*, **23:** 132, 1969.

11. Cleary, D. J.: *Bull. Par. Drug Assoc.*, **23:** 233, 1969.

12. *The pharmacopeia of the United States of America*, 17th revision, Mack Printing Co., Easton, Pa., 1965, p. 809.

13. *Ibid.*, p. 790.

14. Ehrenstein, E.: *op. cit.*, p. 362.

15. *Merck index*, 8th ed., 1968, p. 1281.

16. Martin, E. W. (Ed.): *Husa's pharmaceutical dispensing, op. cit.*

17. *Remington's practice of pharmacy*, ed. Martin and Cook, 12th ed., Mack Publishing Co., Easton, Pa., 1961.

18. Martin, A. N.: *Physical pharmacy*, Lea and Febiger, Philadelphia, 1960, p. 287.

19. Miklos, K.: Conservation of eyedrops, *Br. J. Ophthalmol.*, **15**: 649 (1931)

20. *The Pharmacopeia of the United States of America*, 18th Revision (page proofs), 1970, p. 845.

21. Nair, A. D., and Lach, J. L.: The determination of chlorobutanol in pharmaceuticals by amperometric titration, *J. Am. Pharm. Assoc, Sci. Ed.*, **47**: 46 (1958).

22. Bahal, C. K., and Kostenbauder, H. B.: Interaction of preservatives with macromolecules. V., *J. Pharm. Sci.*, **53**: 1027 (1964).

23. Patel, N. K., and Kostenbauder, H. B.: Interaction of preservatives with macromolecules. I., *J. Pharm. Sci.*, **47**: 289 (1958).

24. Pisano, F. D., and Kostenbauder, H. B.: Interaction of preservatives with macromolecules. II., *J. Pharm. Sci.*, **48**: 310 (1959).

25. Deluca, P. P., and Kostenbauder, H. B.: Interaction of preservatives with macromolecules. IV., *J. Pharm. Sci.*, **49**: 430 (1960).

26. Patel, N. K., and Foss, N. E.: Interaction of some pharmaceuticals with macromolecules. I., *J. Pharm. Sci.*, **53**: 94 (1964).

27. Judis, J.: Studies on the mechanism of action of phenolic disinfectants. I., *J. Pharm. Sci.*, **51**: 261 (1962).

28. Beckett, A. H., and Robinson, A. E.: The inactivation of preservatives by nonionic surface active agents, *SPC*, **31**: 454 (1958).

29. Lachman, L.: The instability of antimicrobial preservatives, *Bull. Par. Drug. Assoc.*, **22**: 127 (1968).

30. *Remington's practice of pharmacy, op. cit.*, p. 638.

31. Foster, J. H. S.: Preservatives of ophthalmic solutions: Part I, *Mfg. Chemist*, **36**: 45 (1965).

32. *Ibid.*

33. *Ibid.*

34. Reigelman, S., Vaughan, D. G., and Okumoto, M.: Antibacterial agents in *Ps. aeruginosa* contaminated ophthalmic solutions, *J. Am. Pharm. Assoc., Sci. Ed.*, **45**: 93 (1956).

35. Runti, C.: Preparation and preservation of *collyria*. I. General part, *Boll. Chim. Pharm.*, **99**: 286 (1960).

36. Foster, J. H. S.: Preservatives of ophthalmic solutions: Part II, *Mfg. Chemist*, **36**: 43 (1965).

37. *Ibid.*

38. Crompton, D. O.: Ophthalmic prescribing, *Australian J. Pharm.*, **43**: 1020 (1962).

39. Appraisal of the safety of chemicals in foods, drugs and cosmetics by Association of Food and Drug Officials of the United States, Topeka, Kansas, 1965, p. 49.

40. Draize, J. H.: Dermal toxicity, *Food-Drug. Cosmet. Law J.*, **10**: 722 (1945).

41. Mullins, J. D.: Plastics for packaging ophthalmic preparations, *Bull. Par. Drug. Assoc.*, **22**: 38 (1968).

42. Beck, R. P.: Criteria for laminar flow devices, *Bull. Par. Drug. Assoc.*, **22**: 222 (1968).

43. 21 CFR Part 133 (28 F.R. 6385 Effective 6/20/63), Drugs, current good manufacturing practice in manufacture, processing, packaging or holding.

44. Dabezies, O. H., Jr.: Contact lenses and their solutions: A review of basic principles, *Eye, Ear, Nose and Throat Monthly*, **45**: 39 (1966).

45. *Ibid.*

46. *Ibid.*

47. Federal insecticide, fungicide and rodenticide act, Section 2a, 2d, and 2n.

48. Lofholm, W.: O-T-C ophthalmic preparations, *J. Am. Pharm. Assoc.*, **NS8** (No. 9): 497 (September 1968).

Chapter 32

FRAGRANCE

MARVIN S. BALSAM

Fragrance is unique in the realm of cosmetics, for it is both an end result in its own right and a quality of products having a diversity of other purposes. Perfume materials of one type or another are used in almost all cosmetics. They constitute the very *raison d'être* of several important products that are sold only or mainly for fragrance (perfumes, toilet waters, colognes, and perhaps bath oils and related bath products). The fragrance is a significant part of many other products, as creams, lotions, face and bath powders, and shaving creams. It plays a minor role in other cosmetics, such as face rouge; it acts primarily as a mask for malodorous ingredients in thioglycolate depilatories and hair-waving lotions. It is usually omitted in a few cosmetics such as eye makeup, many nail products, and some purposely unscented materials for women who tend to be sensitive to perfumes.

In order to impart fragrance to an alcoholic solution or to a cosmetic preparation, the perfumer blends a variety of natural and synthetic raw materials into a complex and usually difficult-to-describe mixture. This blend is known as a "compound" or "perfume compound." Although really a mixture of many components, the well-formulated perfume compound loses the individual characteristics of its constituents and develops a new character.

The blend is also called a perfume, perfume oil, perfume base, essence, or bouquet. In this study the word "perfume" will be used to describe a concentrated alcoholic solution of fragrant materials, sold or ready for sale as a finished product for the consumer; and the base, essence, or aromatic mixture used to impart fragrance to such a perfume or to a cosmetic, soap, or other product, will be called a perfume oil.

A cosmetic may be perfumed conceivably with a single raw material, or with a mixture of two or three such materials, but this is rare; more often the perfume oil contains at least a dozen materials, and it may contain as many as 200 or more. The reasons for such complexity will be discussed. (When the aromatic materials are used to impart flavor, rather than fragrance, as in dentifrices, the number of constituent materials is usually small.)

Raw Materials

Perfume raw materials can be divided into several categories. They are often called "natural" or "synthetic," but such a distinction is arbitrary and makes no provision for many chemical bodies that are not known in pure form in nature, but are isolated from natural products.

The following schematic division is relatively all-inclusive:

1. Plant materials.
   a. Essential oils (obtained by distillation or expression).
   b. Flower oils.
   c. Resins, gums, and exudations.
2. Animal secretions.
3. Chemical substances.
   a. Isolates from plant materials.
   b. Derivatives of plant materials.
   c. Synthetic organic substances.

In some of these categories, which will be described briefly here, there are large numbers of different products, all in frequent commercial use. Thus there are at least 200 essential oils used often, and certainly another 800 which are in use from time to time. There are about a dozen expressed oils, approximately the same number of flower oils, and perhaps a few more resins, gums, and exudations. Some four materials of animal origin are in common usage. Finally, several hundred chemical substances are regular articles being used by the perfume industry, and there are thousands of others at the disposal of the perfumer (1). When one considers that literally millions of different qualitative formulas can be written by using different combinations of these substances, choosing, let us say, between 10 and 50 materials, and when one considers, further, that different fragrances can be obtained by changing the quantitative proportions, one can grasp how limitless are the possibilities for new fragrances and for changes, nuances, and diversifications.

Definitions

The term "essential oil" is often used in the cosmetic and perfume industries as synonymous with perfume oil, base, or "compound." This terminology being incorrect, it will be used only in its scientific sense. Although rather difficult to define, one might start with the suggestion of Haagen-Smit (2), who points out that "in early work . . . the term 'essential' or 'ethereal' oil [is] defined as the oil obtained by the steam distillation of plants." Somewhat more detailed is the definition of Parry (3), who states that for all practical purposes these materials "may be defined as odoriferous bodies of an oily nature obtained almost exclusively from vegetable sources, generally liquid (sometimes semisolid or solid) at ordinary temperatures, and volatile without decomposition." Thousands of essential oils have been investigated, of which, according to Langenau (4), "some 150 to 200 species have been exploited" for commercial purposes.

The essential oils are obtained from various parts of the plant; they come from the seeds (e.g., anise) the wood (e.g., rosewood), and from almost every other part of the plant; cinnamon oil, for example, is obtained from the bark and the leaf. In some instances the oils are made up largely of one material, the odoriferous principle, which is the main substance imparting odor to the plant; an example is oil of clove, of which eugenol constitutes some 90 %. More often, the essential oil is a complex mixture of many different chemical bodies, some of which have been isolated, analyzed, and reconstituted synthetically, and others are still of unknown structure. In some cases a constituent is found only in a single oil, and otherwise unknown in nature, as the tagetone in oil of tagetes. Others, like geraniol and citral, are known to exist in many plant oils.

Essential oils (like other natural products used in perfumery) are subject to the changes of a product made by nature. From one harvest to another, from one climate to the other, the oils are likely to show certain variations. The work of a number of outstanding chemists, investigating essential oils and reporting on botanical origin and physical and chemical properties, has aided in the establishment of standards. Of the early workers, the most important classic contribution was made by Gildemeister and Hoffmann (5), and other important contributions were made by Finnemore (6), Semmler (7), Parry (3), and Charabot and his co-workers (8). Recently an exhaustive and definitive compendium in this field was compiled by Guenther (9), and an excellent contribution has been made in this field by Arctander (10).

Flower oils, or floral oils, are the essences obtained from the plants (generally the flowers) by extraction with volatile or nonvolatile solvents. These products, often referred to in the trade as "absolutes," have been termed "natural perfume materials," so designated in contrast to the essential oils of distillation, because the latter process subjects the oils to the harsh action of heat and water, thus resulting in an oil which may not be that which nature had produced originally. The best known and most commonly used of these materials are jasmin absolute and rose absolute, the latter not to be confused, however, with otto of rose, a product of distillation. Others include tuberose, mimosa, violet leaf, and numerous flowers of lesser interest. The entire field of flower oils was described in considerable detail in an exhaustive study by Naves and Mazuyer (11,12).

The three processes for flower oil production are volatile solvent extraction, immersion in warm fats (maceration), and extraction by cold fats (enfleurage). The historical development of these processes, the yields obtained, the reasons for the choice of process, and the nomenclature of the various materials (concrete, pomade, etc.) are to be found in the Naves-Mazuyer work, cited above.

Another process for essential oil production is expression. It consists in pressing of oil-bearing plant material by mechanical or manual means. It is suitable only for citrus oils, and is used to obtain the oils of lemon, orange, bergamot, grapefruit, lime, tangerine, and other more obscure products. These expressed oils usually contain large proportions of terpenes. In some instances, an oil may be used after the terpenes and even the sesquiterpenes have been partially or completely removed. Guenther (9) states that the terpeneless oils are not only more soluble (in hydroalcoholic solutions) and more stable, but also "much stronger in odor, yet retaining most of the odor and flavor characteristics of the original oil." Nevertheless Langenau (4) finds certain disadvantages, as well as advantages, in the use of terpeneless oils. "The elimination of terpenes and sesquiterpenes removes a part of the characteristic odor and flavor, so that these concentrated products do not display the fine bouquet and freshness of the original oil. However," he goes on to state, "the advantages inherent in the use of such concentrate assure their continued employment in many formulas."

Resins, gums, balsams, and exudations are not so well defined; they are plant materials, including gum styrax, balsam Peru, benzoin, labdanum, myrrh, and others; they are heavy, viscous, highly odorous materials, either obtained as such from the plant or separated from other less valuable materials.

Animal secretions are among the most interesting and valued perfume materials. Although numerous animals secrete odorous materials, and an effort has been made to exploit commercially the products of at least a dozen or more, only four are of importance: castoreum, civet, musk, and ambergris. These materials are dissimilar from each other in many respects, notably odor and chemical constituents. However, all of them are lingering and long-lasting of themselves, and make excellent contributions toward the fixation of perfume. A summary of information about animal materials may be found in *The Science and Art of Perfumery* (13).

Of the chemical bodies used in perfumery, the isolates from plant oils bear the strongest resemblance to the plant materials themselves. In some cases these isolates are the main odorous constituents of the oils. Examples of isolates in common use are eugenol (usually obtained from oil of clove), citral (from lemongrass oil), geraniol (from citronella oil), and others.

Many materials can be obtained from more than one oil; a careful olfactory study will usually reveal to the trained perfumer the source of the material. Thus linalool is present in the oils of rosewood, petitgrain, linaloe seed, bergamot, rose, and jasmin, among others.

Closely related to the isolates are substances derived from such isolates, or directly from the respective essential oils, by chemical reaction. Usually these chemical compounds do not exist as such in nature. These would include products of esterification, such as the formates, acetates, propionates, and other esters of citronellol, linalool, geraniol, rhodinol, terpineol, etc., as linalyl acetate and geranyl acetate, for example. Also products of hydroxylation, such as hydroxycitronellal, and those of cyclization, such as the ionones from citral. To describe these materials as synthetics is not entirely accurate, nor should they be called natural substances. Furthermore, some of them are known in nature (as, e.g., ionone; not methyl ionone, but the oil in which ionone is found is not used as its source material).

The true or total synthetics are the organic chemicals built up from coal tar or petroleum derivatives, or from other basic chemical materials. Within the perfume industry, they are called synthetic aromatic chemicals, but this term is something of a misnomer, since it uses the word "aromatic" not in a technical sense but to convey the idea of giving an aroma. These substances include both aliphatic and aromatic materials, as well as other compounds not readily classified in either of these two groups.

The synthetics include materials known in nature and others unknown in the natural kingdom. Among the former one may cite benzyl acetate, found in jasmin absolute and made synthetically from acetic acid and benzyl alcohol; and linalool, found in bois de rose oil and made synthetically from acetylene and acetone, as well as other starting materials. Examples of widely used products not known in nature are musk ambrette and the other so-called synthetic musks; also cyclamen aldehyde, amyl cinnamic aldehyde, and many others.

There has been an increasing emphasis in the last decade on producing many of these isolates by purely chemical means. In some instances the use of the natural isolate has almost stopped. Of course these newer materials offer the great advantage of product and price stability since they are not dependent on the fluctuations caused by climatic, economic, or political conditions.

The synthetics may be classified in various ways. They are easily classified by their functional groups, as aliphatic alcohols, aliphatic aldehydes, aromatic ketones, macrocyclic lactones, etc. Such a classification aids the organic chemist in the study of the syntheses of such products; but of more concern is whether it can serve the perfumer as a guide to the fragrance type and to the behavior of a substance as a constituent of a perfume oil in a cosmetic preparation. A classification of synthetics and other perfume materials by volatility was suggested by Poucher (14).

For a comprehensive study of organic chemicals used in perfumery, both of plant origin and of total synthesis, the reader is referred particularly to the work of Bedoukian (15). Jacobs (16) has described aromatics used for flavors; *The Givaudan Index* (17) gives constants and specifications for a large number of aromatic substances, and Parry (18) and Poucher (19) describe the use of these materials in perfume. Finally, the works of Moncrieff and Arctander (20,10) are essential in a study of chemicals used in perfumery.

Tinctures, Diluents, Infusions, Resinoids

Most of the materials described above are used in the forms in which they are produced. Whether an essential oil or a synthetic organic chemical, the product may be added to the perfume mixture as such, without the preparation of a solution or other intermediate material. This is not true of some perfume constituents, because of the added value obtained when a so-called tincture is prepared, or because of their physical form, their odor strength, or the expensive nature of the material which dictates its use in small, carefully measured quantities. Thus, for a variety of reasons, some perfume materials are prepared in the form of alcoholic tinctures or infusions, or they are diluted with other materials (e.g., diethyl phthalate or benzyl alcohol).

The preparation of a tincture or dilution can serve several purposes:

1. In the case of animal products such as civet, the tincture undoubtedly improves the odor; it enables the perfumer better to standardize his product, and gives a better material from the viewpoint of workmanship.

2. When the formula calls for a minute amount and the entire formula is being prepared on a small scale, the material in pure form would be difficult to weigh accurately without using a microbalance. Here the use of a 10% solution would simplify the problem.

3. If a substance as powerful as some of the aliphatic aldehydes were to be used, it would be difficult to handle without dilution in a suitable solvent, such as diethyl phthalate, in a predetermined concentration.

4. In the case of resins such as oakmoss, the dissolved material, in a suitable solvent, is easier to standardize from an olfactory viewpoint and to handle in the compounding or blending process.

5. Some materials have increased stability in a dilution, in which they may reach an equilibrium. A notable example is phenyl acetaldehyde, which, in pure form, tends to polymerize, whereas in the form of a 50% solution in diethyl phthalate it is quite stable.

Standardization of Materials

Control of quality of perfume raw materials is based primarily on the olfactory character of the product, and only secondarily on the analyzable physical and chemical characteristics. Standards and specifications of essential oils have been published by Guenther (9). In France a committee of chemists associated with the essential oil industry has published an official compendium of analytical constants (21). For synthetics the chemical and physical constants are found in the works of Bedoukian (15), in *The Givaudan Index* (17), and for those materials widely used in flavors, in the work of Jacobs (16). For numerous essential oils, isolates, and synthetics, physical and chemical constants and standardized analytical procedures have been agreed upon by the Essential Oil Association of the United States.

It is not always possible to correlate chemical purity with olfactory attractiveness. A material may have a higher purity than another grade of the same substance, so far as its chemical analysis is concerned; nevertheless the pure product may not be so "fine" from the viewpoint of value to the perfumer, for any of the following reasons:

1. The nature of the impurity, no matter how small the quantity, may be such as not to affect adversely the odor character.

2. The purification may result in the elimination or the diminution of an "impurity" which was desirable and required in the material; this is particularly true of the isolates, which retain some of the character of the oils from which they have been obtained.

3. There may be present stereoisomers or homologous compounds which affect the odor, but do not show up by the usual analytical methods, although infrared and gas chromatography have overcome this difficulty to a great degree.

Nevertheless, a group of highly purified perfume materials has been described and evaluated by a perfumer (22) who disclosed that they were not interchangeable or indistinguishable from the usual products of commerce, but offered certain possible advantages. If such purified substances are acceptable to the perfumers, then the latter will have assurance of consistency of odor character and freedom of choice of starting material, without fear that there will be changes in the finished product. Although such substances may prove to be more expensive than the less highly purified products, they can also serve to stabilize the market in essential oils, by permitting the manufacturers to switch with ease from one source to another, without changing the olfactory effects obtained.

Before concluding this brief description of the raw materials of perfumery, a word must be added about the so-called specialties. The specialties are materials that are mixtures of essential oils and chemical bodies, and that are used as constituents in the compounding of a finished perfume oil. Actually there is no reason why the perfumer cannot go back to original source materials; however, the specialty has two major advantages:

1. It enables a perfumer to draw upon the experience of another expert or group of experts, rather than to limit himself to his own experience or that of his associates.

2. It enables a perfumer to make minor additions and changes in a perfume without reevaluating the materials in the entire formulation. To elaborate on this point, let us assume that in creating a heavy floral note, one has worked out a product that might be improved with a faint touch of a lilac character; in fact, one wishes to try as little as 0.5 % of a lilac. It might be impracticable to reformulate the entire perfume, adding minute amounts of terpineol, anisic aldehyde, ylang-ylang oil, cinnamic alcohol, phenyl ethyl alcohol, and other common ingredients of a lilac perfume oil. Instead one can employ a previously blended lilac as the ingredient, and this can be either a product of one's own, or that of another producer. If it is one's own lilac, it would be possible to recompute the formula, but as a practical matter, it is sometimes simpler to have formulas within formulas. Thus specialties may aid in saving time, in obtaining smoother blends, and in the judicious use of trace effects.

Many such specialties are those originating with the user, whereas others are often well-known products of the perfume industry. Among the most frequently used specialties are simulations of the floral absolutes, but available at prices far lower than those of the natural products. Other specialties are simulations of some of the expensive essential oils, and finally creations of fantasy, which are not in either of these categories. Examples of a few specialties that have attracted attention and are highly valued in perfumery circles are Wardia, Sophora, and Jasmin du Ballet.

Describing the Perfume Oil

The description of the fragrance that is created, or that is sought, is one of the most difficult and troublesome problems besetting the perfumer. Perfumes that closely simulate the fragrance of an individual known material, usually a flower, are simply described in terms of that material. Thus one speaks of a rose, a lilac, and a lily-of-the-valley perfume.

Fragrances which do not identify themselves with individual flowers are sometimes described in general terms, as floral bouquet or light or heavy floral, or by such words as "oriental," "fantasy perfumes," and "modern blends." These terms are rather general, and do not lend themselves to easy identification or to any marked degree of agreement. Furthermore they lack specificity; a perfumer seeking to create an "oriental" may find himself with his sights set on any one of dozens of fragrances as different from each other as is the rose from the violet, the lilac, and the gardenia, all of which are florals.

The perfumer describes the fragrance, or various characteristics of it, in terms of some of its ingredients, or in terms of the odor of some of these ingredients. Thus a "Farina" type can be described as "citrus," and this refers to the presence of bergamot, lemon, and other citrus notes. Words such as "aldehydic," "woody," "minty," "mossy," and others seem to convey to some degree a uniform and reproducible concept.

All words mentioned hereto are used to describe the type of odor, but not its quality. The use of language to describe the quality of the fragrance is less successful; perfumers in their vocabulary speak of a fragrance as being "round" or "sharp" or they use a variety of other words, but no one has determined whether the same words would be used by different experts examining the same materials.

The failure to communicate with words is one of the chief difficulties in any treatise on perfume formulation. So long as a formula can be described as a lilac, a rose, or a mimosa, for example, the description has some value; and even within one of these floral groups, some subdivisions can be made, as a light rose, a heavy rose, a rose absolute substitute, a sweet rose, a floral-bouqueted rose, a spicy-nuance rose, a tea rose, and others. But as soon as one goes into the realm of fantasy bouquets, there is no method of telling the reader the type of formula to be expected unless one limits the description to such general and not too meaningful terms as "oriental" or "woody" or "mossy."

A group of renowned perfumers, working under the auspices of the American Society of Perfumers, has prepared (as set forth below) a description of the more important fragrance types, as well as definitions for some of the most commonly used descriptive terms in perfumery.

Fragrance Types

1. Oriental. A blend of fragrant complexes culminating in an intense heavy full-bodied fragrance.

2. Cologne blend. Any harmonious combination whose fragrance characteristics are derived from citrus oils.

3. Bouquet. Originally a harmonious combination of two (2) or more floral notes. The term today encompasses other fragrance complexes besides florals. Therefore today a "Bouquet" is a harmonious combination of two (2) or more fragrance complexes.

4. Floral. The fragrance characteristic of an existing known flower type.

5. Chypre. A mossy-woody complex with a characteristic sweet citrus top note, frequently encompassing some floral tones.

6. Fougere. A fragrance combining a dominant sweet note with a mossy, lavender, citrus character.

7. A spice blend. Any fragrance combination falling into either a floral spice such as carnation or having herbal spice characteristics.

8. A wood blend. A fragrance dominated by characteristic woody notes encompassing other scent tones.

9. An aldehydic blend. Blends deriving their specific unique character through the super imposition of certain chemicals called aldehydes.

10. Amber. A heavy full bodied powdery, warm scent tone.

Descriptive Terms

1. Top note. Initial fragrance impression.

2. Body. Main fragrance theme.

3. Dry out. Final phase of the main fragrance.

4. Strength. The relative intensity of a fragrance impression.

5. Lasting qualities. The ability of a fragrance to retain its character, over a given period of time.

6. Diffusion. The ability of a fragrance to radiate from the wearer and permeate the environment.

7. Fixation. The property of fragrance that prolongs the odor life and produces a continuity of odor.

8. Undertones. The subtle characteristics of the fragrance background.

9. Thin. The lack of body and richness.

Life of the Fragrance

The life cycle refers to the various qualities of the perfume as it changes from the moment the bottle is opened and the fragrance reaches the air (and the nose) until the last residual odor has vanished. Ideally, as many would contend, a perfume should be completely unchanging, with the first note the same as the last. This is, however, impossible to achieve except by using a single chemical substance as the perfume, and then the limitations on the fragrance effect are self-defeating. Although measures can be taken to diminish the extent and the rapidity of change (as described in the section on fixation), some alterations do take place.

It is the aim of the perfumer to make such changes artistically acceptable, so that the "life cycle" of the perfume has a unified character throughout, and so that in all its changing ramifications it is recognizable (i.e., characteristic) and acceptable.

The first note that volatizes is referred to as the topnote, and is usually light, quickly disappearing, and often dominated by the lowest-boiling and most volatile materials used. The body of the perfume is its main "life," the note that characterizes it during volatilization. Finally, the residual or lingering note is the end of the life of the fragrance before it completely disappears.

Fixation

There have been several attempts to explain theoretically the phenomenon of fixation; of this literature, the work of Pickthall (23) deserves special attention. Ideally, this author states, the perfumer "should select a range of ingredients, all strong in odor and possessing identical and low vapor pressures." However, when blending a perfume, many substances are used; in fact, some of the natural materials are themselves complex mixtures of substances of varying vapor pressures, and the perfumer could not possibly be restricted to a few materials that might make this "ideal" selection. If the more volatile materials are lost too quickly, Pickthall points out, the balance of the perfume is upset. By use of Raoult's law, Pickthall shows how one liquid affects the rate of volatility of another in a mixture of several substances. By retarding this volatility, such a liquid acts as a fixative.

The fixative, according to Pickthall, should be a material with a low vapor pressure and a low molecular weight, "two properties not often to be found in one chemical substance, especially when oil-soluble."* A compromise is often necessary, in which substances of relatively low volatility and comparatively higher molecular weights are chosen. Furthermore, a fixative should be selected that is "physically attractive for the molecules of the more volatile substances in the perfume." By means of experimental studies, Pickthall demonstrated the existence of fixation and showed, for example, that benzyl benzoate had the power of retarding the rate of evaporation of other more volatile substances.

From a practical point of view, excellent instructions on fixation were given by Poucher (19). He defined the art of fixation as consisting in selecting "those substances which, when blended with the more volatile constituents of a perfume, will prevent their rapid evaporation and at the same time retain the predominating note of their fragrance." In other words, the fixative seeks to "equalize the different rates of evaporation of the various constituents." The fixatives, Poucher points out, may themselves be odorous (whether pleasant or unpleasant), or may have relatively little odor value of their own

* An interesting exception is water, with its low vapor pressure and low molecular weight, but of course lacking other characteristics necessary for use with perfume materials.

(as benzyl benzoate and diethyl phthalate). Poucher then goes on to divide the process of fixation into three stages, as follows:

Pre-fixation, or complete deodorization of the solvent; *blending*, or the addition of modifying fixators during manufacture; *final fixation*, or the admixture of such additional substances as will produce the desired tenacity and distinctiveness, without changing the basic odor.

The first of these processes, or prefixation, is of interest only in the manufacture of alcoholic fragrances, and not in the blending of perfume oils for use in cosmetics. According to Poucher, it is carried out by adding "an aromatic resin extract whose odor is similar to the perfume for which the alcohol is to be used." As an example, he suggests the addition of resins of gum benzoin, balsam tolu, and olibanum to the alcohol, allowing the mixture to stand for 1 month, in order to obtain an alcohol perfectly suitable for perfumes of a verbena or an eau de cologne character. For other types of perfumes, different resins are suggested.

For fixation during blending, Poucher urges the use of aromatic notes of a lasting character, to be introduced with the other blending topnotes; thus, in a violet, he would use costus and ylang-ylang; in lily-of-the-valley, hydroxycitronellal and bois de rose; and in carnation, some heliotropin.

In the final fixation, Poucher prefers a blend of, first, animal fixatives to give warmth and life and to soften the harsh tones that might be present; second, sweet and pleasant notes, usually produced by natural oils, such as bergamot; and, third, the freshness of floral absolutes.

Although the choice of fixative is a question of individual taste, and the determination of the fixative value of various substances in different perfumes almost entirely an empirical matter, it is interesting that Poucher finds that the "essential oil which is preeminent as a fixator for *any* perfume is unquestionably *clary sage*" (emphasis in original).

During recent years there has been a movement away from the use of the so-called animal fixatives, dictated by increasing prices and decreasing availability. Instead, the emphasis has been on many new synthetic materials which have excellent properties of fixation as well as contributing their own pleasantly warm odor.

Building a Perfume

Up to the advent of synthetics in perfumery, the fragrances were simple mixtures of various essential oils and flower oils, together with some animal tinctures. These perfumes, so far as one is able to tell today by reconstructing the formulas (albeit with modern equivalents of the materials) were floral and fresh, but had several disadvantages when compared with the materials of today:

1. The possibilities for a wide range of fragrance effects were most limited; thus the entire field of modern "fantasy" perfumes would be unobtainable.

2. The artistically desirable nuances obtained by the judicious use of small quantities of synthetics were not possible.

3. The entire price structure of perfume materials in modern times would make it uneconomical to employ only natural substances.

4. Synthetics are particularly desirable when the problems of chemical reactivity, solubility, discoloration, and irritating effects are paramount. The use of synthetics in such instances enables one to control better each ingredient entering into the formulation.

Examples of so-called old-fashioned perfume formulations are given in Formulas 1 to 5. They are taken from several of the more popular perfume

Formula 1

Old-fashioned perfume: Esterhazy bouquet

Extrait de fleur d'orange (from pomade)		1 pint
Esprit de rose triple		1 pint
Extract of vetiver		
Extract of vanilla	of each	1 pint
Extract of orris		
Extract of tonquin		
Esprit de neroli		1 pint
Extract of ambergris		$\frac{1}{2}$ pint
Otto of santal		$\frac{1}{2}$ drachm
Otto of cloves		$\frac{1}{2}$ drachm

Formula 2

Old-fashioned perfume: Eau de cologne (first quality)

Spirit (from grape), 60 over proof	6 gal
Otto of neroli, petale	3 oz
Otto of neroli, bigarade	1 oz
Otto of rosemary	2 oz
Otto of orange zeste	5 oz
Otto of citron zeste	5 oz
Otto of bergamot	2 oz

books of the nineteenth century, being the works of Piesse (24), Cooley (25), and Deite (26). They are offered as historical curiosities, and are not to be confused with what might be considered an "old-fashioned" type today, as, for example, a violet or a chypre. Maurer (27) has described the old-fashioned perfumery compounding, particularly of Victorian times, and showed that 13 oils were used to blend an almost complete gamut of floral notes, with a maximum of nine constituents in any single floral.

Formula 3

Old-fashioned perfume: Extrait de millefleurs

Grain-musk (finest)	4 grains
Ambergris (do.)	6 grains
Oil of lemon	3 drachms
Oil of lavender (English) Oil of cloves } of each	2 drachms
Liquid styrax (genuine)	½ drachm
Oil of verbena Oil of pimento Oil of neroli } of each	12 drops (minims)
Rectified spirit	1 pint

Procedure: Macerate in a warm room, with frequent agitation, for a fortnight or three weeks.

Formula 4

Old-fashioned perfume: Bouquet d'amour

Take of

Esprits de rose, jasmin, violette, and cassie	of each, 2 ounces
Essences of musk and ambergris	of each, 1 ounce

Procedure: Mix, and if the liquid be not quite clear, add of strong alcohol, drop by drop, the least quantity sufficient to render it so.

Formula 5

Old-fashioned perfume: Extrait muguet (lily of the valley). Extracts No. 1 from *Pomm. Jonquille* 750 drachms, from *Pomm. Jasmin* 100, from *Pomm. Tubereuse* 200, and from *Pomm. Acacia* and *Pomm. Orange* each 100; bergamot oil 7½ drachms, oil of lemons 2½, angelica oil 3 drops, storax tincture 5 drachms, musk tincture 2½, vanilla tincture 5, ambergris tincture 2, ylang-ylang tincture 100, wintergreen tincture 25, bitter-almond-oil tincture 2½.

Simple Floral Fragrances

Whether derived from perfume formulary, books or from formulas commercially in use, the perfume oil is usually a complex mixture with numerous ingredients. It has been shown by Pickthall (23), however, that it is possible to take a few raw materials and build several floral fragrances by varying the quantitative formulas. Using ten raw materials (including a trace effect of musk ketone), Pickthall blends a simple rose, lilac, lily-of-the-valley, and jasmin.

In Formulas 6 to 11, the author has used eight raw materials to make six different florals. Actually the floral types were chosen because they lend themselves to simple formulas.

Formulas for Simple Floral Fragrances

	6	7	8	9	10	11
	Lilac	Rose	Muguet	Jasmin	Violet	Car-nation
Phenyl ethyl alcohol	30%	35%	15%	5%	20%	25%
Hydroxycitronellal	30	—	45	6	5	5
Geraniol	2	48	20	2	4	5
Amyl cinnamic aldehyde	4	2	5	45	1	1
Benzyl acetate	5	4	5	40	10	3
Ionone	3	4	5	—	60	4
Eugenol	1	2	—	—	—	55
Terpineol	25	5	5	2	—	2

If one were to choose honeysuckle instead of rose, greater complexity would be required. On the other hand, if violet and carnation were not being created, one might drop the ionone and eugenol, reducing the number of constituents to six. Nevertheless, ionone and eugenol being available, they are used to advantage in several of the florals.

The six florals furnish, of course, only a beginning, rather than accomplished examples of well-rounded perfumes. On study of these formulas, it will be seen that actually two or three raw materials were used as the basis for most of the florals, with small proportions of the other ingredients to aid the blend. The lilac is built primarily with terpineol, hydroxycitronellal, and phenyl ethyl alcohol; the rose with phenyl ethyl alcohol and geraniol; the jasmin with amyl cinnamic aldehyde and benzyl acetate.

In making up a formula of this type, or of any other, the grade of material used may play a very important role. In a lilac based to a large extent on terpineol, the difference in grade would almost certainly be discernible to the most casual user. In a text, it is difficult to indicate the grade to be used, except that one should almost always use the more highly refined materials, if cost considerations do not prohibit their use; furthermore, once a grade has been established, it should not vary for the particular formula. A well-written formula should state the grade of the material, and its source. In the case of a complex product such as ionone with its different isomers, the grade would make a much greater difference than in that of, say, amyl cinnamic aldehyde.

The six florals could, of course, be improved if one were to broaden the arsenal of raw materials, and use other naturals, isolates, and synthetics, than the few on which these products were based. In the next series, Formulas 12 to 17, it is shown how an additional 15 raw materials can be employed to improve the character of the six fragrances. Using Formula 6 as 65 parts of a new and improved lilac, a new fragrance is blended, shown as Formula 12. The same is done with the rose, lily-of-the-valley, jasmin, violet, and carnation. The new fragrances vary in total number of constituents from 12 (for the jasmin and the violet) to 17 (for the lilac and carnation). Of the 15

Formulas for Development of Floral Fragrances

	12	13	14	15	16	17
	Lilac	Rose	Muguet	Jasmin	Violet	Carna-tion
Base from corresponding floral perfume (Formulas 6 to 11)	65%	65%	65%	65%	65%	65%
Linalyl acetate	—	1	4	7	7	—
Cinnamic alcohol	11	6	7	3	—	3
Citronellol	—	15	3	—	—	1
Indole 10% in diethyl phthalate	3	1	7	8	—	—
Phenyl acetaldehyde 50% in diethyl phthalate	1	1	—	—	—	—
Phenyl acetaldehyde dimethyl acetal 10% in diethyl phthalate	1	2	—	—	—	—
Geranium oil	—	7	1	—	—	—
Heliotropin	12	—	3	—	7	5
Linalool	4	—	8	7	—	3
Methyl ionone	—	—	—	—	13	4
Ylang ylang oil	1	1	1	7	4	2
Musk ketone	—	—	1	—	3	2
Resin balsam peru	—	—	—	3	—	1
Aubepine (anisic aldehyde)	1	—	—	—	—	—
Isoeugenol	1	1	—	—	1	14

new materials introduced, none has been used in all six fragrances; cinnamic alcohol has been blended into five formulas, and at the other extreme, aubepine (anisic aldehyde) into only one (the lilac).

Thus far, the formulas cited are relatively simple ones, even when the number of ingredients has risen to 17. They are still made up of a few raw materials, and they are still "rough," constituting only a starting point. For a more complex formula, particularly for the blending of a modern fantasy perfume, one has recourse to a larger number of materials, sometimes in small quantities, sometimes as major ingredients. Such widely used oils as lavender and vetiver have not yet been mentioned, or important resins such as oakmoss, or materials indispensable for high-quality fragrances, such as the floral absolutes.

The Study of Published Formulas

Numerous modern formula books have appeared, of which the most important include (but are not limited to) the extremely valuable work of Poucher (19), Gerhardt (28). Mann (29), Cola (30), Jellinek (31), Winter (32,33), and Burger (34), to which one should add the important series of articles by Morel (35), continued under the name of Maurer (36). During recent years publication of this type of formula book has been largely stopped. In the main, this has happened as a result of the increasing complexity of modern perfumes, and also the introduction of many entirely new materials. These books present certain difficulties and drawbacks:

1. The formulas are often based on specialties; since the books have frequently originated in Europe, and the same specialties may not be available in this country, one may not be able to obtain the same product. To obtain an adequate replacement is difficult without studying the specialty itself.

2. The formulas frequently call for uneconomical quantities of floral absolutes; and the use of these formulas, under prevailing economic conditions, would be prohibitive in many instances.

3. There is an unfortunate assortment of excellent, fair, and poor formulations, and one must actually make the product, study it, and test it, before it can be evaluated.

4. The formulas are, of necessity, inadequately described, particularly where the "fantasy perfumes" are involved; hence one does not know where to start when searching for a fragrance of a given character. The experienced perfumer readily recognizes many fragrance types by an examination of the formula, but such a perfumer has less need of the book than has the beginner, who could not be expected to recognize the fragrances without smelling them.

How several authorities have approached one fragrance type is shown in a series of formulas, presented as Formulas 18 to 23. This group of chypre perfume oil formulations appeared in various works. Formula 18 is from Poucher (19), Formula 19 from Cola (30), Formula 20 from Gerhardt (28), Formula 21 from Burger (34), Formula 22 from Winter (32), and Formula 23 from Mann (29). Finally, Formula 24 is a composite that encompasses most of the suggestions and principal characteristics of the six others,

	18	19	20	21	22	23	24
Oakmoss	6.0%	6.0%	12.0%	10.0%	5.0%	3.5%	10%
Patchouli oil	5.0	2.0	3.2	4.0	3.4[1]	1.0	4
Vetiver oil	5.0	6.0[2]	—	—	6.0	0.6	5
Sandalwood oil E.I.	14.0	6.0[3]	12.2	8.0	9.0	4.5	12
Bergamot oil	16.0	22.5	—[4]	20.0[5]	26.0	19.0	22
Methyl ionone	7.0	5.0	1.6[6]	—	6.0	—	4
Lavender oil	—	0.3	—	—	—	—	—
Sweet orange oil	—	—	4.0	5.0	—	—	1
Hydroxycitronellal	5.0	—	—	—	—	—	2
Linalool	—	3.0	16.0[4]	—	4.0	—	4
Linalyl acetate	—	—	20.0[4]	—	—	—	—
Jasmin absolute	5.0	2.0	—	4.0	4.0[7]	1.5	3
Rose absolute	5.0	1.5	—	6.0	4.0[8]	4.5[9]	3
Geranium oil Bourbon	—	—	2.5	2.0	—	—	2
Tuberose absolute	—	—	—	—	—	0.3[4]	—
Carnation specialty	—	—	—	—	3.0	—	—
Iris concrete	—	—	—	—	—	0.2	—
Cassie absolute	1.0	—	—	2.0	—	—	1
Neroli petals	1.0	—	—	2.0	1.2	—	1
Angelica root oil	—	0.5	—	—	—	—	—
Clary sage oil	1.0	3.0	—	—	2.5	—	2
Ylang-ylang oil	—	7.0	—	—	—	—	1
Citronellol	—	—	4.0	—	—	—	—
Benzyl acetate	—	—	—	10.0	—	—	—
Terpineol	—	—	4.8	—	—	—	—
Phenyl ethyl alcohol	—	—	1.5	—	—	—	—
Isobutyl salicylate	9.0	—	—	—	—	—	1
Amyl salicylate	—	—	—	—	1.2	—	—
Eugenol	—	—	—	—	—	—	—
Iseugenol	—	3.5	—	—	—	—	1
Coriander oil	0.5	—	—	—	—	—	—
Estragon oil (tarragon)	0.5	2.5	—	—	1.0	—	1
Cinnamon bark oil	0.2	—	—	—	—	—	—
Cinnamyl acetate	—	2.5	—	—	—	—	—
Sassafras oil	0.2	—	—	—	—	—	—
Methyl salicylate	—	0.2	—	—	—	—	—
p-Methyl acetophenone	—	—	0.4	—	—	—	—

Formulas for Chypre Perfumes

Vanillin	0.5	1.5	3.1	—	1.5	0.7	1
Vanilla resinoid	—	—	—	—	—	0.6	—
Coumarin	—	9.0	2.4	—	9.0	2.0	6
Tonka resinoid	—	—	—	—	1.0	0.6	—
Heliotropin	—	3.5	—	—	4.0	—	2
Nitromusks[10]	4.0	5.0	8.0	15.5	4.5	3.0	8
Aldehyde C-12							
Lauric 10%	0.1	—	—	—	—	—	—
Undecalactone	—	—	0.2	—	—	—	—
Vanilla tincture[11]	—	—	—	—	—	28.0	—
Musk tincture	—	—	—	—	—	20.0	—
Ambreine absolute[12]	3.0	2.5	—	—	—	—	—
Castoreum absolute	1.0	—	—	6.0[13]	0.2[13]	10.0	1
Labdanum	—	—	4.1	5.0	3.0	—	2
Benzoin gum							
resinoid	—	5.0	—	—	—	—	—
Styrax	10.0	—	—	—	—	—	—
Clove resinoid	—	—	—	—	0.5	—	—
Civet purified	—	—	—	0.5	—	—	—

[1] Mixture of oil and resinoid.

[2] Vetiverol instead of vetiver oil.

[3] Santalol instead of sandalwood oil.

[4] Note large percentage of linalool and linalyl acetate in place of bergamot oil.

[5] Formula calls for sesquiterpeneless oil.

[6] Formula calls for ionone 100% instead of methyl ionone.

[7] Given as 2% jasmin absolute and 2% synthetic jasmin.

[8] Given as 1% rose absolute and 3% synthetic rose.

[9] Given as 1% rose absolute and 3.5% synthetic rose.

[10] Nitromusks consist of musk ambrette, musk ketone, and musk xylene (generally called musk xylol), as well as other synthetics available under various trade names, and recommended for use singly or in combinations.

[11] Strength of tinctures not given; it may be assumed that they are 24:128.

[12] Ambreine in Formula 18 is a specialty, the formula of which is given by Poucher.

[13] These are 10% resinoids.

with some adjustments made to avoid conflicting notes and repetitions, and to eliminate specialties. In presenting these formulas, a few minor changes and adjustments have been made, but the versions given here follow in all essential respects those of the original perfumers.

The six creators of these formulas have employed some 54 different raw materials in these formulations. Actually the list can be extended, because several nitromusks are here placed under that single title; vetiver oil and vetiverol are grouped together, as are sandalwood oil and santalol; and natural and imitation (or synthetic) flower oils are likewise placed under a single entry.

Excluding the composite (Formula 24), the formulations include from 15 to 24 different ingredients. A glance at the formulas would quickly indicate that, except for Formula 20, all of these perfume oils would be extremely expensive to manufacture.

The author has arranged these six formulas (and the seventh or composite) by groups of raw materials; this grouping is somewhat arbitrary and in certain instances overlapping. Nevertheless, with such a large number of ingredients with which to cope, it may be helpful to think of such a fragrance in terms of a few groups.

In the first category one finds the materials which form the base, or "the heart" of any chypre note. They are oakmoss, patchouli, vetiver (or vetiverol), sandalwood (or santalol), bergamot, and usually methyl ionone. This small number of materials constitutes from 28 to 55% of each of the formulas. If one should consider that the linalool and linalyl acetate in Formula 20 are present as a replacement for the bergamot, the proportion of the first group of materials in that formula would rise to 65%. Three of the six original formulas contain all of the ingredients in this group; another lacks only the methyl ionone; and still another contains neither the methyl ionone nor the vetiver.

The next group can be considered the supplementary citrus notes, described in this fashion because the main citrus character is imparted by the bergamot. In the new group one finds lavender oil, sweet orange oil, hydroxy-citronellal, linalool, and linalyl acetate. None of the perfumers has found it necessary to use all of these materials, and one has used none.

The constituents following consist of the natural floral notes and simulations thereof. Supplementing these important floral characteristics are the synthetics and isolates making up the following group, materials which contribute to a rose, jasmin, or lilac, but which do not have the depth and roundness of the natural oils of the previous group. Note that only two of the six perfumers drew upon these materials.

The group that follows consists of natural oils, isolates, and synthetics that can be considered to contribute a spicy character to this type of fragrance. Actually, only Formulas 18 and 19 call these materials into play to any considerable extent; in Formulas 20 and 22 they are little used; and in Formulas 21 and 23 they are completely omitted. Certainly it is here that one can trace the major differences in the odor character of the various chypre notes created by these perfumers.

Finally one comes to the resins and tinctures, to the fixative notes, to the sweet, the musky, and the long-lasting contributions to this fragrance. Natural and synthetic musks; such resins as labdanum, benzoin, and styrax;

vanillin and vanilla; coumarin and tonka—these are among the materials chosen by various perfumers, to blend with the oakmoss, and to give depth, strength, and lasting value to this oil.

What can one learn from the study of a group of formulas of this type? First, how one fragrance is approached by several different perfumers, the degrees of similarity and dissimilarity among them, the peculiarities of the approaches. Then, if a young perfumer should wish to formulate each of these perfumes in the laboratory, smelling each raw material as he uses it, and continuously smelling the mixture as it is being compounded in order to detect the constant changes that it undergoes as each material is added, he can find the specific contributions to the complex fragrance that each constituent makes. Now, with the completed fragrances at hand, their study will indicate not merely which is "best" or most to one's liking, but which are the points of desirability to be enhanced and which are the less desirable notes to be reduced or eliminated. There may be something in each formula that is desirable, and something in each that makes it jarring; by compounding each with great care, the perfumer may be able to "put his nose" on the notes that are to be brought out and the ones to be rejected.

The formulas will, of course, require revision to suit the needs of a company or a client. Except for Formula 20, all of these perfume oils would be quite expensive for most purposes. Some contain "synthetic" essential oils or specialties, and these would have to be selected. In addition, other revisions may be necessary, as will be seen below in the study of another fragrance note.

It is suggested that the formula books can be helpful only if they are used as a point of departure, by grouping together several related formulas, as has been done here with the chypre note, and making a study of that odor type, both on paper and in the laboratory.

Revision and Adaptation

Let us now suppose that a perfume has been formulated and that it is satisfactory for a specific purpose, but requires a series of revisions. The original perfume oil is shown as Formula 25, which is a composite type that would probably best be described as having an oriental-citrus character. The progressive revisions required may be for the purpose of lowering the price so that it will be usable in a toilet water or a cosmetic, or for the purpose of producing a difference in the fragrance without making a fundamental change (that is, to diminish or to enhance a single note); or the formula may have to be altered to accommodate the use of the perfume under conditions in which it might otherwise cause irritation, discoloration, or other ill effects when incorporated in a given preparation.

The revision of a formula for the sake of lowering the cost may best be handled by a study of the fluctuating markets. On studying the formula,

Formulas for a Series of Related Perfume Oils

	25	26	27	28	29	30	31	32	33	34	35
Orange oil	7.0%	7.0%	13.0%	15.0%	15.0%	7.5%	15.0%	13.0%	5.0%	2.0%	1.0%
Bergamot oil	30.0	30.0	—	—	—	31.5	—	—	26.0	17.0	9.0
Lemon oil	4.0	4.0	13.0	14.0	—	4.0	14.0	13.0	3.0	2.0	1.0
Lemon replacement	—	—	—	—	14.0	—	—	—	—	—	—
Linalyl acetate	—	—	15.0	—	—	—	15.0	14.5	7.0	15.0	13.0
Terpinyl acetate	—	—	—	17.0	17.0	—	—	5.0	—	—	—
Lime oil	—	—	0.5	0.5	0.5	—	0.5	0.5	—	—	—
Sandalwood oil	6.0	6.0	2.0	—	—	6.5	3.5	6.0	7.5	5.0	6.0
Balsam copaiba	—	—	2.0	2.5	2.5	—	2.0	—	—	—	—
Guaiacwood oil	—	—	2.0	2.5	2.5	—	2.0	—	—	—	—
Patchouli oil	6.0	6.0	2.0	—	—	6.5	2.5	6.0	7.5	5.0	6.0
Cedarwood oil	—	—	2.5	5.0	5.0	—	3.0	—	—	—	—
Lavender oil	6.0	6.0	6.0	—	—	6.5	—	5.5	5.5	2.0	1.0
Lavandin oil	—	—	—	6.5	6.5	—	6.5	—	—	—	—
Rose de mai absolute	0.5	—	—	—	—	—	—	—	—	—	—
Rose specialty	3.0	3.5	3.5	3.5	4.0	3.5	3.5	3.5	3.5	6.0	6.0
Jasmin absolute	2.0	—	—	—	—	—	—	—	—	—	—
Jasmin specialty	—	2.0	2.0	2.0	2.0	2.0	2.0	2.0	1.5	6.0	8.0
Resin castoreum	0.5	0.5	0.5	0.5	0.5	0.5	0.5	0.5	0.5	0.3	0.3
Resin benzoin	2.0	2.0	2.0	2.0	2.0	—	—	2.0	3.0	2.0	2.0
Resin opoponax	3.0	3.0	3.0	3.0	3.0	3.0	3.0	3.0	3.5	1.5	1.0

Ingredient	1	2	3	4	5	6	7	8	9	10	11
Resin styrax	—	—	—	—	—	0.5	0.5	—	—	—	—
Tincture musk Tonquin, 4:128	5.0	2.5	—	—	—	5.5	—	5.0	2.5	5.0	5.0
Amber synthetic	—	—	—	0.5	0.5	—	—	—	—	—	—
Tincture vanilla bean, 16:128	3.5	3.5	3.5	—	—	—	—	3.5	3.5	3.0	3.0
Vanillin	1.0	1.0	1.0	2.0	2.0	—	—	1.0	1.5	0.5	1.0
Coumarin	7.0	7.0	7.0	7.5	8.0	9.5	9.5	7.0	8.5	5.0	5.0
Ethylene brassylate, 10%	—	2.5	5.0	2.5	1.0	5.0	10.5	—	2.5	1.0	1.0
Tincture ambergris, 4:128	2.5	2.5	2.5	—	—	2.5	—	2.5	2.5	2.5	2.5
Musk ketone	3.0	3.0	3.0	—	—	—	—	—	—	—	—
Musk ambrette	2.0	2.0	2.0	—	—	—	—	—	—	—	—
Musk xylol	—	—	—	5.5	5.5	—	—	—	—	—	—
Phenyl ethyl alcohol	1.8	1.8	1.8	1.8	1.8	2.0	2.0	1.8	1.8	2.0	2.0
Linalool	3.0	3.0	3.5	4.5	5.0	3.0	3.5	3.0	2.5	6.0	5.0
Indole, 10%	0.5	0.5	0.5	0.5	0.5	—	—	0.5	0.5	1.0	0.5
Methyl anthranilate	0.2	0.2	0.2	0.2	0.2	—	—	0.2	0.2	0.2	0.2
Hydroxycitronellal	—	—	—	—	—	—	—	—	—	4.5	4.0
Methyl ionone	—	—	—	—	—	—	—	—	—	3.0	5.0
Aldehyde C-10, 10%	—	—	—	—	—	—	—	—	—	—	2.0
Aldehyde C-11, 1%	0.5	0.5	1.0	1.0	1.0	0.5	1.0	1.0	0.5	0.5	2.0
Aldehyde C-12 lauric, 10%	—	—	—	—	—	—	—	—	—	—	2.0
Ylang-ylang oil	—	—	—	—	—	—	—	—	—	2.0	5.0
Styrallyl acetate	—	—	—	—	—	—	—	—	—	—	0.5

it becomes apparent where the greatest contribution to its cost is found; the costly materials may be able to be replaced, in whole or in part, by other natural oils, synthetics, or specialties whose fragrance character is sufficiently similar to permit such substitution.

Revision of a formula because of chemical incompatibility is often a matter of isolating the ingredient or ingredients causing such difficulty. For this purpose, one has recourse to a wealth of literature and a cumulative experience, as well as knowledge of the chemistry of the raw materials and of the cosmetic preparation in which the oil is to be incorporated. For example, in a suntan preparation, one would not wish to use oil of bergamot because of its photosensitizing potential. In a white cream, one would use indole only after thorough tests had indicated that it will have no discoloring effect. In a cold wave, one would avoid certain aldehydes because of their instability in the presence of strong reducing agents (thioglycolate).

Yet, except for the danger of irritation, all of these rules are far too broad and too general, for they do not take into consideration the activity of these materials under the specific conditions of use, the fact that they are often used to excellent advantage only in trace amounts, and the inhibiting effect of other perfume materials on the behavior of a single ingredient. For example, the warning that methyl anthranilate is not usable in a white soap may be true as a general statement, but its judicious use is not impossible under proper conditions. The writer is not impressed by the wisdom of following a chart warning against the use of individual materials for certain purposes, like that by numerous authors who have contributed guides in the literature, except as rules to be disregarded as often as they are obeyed. As an instance, one such list finds hydroxycitronellal "definitely unusable" in cream, lipsticks, massage oils, brilliantines, hair oils, face lotions, and liquid shampoos. Yet it can be said from experience that the proper use of this aromatic material in perfume oils for such purposes, even when playing a major part, has brought gratifying results.

To return to Formula 25 and to its modifications, these revisions have been made for the purpose of bringing the cost down, adapting the fragrance for toilet water and cosmetic creams, making it nondiscoloring, or bringing out certain desired olfactory effects. The replacements that might be made would depend on the taste and discretion of the perfumer, on the notes that are found most pleasing, as well as on the fluctuations in the prices of raw materials, and other considerations.

On the basis of average prices prevailing in the American market in 1969, the cost of the raw materials necessary for the manufacture of 1 lb of these oils varies between about \$3 and \$25/lb.

The first fragrance in this series, Formula 25, is intended for use in perfume. The second fragrance is for use in toilet water, and the revision should therefore be based almost exclusively on cost requirements, because it is usually desired that the toilet water and the perfume have the same fragrance, except for strength.

The degree of change in adapting an oil for use in toilet water is almost always determined by economic considerations. In this instance, the absolutes were omitted (rose de mai and jasmin) and specialties substituted for them. The nature of the specialty chosen for this purpose is best determined by the perfumer, and is not indicated in the formula. One other change will be noted: the use of a macrocyclic musk (ethylene brassylate) as a partial replacement for tincture of Tonquin musk. As a result of these replacements, the new perfume oil is about half the cost of the old; yet it should be very similar in odor character, although not indistinguishable from the original.

In the next perfume group, Formulas 27 and 28, the perfumer adapts his oils for use in cosmetic products. Again he is faced with a high cost, and seeks to bring it down, emphasizing the replacement of certain of the higher-priced essential oils. Bergamot oil (very high in the mid-1960's) is eliminated, and compensated for with linalyl acetate and with the effects of other citrus oils. The proportion of sandalwood oil is diminished, as is that of patchouli oil; and in their place one introduced such notes as balsam, guaiacwood, and cedarwood. The macrocyclic musk now completely replaces the Tonquin musk.

These revisions for the purpose of lowering costs and making the product usable for perfuming cosmetics might continue indefinitely; however, each such revision sacrifices some of the olfactory character of the original oil. The process of lowering the cost is concluded in Formulas 28 and 29. The linalyl acetate, which has been introduced to take the place of bergamot oil, now gives way to terpinyl acetate. Sandalwood and patchouli oils, already partially replaced, are now completely eliminated. Tincture of vanilla bean is deleted in favor of vanillin; musk xylol is used as a substitute for other musks; oil of lavender is sacrificed in favor of oil of lavandin; even lemon oil is replaced in Formula 29, in which a mixture of lemon terpenes and citral is used.

To adapt the original perfume oil for use in cosmetic products, one may find it necessary to remove materials that tend to cause discoloration. It is possible that in certain cosmetics all of the oils, from Formulas 25 to 29, would discolor. Formulas 30 and 31 show how such a problem might be approached. In Formula 30 the floral absolutes have been eliminated (for reasons of economy); now the vanilla bean tincture and the vanillin must be removed, the synthetic musks eliminated, and no methyl anthranilate and indole are used. To compensate for the loss of vanillin, the proportion of coumarin is increased; and to retain the musky notes, both musk Tonquin and ethylene brassylate are used. In Formula 31 all of the discoloring agents that had been discarded in the previous formula are omitted again; in addition, some economies have been effected, in a manner similar to some of the earlier revisions.

Formula 32 is an example of an oil worked out for a specific purpose: the removal of the bergamot oil, because of the photosensitization potential of that material. To compensate for this loss, orange, lemon, and lime oils, linalyl acetate, and lavender oil are either introduced into the formula or their proportions are increased.

In the last three fragrances, the changes are intended to enhance a given olfactory note or to introduce a new one. Formula 33 is heavier and sweeter than the previous oils. This heaviness comes out with increased proportions of sandalwood and patchouli oils; the proportions of benzoin and opoponax are higher, as are those of vanillin and coumarin. These are not drastic changes, because the basic character was meant to remain intact.

In Formula 34, nuances in the fragrances are created by the addition of new materials that had not previously appeared in the formulas. They are not radically different from other ingredients, for they must blend in and become a part of the entire fragrance. Thus among other changes this formula sees, for the first time in the series, some hydroxycitronellal, which blends very well with the linalool and the citrus notes; some methyl ionone, effective with the woody and floral notes; and ylang-ylang oil, which is excellent with a jasmin character. To work these additions into the fragrance, the jasmin and rose specialties are increased, the indole content is made somewhat higher, and the linalool content brought up to a level greater than in any of the previous versions. To accommodate these increases, there are corresponding decreases in several materials, such as bergamot, orange, and lemon, among others.

A last version is offered in Formula 35. By the introduction of new materials, a "modern, sophisticated" tone is given to the perfume. Some fatty aldehydes are introduced (hitherto there was but a trace effect), and the floral concept is given a new direction with styrallyl acetate. The methyl ionone and the ylang-ylang oil, previously shown in Formula 34, are strengthened.

What has been done here with one type of fragrance could be repeated almost indefinitely, making variations, revisions, and replacements, and bringing out different effects, introducing new ones, and eliminating old ones. A similar process can be repeated for literally hundreds of different types of perfumes, each of which may be carried through a long series of revisions.

In the production of the perfume oil, a process generally called compounding, it is usual practice for the crystalline or solid ingredients to be weighed out first, and then to follow them with the liquid materials. This mixture is then agitated, usually by a suitable stirrer, until complete solution is accomplished. Care must be taken not to agitate too vigorously, thereby introducing an excess amount of air. The mixture will sometimes require gentle heating to bring the crystals into solution. It is important, however, that the heating should not be too severe, for this may affect the perfume. Usually the best way to do this is to heat the crystals with some portion of compound that would not be affected by heating, and then add this melted mixture to the balance of the completed compound. During the compounding, extreme precautions must be taken to prevent contamination of any of the perfume materials.

Solubilized Perfume

In the series of formulas shown, there are no water-soluble or solubilized perfume oils. Strictly speaking, there are few perfume oils that are water-soluble, although many are water-solubilized. Although some perfume materials exhibit slight solubility in water, most materials have very poor solubility, not only in water but even in hydroalcoholic solutions of less than 65 % alcohol content.

The ideal water-soluble perfume would consist of a mixture of perfume materials soluble to at least 1 % in water, and preferably higher. All aromatic research thus far has led to very few organic compounds having interesting odors that are soluble in water, even to this limited extent. The outstanding exception is probably phenyl ethyl alcohol, for which a solubility as high as 2 % in water has been claimed with regard to some grades, although the usual commercial grades are soluble only to about 1 % or less. Various alcohols, some terpeneless oils, and a few other perfume materials exhibit solubility to the extent of 0.1 to 1.0 %.

All perfume oils can be solubilized; some require a smaller percentage of solubilizer in proportion to the oil than do others. A perfume oil containing resins, terpenes, and certain crystalline materials will be more difficult to solubilize, and the percentage of solubilizer to oil will have to be relatively high. A perfume oil, on the other hand, based largely on terpeneless oils, alcohols, compounds of low molecular weight, and polar compounds, will require a smaller percentage of solubilizer when used in water or in hydro-alcoholic solutions.

Several other factors influence the choice of solubilizer and the proportion of solubilizer to perfume oil. These include the nature of the vehicle (and particularly whether distilled water or ordinary tap water is to be used) and the amount of fragrance required (or the proportion of aromatic materials needed in the solution). Solubilizers that have been suggested for perfume oils include soap, synthetic detergents, and many commonly used emulsifiers. There are several polyglycols (propylene and dipropylene) that are frequently used; also surfactants such as polyoxyethylene derivatives of sorbitan fatty acid esters and sorbitan fatty acid esters. These products are available under such trade names as Tweens and Spans.

From the above, it will be seen that one preferably should not create a formula calling for a given amount of a particular solubilizer, and the remaining percentage of perfume oils. Although such products are being offered to industry under the name of water-soluble perfumes, it is preferable that they be made to order for the specific vehicle and perfume oil involved.

Incorporation of Perfume

After the fragrance has been formulated and compounded, it is ready to be added to the cosmetic preparation. Unless the perfume is properly incorporated, all of the care taken in its blending and selection can go for naught.

Generally speaking, the perfume oil is best incorporated into a cosmetic product at the earliest period in manufacture at which this can be done while subjecting the perfume oil to the least possible amount of heat, and at the latest period at which it can be done without risking the possibility of poor distribution.

In other words, one must avoid undue subjection of the perfume oil to heat, because the oil, or some of its constituents, may break down or be lost by evaporation. For this reason the perfume oil is often incorporated into a cream or lotion during the cooling stage, e.g., after the emulsion has been formed, and as the material is cooling and has reached about 45°C. On the other hand, it would be inadvisable, as a rule, to add the perfume to the oil-soluble part of the formula and then form the emulsion, for this process may require heating to 85 or 90°C as well as the subjection of the perfume to the chemicals used in emulsification.

Particularly in a solid or semisolid preparation (mug shaving soap, for example), the distribution of the perfume oil is most important. A poor distribution means that part of the product might be underperfumed, hence defeating the purpose of the fragrance; another part might be overperfumed, which is not only aesthetically undesirable, but which may have certain deleterious effects not caused by the smaller proportion.

Problems in Formulation

In the formulation of a perfume, one must take into consideration primarily the type of fragrance desired; that is to say, is the perfumer aiming to work along the line of some previously known or previously formulated material, and if the latter, with what changes, modifications, and improvements? From his previous knowledge, from his own formula books, and from his acquaintance with perfume raw materials, he can have a fairly accurate, although rough, concept of what the basic elements might be, in order to obtain the fragrance which he seeks. Then the improvement of that material, its being "finished" or "rounded out," and its meeting the other necessary requirements or specifications, constitute the main problems facing him.

In addition, he is faced with several other problems, one of which is cost. A perfume in almost any fragrance category can be formulated in a wide variety of price ranges, using the least expensive materials, in which costs can be kept low, to the most expensive absolutes, which could make the perfume cost a hundred dollars or more per lb. The price range within which a perfumer is formulating determines the materials with which he will work, and the quantities of those materials that he will allow himself to use; of course, this will affect the quality of the product.

The perfumer must likewise evaluate his formula from the point of view of the end use to which the fragrance is to be put, because this likewise affects the selection of the raw materials to be employed. Acidity or alkalinity of the product in which the perfume is to be used, the solubility of the perfume oils in the product, the tendency to discoloration of a white product or color change in a colored product, the tendency to irritate, and the presence in the formula of reactive materials—these are among the problems requiring study, not only in the formulation of the perfume, but in its adaptation for a specific cosmetic.

Of these problems, perhaps none is quite so important as the elimination of any tendency of the perfume materials to cause irritation or sensitization. This problem was investigated with thoroughness by Klarmann (37), who points out that "perfume hypersensitivity due to sensitization is both extremely rare and readily gotten rid of." Indispensable for further study is the compilation of Greenberg and Lester (38), in which almost every published reference to the sensitizing effects of perfume materials (as well as other cosmetic ingredients) is listed. However, the elimination of materials cited in the Greenberg-Lester study is neither necessary nor practical, because in many instances the only reference is to a single case of allergy in a material that has been used in literally thousands of formulas. Furthermore, such elimination would not in and of itself preclude the possibilities of sensitization due to perfume, because, as Klarmann (37) points out, "hardly anything is known concerning the existence of sensitizing synergisms or antagonisms of groups of perfume materials."

The entire question of "synergisms or antagonisms of groups of perfume materials" must enter into any consideration of problems other than sensitization, e.g. discoloration, effect on emulsion stability, and others. Most of the studies of the effect of perfumes on these important properties of cosmetics have been concerned with single ingredients. Even when the results of such studies indicated the inadvisability of using certain ingredients for a specific type of cosmetic, it was not at all proved that when the particular perfume raw material was incorporated into a formula in which it was blended with numerous other substances, it continued to have any ill effects. This is not to state that the investigations of the behavior of individual perfume raw materials are without value, but the information can only be used as a vague indication of what products to include and exclude from a formula, rather than as a rigid rule.

Some of the more important problems involved in the perfuming of cosmetics have been reviewed by McDonough (39) and by Morel (40). The behavior of perfume materials in emulsions, particularly their effect on the stability of emulsions, has been studied by Pickthall (41), Karas (42), and Wynne (43), whose findings are at variance with those of Karas. Wynne recognized the necessity of studying complex blends as well as individual ingredients, but the formulas of the perfume oils under investigation are not disclosed. Other relevant studies of perfume materials under specific conditions have been made by Wight, Tomlinson, and Kirmeier (44), who examined the behavior of aromatics in polyethylene containers; by Kilmer (45), who studied the problems involved in the perfuming of soaps and synthetic detergents; by Foresman and Pantaleoni (46), who considered perfume constituents of aerosols; and by Sagarin and Balsam (47), who studied these materials in thioglycolate hair-waving preparations.

Special steps are necessary in the formulation of a well-balanced fragrance for use in a product whose perfuming poses great difficulty. Even with the wealth of experience at his disposal, the perfumer may encounter a problem which for some reason or other does not fit in with his or his company's background. Remarkably, this occurs more often than might be supposed. The perfumer should have on hand some of the unperfumed product, or some knowledge of how to simulate this product. He should know, through preliminary experimentation, in what direction his best efforts lie. A thorough knowledge of the chemistry of the unperfumed product, as well as of the perfume materials, is also a must. The next phase may require hundreds of test samples of individual perfume materials in the product. These samples are exposed to elevated oven temperatures, to approximate an aging process. Next, these many test samples are examined to determine what odor changes have occurred. Also noted will be the strength or coverage that the perfume material exhibits. Discoloration, or in some cases bleaching, must also be examined. At this point a "picture" begins to emerge to the experienced perfumer. He can now put together a perfume that has a good chance of working. The new mixture or mixtures are again subject to the testing outlined above, and finally a finished, polished perfume or perfumes are produced. Through this whole process, the perfumer is constantly balancing his formulation. He makes substitutions where he can. He may decide to leave in a material which discolors slightly. He uses it in a small quantity where he feels its value to the finished product warrants its inclusion. During this whole process there is, in back of his mind, a theme or type for the finished perfume. This may be of his own choosing, or it may be imposed by others. Concurrent with this whole program, numerous other chemical tests may be employed, other than odor testing alone.

Use of Instrumentation

Several analytical tools have become increasingly important in the field of perfumery, such as gas chromatography and infrared spectroscopy. To a lesser degree ultraviolet spectroscopy, mass spectroscopy, and NMR (nuclear magnetic resonance) have assumed some importance. Of course one might surmise that these instruments would be useful in the manufacture of perfume raw material, as they are to other chemical manufacturers, for process and quality control. However, in perfumery they take on a different significance. The perfumer has always been obsessed with the dream of duplicating many of the essential oils found in nature, particularly those such as jasmin and rose absolute. With the use of these instruments the realization of this dream has during recent years been fulfilled to some degree. Furthermore, the everyday work of raw material evaluation and duplication is greatly facilitated.

Of these instruments, the one most used by the perfumer is gas chromatography. In fact, no well-equipped laboratory would be found without at least one such device. It might almost be viewed as an extension of the perfumer's nose. Of the various types of detectors used in gas chromatography, this author believes that thermal conductivity is the best all-round type for use by the perfumer. It is not as sensitive as some of the others, but sufficient material is injected with this type of detection system so that the individual components passing through the outlet may be smelled by the perfumer standing close, and identified. This technique of smelling and identifying the components is used in conjunction with the conventional one where a graph is obtained. The area under the curve would provide the quantity of the component, and the time it takes passing through the column identifies it. This method of working closely with gas chromatography helps the perfumer enormously.

Alcoholic Fragrance Solutions

The most popular fragrance products on the market are in the form of alcoholic solutions. Such solutions are handled under different designations, such as perfume, toilet water or eau de toilette, cologne or eau de cologne; less frequently one encounters a variety of other fanciful names, such as eau de parfum, essence, fragrant water, etc. In Europe the word "lotion" is synonymous with "toilet water" as used in America (and as defined below), but this meaning of lotion has never been adopted in America.

By and large, the term "perfume" is meant to connote an alcoholic solution in which the content of the perfume oil is relatively high and the odor strength and lasting power relatively good. There exists, however, neither an official ruling nor an accepted trade standard relative to the percentage of perfume oil that should be used in order to warrant description of a given solution by the word "perfume." As a matter of fact, merely to cite the number of fl oz of odorous concentrate per gal of alcohol would be quite inadequate, because this would be based on the false assumption that the same percentage of various oils would give similar odor strengths. It is entirely possible that the strength and lasting power obtained by one type with 16 fl oz of perfume oil per gal would be comparable to that obtained with 24 fl oz of another oil.

It is therefore a matter of individual judgment for each manufacturer to decide the percentage of perfume oil to be used in the perfume. This decision must be based above all else on the type of fragrance and the effects obtained in various strengths; the study of such effects can only be empirical. No doubt, costs must enter into consideration, and particularly the cost of the perfume oil and the price at which the finished perfume is to be sold. Finally, the nature of the prospective clientele is an important factor in deciding the strength of the oil to be used.

It is general trade practice, in the better perfumes, to use between 20 and 24 oz of oil/gal of alcohol. These are not, however, upper and lower limits. Perfumes containing as high as 36 oz of oil/gal are known on the American market; other products, labeled and sold as perfumes, often contain between 10 and 16 oz/gal.

The more dilute fragrance solution, generally made with up to 8 oz of oil/gal, is known as toilet water or eau de cologne; in America these designations have sometimes been used interchangeably. In the case of some houses handling both a toilet water and a cologne, the former usually refers to the stronger of the two solutions. Historically, however, the term "eau de cologne" was meant to refer to a specific type of fragrance with a citrus note and a refreshing character, rather than to any fragrance type that happened to be in a form more dilute than the perfume. This European definition of eau de cologne is still adhered to by some houses; their colognes continue to have the top notes of a bergamot, lemon, and lavender that were historically associated with the word "cologne."

Again there are no legal provisions that govern the use of the words "toilet water" and "cologne," so far as concentration of perfume oil in the alcohol is concerned. The main delimiting factors are fragrance effects and costs. Inasmuch as colognes and toilet waters are generally sold in relatively large containers at prices far below those of the true perfumes, it is economically impossible to use large percentages of perfume oils in colognes. However, fragrance strength is not sought in the same manner as in a perfume.

With this in mind, it can be said that most toilet waters and colognes contain between 3 and 6 oz of perfume oil/gal of alcohol, with the limits lowered in some instances to 2 oz, and raised in others to 8 oz.

Another major difference between the perfume and the more dilute form of fragrance (whether cologne or toilet water) is the concentration (or proof) of the alcohol used. A perfume with 16 to 24 oz of oil/gal may contain only a very low percentage of water in order to remain clear and to have the oil go readily into solution. For this purpose, 95% alcohol is usually employed. When the percentage of oil is lowered, a weaker alcohol may also be used, usually going down to 80 or 85%, and sometimes as low as 75%. The amount of water present in the alcohol will be determined by the type of oil being used and its solubility characteristics, the percentage of perfume oil/gal of alcohol, the demands of the formulator in the matter of costs, and the fragrance effect obtained with varying percentages of water.

In addition to the obvious saving in cost, it is felt by many that a hydroalcoholic perfume solution has more "lift" and a greater freshness when applied to the skin than the solution in 95% alcohol. If solubility is a critical problem in a low-proof alcoholic solution, this can be met in various ways. One can decrease the amount of water, decrease the percentage of perfume oil, or revise the formula to diminish the percentage of resins, crystals, terpenes, and other difficultly soluble materials in the perfume. It is always advisable to use distilled or deionized water instead of tap water, a practice which will sometimes diminish the difficulties arising as a result of solubility problems. The use of a solubilizer, however, for a hydroalcoholic solution, may not always be practical, because of foam, "sticky feel," or cloudiness that the solubilizer may cause.

The ethyl alcohol used for perfumes and other fragrance and cosmetic purposes is denatured, and is known as specially denatured alcohol or SDA. There are numerous denaturants, several of which have been approved by the government for perfumes, toilet waters, and colognes. As a practical matter, two SDA formulas are most frequently used, SDA 39-C and 40. The former is used mostly for perfumes, the latter usually for toilet waters, colognes, after shave solutions, and other fragrance products requiring alcohol. Of all the denatured alcohols, SDA 39-C is considered by most perfumers to have the finest odor, the one in which the denaturant interferes least with the fragrance. However, some fragrance manufacturers use SDA 39-C and 40 interchangeably, although not all perfumers approve of this practice. The difference between the two alcohols is only slightly perceptible, and approaches the point of being negligible when a perfume oil is added; however, it seems best to refrain from changing from one denatured alcohol to another, even if government approval is available.

All alcoholic solutions must be approved by the government before they can be manufactured. The government body having jurisdiction is the Alcohol, Tobacco, and Firearms Division of the United States Department of the Treasury. If a formula calls for the use of SDA 39-C and has been approved specifically in that manner, one cannot of course substitute SDA 40 without official permission. The Alcohol Division will not approve a formula if, in its opinion, the quantity and character of the perfume oil used make it possible readily to remove this oil and to render the alcohol potable.

The manufacture of an alcoholic fragrance solution is a delicate operation requiring careful selection of equipment and understanding of other critical factors. These are described separately in the following chapter.

REFERENCES

1. Arctander, S.: *Perfume & flavor chemicals (aroma chemicals)*, 2 Vols., published by author, 1969.

2. Haagen-Smit, A. J.: The Chemistry, origin and function of essential oils in plant life, in Guenther, E.: *The essential oils*, Vol. 1, Van Nostrand, New York, 1948, p. 17.

3. Parry, E. J.: *The chemistry of essential oils and artificial perfumes*, 4th ed., Scott, Greenwood, London, 1922.

4. Langenau, E. E.: "Oils, essential" in *Encyclopaedia of chemical technology*, Vol. IX, Interscience Publishers, New York-London, 1966, p. 569.

5. Gildemeister, E., and Hoffmann, F.: *The volatile oils*, trans. by E. Kremers, 3 vols., Longmans, Green, London, 1913–1922.

6. Finnemore, H.: *The essential oils*, E. Benn, London, 1926.

7. Semmler, F. W.: *Die ätherischen Öle*, 4 vols., Veit, Leipzig, 1906–1907.

8. Charabot, E. T., Dupont, J., and Pillet, L.: *Les huiles essentielles et leurs principaux constituants*, C. Béranger, Paris, 1899.

9. Guenther, E.: *The essential oils*, 6 vols., Van Nostrand, New York, 1948–1952.

10. Arctander, S.: *Perfume and flavor materials of natural origin*, published by author, 1960.

11. Naves, Y. R., and Mazuyer, G.: *Les parfums naturels*, Gauthier-Villars, Paris, 1939.

12. Naves, Y. R., and Mazuyer, G.: *Natural perfume materials*, trans. by E. Sagarin, Reinhold, New York, 1947.

13. Sagarin, E.: *The science and art of perfumery*, 2nd ed., Greenberg, New York, 1955.

14. Poucher, W. A.: A classification of odours and its uses, *JSCC*, **6**: 80 (1955).

15. Bedoukian, P.: *Perfumery synthetics and isolates*, Van Nostrand, New York, 1951.

16. Jacobs, M. B.: *Synthetic food adjuncts*, Van Nostrand, New York, 1947.

17. *The Givaudan index*, Givaudan-Delawanna, Inc., New York, 1949.

18. Parry, E. J.: *Parry's cyclopedia of perfumery*, 2 vols., Blakiston, Philadelphia, 1925.

19. Poucher, W. A.: *Perfumes, cosmetics and soaps*, 3 vols., 6th ed., Van Nostrand, New York, 1942.

20. Moncrieff, R. W.: *The chemistry of perfumery materials*, United Trade Press, London, 1949.

21. Igolen, G., Crabalona, L., Daumas, J., Benezet, L., and Telsseire, P.: Méthodes et constantes analytiques des huiles essentielles, adoptées par une commission de normalisation, Syndicat Fabricants et Importeurs Huiles Essentielles et Produits Aromatiques Naturels, Grasse 1954, through Bedoukian, P. Z.: Progress in perfumery materials, Part II., *Am. Perf.*, **67**: 44 (March 1956).

22. Charpy, J. F.: A perfumer evaluates pure perfumery chemicals, *SPC*, **28**: 1042 (1955).

23. Pickthall, J.: Talking of perfumes again, *JSCC*, **5**: 182 (1954).

24. Piesse, G. W. S.: *The art of perfumery*, 2nd ed., Longman, Brown, Green, Longmans, and Roberts, London, 1856.

25. Cooley, A. J.: *The toilet in ancient and modern times*, Lippincott, Philadelphia, 1873.

26. Deite, C.: *A practical treatise on the manufacture of perfumery*, trans. W. T. Brannt, Baird, Philadelphia, 1892.

27. Maurer, E. S.: Simple floral perfumes, *JSCC*, **4**: 179 (1953).

28. Gerhardt, O.: *Das Komponieren in der Parfumerie*, Akademische Verlagsgesellschaft M.b.H., Leipzig, 1931.

29. Mann, H.: *Die moderne Parfumerie*, Julius Springer, Vienna, 1932.

30. Cola, F.: *Le livre du parfumeur*, Imprimerie des Etablissements Casterman, Tournai (Belgium) and Paris, 1931.

31. Jellinek, P.: *The practice of modern perfumery*, trans. A. J. Krajkeman, Interscience Publishers, New York-London, 1954.

32. Winter, F.: *Riechstoffe und Parfumerungstechnik*, Julius Springer, Vienna, 1933.

33. Winter, F.: *Handbuch der gesamten Parfumerie und Kosmetik*, Springer-Verlag, Vienna, 1949.

34. Burger, A. M.: *Leitfaden der modernen Parfumerie*, Walter de Gruyter and Co., Berlin and Leipzig, 1930.

35. Morel, C.: The essentials of perfume compounding, *SPC*, 1952–1954.

36. Maurer, E. S.: The essentials of perfume compounding, *SPC*, 1955.

37. Klarmann, E. G.: Perfume and the skin, *Am. perf.*, **64**: 425 (1954).

38. Greenberg, L. A., and Lester, D.: *Handbook of cosmetic materials*, Interscience Publishers, New York-London, 1954.

39. McDonough, E. G.: Problems in perfuming cosmetics, *Am. perf.*, **55**: 205 (1950).

40. Morel, C.: Perfuming cosmetics and toilet preparations, *SPC*, **19**: 917 (1946).

41. Pickthall, J.: The effect of perfumery chemicals on emulsified products, *SPC*, **28**: 69 (1955).

42. Karas, S. A.: The effect of some aromatic chemicals and essential oils upon the stability of cosmetic emulsions, *JSCC*, **1**: 374 (1949).

43. Wynne, W.: Effect of perfume oils on emulsions, *Am. perf.*, **54**: 381 (1949).

44. Wight, C. F., Tomlinson, J. A., and Kirmeier, S.: Difficulties encountered in the use of polyethylene as a packaging material, *Proc. Sci. Sec. TGA*, **19**: 30 (1953).

45. Kilmer, E. D.: Some aspects of soap and detergent perfumery, *Proc. Sci. Sec. TGA*, **21**: 20 (1954).

46. Foresman, R. A., and Pantaleoni, R.: A study of odor stability on aerosol perfuming, *Proc. Sci. Sec. TGA*, **21**: 36 (1954).

47. Sagarin, E., and Balsam, M. S.: The behavior of perfume materials in thioglycolate hairwaving preparations, *JSCC*, **7**: 480 (1956).

Chapter 33

EMULSIFIED AND SOLID FRAGRANCES

Robert F. Schuler

The main products of the fragrance industry have always been concentrated essences in the form of clear alcoholic liquids, commonly known as perfumes, colognes, and toilet waters. From time to time, however, the industry has tried to expand the overall sales volume of fragrances by introducing other forms. Among these are the emulsified and solid forms of fragrance.

Emulsified cream products, in particular, attained considerable popularity during World War II, when limitations were placed on the use of alcohol for colognes and perfumes. The demand for fragrance products became overwhelming during the war years, and the only way to supply the need was to formulate emulsified cream colognes without alcohol. These products were successfully marketed then, but most were discontinued as soon as alcohol became freely available at the end of the war.

During the 1950's, however, emulsified fragrances appeared again and attained some popularity. They did not contain alcohol, even though alcohol was freely available. This was probably due to the difficulty of formulating alcohol into stable liquid emulsions, which would not separate into layers a short time after being made (14, 17). During the 1960's emulsified fragrances achieved considerable acceptance, when it became technically possible to formulate stable emulsified fragrances with high alcohol contents (12).

The 1960's also witnessed the introduction of emulsified fragrances in aerosol forms: puffs and foams of colognes and perfumes. The popularity of this type of product has not yet been established. Only time will reveal the extent of acceptance of these emulsified fragrances. Their formulations will not be found in this chapter; they are covered in the chapter on aerosol cosmetics.

Fragrances in solid form have been marketed for over 60 years, but have never attained wide consumer acceptance. As early as 1915, a solid cologne was sold in England. About 1932, one was marketed in this country. During the 1950's solid stick colognes became quite popular, but sales during the 1960's did not increase.

Emulsified Fragrances

The emulsified fragrances are called by various names: cream sachet, liquid sachet, lotion sachet, liquid skin sachet, liquid cream sachet, cream lotion sachet, perfume cream sachet, and others. The alcoholic emulsified fragrances have been given such names as veils of perfume, toilet water, or cologne, silks, or skin balms. These products are used in the same way as conventional perfumes and colognes—a dab behind each ear, on the wrist, temple, at the crook of the elbow, or behind the knee—as well as in the form of body or hand lotions.

The emulsified forms of fragrances have many appealing characteristics:

1. Attractive translucent or opaque appearance.
2. Rich, viscous consistency.
3. Long-lasting quality, because of the fixative action of the waxes and/or film formers in the formulation.
4. Skin-softening and moisturizing action.
5. Freedom from alcohol sting, when alcohol is not used.
6. Cooling, quick-drying action, when alcohol is used.
7. Extra "lift" to the fragrance, when alcohol is used.

Emulsified fragrances are highly perfumed emulsions of semisolid or liquid consistency. They usually contain from 1 to 10 % of a given perfume oil, depending on whether the intensity of a toilet water or a perfume fragrance is desired.

The perfume oils used in these preparations are artistically blended combinations of natural essential oils, substances isolated from essential oils, and aromatic organic chemicals of synthetic origin. The more expensive perfume oils also contain floral extractives, resinoids, and animal fixatives.

The materials used in the perfume oils must be chosen with special care to avoid components that change in the presence of water. Aromatic oils with alcohol and ether groups are stable under most conditions, but aldehydes, ketones, and esters may hydrolyze, polymerize, or become oxidized. In particular, they are likely to decompose in an aqueous alkaline medium required by some emulsifying agents (11).

Even after the judicious selection of the ingredients for the perfume oil, it is necessary to observe the finished preparation closely for deterioration of the fragrance. Unfortunately, there are no reliable accelerated methods of ascertaining these changes; the finished product should be aged at ordinary temperatures for at least 6 months. Accelerated test results, at an elevated temperature of 45°C, may be misleading, since practically all perfumes deteriorate somewhat under this condition. An unusually rapid breakdown of fragrance at 45°C, however, is an indication of probable poor future stability at room temperature. It is important that the perfume compound have sufficient stability to withstand the heat applied during the manufacture of the emulsified fragrance.

The main problem in formulating emulsified fragrances is that of producing a stable emulsion. This is rendered more difficult than usual by the high concentration of perfume oil used. In addition to stability, the product must not be greasy upon application to the skin and must not leave an oily residue. It should, of course, have the elegance of appearance expected of any cosmetic preparation (13).

Among the commonly used emulsifying agents are the following:

Anionics

1. Carboxy vinyl polymers neutralized with alkalis. Carbopol 934, 940, or 941, neutralized with diethanolamine, triethanolamine, monoisopropanolamine, diisopropanolamine, triethylamine, triamylamine, or dodecylamine.

2. Fatty acid soaps. Potassium and sodium stearates for semisolid emulsions; triethanolamine stearate, 2-methyl-2-amino-1,3-propanediol stearate for liquid emulsions, etc.

3. Sulfated alcohols. Sodium lauryl sulfate, ammonium lauryl sulfate, sodium stearyl sulfate for both semisolid and liquid emulsions.

4. Oleyl ether phosphates. Polyoxyethylene (10) oleyl ether phosphate (Crodafos N-10 neutral).

Nonionics

1. Polyhydric alcohol esters. Glyceryl monostearate, propylene glycol stearate, diethylene glycol monostearate, etc.

2. Polyethylene oxide ethers. Polyoxyethylene (3) oleyl alcohol (Volpo 3), polyoxyethylene (10) oleyl alcohol (Volpo 10).

3. Polyethoxylated sorbitan esters. Polyoxyethylene sorbitan monostearate (Tween 60), polyoxyethylene sorbitan monolaurate (Tween 20), polyoxyethylene sorbitan monooleate (Tween 80), etc.

4. Sorbitan esters. Sorbitan monostearate (Arlacel 60), sorbitan monolaurate (Arlacel 20), sorbitan monooleate (Arlacel 80), sorbitan sesquioleate (Arlacel 83), etc.

5. Liquid solutions of lanolin alcohols. Mixtures of cholesterol and other free sterols in a liquid hydrocarbon base (Amerchol L-101, Nimlesterol, etc.).

6. Polyethylene glycol esters. Polyethylene glycol 200, 300, 400, 600, and 1000 mono- and dilaurates, oleates, stearates, etc.

7. Acetylated polyoxyethylene derivates of lanolin. Acetylated polyoxyethylene (10) derivate of lanolin (Solulan 98, etc.).

8. Nonylphenoxypoly (ethyleneoxy) ethyl alcohols. Nonylphenoxypoly (ethyleneoxy) (9-10) ethanol (Igepal CO-630, etc.).

9. Polyethylene glycol ethers of lanolin alcohol. Polyethylene (5) glycol ether of lanolin (Polychol 5, etc.).

10. Ethylene oxide condensate of a propylene oxide-propylene glycol condensate (Pluronic P123, etc.).

11. Ethoxylated cholesterol.

Cationics

1. Cetyl trimethyl ammonium bromide, diisobutyl phenoxy ethoxy ethyl dimethylbenzyl ammonium chloride (Hyamine 1622), alkyl (C_{10}-C_{14}) dimethylbenzyl ammonium chloride USP, N(stearoyl colamino formyl methyl) pyridinium chloride, cetyl pyridinium chloride, etc. These are reacted with anionics such as a fatty acid on a molecule-to-molecule basis to form large molecules which behave as emulsifiers.

The most versatile emulsion systems for providing stable emulsified fragrances—with and without alcohol—are those formed with a trio of water-soluble resins: Carbopol 934, 940, and 941 (1,2). Carbopol resins are carboxy vinyl polymers of high molecular weight. They are supplied as dry, fluffy powders in acid form and, after dispersing in water, must be neutralized with alkali to develop viscosity. Selection of a suitable neutralizing agent is the key to successful formulation of emulsified fragrances. Amines, in particular, are useful because they enhance oil and alcohol solubility of Carbopol resins.

Component	Suggested concentration ranges
Carbopol 941, 940	0.1 to 0.5%
Carbopol 934	0.1 to 0.3
Emollient oil (isopropyl myristate, etc.)	0.0 to 10.0
Perfume	0.1 to 10.0
Alcohol	10.0 to 50.0
Neutralizing agent (amines)	0.1 to 0.5
Water	q.s. 100.00

Some general guidelines for formulating Carbopol-emulsified fragrances are shown in the accompanying chart. The Carbopol dispersion may be prepared beforehand at a 0.5 % level. It should, however, be protected from microbial growth, from metallic ions, and also from light to prevent depolymerization and subsequent loss of viscosity. In fact, the use of antimicrobials, chelating agents, and ultraviolet light absorbers in the Carbopol dispersion, as well as in the finished formulations, may be desired to prevent mold growth and depolymerization catalyzed by light and trace metals.

A number of formulations are presented here as illustrations of emulsified fragrances in semisolid and liquid consistencies. Formulas 1 to 21 are suggested starting points. Opaque lotions are shown in Formulas 1, 2, 10, and 19 to 21. Pearly lotions are given in Formulas 3 to 6 and 11 and 12. Formulas 7 and 8 are pearlescent lotions, and Formula 9 is a translucent lotion. Opaque creams are given in Formulas 13 through 18.

Formulas for Alcoholic Fragrance Emulsions
*(Opaque Lotions)**

	1	2
Carbopol 934 resin	0.3%	—
Carbopol 941 resin	—	0.4%
Water, deionized	65.4	45.2
Isopropyl lanolate	1.0	1.0
Polyoxyethylene (40) stearate	3.0	3.0
Triethanolamine	0.3	0.4
Ethyl alcohol	30.0	50.0
Perfume	q.s.	q.s.

Procedure: In a stainless steel steam-jacketed kettle, add the Carbopol resin slowly to the water at room temperature with rapid agitation, until complete dispersion results, and then heat to 75°C. In another similar kettle, melt the isopropyl lanolate and polyoxyethylene (40) stearate and adjust to 75°C. Add the Carbopol dispersion to the oil phase and mix for 5 min. Mix in the triethanolamine and stir while cooling to 38°C. Add the alcohol with the perfume previously dissolved in it, and stir until the batch reaches 30°C.

* Ref. 3.

*Formula 3**

Carbopol 941 resin	0.3 to 0.5%
Water, deionized	40.3 to 38.1
Ethyl alcohol	44.3
Diisopropanolamine, 10%, in water	3.0 to 5.0
Pearl essence	0.1
Solulan 98	3.0
Perfume	4.0
Igepal CO-630	5.0

Procedure: In a stainless steel mixing tank, add the Carbopol 941 resin slowly to the water with rapid agitation until complete dispersion results. Add the alcohol and follow with the diisopropanolamine solution. Next mix in the pearl essence previously dispersed in the Solulan 98. Finally, add the perfume previously combined with the polyoxyethylated (9-10) nonylphenol (Igepal CO-630). Avoid aeration by the use of a suitable mixing unit and by carefully controlling its speed.

* Ref. 4.

Formulas for Alcoholic Fragrance Emulsions (Pearly Lotions)*

	4	5
Carbopol 941 resin	0.3%	0.25%
Water, deionized	42.2	44.15
Ethyl alcohol	50.0	50.00
Triethanolamine, 10%, in water	3.0	2.50
Pearl essence	1.5	0.10
Solulan 98	3.0	3.00
Perfume	q.s.	q.s.

Procedure: In a stainless steel mixing tank, add the Carbopol 941 slowly to the water with rapid agitation until complete dispersion results. Mix in the alcohol thoroughly with the perfume previously dissolved in it. Add the triethanolamine slowly and mix well. Blend the pearl essence with the Solulan 98 and add to the batch while mixing. (The Carbopol and triethanolamine levels may be adjusted to compensate for the viscosity imparted by the perfume.)

* Ref. 3.

Formula 6*

Natrosol 250 HR	0.10%
Carbopol 941 resin	0.15
Water, deionized	43.15
Diisopropyl adipate	2.00
Ethyl alcohol	50.00
Triethanolamine, 10%, in water	1.50
Solulan 98	3.00
Pearl essence	0.10
Perfume	q.s.

Procedure: Prepare the Natrosol dispersion beforehand by heating a portion of the water to 80°C and adding the Natrosol with rapid mixing. Disperse the Carbopol 941 thoroughly with high-speed agitation in the balance of the water at room temperature. Add the Natrosol dispersion to the Carbopol 941 dispersion, and mix until uniform. Dissolve the perfume and the diisopropyl adipate in the alcohol and add to the combined gum dispersions. Add the triethanolamine solution slowly, but with sufficient agitation to avoid localized neutralization of the Carbopol. Blend the Solulan 98 with the pearl essence and mix into the batch until uniform.

* Ref. 3.

Formulas for Alcoholic Fragrance Emulsions (Pearlescent) *

	7	8
Volpo 10	3.0%	1.5%
Polychol 5	—	1.5
Isopropyl myristate	5.0	5.0
Perfume	1.0	1.0
Water, deionized	34.0	32.0
Carbopol 940 resin, 1.2%		
dispersion	25.0	27.0
Ethyl alcohol	32.0	32.0
Diethanolamine	q.s. to pH 6.4	

Procedure: In a stainless steel tank, mix together the Volpo 10, Polychol 5, and the isopropyl myristate. Blend in the perfume and follow with the water. Mix in the Carbopol 940 dispersion with vigorous agitation. Finally, add the diethanolamine previously dissolved in the alcohol.

* Ref. 6.

Formula 9 *

Volpo 3	1.0%
Crodafas N-10 neutral	1.0
Water, deionized	36.0
Perfume	2.0
Carbopol 941 resin, 2%	
dispersion	20.0
Ethyl alcohol,	40.0
Triethanolamine	q.s. to pH 6.5

Procedure: In a stainless steel tank, mix together the Volpo 3, Crodafas N-10, and the water. Blend in the perfume and add the Carbopol 941 dispersion, with constant mixing. Finally, add the triethanolamine previously dissolved in the alcohol.

* Ref. 6.

Formula 10 *

Carbopol 934 resin	0.35%
Water, deionized	74.48
Ceraphyl 28	0.87
Cerasynt 840	3.94
Ethyl alcohol	17.50
Perfume	2.62
Monoisopropanolamine	0.24

Procedure: In a stainless steel mixing kettle, add the Carbopol 934 slowly to the water with rapid agitation until complete dispersion results and then heat to 70 to 75°C. In a separate container, mix the Ceraphyl 28 and Cerasynt 840, and heat to 70 to 75°C. Add the lactate-monostearate blend to the Carbopol dispersion with continuous mixing until the temperature reaches 50°C. Finally, mix in the previously blended alcohol, perfume, and monoisopropanolamine.

* Ref. 7.

Formula 11*

Carbopol 941 resin	0.30%
Ethyl alcohol	29.70
Uvinul D-50	0.07
Ceraphyl 50	2.50
Diisopropyl adipate	2.50
Perfume	1.50
Water	62.83
Diethanolamine	0.30
Pearl essence	0.30

Procedure: In a stainless steel mixing tank, disperse the Carbopol 941 thoroughly in one-half of the alcohol. In a separate vessel, dissolve the Uvinul D-50, Ceraphyl 50, diisopropyl adipate, and perfume in the remainder of the alcohol, and add to the Carbopol dispersion. Then add one-third of the water, with rapid mixing. In another container, combine the remainder of the water, diethanolamine, and pearl essence, with continuous mixing to maintain the pearl essence in suspension. Finally, add this mixture to the other mixture slowly with continuous high-speed propeller agitation.

   * Ref. 7.

Formula 12*

Carbopol 940 resin	0.15%
Water, deionized	46.12
Ethyl alcohol	50.00
Diisopropanolamine, 10%, in water	1.20
Diisopropyl adipate	2.50
Pearl essence	0.03
Perfume	q.s.

Procedure: In a stainless steel tank, add the Carbopol 940 slowly to the water at room temperature with rapid agitation until complete dispersion results. Mix in the alcohol with the perfume previously dissolved in it. Add the diisopropanol-amine slowly and mix well. Blend the pearl essence with the diisopropyl adipate and mix into the batch.

   * Ref. 5.

Formula 13*

Glyceryl monostearate, self-emulsifying	10%
Cetyl alcohol	3
Water, deionized	16
Carbopol 940 resin, 2% dispersion	62
Perfume	8
Triethanolamine	1

Procedure: In a stainless steel cream-mixing kettle, heat the glyceryl monostearate and cetyl alcohol to 70°C. Heat the water to 75°C and add to the glyceryl monostearate-cetyl alcohol with mixing until a uniform emulsion is formed. Mix in the Caropol dispersion thoroughly and start cooling. At 40°C, add the perfume. Finally, add the triethanolamine and mix slowly until a cream is formed.

   * Ref. 4.

Formula 14*

Glyceryl monostearate, acid-stable, self-emulsifying	19.50%
Spermaceti	6.50
Propyl *p*-hydroxybenzoate	0.10
Water, deionized	59.25
Glycerol	6.50
Methyl *p*-hydroxybenzoate	0.15
Perfume	8.00

Procedure: In a stainless steel cream-mixing kettle, heat the glyceryl monostearate, spermaceti, and propyl *p*-hydroxybenzoate to 80°C. In a separate container, heat the water, glycerol, and methyl *p*-hydroxybenzoate to 80°C. Add the water phase to the oil phase at 80°C with constant mixing and allow to cool. At 35°C, add the perfume and continue mixing until the cream has nearly cooled to room temperature.

   * Ref. 8.

Formula 15*

Glyceryl monostearate, self-emulsifying	10%
Pluronic P123	16
Propylene glycol	6
Lanolin, anhydrous	2
Stearic acid, triple-pressed	3
Perfume	4
Water, deionized	59
Preservative	q.s.

Procedure: In a stainless steel cream-mixing kettle, mix together all of the components, except the perfume, and warm to 60°C. Allow to cool to 45°C, and mix in the perfume and preservative. Pour into containers, and allow to cool and set up in cream form.

   * Ref. 9.

Formula 16

Stearic acid, triple-pressed	18.0%
Cetyl alcohol	0.5
Water, deionized	70.2
Propylene glycol	5.0
Methyl *p*-hydroxybenzoate	0.1
Potassium hydroxide	1.2
Perfume	5.0

Procedure: Melt the stearic acid and cetyl alcohol in a stainless steel steam-jacketed kettle at 70°C. In a similar kettle, dissolve the methyl *p*-hydroxybenzoate, potassium hydroxide, and propylene glycol in the water at 70°C with agitation. Add the oil phase to the water phase with constant mixing by means of a slowly moving paddletype agitator. Allow to cool with continued agitation, and at 45°C add the perfume slowly. Continue mixing, and cool rapidly by circulating cold water through the jacket of the kettle, until the batch has reached room temperature and has set in semisolid form. Pack into containers by means of a cream-filling machine commonly used for vanishing-type creams.

Formula 17

Beeswax, white	12.0%
Sorbitan monostearate	5.0
Water, deionized	66.0
Propylene glycol	5.9
Polyoxyethylene sorbitan monostearate	5.0
Methyl *p*-hydroxybenzoate	0.1
Sodium borate	1.0
Perfume	5.0

Procedure: Same as for Formula 16, except that the oil phase in this formulation consists of beeswax and sorbitan monostearate.

Formula 18

Cetyl alcohol	5.0%
Stearyl alcohol	5.0
Spermaceti	5.0
Water, deionized	73.0
Propylene glycol	5.9
Sodium lauryl sulfate	1.0
Methyl *p*-hydroxybenzoate	0.1
Perfume	5.0

Procedure: Same as for Formula 16, except that the oil phase in this formulation consists of cetyl alcohol, stearyl alcohol, and spermaceti, and the mixing temperature is 65°C.

Formula 19

Stearic acid, triple-pressed	2.0%
Cetyl alcohol	0.5
Water, deionized	79.5
Propylene glycol	5.9
Polyethylene glycol 400 monostearate	6.0
Triethanolamine	1.0
Methyl *p*-hydroxybenzoate	0.1
Perfume	5.0

Procedure: Melt the stearic acid and cetyl alcohol in a stainless steel steam-jacketed kettle at 60°C. In a similar kettle, heat the water, the propylene glycol, polyethylene glycol 400 monostearate, triethanolamine, and methyl *p*-hydroxybenzoate to 60°C with agitation. Add the oil phase at 60°C to the water phase at 60°C with rapid agitation. Allow to cool with continued agitation, and at 45°C add the perfume slowly. Continue mixing and cool rapidly by circulating cold water through the jacket of the kettle until the batch reaches room temperature. Fill into bottles with the usual liquid-filling equipment.

Formula 20

Stearic acid, triple-pressed	2.5%
Glyceryl monostearate, self-emulsifying	1.0
Amerchol L-101	1.0
Water, deionized	83.8
Propylene glycol	4.0
Triethanolamine	0.1
Methyl *p*-hydroxybenzoate	0.1
Veegum HV	0.5
Perfume	7.0

Procedure: Same as for Formula 19, except that the oil phase consists of stearic acid, glyceryl monostearate, and Amerchol L-101, and the mixing temperature is 80°C. The Veegum HV is most conveniently handled when it is added to the aqueous phase in the form of a previously prepared 5% dispersion in water.

Formula 21*

Water, deionized	78.24%
Glycerol	10.00
Methyl *p*-hydroxybenzoate	0.10
Cetyl trimethyl ammonium bromide	0.16
Pectin, citrus N.F.	1.50
Perfume	10.00

Procedure: In a stainless steel steam-jacketed kettle, heat the water, glycerol, methyl *p*-hydroxybenzoate, and cetyl trimethyl ammonium bromide to 75°C with high-speed mixing (400 rpm propeller-type blade). Slowly sprinkle in the pectin and continue mixing until the dispersion is complete. Allow to cool to 35°C, add the perfume slowly, and continue mixing until the temperature drops to 28°C. Pass through a colloid mill at an opening of 0.010 in.

* Ref. 10.

The formulation of stable emulsified fragrances is a more difficult task than is encountered with other kinds of cosmetic emulsions. As mentioned previously, this is due to the high concentration of perfume used. The formulations should be kept as simple as possible—with perfume, emulsifying agent, and alcohol or a polyol as the essential ingredients.

The concentration of the perfume oil varies from 1 to 10%, depending on whether the product is intended to approximate a toilet water or a perfume. The lowest possible concentration should be used, consistent with the formulation of a product with the desired intensity of fragrance. The higher the concentration of perfume, the more difficult the stabilization of the emulsion.

The perfume is added to the batch at the lowest possible temperature during the manufacturing process, because of deterioration of the fragrance when subjected to elevated temperatures. A temperature of 45°C has been found satisfactory in many formulations. Immediately after the perfume has become mixed with the product, the batch should be cooled rapidly by circulating cold water through the jacket of the kettle to minimize the loss of fragrance.

Most emulsions are formed at temperatures of 60 to 80°C, and the perfume is added after the emulsion has been completed. A stable emulsion is sometimes difficult to obtain with certain perfumes when they are added after the emulsion has been formed. In such a case, the perfume is added to the oil phase and incorporated in the product during the emulsification step. When this procedure is followed, the product should be observed carefully for possible deterioration of the fragrance caused by the heat of processing.

There is considerable variation in the effect of different perfume oils on the stability of emulsions. Some perfume oils will not cause "creaming" or "separation" of emulsions, even in the high concentrations used in fragrance emulsions. Others will cause such emulsions to "separate" or "cream" within a few hours.

Unfortunately, there is no available comprehensive study of the effect of perfume ingredients on emulsion stability. A stable emulsion, therefore, is obtained only by trial and error. Various perfume oils, or different emulsifying agents with the same perfume oil, should be tried until a stable emulsion results.

Wynne (18) described the effects of certain perfume materials on emulsion stability. He studied the effects of the following for 3 months at a concentration of 1%.

1. Synthetic aromatics. Terpineol extra, phenyl ethyl alcohol, geraniol pure, hydroxycitronellal, and amyl cinnamic aldehyde.

2. Essential oils. Rose de mai absolute, geranium Bourbon, and lavender (50% ester content).

3. Compounded perfume oils. A modern bouquet, a lilac type, a light floral type, and two of a rose character.

Wynne's results are summarized briefly as follows:

1. Triethanolamine stearate emulsion. Terpineol was the only material that caused this type of emulsion to separate.

2. Sorbitan monostearate-polyoxyethylene sorbitan monolaurate emulsion. None of the materials tested affected the emulsion.

3. Aminoglycol stearate emulsion. One of the rose perfume oils caused this emulsion to separate.

4. Potassium stearate-quince seed mucilage emulsion. Separation was caused by geraniol pure, terpineol extra, and a light floral perfume oil.

The use of ethyl alcohol and/or polyhydric alcohols is recommended in the formulation of emulsified fragrances. They serve the useful purpose of being (a) antifreeze agents (thus preventing solidification of the product in cold weather with consequent breaking of the bottle, or separation of the emulsion), and (b) solvents for the perfume oil. The commonly used polyols are Carbitol, glycerol, polethylene glycols, propylene glycol, sorbitol, and 1,3-butylene glycol.

It is sometimes desirable to impart color to an emulsified fragrance product to improve its appearance. Before color is added, the formulations range from white, ivory, or pale yellow, to brown, depending on the color of the perfume oil. The addition of a certified cosmetic color greatly enhances the cosmetic elegance of many emulsified fragrance products. Attractive, stable pastel shades can be obtained with the following dyes, alone or in combination with each other:

Pink: D&C Red. No. 19.
Green or blue: D&C Green No. 5.
Yellow: FD&C Yellow No. 5.

These dyes are among the most stable. Others are likely to be unstable.

The final product should be observed carefully for stability throughout a 6-month period and preferably longer before it is recommended for manufacture and sale. The samples for stability studies should be placed in the package intended for the final product. Samples should then be stored in a dark cabinet as reference samples, and others in an oven at 45°C. If the product is packaged in a transparent container, a third group of samples should be placed on an open shelf, exposed to diffused light, with a fourth lot of samples on a windowsill for exposure to direct sunlight. Finally, samples should be placed in a refrigerator (5°C) and others in the deep-freeze section (−10°C) to determine the effect of cold temperatures on the product. After 24 hr the cold samples are allowed to stand at room temperature for another 24 hr, and reexamined.

The following possible deficiencies should be looked for in all the stability samples:

1. Water separation—the formation of a clear, liquid lower layer.

2. Creaming—the formation of an opaque layer near the surface, somewhat similar in appearance to that of the cream layer that rises to the surface of nonhomogenized milk.

3. Oil separation—the separation of oil droplets, or a clear oil layer on the surface.

4. Fragrance changes—determined by olfactory examination.

5. Color changes—observed visually.

6. Viscosity changes—some formulations become too viscous to flow through the opening of the bottle; others decrease in viscosity to a watery consistency.

An emulsion that reveals no "separation" or "creaming" for 3 weeks at 45°C will in all likelihood retain its stability at ordinary temperatures. Furthermore, an emulsion that successfully withstands five changes from 24 hr at 45°C to 24 hr at −10°C should remain in a stable condition for at least 2 years at ordinary temperatures.

The transition from laboratory batches of emulsified fragrances to large plant-scale batches frequently creates stability problems. Therefore final stability observations should be made on samples obtained from a pilot-plant batch produced in the same equipment as that used for regular plant production. During the manufacture of pilot-plant batches, the most favorable temperature conditions and rates of stirring should be carefully determined, because some formulations may "cream" or "separate," when even slight deviations are made from optimum conditions.

The packaging of emulsified fragrances presents no difficult problems. These preparations are packaged in glass with pulp tinfoil, cork tinfoil, polyethylene, or 14B Armstrong white rubber as the cap liners. Liners with Vinylite or Saran facings should be carefully tested before being used, because emulsified fragrances sometimes cause separation of these facings from the cork or pulp backing. Polyethylene bottles should not be used, because some perfume oils diffuse through the sides of the container with a consequent loss and distortion of the fragrance.

Emulsified fragrances in the form of semisolid and liquid cream perfumes have been accepted favorably by the public, and have become firmly established as members of the family of fragrance products. They have not, however, attained the popularity of clear, alcoholic, liquid fragrance preparations.

Solid Fragrances

The most popular forms of solid fragrances are the cologne sticks or "frozen" colognes. As their names imply, they are solid forms of cologne and appear on the market usually in the shape of small molded cylinders, wrapped in foil and packaged in airtight containers. They are used in the same manner as regular colognes.

Solid colognes offer the consumer many practical advantages. They are usually reasonably priced, easily applied, spillproof as well as leakproof, which permits them to be carried in handbags, suitcases, or glove compartments of cars without danger of staining other items in the same area (25).

Solid colognes are essentially liquid colognes which have been solidified with a gellant, and are usually composed of the following materials:

Perfume	2 to 5%
Ethyl alcohol SD-39C or -40	85 to 90
Solidifying ingredient	6 to 5
Polyhydric alcohol	0 to 5
Water	5 to 10

The key to a well-formulated cologne stick is the solidifying agent. Hard soaps are used in modern solid colognes. Sodium stearate is the one generally selected (19, 20). It may be made *in situ* in liquid cologne from stearic acid and sodium hydroxide or by the direct addition of sodium stearate. The use of already-made sodium stearate is recommended, because it simplifies the production procedure (21).

If sodium stearate is made during the manufacturing process, triple-pressed stearic acid or stearic acid manufactured from hydrogenated oils is used (22). The higher the palmitic acid content of stearic acid, the more transparent is the stick, whereas a higher stearic acid content yields a stiffer and more opaque product. It is also important that the stearic acid contain a minimum amount of unsaponifiable matter, so that there will be no impairment of its gelating efficiency. The presence of oleates also interferes with gelation. The hardness of a cologne stick is determined by the quantity of sodium stearate in the formula, increasing amounts producing a harder stick.

The inclusion of potassium chloride, castor oil, or castor oil fatty acids has been recommended for the preparation of clear, transparent gels, but castor oil and its fatty acids tend to reduce the gel strength. Other solidifying ingredients mentioned in the literature are rosin soaps, candelilla wax, beeswax, carnauba wax, potassium diacetone fructose sulfate, acetanilide, calcium acetate, and ethyl cellulose (24).

In addition to a solidifying agent, a cologne stick should contain a non-volatile solvent in the form of a polyhydric alcohol as a plasticizer (*a*) to prevent the stick from being too brittle, (*b*) to prevent the film from drying too rapidly on the skin during application, and (*c*) to prevent the deposition on the skin of a white powdery layer of sodium stearate. The polyol also serves as a solvent for the gelating agent, so that higher concentrations may be used in the cologne stick to render it more rigid to wilting summer heat.

The commonly used polyols are Carbitol, propylene glycol, sorbitol, glycerol, polyethylene glycol 300 or 400, 1,3-butylene glycol, and 2-ethyl-1,3-hexanediol. Fatty acid esters such as isopropyl palmitate or myristate are also used (24, 26, 28, 29).

Small amounts of water are used in cologne stick formulations, mainly to dissolve the sodium hydroxide, if the sodium stearate is made during the manufacturing process. Concentrations of water above 10% should be

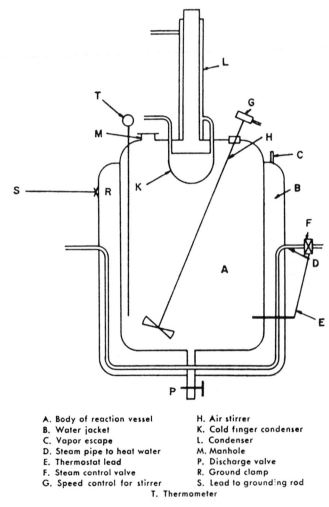

A. Body of reaction vessel   H. Air stirrer
B. Water jacket   K. Cold finger condenser
C. Vapor escape   L. Condenser
D. Steam pipe to heat water   M. Manhole
E. Thermostat lead   P. Discharge valve
F. Steam control valve   R. Ground clamp
G. Speed control for stirrer   S. Lead to grounding rod
                  T. Thermometer

Fig. 1.   Installation of commercial production of cologne sticks. From Fishback (27). Courtesy Toilet Goods Association.

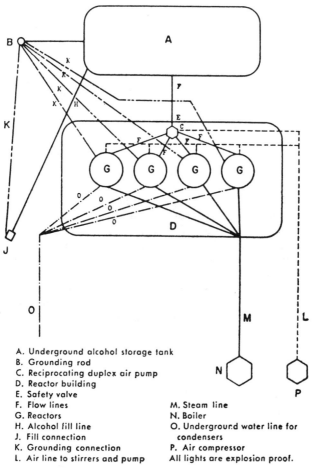

A. Underground alcohol storage tank
B. Grounding rod
C. Reciprocating duplex air pump
D. Reactor building
E. Safety valve
F. Flow lines
G. Reactors
H. Alcohol fill line
J. Fill connection
K. Grounding connection
L. Air line to stirrers and pump

M. Steam line
N. Boiler
O. Underground water line for
   condensers
P. Air compressor
All lights are explosion proof.

Fig. 2. Flow sheet for manufacture of cologne sticks. From Fishback (27). Courtesy Toilet Goods Association.

avoided because of the possibility of forming unsightly white spots of sodium stearate or stearic acid crystals. Preferably, one uses no more than 5 % water (23).

The manufacture of solid cologne is best accomplished in a stainless steel steam-jacketed kettle. The kettle should have a tightly fitted cover with an opening for a reflux condenser, one for an agitator shaft, and another for the addition of materials to the batch. Explosion-proof or air motors should be used to minimize the danger of explosion and fire. All equipment should, of course, be properly grounded. Figures 1 and 2 from Fishback (27) represent a typical installation for the commercial production of cologne sticks.

The manufacturing procedure is best described by the presentation of two typical cologne stick formulations by Fishback (27), Formulas 22 and 23, and in Formula 24 by Osipow (30).

Formula 22

Ethyl alcohol	81%
Carbitol	3
Glycerol	3
Water, deionized	5
Sodium stearate	6
Perfume	2
Color solution	q.s.

Procedure: Place all the ingredients, except the perfume oil, in a closed stainless steel kettle and heat with constant stirring to reflux temperature. (Considerable time is required to dissolve the sodium stearate completely.) When the sodium stearate is completely dissolved, add the perfume and color, and pour into molds at 65°C. (Aluminum molds have been found satisfactory for this purpose.) Allow to cool slowly at room temperature. (Rapid cooling may cause the formation of a hollow core in the center of the stick because of excessively rapid contraction. It may also stimulate greater crystallization, resulting in a more opaque stick. A slow rate of cooling produces a more transparent product.) Wrap in foil and place in airtight containers.

Experimental laboratory batches may be prepared in the same manner with ordinary laboratory glassware: an Erlenmeyer flask fitted with a cork stopper and a 1-ft length of glass tubing as an air-cooled condenser. The ingredients are added to the Erlenmeyer flask in the same manner as for the

Formula 23

Ethyl alcohol	80.61%
Carbitol	3.00
Glycerol	3.00
Stearic acid	5.55
Water, deionized	5.00
Sodium hydroxide	0.84
Perfume	2.00
Color solution	q.s

Procedure: Heat the alcohol, Carbitol, stearic acid, and glycerol with constant stirring at 70°C in a closed stainless steel kettle. Slowly add the sodium hydroxide, previously dissolved in water. Mix vigorously for about 15 min. Remove a sample and titrate for excess alkalinity or acidity. Adjust the batch for an excess of 5% of stearic acid. Add the perfume oil and color solution and pour into molds at 65°C. Allow to cool slowly. Wrap in foil and place in airtight containers.

Formula 24

Ethyl alcohol	78.0%
Sodium stearate	5.5
1,3-Butylene glycol	5.0
Water, deionized	11.5
Perfume	q.s.

Procedure: Same as for Formula 22.

manufacturing batches described above. A steam bath is used for heating, and the turbulence caused by the boiling alcohol provides enough agitation to mix the ingredients thoroughly. A small amount of alcohol is lost through the air condenser, but this is compensated for by adding a comparable amount of alcohol, just before pouring the batch into molds.

The most important features to be considered in the formulation of a cologne stick are (a) the translucency in appearance, (b) the ease of application, (c) the cooling action on the skin when used, and (d) the fragrance. The translucency is affected by the amount of soap in the formula, the palmitic acid content of the stearic acid used, and the rate of cooling of the molded sticks. The most translucent product is produced when the soap content is at a minimum, when the amount of palmitic acid in the stearic acid is at a maximum, and when the rate of cooling is slow. If a more opaque product is desired, zinc stearate may be added to the preparation.

The ease of application to the skin is favored by the increase in content of polyhydric alcohol as a plasticizer. Maximum quantities should be used, short of the concentrations that retard rapid drying and that reduce the cooling action of the product.

The fragrance of a cologne stick may not prove so satisfactory as that of the corresponding liquid cologne, even when the same perfume oil is used in both compositions. The top note of the cologne stick may not have the delicacy of that of the liquid cologne because of the odor imparted by the soap in the formulation. This is particularly true of light floral compounds. The lasting quality on the skin of the fragrance of a cologne stick, however, is usually greater than that of the corresponding liquid cologne.

The perfume oil for a cologne stick should therefore be selected with great care. Even though a perfume oil has been used successfully in a liquid cologne, it should be studied for stability in the cologne stick form, and modifications made, if necessary, to attain the maximum stability of the fragrance. The alkaline *p*H of a cologne stick and the heat used during the manufacturing process could conceivably affect the stability of the perfume oil.

The packaging of solid colognes involves two important considerations: (a) The wrapping material must be resistant to corrosion by the stick (in which the alcohol is the most likely corroding agent), and (b) the container closure must be absolutely tight to prevent the evaporation of the alcohol. One of the most satisfactory wrapping materials is tinfoil. Another suitable material is aluminum foil with an overall coating of vinyl lacquer.

Glass containers are in common use for the packaging of cologne sticks. A satisfactory closure is obtained with a polyethylene plug that fits inside the mouth of the container, or a polyethylene cap fitting over the mouth of the bottle. Aluminum metal containers with a lacquer coating are also recommended for cologne sticks, with polyethylene plugs or caps as closures. Screw caps are not recommended as closures, because they do not provide a tight enough seal and result in a heavy volume of returned goods in the form of shrunken cologne sticks.

Plastic cases have been developed with swivel or "push-up" devices for convenient handling. Spiral-wound paper tubes with a "push-up" bottom have also been used as containers. These containers have the advantage of eliminating the molding operation, because the tube serves as a mold into which the hot cologne mix is poured. Before a container is adopted, however, careful studies of the alcohol evaporation losses should be made.

Cologne sticks, by virtue of their convenience, have made a place for themselves among the established fragrance products. They enjoy some consumer acceptance, but have not attained the popularity of clear, alcoholic, liquid fragrance preparations.

Solid fragrances other than alcoholic solid preparations have been marketed from time to time but have never become popular. Transparent fragrance sticks that will not dry out can be made with propylene glycol, polyethylene glycol 300 or 400, or 2-ethyl-1,3-hexanediol. A typical example is shown in Formula 25.

Formula 25*

Sodium stearate USP	15%
Water	5
Propylene glycol	60
Perfume	20

* Ref. 31.

Procedure: Heat the sodium stearate, water, and propylene glycol with constant stirring in a stainless steel kettle. When the sodium stearate has been completely dissolved, lower the temperature to 75°C. To this solution, add the perfume which has already been preheated to almost 75°C. Stir constantly and pour into molds.

Perfume oils dissolved in a combination of waxes yield a solidified form of fragrance. Typical examples are shown in Formulas 26 to 31.

Formulas for Solid Fragrances

	26	27	28*	29*	30†	31‡
Paraffin wax, m.p. 71 to 74°C	20%	56%	60%	—	—	—
Petrolatum, white	65	20	—	—	—	—
Perfume	15	17	15	20%	25%	25.0%
Spermaceti	—	4	—	—	—	—
Glycerol	—	3	—	—	—	—
Carnauba wax	—	—	3	—	—	27.5
Ozokerite, white	—	—	7	8	—	—
Mineral oil	—	—	15	12	5	—
Lanolin, anhydrous	—	—	—	8	—	—
Microcrystalline wax	—	—	—	30	—	—
Isopropyl myristate	—	—	—	12	—	—
Glyceryl monostearate	—	—	—	—	70	—
Benzyl benzoate	—	—	—	—	—	47.5
Beeswax	—	—	—	10	—	—

* Ref. 24.
† Ref. 31.
‡ Ref. 4.

Procedure: A general procedure for manufacturing these wax-based formulations involves (*a*) melting the components together at the lowest possible temperature, (*b*) adding the perfume, and (*c*) pouring into molds before solidification begins.

Among the more popular forms of wax-based solid fragrances are perfumed candles. A properly formulated product will fill a room with a gentle fragrance after burning for about 20 min. A candle of 4 to 7 oz in weight should last for over 30 hr, depending on the size and burning quality of the wick. It is essential to avoid a heavy flame, which burns too fast, smokes badly, and distorts the fragrance.

The early New England settlers (32), particularly in the Cape Cod area, prepared their own perfumed candles by boiling bayberries (*Myrica cerifera*) in rainwater and then skimming the wax from the surface. After refining, the resulting material has a pleasant odor, burns with a clear white flame, and imparts an aromatic, balsam-like fragrance to the surroundings.

Other fragrances are now being used in scented candles. Among them are reproductions of fragrances found in nature, such as hollyberry, pine, and floral types. During the 1960's, candles appeared on the market containing more sophisticated modern fragrance types.

The ideal fragrance candle should contain the best features of a conventional candle, combined with a suitable perfume oil. The perfume oil must be soluble in the wax candle formulation to avoid bleeding and should not retard or accelerate the burning rate of the flame of the finished candle.

The candle should be studied (a) for length of burning time, (b) to determine whether the heat vaporizes the fragrance at a constant rate, (c) to determine whether dyes used for the candle's color bleed or discolor, and (d) for shelf-life stability.

Waugh (32) describes the following as typical ingredients of an incense-type perfume oil:

> Essential oils: Oils of sandalwood, patchouli, expressed limes and geranium (Algerian).
> Aromatic chemicals: Geraniol and terpineol.
> Fixatives of plant origin: Peru balsam, resinoid myrrh, resinoid olibanum, and oakmoss extractives.
> Fixatives of animal origin: Civet and castoreum.
> Fixatives of chemical nature: Synthetic musk.

Typical perfumed candle formulations are shown in Formulas 32 and 33.

Formulas for Fragrance Candles

	32	*33*
Stearic acid, double-pressed	20%	30%
Paraffin wax, m.p. 137°F.	75	65
Perfume	5	5

Procedure: Melt the stearic acid and paraffin wax at the lowest possible temperature, stir in the perfume, and pour into molds into which wicks have already been placed.

New forms of fragrance, as they appear on the marketplace, will continue to attract some consumers. Perfume cream sachets, cologne veils, perfume silks, cologne puffs, perfume foams, stick colognes, and perfumed candles will enjoy periods of popularity from time to time. However, in the long run these forms, in all probability, will never begin to approach the popularity of clear alcoholic fragrance preparations as packaged in conventional glass bottles or in aerosol containers.

REFERENCES

1. Kalish, J.: Alcoholic emulsions, *DCI*, **97**: 665 (1965).
2. B. F., Goodrich, Chemical Company: *Service Bulletin GC-36*, revised, Carbopol water-soluble resins.
3. American Cholesterol Products, Inc., Handbook, 1968.
4. International Flavors and Fragrances, Inc.: Personal communication.
5. B. F. Goodrich, *Chemical company formulary issue No. 4*.
6. Croda, Inc.: *Cosmetic and pharmaceutical formulary*, 1970, p. 21.
7. Van Dyk and Company, Inc.: *Technical Bulletin*, 1968.
8. Stabilized emulsions, *Schimmel Briefs*, 1964.
9. Wyandotte Chemicals Corporation Technical Data Sheet 0-74.

10. Manchey, L. L., and Schneller, G. H.: U. S. Pat. 2,372,159 (1945).

11. Spotlight on sachets, *Givaudanian*, May 1965.

12. Emulsions suitable for liquid and cream colognes, *Schimmel Briefs*, No. 91, 1942.

13. Hilfer, H.: Cream sachets, *DCI*, **62**: 615 (1948).

14. Cosmetic sachet powders and creams, *Schimmel Briefs*, No. 208, 1952.

15. Hilfer, H.: Cream sachets, *DCI*, **70**: 472 (1952).

16. Liquid cream sachet (questions and answers, No. 1035), *Am. Perf.*, **62**: 263 (1953).

17. Expanding fragrance, *DCI*, **74**: 511 (1954).

18. Wynne, W.: Effect of perfume oils on emulsions, *Am. Perf.*, **54**: 381 (1949).

19. Sluijs, K.: Solid colognes and perfumes, *DCI*, **64**: 105 (1949)

20. Solid colognes (readers' questions), *DCI*, **65**: 575 (1949).

21. deNavarre, M. G.: How to make solid colognes, *Am. Perf.*, **56**: 289 (1950).

22. Hilfer, H.: Solidified fragrances, *DCI*, **66**: 522 (1950).

23. deNavarre, M. G.: Desiderata (stick cologne), *Am. Perf.*, **57**: 353 (1951).

24. Jannaway, S. P.: *Alchemist*, **5**: 67 (1951).

25. What is behind the phenomenal sales increase in solid colognes? *Givaudanian*, April–May 1952, p. 3.

26. Hilfer, H.: Solidified cosmetics, *DCI*, **73**: 326 (1953).

27. Fishback, A. L.: Cologne sticks and related products—their formulation, manufacture and analysis, *Proc. Sci. Sec. TGA*, **20**: 20 (1953).

28. Pears, G. S.: Isopropyl myristate, *PEOR*, **44**: 84 (1953).

29. Cologne sticks (readers' questions), *DCI*, **74**: 873 (1954).

30. Osipow, L. I., Marra, D., and Resnansky, N.: 1,3-Butylene glycol in cosmetics, Part III, *DCI*, **103**: 75 (1968).

31. Givaudan-Delawanna, Inc.: Personal communication.

32. Waugh, T.: Perfuming of candles, *Am. Perf.*, **83**: 39 (July 1968).

AUTHOR INDEX

SUBJECT INDEX

INDEX OF TRADE NAMES

*Registered trademark, United States Patent Office.

Crodafos N.10 Neutral (Croda), oleyl ether phosphate

Crodamol (Croda), isopropyl myristate

*Darvan #1** (Vanderbilt), dispersing agent

DCMX, dichlorometaxylenol

DDA, dodecylamine

Deltyl Extra* (Givaudan), isopropyl esters of saturated fatty acids

*Deriphat 160C** (General Mills), partial sodium salt of N-lauryl-β-iminodipropionic acid

Dicrylan 325-50* (Ciba-Geigy), acrylic copolymer dispersion

Drewmulse 1128* (Drew), acid-stable glyceryl monostearate

*Dry-Flo** (National Starch), modified corn starch

Duponol C* (du Pont, Dyes and Chemicals Division), sodium lauryl sulfate USP

Duponol WA Paste* (du Pont, Dyes and Chemicals Division), aqueous form of sodium lauryl sulfate

Duponol WAQ* (du Pont, Dyes and Chemicals Division), aqueous form of sodium lauryl sulfate

Duponol WAT* (du Pont, Dyes and Chemicals Division), an alkylalkylolamine sulfate

E-607 and E-607S (Witco), see *Emcol E-607* and *Emcol E-607S*

EDTA, ethylene diamine tetraacetic acid

Emcol E-607 (Witco), lapyrium chloride

Emcol E-607S (Witco), stearoyl analog of *Emcol E-607*

Emcol MAS (Witco), a fatty amide

Emcol 14 (Witco), polyglycerol fatty acid ester

Emulphor EL-719* and *VN-430* (GAF), dispersing and emulsifying agents

Emulsynt 610-A (Van Dyk), polyethylene glycol 400 monolaurate

Emulsynt 1055 (Van Dyk), polyoxyalkylene oleatelaurate

Escalol 106* and *Escalol* 506* (Van Dyk), synscreens; the former is glyceryl *p*-aminobenzoate; latter is amyl *p*-dimethyl aminobenoate

Ethomeen TD/25* (Armak), ethoxylated aliphatic amine

Ethoxylan 100* (Malmstrom), water- and oil-soluble lanolin in 100% active form; also available as *Ethoxylan* 50*, a 50% aqueous solution

*Ethoxylols** (Malmstrom), nonionic emulsifiers; *Ethoxylol* 5* is reaction product of lanolin alcohols with 5 mols of ethylene oxide; *Ethoxylol 16* with 16 mols of ethylene oxide; *Ethoxylol 16R* These products were formerly known as *Nimcolan,* * *Nimcolan S,* * and *Super Nimcolan S,* * respectively

*Ethylan** (Robinson-Wagner), liquid lanolin ester

*Eucerin** (Duke), lanolin absorption base in W/O emulsion

Fifty Super, see *Argonol 50 Super*

Fluilan (Croda), liquid lanolin

Foamole L (Van Dyk), linoleic alkanolamide

Forlan (R.I.T.A.), solid absorption base

*Freon** (du Pont, Freon Products Division), fluorinated hydrocarbon propellants

*G-4** (Givaudan), dichlorophene

*G-11** (Givaudan), hexachlorophene

Gafanol E-550B* (GAF), polyethylene glycol

Gantrez AN-119 ester, ES-225,* and *ES-425* (GAF), film formers

*Genetron** (Allied), fluorinated hydrocarbon propellants

Gerstoffen S Wax, see *Hoechst Wax S*

*Giv-Tan F** (Givaudan), sunscreening agent

Hartolans (Croda), lanolin alcohols

Hodag CSA-80 (Hodag), nonionic fatty ester

Hoechst Wax S (American Hoechst), acid wax derived from montan wax

Hostaphat KO 280* (American Hoechst), emulsifier

Hostaphat KO 380* (American Hoechst), triester emulsifier

Hyamine 10-X* and *Hyamine* 1622* (Rohm and Haas), synthetic organic bactericides and disinfectants

*Hychol** (Robinson-Wagner), anhydrous lanolin

Igepal CO-430* and *CO-630* (GAF), alkyl phenoxyethylene ethanols

Igepon A, T, AC-78,* and *TC-42* (GAF), anionic surfactants

IMP, insoluble sodium metaphosphate

IPC, isopropylcatechol

Isocreme (Croda), absorption base

Iso-Lan* (Goldschmidt), semi-liquid lanolin absorption base and emollient moisturizer

Isopar,* (Humble), isoparaffinic saturated water-white hydrocarbons

Isopropylan* and *Isopropylan* 50* (Robinson-Wagner), liquid lanolin esters

Keltrol* (Kelco), xanthan gum product

Kelzan (Kelco), xanthan gum product; see also *Keltrol**

Kessco* PEG-600 distearate (Armak), fatty acid ester

Kohnstamm Tablet Brown (Kohnstamm), a blend of FDA approved synthetic iron oxides

Lamepon* (Stepan), see *Maypon*

Lanacet (Malmstrom), acetylated lanolin

Lanacid (Malmstrom), mixture of naturally occurring lanolin fatty acids

Lanesta (Westbrook-Marriner), isopropyl ester of lanolin fatty acids

Laneto* 50 and *Laneto* 100* (R.I.T.A.), completely water-soluble lanolin

Lanexol (Croda), alkoxylated liquid lanolin derivative

Lanfrax* WS 55 (Malmstrom), ethoxylated water-soluble lanolin wax

Lanogel* (Robinson-Wagner), water-soluble lanolin

Lanogene* (Robinson-Wagner), lanolin oil

Lanoil (Lanaetex), liquid lanolin

Lanosol* (Robinson-Wagner), lanolin suspension

Lantrol* (Malmstrom), oil-soluble liquid fraction of lanolin

Lantrol* A WS (Malmstrom), alkoxylated version of *Lantrol**

Lathanol LAL* (Stepan), sodium lauryl sulfoacetate

Lipal* 4MA (Drew), ethoxylated myristyl alcohol

Lipal* 20-OA (Drew), ethoxylated oleyl alcohol

Lipal* 15CSA (Drew), ethoxylated cetyl-stearyl alcohol

Liquid Base (Croda), special *Hartolan* solution

Loramine OM-101 (Dutton & Reinisch), a complex fatty alkylolamide

Maypon* UD (Stepan), potassium salt of condensation product of undecylenic acid with a complex of polypeptides and amino acids derived from collagen protein

Maypon* 4C (Stepan), potassium salt of condensation product of coconut fatty acids with a complex of polypeptides and amino acids derived from collagen protein

Maypon* 4CT (Stepan), triethanolamine salt of condensation product of coconut fatty acids with a complex of polypeptides and amino acids derived from collagen protein

MBE, monobenzyl ether

MEA, monoethanolamine

Mearlin-AC (Mearl), see *Timica*

Merpol* OJ (du Pont, Dyes and Chemicals Division), see *Alkanol* OJ*

Metadelphene* (Hercules), grade of N,N-diethyl-*m*-toluamide

Methocel* 60HG (Dow), hydroxypropyl-methylcellulose

Methyl Cellosolve* (Union Carbide), ethylene glycol monomethyl ether

Methyl Cellosolve* Acetate (Union Carbide), ethylene glycol monomethyl ether acetate

Methyl Paraben, methyl *p*-hydroxybenzoate

Methyl Parasept* (Heyden), methyl *p*-hydroxybenzoate

Mibiron* (Rona), bismuth oxychloride coated mica, pearl pigment

Miranols* (Miranol), amphoteric surface-active agents; grades include *Miranol* HM Conc., Miranol* DM, Miranol* C2M Conc., Miranol* MM Conc*

Modulan* (Amerchol), acetylated lanolin

Monosulph* (Nopco), sulfated castor oil

MP-10 (Rona), see *Timiron**

Myrj* 52 (ICI-Atlas), polyoxyl 40 stearate USP (polyoxyethylene 40 stearate)

Myrj* 52 (ICI-Atlas), polyoxyethylene 50 stearate

Myvacet* Type 9-40 (Eastman), acetylated monoglyceride with minimum 90% mono-ester content

*Vancide** (Vanderbilt), bacteriostat and fungistat

*Vanseal CS** (Vanderbilt), cocayl sarcosinate

*Veegum** (Vanderbilt), magnesium aluminum silicate

*Versene** *100* (Dow), sodium salt of EDTA

*Vicryl** *XOR-63* (Sherwin Williams), terpolymer resin

*Viscolan** (Amerchol), dewaxed lanolin

Volpo 3, Volpo 10, and *Volpo 20* (Croda), ethoxylated oleyl alcohols; known in United Kingdom as *Volpo N.3, Volpo N.10,* and *Volpo N.20*

*Vulca** *90* (National Starch), cross-linked corn starch

*Wecobee** *R* (Drew), hydrogenated coconut oil

*WSP-X 250** (Wilson), collagen-derived protein hydrolysate

*Zephiran** (Winthrop), alkyl dimethyl benzyl ammonium chloride

SUPPLIERS

Allied Chemical Corp., Morristown, N.J.
Amerchol, Edison, N.J.
American Cyanamid Co., Wayne, N.J.
American Hoechst Corp., Mt. Holly, N.C.
American Lecithin Co., Atlanta, Ga.
Arco Chemical Co., Philadelphia, Penna.
Armak Co., Chicago, Ill.

Cabot Corp., Boston, Mass.
Ciba-Geigy Corp., Summit, N.J.
Colgate-Palmolive Co., Jersey City, N.J.
Croda, Inc., New York, N.Y.

Dow Chemical Corp., Midland, Mich.
Dragoco, Inc., Totowa, N.J.
Drew, Division of Pacific Vegetable Oil Corp., Dover, N.J.
Duke Laboratories, South Norwalk, Conn.
E. I. du Pont de Nemours & Co., Inc., Wilmington, Del.
Dutton & Reinisch Ltd., London, England

Eastman Chemical Products., Inc., Kings-port, Tenn.

Firmenich Inc., New York, N.Y.

GAF Corporation, New York, N.Y.
Geigy Chemical Corp., Ardsley, N.Y.
General Electric Co., Waterford, N.Y.
General Mills & Chemicals, Inc., Minneapolis, Minn.
Givaudan Corp., Clifton, N.J.
Goldschmidt Chemical, Division of Wilson Pharmaceutical & Chemical Corp., New York, N.Y.
B. F. Goodrich Chem. Co., Cleveland, O.

Hercules Inc., Wilmington, Del.
Heyden Chemical Corp., New York, N.Y.
Hodag Chemical Corp., Skokie, Ill.
Humble Oil & Refining Co., Houston, Tex.

ICI-Atlas: ICI America Inc., Atlas Chemicals Division, Wilmington, Del.

Kelco Co., New York, N.Y.
H. Kohnstamm & Co., Inc., Brooklyn, N.Y.

Lanaetex Products, Inc., Elizabeth, N.J.

Malmstrom Chemical Corp., Linden, N.J.
Mearl Corp., New York, N.Y.
Miranol Chemical Co., Inc., Irvington, N.J.
Monsanto Chemical Co., St. Louis, Mo.

National Starch and Chemical Corp., Plainfield, N.J.
Nopco Chemical Division, Shamrock Chemical Co., Morristown, N.J.

Onyx Chemical Co., Jersey City, N.J.

Pfizer Chemicals Division, Greensboro, N.C.
Philadelphia Quartz Co., Philadelphia, Penna.

Reheis Chemical Co., Chicago, Ill.
R.I.T.A. Chemical Corp., Chicago, Ill.
Robeco Chemicals, Inc., New York, N.Y.
A.H. Robins Co., Richmond, Va.
Robinson-Wagner Co., Inc., Mamaroneck, N.Y.
Rohm and Haas Co., Philadelphia, Penna.
Rona Pearl Co., Bayonne, N.J.

The Editors

Marvin S. Balsam, founder and president of Standard Aromatics, Inc.; B.A. Columbia College, 1948, with additional work in Chemical Engineering and Business Administration; member of the American Society of Perfumers and the Society of Cosmetic Chemists.

S. D. Gershon, technical planning director, Lever Brothers Company; Ph.C., University of Illinois, 1930; B.S., M.S., and Ph.D., University of Chicago, 1934, 1935, and 1938, respectively; assistant professor of chemistry, University of Illinois, 1930-1943; with Lever Brothers, 1943 to date; honors include Leo Mrazek Chemistry Prize, Fairchild Scholar, P. Mandabach Award in Materia Medica, and Medal Award of Society of Cosmetic Chemists; president, SCC, 1952; member, editorial committee, *Journal of Society of Cosmetic Chemists*; editorial advisory board, Oral Research Abstracts; member of praesidium, International Federation of Societies of Cosmetic Chemists; author of numerous publications and patents in carbohydrates, oral products, hair products, and related preparations.

M. M. Rieger, Warner-Lambert Company; B.S., University of Illinois, 1941; M.S., University of Minnesota, 1942; Ph.D., Physical Organic Chemistry, University of Chicago, 1948; University of Chicago Fellowship, 1947; Sigma Xi; C.I.B.S. award, 1960; formerly associated with Lever Brothers Company; author of numerous publications and holder of patents in cosmetics and toiletries; member, Codex Subcommittee of Cosmetics, Toiletries and Fragrance Association; editor, *Journal of the Society of Cosmetic Chemists*, 1963-1967; president, Society of Cosmetic Chemists, 1972.

Edward Sagarin, Ph.D., New York University, 1966; during his career in cosmetic and perfume industry, was associated with Givaudan Corporation, then with Standard Aromatics; author of *The Science and Art of Perfumery*; a few years after publication of the first edition of *Cosmetics: Science and Technology*, returned to school to study sociology, and obtained degree; has written extensively in sociology, particularly on crime and deviant behavior, presently teaching in that field.

S. J. Strianse, vice-president in charge of research and development and member of Board of Directors, Yardley of London, Inc. (subsidiary of British-American Tobacco Co.); B.S., Long Island University, 1941; formerly with Shulton, Vick Chemical Company, Richard Hudnut; president, Society of Cosmetic Chemists, 1957; member, Praesidium and Council, International Federation of Societies of Cosmetic Chemists and President of Federation, 1963-1964; author of numerous patents and papers, and recipients of plaque awards from Society of Cosmetic Chemists and International Federation of Societies of Cosmetic Chemists.

The Contributors

Marvin S. Balsam (*Fragrance*): *see* The Editors.

Richard H. Barry, Ph.D. (*Depilatories*), director of research, Sauter Laboratories, division of Hoffmann-La Roche; formerly vice-president and technical director, Union Pharmaceutical Co., subsidiary of Schering Corp.

Saul A. Bell, Phar. D. (*Preshave and Aftershave Preparations*), section head, Proprietary Products Development, Chesebrough-Pond's; formerly technical director, Seaforth Division, Vick Chemical; past chairman, New York Chapter, Society of Cosmetic Chemists.

Frank J. Berger (*Hair Conditioners, Lacquers, Setting Lotions, and Rinses*), director of research and development, Wella Corp.; former member, Program Committee, Society of Cosmetic Chemists.

William C. Doviak, M.S. (*Nail Lacquers and Removers*); president, Lacquerite Division, Supronics Corp.; formerly director of research and development, Charter Industries.

S. D. Gershon (*Permanent Waving*): *see* The Editors.

M. A. Goldberg, Ph.D. (*Permanent Waving*), technical services manager, Research and Development Division, Lever Bros.; formerly research chemist, Lady Esther, Ltd.

George G. Kolar, B.S. (*Hair Straighteners*), president, Kolar Laboratories; past president, Society of Cosmetic Chemists and former chairman of its Chicago section.

Edward W. Lang, B.S. (*Shampoos*), product development engineer, Toilet Goods Division, Procter & Gamble.

Richard K. Lehne, Ph.D. (*Hair-Grooming Preparations*), director of research and development, Church and Dwight Co.; former director of international research and development, Consumer Products Division, American Cyanamid; formerly with Mennen Co., Colgate-Palmolive, and Wildroot.

George H. Megerle, Ph.D. (*Hair Conditioners, Lacquers, Setting Lotions, and Rinses*), research chemist, Wella Corp.

Aaron Miller, B.S. (*Hair Straighteners*), technical director, Kolar Laboratories.

John D. Mullins, Ph.D. (*Eye Lotions*), director of development for science and technology, Alcon Laboratories; formerly assistant director, Product Development, Mead Johnson Research Center, and research associate, Merck Sharp & Dohme.

Sophie Plechner, Ph.D. (*Antiperspirants and Deodorants*), manager, Technical Services International, Carter-Wallace; formerly director and president, Society of Cosmetic Chemists; medal award winner of Society, 1965; former chairman of Scientific Section, Toilet Goods Association, and member of its Scientific Advisory Committee, 1947-68.

Donald H. Powers, Ph.D., deceased (*Shampoos*), was one of the members of the Editorial Board of the first edition of Cosmetics; Science and Technology. At the time, he was director of research, Lambert-Hudnut Division, Warner-Lambert Pharmaceutical Corp.; had been director of textile chemical department, Monsanto Chemical Co.; was a former president of Society of Cosmetic Chemists, former chairman of the Scientific Section of Toilet Goods Association, and was a member of the Advisory Board of National Research Council of U.S. Army Quartermaster Corps.

M. M. Rieger (*Permanent Waving*): *see* The Editors.

Morris J. Root, M.S. Ch.E. (*Aerosol Cosmetics*), vice-president of research and development, Barr-Stalfort Co. (Division of Pittway Corp.); formerly, assistant research director, Rayette Co.; formerly president, Society of Cosmetic Chemists, and former chairman of its Midwest Chapter.

Robert E. Sauté, Ph.D. (*Bath Preparations*), vice-president and technical director, Vanda Beauty Counselor; former director of research and

690

development, Toiletries Division, Gillette; previously, director of product and process development, Avon.

Warren R. Schubert, B.S. (*Shaving Preparations: Soaps, Creams, Oils, and Lotions*), senior research chemist, Toilet Soaps Section, Colgate-Palmolive Co.

Robert F. Schuler, Sc.D. (*Emulsified and Solid Fragrances*), group manager, Toiletries Division, Gillette Co.; formerly section head, Chesebrough-Pond's, and technical director, Prince Matchabelli.

Stephen Shernov, B.S. (*Aerosol Hair Products*), aerosol group leader, Faberge, Inc.; formerly aerosol chemist, International Flavors & Fragrances; member CSMA Committee for Aerosol Labeling.

Neil D. Stiegelmeyer, B.S.Ch.E. (*Shampoos*), science teacher, Beechwood High School, Ft. Mitchell, Kent.; formerly with Procter & Gamble.

Leonard J. Viola, M.S. (*Fingernail Elongators and Accessory Nail Preparations*), group leader, research and development, Pharmacy Dept., Vick Chemical; formerly with Caryl Richards and Warner-Lambert.

Florence E. Wall, M.A. (*Bleaches, Hair Colorings, and Dye Removers*), technical writer and editor; formerly, director of trade education, Inecto, Inc.; lecturer on cosmetology, School of Education, New York University, and at other colleges; author *Principles and Practice of Beauty Culture*, and other volumes; medalist, Society of Cosmetic Chemists; honorary member, American Institute of Chemists.